全栈开发技术丛书

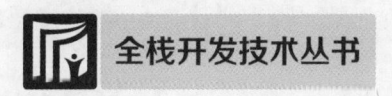

SSM+Spring Boot+Vue.js 3 全栈开发从入门到实战

（微课视频版）（第2版）

陈恒 主编
蒋伟 赵璘 赵志方 孙云浩 副主编

清华大学出版社
北京

内 容 简 介

本书系统地介绍SSM、Spring Boot、MyBatis-Plus和Vue.js 3的重要内容，分三个阶段：第一阶段为SSM框架整合开发(第1~4章)，内容包括Spring、Spring MVC、MyBatis以及基于SSM+JSP的案例开发；第二阶段为Spring Boot框架开发(第5~11章)，内容包括Spring Boot的入门、核心、Web开发、数据访问、安全控制、异步消息以及基于Spring Boot＋MyBatis＋Thymeleaf的案例开发；第三阶段为Vue.js 3前端框架开发(第12~16章)，内容包括Vue.js基础、Vue.js进阶、MyBatis-Plus、单元测试以及基于Spring Boot＋Vue.js 3＋MyBatis-Plus的案例开发。

本书内容不仅有框架知识的介绍，还有大量的实例和案例，这些实例和案例侧重实用性，通俗易懂。读者通过本书可以快速掌握SSM、Spring Boot和Vue.js 3的基础知识、编程技巧以及完整的开发体系，为大型项目开发打下坚实的基础。

本书可作为高等院校计算机及相关专业的教材或教学参考书，也可作为Java技术的培训教材。

版权所有，侵权必究。举报：010-62782989，beiqinquan@tup.tsinghua.edu.cn。

图书在版编目(CIP)数据

SSM＋Spring Boot＋Vue.js 3全栈开发从入门到实战：微课视频版/陈恒主编. -- 2版.
北京：清华大学出版社，2025.5. -- (全栈开发技术丛书). -- ISBN 978-7-302-69133-4
Ⅰ.TP393.092.2
中国国家版本馆CIP数据核字第2025ED5906号

策划编辑：魏江江
责任编辑：王冰飞
封面设计：刘 键
责任校对：时翠兰
责任印制：杨 艳

出版发行：清华大学出版社
网　　址：https://www.tup.com.cn，https://www.wqxuetang.com
地　　址：北京清华大学学研大厦A座　　邮　编：100084
社 总 机：010-83470000　　邮　购：010-62786544
投稿与读者服务：010-62776969，c-service@tup.tsinghua.edu.cn
质量反馈：010-62772015，zhiliang@tup.tsinghua.edu.cn
课件下载：https://www.tup.com.cn，010-83470236

印 装 者：三河市龙大印装有限公司
经　　销：全国新华书店
开　　本：185mm×260mm　　印 张：30.5　　字 数：799千字
版　　次：2022年3月第1版　2025年7月第2版　　印 次：2025年7月第1次印刷
印　　数：17501~19000
定　　价：99.80元

产品编号：107803-01

前言

本书适合具有 Java 编程基础和一定 Java Web 相关知识的读者学习。

本书是第 2 版,使用 Spring Framework 6.1.5＋MyBatis 3.5.14＋Spring Boot 3.2.4＋Vue.js 3.4.21＋MyBatis-Plus 3.5.5 编写实例。与第 1 版相比,本书删除了监控 Spring Boot 应用、人事管理系统的设计与实现等内容,增加了 MyBatis-Plus 基础知识。

本书系统地介绍 SSM、Spring Boot、MyBatis-Plus 和 Vue.js 3 的重要内容,分三个阶段:第一阶段为 SSM 框架整合开发(第 1～4 章),内容包括 Spring、Spring MVC、MyBatis 以及基于 SSM＋JSP 的案例开发;第二阶段为 Spring Boot 框架开发(第 5～11 章),内容包括 Spring Boot 的入门、核心、Web 开发、数据访问、安全控制、异步消息以及基于 Spring Boot＋MyBatis＋Thymeleaf 的案例开发;第三阶段为 Vue.js 3 前端框架开发(第 12～16 章),内容包括 Vue.js 基础、Vue.js 进阶、MyBatis-Plus、单元测试以及基于 Spring Boot＋Vue.js 3＋MyBatis-Plus 的案例开发。本书的重点不是框架知识的简单介绍,而是精心设计的大量实例和案例。读者通过本书可以快速掌握 SSM、Spring Boot、Vue.js、MyBatis、MyBatis-Plus 等框架的实践应用,提高 Java EE 应用的开发能力。

全书共 16 章,具体如下。

第一阶段:SSM 框架整合开发

第 1 章:Spring,包括 Spring 开发环境的构建、Spring IoC、Spring AOP、Spring Bean 以及 Spring 的数据库编程等内容。

第 2 章:Spring MVC,包括 Spring MVC 的工作原理、Spring MVC 的工作环境、基于注解的控制器、表单标签库与数据绑定以及 Spring MVC 的基本配置等内容。

第 3 章:MyBatis,包括 MyBatis 的工作原理、SSM 框架整合开发、核心配置文件、SQL 映射文件、级联查询、动态 SQL 以及 MyBatis 的缓存机制等内容。

第 4 章:名片管理系统的设计与实现(SSM＋JSP),本章内容是对第 1～3 章学习的巩固。

第二阶段:Spring Boot 框架开发

第 5 章:Spring Boot 入门,包括 Spring Boot 特性、Maven 简介以及使用 IntelliJ IDEA 快速构建 Spring Boot 应用等内容。

第 6 章:Spring Boot 核心,包括核心注解、基本配置、自动配置原理以及条件注解等内容。

第 7 章:Spring Boot 的 Web 开发,包括 Spring Boot 的 Web 开发支持、Thymeleaf 视图模板引擎技术、JSON 数据交互、文件上传与下载、异常统一处理以及对 JSP 的支持等内容。

第 8 章:Spring Boot 的数据访问,包括 Spring Data JPA、Spring Boot 整合 MyBatis、REST、MongoDB、Redis、数据缓存 Cache 技术等内容。

第 9 章:电子商务平台的设计与实现(Spring Boot＋MyBatis＋Thymeleaf),本章内容是对第 5～8 章学习的巩固。

第 10 章:Spring Boot 的安全控制,包括 Spring Security 快速入门以及基于 Spring Data

JPA 的 Spring Boot Security 操作实例等内容。

第 11 章：Spring Boot 的异步消息，包括消息模型、JMS 与 AMQP 企业级消息代理、Spring Boot 对异步消息的支持以及异步消息通信实例等内容。

第三阶段：Vue.js 3 前端框架开发

第 12 章：Vue.js 基础，包括 Vue.js 的安装、生命周期、插值与表达式、计算属性和监听器、内置指令、组件、自定义指令、响应性以及 setup 组件选项等内容。

第 13 章：Vue.js 进阶，包括 Vue Router 的基本用法与高级应用、setup 语法糖以及 Element Plus UI 组件库等内容。

第 14 章：MyBatis-Plus，包括 Spring Boot 整合 MyBatis-Plus、MyBatis-Plus 注解、CRUD 接口以及条件构造器等内容。

第 15 章：Spring Boot 单元测试，包括 JUnit 5 注解、JUnit 5 断言、单元测试用例以及使用 Postman 测试 Controller 层等内容。

第 16 章：电子商务平台的设计与实现（Spring Boot＋Vue.js 3＋MyBatis-Plus），本章内容是对本书整体学习的巩固。

为便于教学，本书提供丰富的配套资源，包括教学大纲、教学课件、电子教案、程序源码、习题答案、在线作业和微课视频。

资源下载提示

课件等资源：扫描封底的"图书资源"二维码，在公众号"书圈"下载。

素材（源码）等资源：扫描目录上方的二维码下载。

在线自测题：扫描封底的作业系统二维码，再扫描自测题二维码，可以在线做题及查看答案。

微课视频：扫描封底的文泉云盘防盗码，再扫描书中相应章节的视频讲解二维码，可以在线学习。

本书是辽宁省一流本科课程"工程项目实训"以及辽宁省普通高等学校一流本科教育示范专业"大连外国语大学计算机科学与技术专业"的建设成果。

本书的出版得到清华大学出版社相关人员的大力支持，在此表示衷心感谢。同时，本书编者参阅了相关书籍、博客以及部分官网资源，在此对这些资源的贡献者与分享者深表感谢。由于前端框架技术发展迅速，持续改进与优化，加上编者水平有限，书中难免会有不足之处，敬请各位专家和读者批评指正。

编 者

2025 年 3 月

目录

扫一扫
源码下载

第一阶段　SSM 框架整合开发

第 1 章　Spring

1.1 Spring 概述 2
　　1.1.1 Spring 的由来 2
　　1.1.2 Spring 的体系结构 2
1.2 Spring 开发环境的构建 4
　　1.2.1 配置 IntelliJ IDEA 的 Web 服务器 4
　　1.2.2 Spring 的下载及目录结构 6
　　1.2.3 第一个 Spring 入门程序 6
1.3 Spring IoC 9
　　1.3.1 Spring IoC 的基本概念 9
　　1.3.2 Spring 的常用注解 10
　　1.3.3 基于注解的依赖注入 11
1.4 Spring AOP 13
　　1.4.1 Spring AOP 的基本概念 13
　　1.4.2 基于注解开发 AspectJ 15
1.5 Spring Bean 19
　　1.5.1 Bean 的实例化 19
　　1.5.2 Bean 的作用域 21
　　1.5.3 Bean 的初始化和销毁 22
1.6 Spring 的数据库编程 24
　　1.6.1 Spring JDBC 的 XML 配置 24
　　1.6.2 Spring JdbcTemplate 的常用方法 24
　　1.6.3 基于@Transactional 注解的声明式事务管理 28
　　1.6.4 如何在事务处理中捕获异常 30
本章小结 31
习题 1 31

第 2 章　Spring MVC

- 2.1 Spring MVC 的工作原理 …… 32
- 2.2 Spring MVC 的工作环境 …… 33
    - 2.2.1 Spring MVC 所需要的 JAR 包 …… 33
    - 2.2.2 使用 IntelliJ IDEA 开发 Spring MVC 应用 …… 34
- 2.3 基于注解的控制器 …… 39
    - 2.3.1 Controller 注解类型 …… 39
    - 2.3.2 RequestMapping 注解类型 …… 39
    - 2.3.3 编写请求处理方法 …… 40
    - 2.3.4 Controller 接收请求参数的常见方式 …… 41
    - 2.3.5 重定向与转发 …… 46
    - 2.3.6 应用@Autowired 进行依赖注入 …… 47
    - 2.3.7 @ModelAttribute …… 49
- 2.4 表单标签库与数据绑定 …… 50
    - 2.4.1 表单标签库 …… 50
    - 2.4.2 数据绑定 …… 53
- 2.5 拦截器 …… 58
    - 2.5.1 拦截器的定义 …… 58
    - 2.5.2 拦截器的配置 …… 59
    - 2.5.3 拦截器的执行流程 …… 60
- 2.6 文件的上传 …… 63
- 本章小结 …… 67
- 习题 2 …… 67

第 3 章　MyBatis

- 3.1 MyBatis 简介 …… 68
- 3.2 MyBatis 的环境构建 …… 68
- 3.3 MyBatis 的工作原理 …… 69
- 3.4 MyBatis 的核心配置 …… 70
- 3.5 使用 IntelliJ IDEA 开发 MyBatis 入门程序 …… 71
- 3.6 SSM 框架整合开发 …… 75
    - 3.6.1 相关 JAR 包 …… 75
    - 3.6.2 MapperScannerConfigurer 方式 …… 75
    - 3.6.3 整合示例 …… 76
    - 3.6.4 SqlSessionDaoSupport 方式 …… 81
- 3.7 使用 MyBatis Generator 插件自动生成映射文件 …… 85
- 3.8 映射器概述 …… 86

3.9 <select> 元素 ·········· 87
   3.9.1 使用 Map 接口传递参数 ·········· 88
   3.9.2 使用 Java Bean 传递参数 ·········· 90
   3.9.3 使用@Param 注解传递参数 ·········· 91
   3.9.4 <resultMap> 元素 ·········· 92
   3.9.5 使用 POJO 存储结果集 ·········· 92
   3.9.6 使用 Map 存储结果集 ·········· 93
3.10 <insert>、<update>、<delete> 以及 <sql> 元素 ·········· 94
   3.10.1 <insert> 元素 ·········· 94
   3.10.2 <update> 与 <delete> 元素 ·········· 96
   3.10.3 <sql> 元素 ·········· 96
3.11 级联查询 ·········· 97
   3.11.1 一对一级联查询 ·········· 97
   3.11.2 一对多级联查询 ·········· 100
   3.11.3 多对多级联查询 ·········· 103
3.12 动态 SQL ·········· 106
   3.12.1 <if> 元素 ·········· 106
   3.12.2 <choose>、<when>、<otherwise> 元素 ·········· 107
   3.12.3 <trim> 元素 ·········· 108
   3.12.4 <where> 元素 ·········· 109
   3.12.5 <set> 元素 ·········· 110
   3.12.6 <foreach> 元素 ·········· 111
   3.12.7 <bind> 元素 ·········· 112
3.13 MyBatis 的缓存机制 ·········· 113
   3.13.1 一级缓存（SqlSession 级别的缓存）·········· 113
   3.13.2 二级缓存（Mapper 级别的缓存）·········· 114
本章小结 ·········· 116
习题 3 ·········· 116

第4章 名片管理系统的设计与实现（SSM＋JSP）

4.1 系统设计 ·········· 117
   4.1.1 系统功能需求 ·········· 117
   4.1.2 系统模块划分 ·········· 117
4.2 数据库设计 ·········· 118
   4.2.1 数据库概念结构设计 ·········· 118
   4.2.2 数据库逻辑结构设计 ·········· 118
4.3 系统管理 ·········· 119
   4.3.1 所需 JAR 包 ·········· 119
   4.3.2 JSP 页面管理 ·········· 119
   4.3.3 包管理 ·········· 119

		4.3.4 配置管理	120
4.4	组件设计		120
	4.4.1	工具类	120
	4.4.2	统一异常处理	120
	4.4.3	验证码	121
4.5	名片管理		121
	4.5.1	领域模型与持久化实体类	121
	4.5.2	Controller 实现	121
	4.5.3	Service 实现	123
	4.5.4	Dao 实现	125
	4.5.5	SQL 映射文件	125
	4.5.6	添加名片	126
	4.5.7	名片管理主页面	127
	4.5.8	修改名片	127
	4.5.9	删除名片	128
4.6	用户相关		128
	4.6.1	领域模型与持久化实体类	128
	4.6.2	Controller 实现	128
	4.6.3	Service 实现	129
	4.6.4	Dao 实现	130
	4.6.5	SQL 映射文件	130
	4.6.6	注册	130
	4.6.7	登录	131
	4.6.8	修改密码	131
	4.6.9	安全退出	131
本章小结			131
习题 4			131

第二阶段　Spring Boot 框架开发

第 5 章　Spring Boot 入门

5.1	Spring Boot 概述	134
	5.1.1 什么是 Spring Boot	134
	5.1.2 Spring Boot 的优点	134
	5.1.3 Spring Boot 的主要特性	134
5.2	第一个 Spring Boot 应用	135
	5.2.1 Maven 简介	135
	5.2.2 Maven 的 pom.xml	135
	5.2.3 使用 IntelliJ IDEA 快速构建 Spring Boot 应用	136

本章小结 ·· 139

习题 5 ·· 139

第 6 章　Spring Boot 核心

6.1 Spring Boot 的基本配置 ·· 140

　6.1.1 启动类和核心注解@SpringBootApplication ·· 140

　6.1.2 Spring Boot 的全局配置文件 ·· 141

　6.1.3 Spring Boot 的 Starters ·· 141

6.2 读取应用配置 ·· 142

　6.2.1 Environment ·· 142

　6.2.2 @Value ··· 143

　6.2.3 @ConfigurationProperties ··· 143

　6.2.4 @PropertySource ··· 144

6.3 日志配置 ·· 145

6.4 Spring Boot 的自动配置原理 ·· 147

6.5 Spring Boot 的条件注解 ·· 148

　6.5.1 条件注解 ··· 149

　6.5.2 自定义条件 ··· 149

　6.5.3 自定义 Starters ··· 151

本章小结 ·· 155

习题 6 ·· 155

第 7 章　Spring Boot 的 Web 开发

7.1 Spring Boot 的 Web 开发支持 ··· 157

7.2 Thymeleaf 模板引擎 ·· 157

　7.2.1 Spring Boot 的 Thymeleaf 支持 ··· 158

　7.2.2 Thymeleaf 基础语法 ·· 159

　7.2.3 Thymeleaf 的常用属性 ·· 161

　7.2.4 用 Spring Boot 与 Thymeleaf 实现页面信息的国际化 ····························· 165

　7.2.5 Spring Boot 与 Thymeleaf 的表单验证 ··· 168

　7.2.6 基于 Thymeleaf 与 BootStrap 的 Web 开发实例 ······································ 171

7.3 用 Spring Boot 处理 JSON 数据 ·· 174

　7.3.1 JSON 数据结构 ··· 174

　7.3.2 JSON 数据转换 ··· 175

7.4 Spring Boot 文件的上传与下载 ·· 178

7.5 Spring Boot 的异常统一处理 ·· 182

　7.5.1 自定义 error 页面 ··· 183

　7.5.2 @ExceptionHandler 注解 ·· 185

7.5.3 @ControllerAdvice 注解 ·············· 186

7.6 Spring Boot 对 JSP 的支持 ·············· 187

本章小结 ·············· 190

习题 7 ·············· 190

第 8 章 Spring Boot 的数据访问

8.1 Spring Data JPA ·············· 191
    8.1.1 Spring Boot 的支持 ·············· 192
    8.1.2 简单条件查询 ·············· 192
    8.1.3 关联查询 ·············· 199
    8.1.4 @Query 和 @Modifying 注解 ·············· 213
    8.1.5 排序与分页查询 ·············· 214

8.2 REST ·············· 217
    8.2.1 REST 简介 ·············· 217
    8.2.2 Spring Boot 整合 REST ·············· 219
    8.2.3 Spring Data REST ·············· 220
    8.2.4 REST 服务测试 ·············· 221

8.3 MongoDB ·············· 225
    8.3.1 安装 MongoDB ·············· 226
    8.3.2 Spring Boot 整合 MongoDB ·············· 226
    8.3.3 增、删、改、查 ·············· 227

8.4 Redis ·············· 230
    8.4.1 安装 Redis ·············· 230
    8.4.2 Spring Boot 整合 Redis ·············· 232
    8.4.3 使用 StringRedisTemplate 和 RedisTemplate ·············· 233

8.5 数据缓存 Cache ·············· 236
    8.5.1 Spring 缓存支持 ·············· 236
    8.5.2 Spring Boot 缓存支持 ·············· 238
    8.5.3 使用 Redis Cache ·············· 241

8.6 Spring Boot 整合 MyBatis ·············· 242

本章小结 ·············· 245

习题 8 ·············· 245

第 9 章 电子商务平台的设计与实现（Spring Boot ＋ MyBatis ＋ Thymeleaf）

9.1 系统设计 ·············· 246
    9.1.1 系统功能需求 ·············· 246

		9.1.2 系统模块划分 ·································	246
9.2	数据库设计 ·····························		247
	9.2.1	数据库概念结构设计 ·····················	247
	9.2.2	数据逻辑结构设计 ······················	248
	9.2.3	创建数据表 ·························	250
9.3	系统管理 ······························		250
	9.3.1	添加相关依赖 ························	250
	9.3.2	HTML 页面及静态资源管理 ················	250
	9.3.3	应用的包结构 ························	251
	9.3.4	配置文件 ···························	252
9.4	组件设计 ······························		252
	9.4.1	管理员登录权限验证 ·····················	252
	9.4.2	前台用户登录权限验证 ····················	253
	9.4.3	验证码 ·····························	253
	9.4.4	统一异常处理 ························	253
	9.4.5	工具类 ·····························	254
9.5	后台管理子系统的实现 ·······················		254
	9.5.1	管理员登录 ··························	254
	9.5.2	类型管理 ···························	255
	9.5.3	添加商品 ···························	258
	9.5.4	查询商品 ···························	260
	9.5.5	修改商品 ···························	262
	9.5.6	删除商品 ···························	263
	9.5.7	查询订单 ···························	263
	9.5.8	用户管理 ···························	264
	9.5.9	按月统计 ···························	265
	9.5.10	按类型统计 ·························	266
	9.5.11	安全退出 ···························	267
9.6	前台电子商务子系统的实现 ·····················		268
	9.6.1	导航栏及首页搜索 ······················	268
	9.6.2	推荐商品及最新商品 ·····················	269
	9.6.3	用户注册 ···························	271
	9.6.4	用户登录 ···························	272
	9.6.5	商品详情 ···························	273
	9.6.6	收藏商品 ···························	274
	9.6.7	购物车 ·····························	275
	9.6.8	下单 ·······························	277
	9.6.9	个人信息 ···························	279
	9.6.10	我的收藏 ···························	280
	9.6.11	我的订单 ···························	281
本章小结 ·······································			282
习题 9 ···			282

第 10 章　Spring Boot 的安全控制

10.1 Spring Security 快速入门 ... 283
10.1.1 什么是 Spring Security ... 283
10.1.2 Spring Security 的用户认证 ... 283
10.1.3 Spring Security 的请求授权 ... 284
10.1.4 Spring Security 的核心类 ... 285
10.1.5 Spring Security 的验证机制 ... 287
10.2 Spring Boot 的支持 ... 287
10.3 实际开发中的 Spring Security 操作实例 ... 287
本章小结 ... 296
习题 10 ... 296

第 11 章　Spring Boot 的异步消息

11.1 消息模型 ... 297
11.1.1 点对点式 ... 297
11.1.2 发布/订阅式 ... 297
11.2 企业级消息代理 ... 297
11.2.1 JMS ... 297
11.2.2 AMQP ... 299
11.3 Spring Boot 的支持 ... 301
11.3.1 JMS 的自动配置 ... 301
11.3.2 AMQP 的自动配置 ... 301
11.4 异步消息通信实例 ... 301
11.4.1 JMS 实例 ... 301
11.4.2 AMQP 实例 ... 305
本章小结 ... 309
习题 11 ... 309

第三阶段　Vue.js 3 前端框架开发

第 12 章　Vue.js 基础

12.1 网站交互方式 ... 312
12.1.1 多页应用程序 ... 312
12.1.2 单页应用程序 ... 313
12.2 MVVM 模式 ... 314

12.3	Vue.js 是什么	314
12.4	安装 Vue.js	315
	12.4.1　本地独立版本方法	315
	12.4.2　CDN 方法	315
	12.4.3　NPM 方法	315
	12.4.4　命令行工具（CLI）方法	315
12.5	第一个 Vue.js 程序	315
	12.5.1　安装 Visual Studio Code 及其插件	315
	12.5.2　创建第一个 Vue.js 应用	316
	12.5.3　声明式渲染	317
	12.5.4　Vue.js 的生命周期	319
12.6	插值与表达式	320
	12.6.1　文本插值	320
	12.6.2　原始 HTML 插值	321
	12.6.3　JavaScript 表达式	321
12.7	计算属性和监听器	321
	12.7.1　计算属性 computed	321
	12.7.2　监听器属性 watch	324
12.8	内置指令	325
	12.8.1　v-bind 指令	325
	12.8.2　条件渲染指令 v-if 和 v-show	326
	12.8.3　列表渲染指令 v-for	328
	12.8.4　事件处理	329
	12.8.5　表单与 v-model	330
	12.8.6　实战: 购物车实例	333
12.9	组件	335
	12.9.1　组件的注册	335
	12.9.2　使用 props 传递数据	337
	12.9.3　组件的通信	339
	12.9.4　动态组件与异步组件	344
	12.9.5　实战: 正整数数字输入框组件	345
12.10	自定义指令	347
	12.10.1　自定义指令的注册	347
	12.10.2　实战: 实时时间转换指令	348
12.11	响应性	350
	12.11.1　什么是响应性	350
	12.11.2　响应性原理	351
12.12	setup 组件选项	351
	12.12.1　setup 函数的参数	351
	12.12.2　setup 函数的返回值	353
	12.12.3　使用 ref 创建响应式引用	354
	12.12.4　在 setup 内部调用生命周期钩子函数	355

本章小结 ………………………………………………………………………………………… 355
习题 12 …………………………………………………………………………………………… 355

第 13 章　Vue.js 进阶

13.1　单文件组件与 webpack ……………………………………………………………… 359
13.2　安装 Node.js 和 NPM ………………………………………………………………… 360
　　13.2.1　安装 Node.js ………………………………………………………………… 360
　　13.2.2　NPM 的常用命令 …………………………………………………………… 360
13.3　Vue Router …………………………………………………………………………… 363
　　13.3.1　Vue Router 的安装 ………………………………………………………… 363
　　13.3.2　Vue Router 的基本用法 …………………………………………………… 367
　　13.3.3　Vue Router 的高级应用 …………………………………………………… 371
　　13.3.4　路由钩子函数 ……………………………………………………………… 383
　　13.3.5　路由元信息 ………………………………………………………………… 386
　　13.3.6　登录权限验证实例 ………………………………………………………… 387
13.4　setup 语法糖 …………………………………………………………………………… 390
　　13.4.1　属性与方法的绑定 ………………………………………………………… 390
　　13.4.2　路由 ………………………………………………………………………… 391
　　13.4.3　组件传值 …………………………………………………………………… 393
13.5　Element Plus UI 组件库 ……………………………………………………………… 395
　　13.5.1　Element Plus 的安装 ………………………………………………………… 395
　　13.5.2　Element Plus 组件的介绍 …………………………………………………… 396
　　13.5.3　Element Plus 组件的应用 …………………………………………………… 396
　　13.5.4　按需引入 Element Plus ……………………………………………………… 398
本章小结 ………………………………………………………………………………………… 400
习题 13 …………………………………………………………………………………………… 400

第 14 章　MyBatis-Plus

14.1　MyBatis-Plus 简介 …………………………………………………………………… 401
14.2　Spring Boot 整合 MyBatis-Plus ……………………………………………………… 401
14.3　MyBatis-Plus 注解 …………………………………………………………………… 404
14.4　CRUD 接口 …………………………………………………………………………… 407
14.5　条件构造器 …………………………………………………………………………… 413
本章小结 ………………………………………………………………………………………… 418
习题 14 …………………………………………………………………………………………… 418

第15章　Spring Boot 单元测试

15.1　JUnit 5 ·· 419
　　15.1.1　JUnit 5 简介 ··· 419
　　15.1.2　JUnit 5 注解 ··· 419
　　15.1.3　JUnit 5 断言 ··· 421
15.2　单元测试用例 ·· 422
　　15.2.1　测试环境的构建 ··· 422
　　15.2.2　测试 Mapper 接口 ··· 424
　　15.2.3　测试 Service 层 ·· 425
15.3　使用 Postman 测试 Controller 层 ·· 426
本章小结 ·· 429
习题 15 ·· 429

第16章　电子商务平台的设计与实现（Spring Boot ＋ Vue.js 3 ＋ MyBatis-Plus）

16.1　使用 IntelliJ IDEA 构建后端系统 ·· 430
　　16.1.1　创建 Spring Boot Web 应用 ·· 430
　　16.1.2　修改 pom.xml ··· 430
　　16.1.3　配置数据源等信息 ··· 431
　　16.1.4　创建持久化实体类 ··· 431
　　16.1.5　创建 Mapper 接口 ··· 431
　　16.1.6　创建业务层 ·· 433
　　16.1.7　创建控制器层 ·· 441
　　16.1.8　创建跨域响应头设置过滤器 ······································· 445
　　16.1.9　创建工具类 ·· 445
　　16.1.10　MyBatis-Plus 分页插件、Redis 以及 Token 签名配置 ·········· 446
16.2　使用 Vue CLI 构建前端系统 ··· 446
　　16.2.1　使用 Vue CLI 构建前端项目 ebusiness-vue ··············· 446
　　16.2.2　安装 Element Plus 和 @element-plus/icons-vue ········ 446
　　16.2.3　安装 ECharts ··· 447
　　16.2.4　安装 Axios 模块并设置跨域访问 ······························ 447
　　16.2.5　管理员登录组件 ··· 447
　　16.2.6　后台管理主界面组件 ··· 449
　　16.2.7　商品类型管理组件 ··· 450
　　16.2.8　商品管理组件 ·· 450
　　16.2.9　订单管理组件 ·· 450
　　16.2.10　销量统计（按月）组件 ··· 453
　　16.2.11　订单统计（按类型）组件 ··· 455

16.2.12	前端首页组件 ……………………………………………………	457
16.2.13	用户注册组件 ……………………………………………………	459
16.2.14	用户登录组件 ……………………………………………………	459
16.2.15	个人信息组件 ……………………………………………………	459
16.2.16	商品详情组件 ……………………………………………………	460
16.2.17	我的购物车组件 …………………………………………………	463
16.2.18	我的订单组件 ……………………………………………………	466
16.2.19	我的收藏组件 ……………………………………………………	466
16.2.20	订单确认组件 ……………………………………………………	466
16.2.21	配置路由 …………………………………………………………	467

16.3 测试运行 ……………………………………………………………………… 469
本章小结 …………………………………………………………………………… 469
习题 16 ……………………………………………………………………………… 469

第一阶段　SSM 框架整合开发

第 1 章 Spring

学习目的与要求

本章重点讲解 Spring 的基础知识。通过本章的学习，了解 Spring 的体系结构，理解 Spring IoC 与 AOP 的基本原理，了解 Spring Bean 的生命周期、实例化以及作用域，掌握 Spring 的事务管理。

本章主要内容

- Spring 开发环境的构建
- Spring IoC
- Spring AOP
- Spring Bean
- Spring 的数据库编程

Spring 是当前主流的 Java 开发框架，为企业级应用开发提供了丰富的功能。掌握 Spring 框架的使用，已是 Java 开发者必备的技能之一。本章将学习如何使用 IntelliJ IDEA 开发 Spring 程序，不过在此之前需要构建 Spring 的开发环境。

1.1 Spring 概述

1.1.1 Spring 的由来

Spring 是一个轻量级 Java 开发框架，最早由 Rod Johnson 创建，目的是解决企业级应用开发的业务逻辑层和其他各层的耦合问题。它是一个分层的 Java SE/EE full-stack（一站式）轻量级开源框架，为开发 Java 应用程序提供全面的基础架构支持。Spring 负责基础架构，因此 Java 开发者可以专注于应用程序的开发。

Spring Framework 6.0 于 2022 年 11 月正式发布，这是 2023 年及以后新一代框架的开始，包含 OpenJDK 和 Java 生态系统中当前和即将到来的创新。Spring Framework 6.0 作为重大更新，要求使用 Java 17 或更高版本，并且已迁移到 Jakarta EE 9+（在 Jakarta 命名空间中取代了以前基于 Javax 的 API），以及对其他基础设施的修改。基于这些变化，Spring Framework 6.0 支持最新的 Web 容器，如 Tomcat 10，以及最新的持久性框架 Hibernate ORM 6.1。这些特性仅可用于 Servlet API 和 JPA 的 Jakarta 命名空间变体。

在基础架构方面，Spring Framework 6.0 引入了 Ahead-Of-Time（AOT）转换的基础以及对 Spring 应用程序上下文的 AOT 转换和相应的 AOT 处理支持的基础，能够事先将应用程序中或 JDK 中的字节码编译成机器码。在 Spring Framework 6.0 中还有许多新功能和改进可用，例如 HTTP 接口客户端、对 RFC 7807 问题详细信息的支持以及 HTTP 客户端的基于 Micrometer 的可观察性。

1.1.2 Spring 的体系结构

Spring 的功能模块被有组织地分散到约 20 个模块中，这些模块分布在核心容器（Core Container）层、数据访问/集成（Data Access/Integration）层、Web 层以及面向切面的编程

（Aspect Oriented Programming，AOP）模块、植入（Instrumentation）模块、消息传输（Messaging）和测试（Test）模块中，如图 1.1 所示。

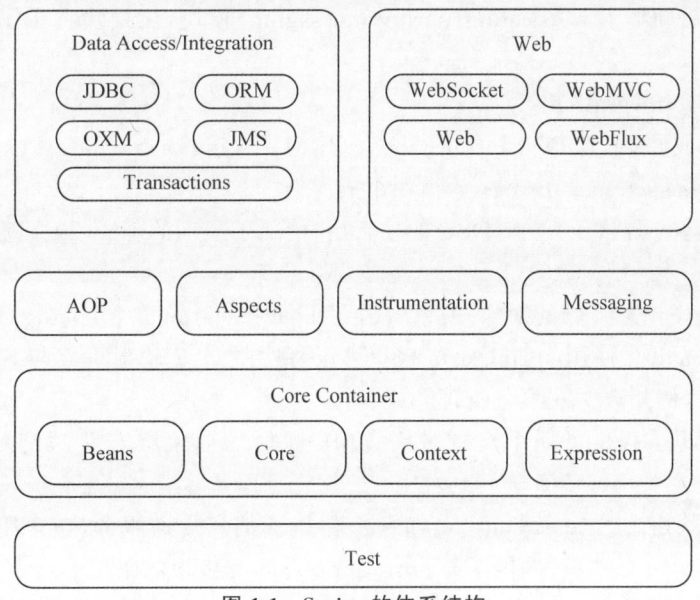

图 1.1 Spring 的体系结构

❶ Core Container

Spring 的 Core Container 是其他模块建立的基础，由 Beans（spring-beans）、Core（spring-core）、Context（spring-context）和 Expression（spring-expression，Spring 表达式语言）等模块组成。

spring-beans 模块：提供了 BeanFactory，是工厂模式的一个经典实现，Spring 将管理对象称为 Bean。

spring-core 模块：提供了框架的基本组成部分，包括控制反转（Inversion of Control，IoC）和依赖注入（Dependency Injection，DI）功能。

spring-context 模块：建立在 spring-beans 和 spring-core 模块基础之上，提供一种框架式的对象访问方式，是访问定义和配置的任何对象媒介。

spring-expression 模块：提供了强大的表达式语言支持运行时查询和操作对象图。这是对 JSP 2.1 规范中规定的统一表达式语言（Unified EL）的扩展。该语言支持设置和获取属性值、属性分配，方法调用，访问数组、集合和索引器的内容，逻辑和算术运算，变量命名以及从 Spring 的 IoC 容器中以名称检索对象。它还支持列表投影、选择以及常见的列表聚合。

❷ AOP 和 Instrumentation

在 Spring 框架中与 AOP 和 Instrumentation 相关的模块有 AOP（spring-aop）模块、Aspects（spring-aspects）模块以及 Instrumentation（spring-instrument）模块。

spring-aop 模块：提供了一个符合 AOP 要求的面向切面的编程实现，允许定义方法拦截器和切入点，将代码按照功能进行分离，以便于干净地解耦。

spring-aspects 模块：提供了与 AspectJ 的集成功能，AspectJ 是一个功能强大且成熟的 AOP 框架。

spring-instrument 模块：提供了类植入（Instrumentation）支持和类加载器的实现，可以在特定的应用服务器中使用。Instrumentation 提供了一种虚拟机级别支持的 AOP 实现方式，使得开发者无须对 JDK 做任何升级和改动就可以实现某些 AOP 的功能。

❸ Messaging

Spring 4.0 以后新增了 Messaging（spring-messaging）模块，该模块提供了对消息传递体系结构和协议的支持。

❹ Data Access/Integration

数据访问/集成层由 JDBC（spring-jdbc）、ORM（spring-orm）、OXM（spring-oxm）、JMS（spring-jms）和 Transactions（spring-tx）模块组成。

spring-jdbc 模块：提供了一个 JDBC 的抽象层，消除了烦琐的 JDBC 编码和数据库厂商特有的错误代码解析。

spring-orm 模块：为流行的对象-关系映射（Object-Relational Mapping）API 提供集成层，包括 JPA 和 Hibernate。使用 spring-orm 模块可以将这些 O-R 映射框架与 Spring 提供的所有其他功能结合使用，例如声明式事务管理功能。

spring-oxm 模块：提供了一个支持对象/XML 映射的抽象层实现，如 JAXB、Castor、JiBX 和 XStream。

spring-jms 模块（Java Messaging Service）：指 Java 消息传递服务，包含用于生产和使用消息的功能。自 Spring 4.1 之后，提供了与 spring-messaging 模块的集成。

spring-tx 模块（事务模块）：支持用于实现特殊接口和所有 POJO（普通 Java 对象）类的编程和声明式事务管理。

❺ Web

Web 层由 Web（spring-web）、WebMVC（spring-webmvc）、WebSocket（spring-websocket）和 WebFlux（spring-webflux）模块组成。

spring-web 模块：提供了基本的 Web 开发集成功能。例如，多文件上传功能、使用 Servlet 监听器初始化一个 IoC 容器以及 Web 应用上下文。

spring-webmvc 模块：也称为 Web-Servlet 模块，包含用于 Web 应用程序的 Spring MVC 和 REST Web Services 实现。Spring MVC 框架提供了领域模型代码和 Web 表单之间的清晰分离，并与 Spring Framework 的所有其他功能集成，本书后续章节将会详细讲解 Spring MVC 框架。

spring-websocket 模块：Spring 4.0 后新增的模块，它提供了 WebSocket 和 SockJS 的实现，主要是与 Web 前端的全双工通信的协议。

spring-webflux 模块：spring-webflux 是一个新的非堵塞函数式 Reactive Web 框架，可以用来建立异步、非阻塞、事件驱动的服务，并且扩展性非常好（该模块是 Spring 5 新增的模块）。

❻ Test

Test（spring-test）模块支持使用 JUnit 或 TestNG 对 Spring 组件进行单元测试和集成测试。

1.2 Spring 开发环境的构建

在使用 Spring 框架开发应用之前，应先搭建其开发环境。

▶ 1.2.1 配置 IntelliJ IDEA 的 Web 服务器

IntelliJ IDEA 是一款被人们广泛认可的 Java 编程语言的集成开发环境，由 JetBrains 公司开发。IntelliJ IDEA 的强大之处在于其智能代码助手、代码自动提示、重构、Java EE 支持、各类版本工具(如 git、svn 等)、JUnit、CVS 整合、代码分析、创新的 GUI 设计等功能。无论是初学者还是经验丰富的开发者,IntelliJ IDEA 都能够帮助他们提高工作效率,并提升代码质量。

IntelliJ IDEA 提供两个版本,分别为旗舰版(Ultimate)和社区版(Community)。旗舰版需要付费,社区版是免费、开源的。建议读者使用 edu 邮箱或学信网信息注册 IntelliJ IDEA 的旗舰版(每年注册一次),为以后的 Java 企业级开发奠定基础。

IntelliJ IDEA 的下载地址是 https://www.jetbrains.com/idea/download/。本书下载的是 IntelliJ IDEA 旗舰版,文件是 ideaIU-2023.2.1.exe,双击该文件并接受所有默认选项即可完成 IntelliJ IDEA 的安装。虽然 IntelliJ IDEA 自带 OpenJDK,但建议读者在使用前先安装 JDK 和 Web 服务器。

❶ 安装 JDK

登录 Oracle 官方网站(http://www.oracle.com/technetwork/java)根据操作系统的位数下载相应的 JDK,例如 64 位的系统使用 64 位的 JDK。本书采用的 JDK 是 jdk-20_windows-x64_bin.exe。

JDK20 的安装程序默认将常用的开发工具(包括 java.exe、javac.exe、javaw.exe 以及 jshell.exe)自动复制到 C:\Program Files\Common Files\Oracle\Java\javapath 目录中,并将该目录自动添加到 Path 环境变量中,如果仅需要使用这些常用的开发工具,不再需要配置环境变量。

❷ Web 服务器

登录 Apache 软件基金会的官方网站(http://jakarta.Apache.org/tomcat),下载 Tomcat 10.0 的免安装版(本书采用 apache-tomcat-10.0.23-windows-x64.zip)。在登录网站后,首先在 Download 中选择 Tomcat 10,然后在 Binary Distributions 的 Core 中选择相应版本。

在安装 Tomcat 之前需要先安装 JDK 并配置系统环境变量 Java_Home。将下载的 apache-tomcat-10.0.23-windows-x64.zip 解压到某个目录下,解压缩后将出现如图 1.2 所示的 Tomcat 目录结构。

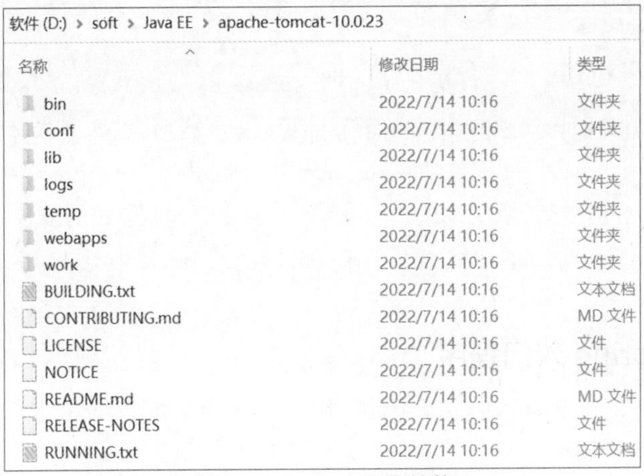

图 1.2 Tomcat 目录结构

执行 Tomcat 根目录下 bin 文件夹中的 startup.bat 来启动 Tomcat 服务器。执行 startup.bat 启动 Tomcat 服务器会占用一个 MS-DOS 窗口,如果关闭当前 MS-DOS 窗口将关闭 Tomcat 服务器。

Tomcat 服务器启动后,在浏览器的地址栏中输入"http://localhost:8080",将出现如图 1.3 所示的 Tomcat 测试页面。

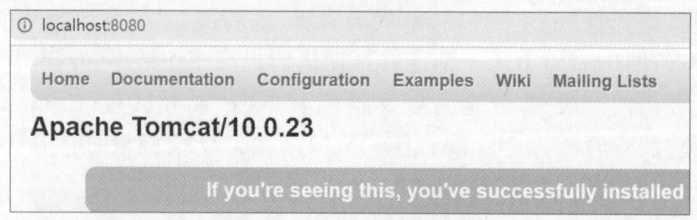

图 1.3　Tomcat 测试页面

❸ 集成 Tomcat

启动 IntelliJ IDEA，选择 File→Settings 菜单项，在弹出的对话框中选择 Build，Execution，Deployment→Application Servers，然后单击＋，选择 Tomcat Server，弹出如图 1.4 所示的 Tomcat Server 界面，在此选择 Tomcat 目录。

图 1.4　Tomcat 配置界面

在图 1.4 中单击 OK 按钮，即可完成 Tomcat 配置。

至此，可以使用 IntelliJ IDEA 创建 Java Web 应用，并在 Tomcat 下运行。

▶1.2.2　Spring 的下载及目录结构

在使用 Spring 框架开发应用程序时，需要引用 Spring 框架自身的 JAR 包。Spring 框架的 JAR 包可以从 Maven 中央库（https://mvnrepository.com/）获得，在编写本书时，Spring 框架的最新版本是 6.1.5。

在 Spring 的 JAR 包中有 4 个基础包，即 spring-core-6.1.5.jar、spring-beans-6.1.5.jar、spring-context-6.1.5.jar 和 spring-expression-6.1.5.jar，分别对应 Spring 核心容器的 4 个模块，即 spring-core 模块、spring-beans 模块、spring-context 模块和 spring-expression 模块。

对于 Spring 框架的初学者，在开发 Spring 应用时，只需要将 Spring 的 4 个基础包和 Spring Commons Logging Bridge 对应的 JAR 包 spring-jcl-6.1.5.jar 复制到 Web 应用的 WEB-INF/lib 目录下即可。

▶1.2.3　第一个 Spring 入门程序

【例 1-1】　通过一个简单的入门程序向读者演示 Spring 框架的使用过程。

其具体实现步骤如下。

❶ 使用 IntelliJ IDEA 创建模块并导入 JAR 包

首先使用 IntelliJ IDEA 创建一个名为 ch1 的项目，如图 1.5 所示。

然后在项目 ch1 中创建一个名为 ch1_1 的模块，如图 1.6 所示。

接着右击模块名 ch1_1，选择 Add Frameworks Support 菜单项（如果找不到 Add Frameworks Support 菜单项，选中模块名双击 Shift 键，在 Actions 选项中打开 Add Frameworks Support），给模块 ch1_1 添加 Web Application，如图 1.7 所示。

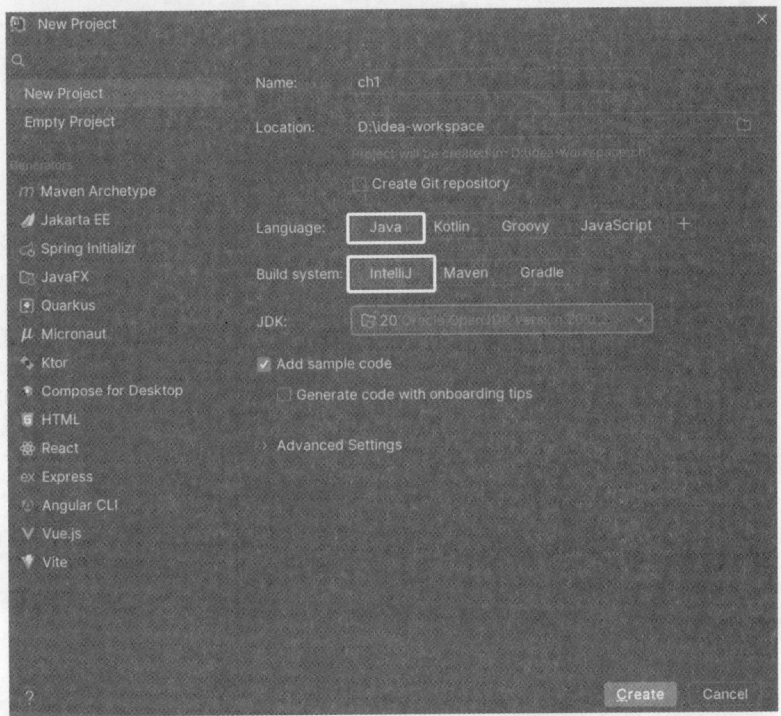

图1.5 使用IntelliJ IDEA创建项目

图1.6 使用IntelliJ IDEA创建模块

最后将Spring的4个基础包和Spring Commons Logging Bridge对应的JAR包spring-jcl-6.1.5.jar复制到ch1_1的WEB-INF/lib目录中。具体做法是：在WEB-INF目录下创建lib目录，将JAR包复制到lib中，然后右击选择Add as Library，添加为模块依赖，如图1.8所示。

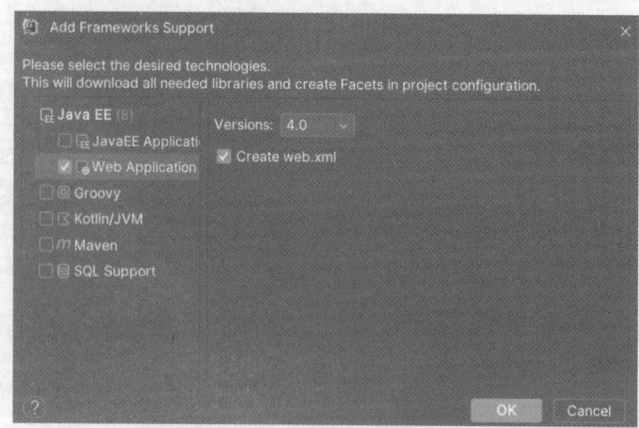

图 1.7　给模块添加 Web Application

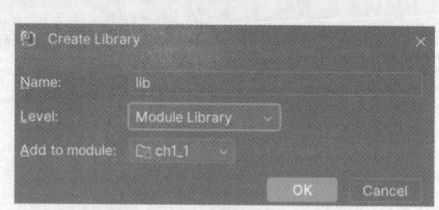

图 1.8　给模块添加依赖

注意：在讲解 Spring MVC 框架前，本书的实例并没有真正运行 Java Web 应用。创建 Java Web 应用的目的是方便添加相关 JAR 包。

❷ 创建接口 TestDao

Spring 解决的是业务逻辑层和其他各层的耦合问题，因此它将面向接口的编程思想贯穿整个应用系统。

在模块 ch1_1 的 src 目录下创建一个 dao 包，并在 dao 包中创建接口 TestDao，在接口中定义一个 sayHello()方法，代码如下：

```java
package dao;
public interface TestDao {
    public void sayHello();
}
```

❸ 创建接口 TestDao 的实现类 TestDaoImpl

在 dao 包下创建 TestDao 的实现类 TestDaoImpl，代码如下：

```java
package dao;
public class TestDaoImpl implements TestDao{
    @Override
    public void sayHello() {
        System.out.println("Hello, Study hard!");
    }
}
```

❹ 创建配置文件 applicationContext.xml

在模块 ch1_1 的 src 目录下创建 Spring 的配置文件 applicationContext.xml，并在该文件中使用实现类 TestDaoImpl 创建一个 id 为 test 的 Bean，代码如下：

```xml
<?xml version="1.0" encoding="UTF-8"?>
<beans xmlns="http://www.springframework.org/schema/beans
```

```
    xmlns:xsi="http://www.w3.org/2001/XMLSchema-instance"
    xsi:schemaLocation="http://www.springframework.org/schema/beans
       http://www.springframework.org/schema/beans/spring-beans.xsd">
    <!-- 将指定类 TestDaoImpl 配置给 Spring,让 Spring 创建其实例 -->
    <bean id="test" class="dao.TestDaoImpl" />
</beans>
```

注意：配置文件的名称可以自定义，但习惯上命名为 applicationContext.xml，有时候也命名为 beans.xml。有关 Bean 的创建将在本书后续章节详细讲解，这里读者只需了解即可。另外，配置文件信息不需要读者手写，可以从 Spring 的帮助文档中复制（使用浏览器打开 https://docs.spring.io/spring-framework/docs/current/reference/html/core.html#spring-core，在 1.2.1 Configuration Metadata 下即可找到配置文件的约束信息）。

❺ 创建测试类

在模块 ch1_1 的 src 目录下创建一个 test 包，并在 test 包中创建 Test 类，代码如下：

```
package test;
import org.springframework.context.ApplicationContext;
import org.springframework.context.support.ClassPathXmlApplicationContext;
import dao.TestDao;
public class Test {
    public static void main(String[] args) {
        ApplicationContext appCon = new ClassPathXmlApplicationContext("applicationContext.xml");
        //从容器中获取 test 实例
        TestDao tt = appCon.getBean("test", TestDao.class);          //test 为配置文件中的 id
        tt.sayHello();
    }
}
```

在执行上述 main() 方法后，将在控制台中输出"Hello，Study hard!"。在上述 main() 方法中并没有使用 new 运算符创建 TestDaoImpl 类的对象，而是通过 Spring IoC 容器来获取实现类对象，这就是 Spring IoC 的工作机制（1.3 节将详细讲解 Spring IoC 的工作机制）。

1.3 Spring IoC

▶ 1.3.1 Spring IoC 的基本概念

控制反转（IoC）是一个比较抽象的概念，是 Spring 框架的核心，用来消减计算机程序的耦合问题。依赖注入（DI）是 IoC 的另外一种说法，只是从不同的角度描述相同的概念。下面通过实际生活中的一个例子解释 IoC 和 DI。

当人们需要一件东西时，第一反应就是找东西。例如想吃面包，在没有面包店的情况下，最直观的做法可能是自己按照口味制作面包，也就是需要主动制作面包。然而时至今日，各种面包店盛行，如果自己不想制作面包，可以把自己的口味告诉店家，由店家制作，然后购买。注意这里自己并没有制作面包，而是由店家制作，但是面包完全符合自己的口味。

上面只是举了一个非常简单的例子，但包含了控制反转的思想，即把制作面包的主动权交给店家。下面通过面向对象编程思想继续探讨这两个概念。

当某个 Java 对象（调用者，比如您）需要调用另一个 Java 对象（被调用者，即被依赖对象，比如面包）时，在传统编程模式下，调用者通常会采用"new 被调用者"的代码方式来创建对象（比如您自己制作面包）。这种方式会增加调用者与被调用者之间的耦合性，不利于后期代码的升级与维护。

当 Spring 框架出现后，对象的实例不再由调用者来创建，而是由 Spring 容器（比如面包店）来创建。Spring 容器会负责控制程序之间的关系（比如面包店负责控制您与面包的关系），而不是由调用者的程序代码直接控制。这样，控制权由调用者转移到 Spring 容器，控制权发生了反转，这就是 Spring 的控制反转。

从 Spring 容器角度来看，Spring 容器负责将被依赖对象赋值给调用者的成员变量，相当于为调用者注入它所依赖的实例，这就是 Spring 的依赖注入，主要目的是解耦，体现一种"组合"的理念。

综上所述，控制反转是一种通过描述（在 Spring 中可以是 XML 或注解）并通过第三方去产生或获取特定对象的方式。在 Spring 中实现控制反转的是 IoC 容器，其实现方法是依赖注入。

▶1.3.2 Spring 的常用注解

在 Spring 框架中，尽管使用 XML 配置文件可以很简单地装配 Bean，但如果应用中有大量的 Bean 需要装配，会导致 XML 配置文件过于庞大，不方便以后的升级和维护。因此，更多的时候推荐开发者使用注解（annotation）的方式去装配 Bean。

需要注意的是，基于注解的装配需要使用＜context:component-scan＞元素或@ComponentScan 注解定义包（注解所在的包）扫描的规则，然后根据定义的规则找出哪些类（Bean）需要自动装配到 Spring 容器中，然后交由 Spring 进行统一管理。

Spring 框架基于 AOP 编程（面向切面编程）实现注解解析，因此在使用注解编程时需要导入 spring-aop-6.1.5.jar 包。

❶ 声明 Bean 的注解

1）@Component

该注解是一个泛化的概念，仅表示一个组件对象（Bean），可以作用在任何层次上，没有明确的角色。

2）@Repository

该注解用于将数据访问层（DAO）的类标识为 Bean，即注解数据访问层 Bean，其功能与@Component()相同。

3）@Service

该注解用于标注一个业务逻辑组件类（Service 层），其功能与@Component()相同。

4）@Controller

该注解用于标注一个控制器组件类（Spring MVC 的 Controller），其功能与@Component()相同。

❷ 注入 Bean 的注解

1）@Autowired

该注解可以对类的成员变量、方法及构造方法进行标注，完成自动装配的工作。通过使用@Autowired 来消除 setter 和 getter 方法。默认按照 Bean 的类型进行装配。

2）@Resource

该注解与@Autowired 的功能一样，区别在于该注解默认是按照名称来装配注入的，只有当找不到与名称匹配的 Bean 时才会按照类型来装配注入；而@Autowired 默认按照 Bean 的类型进行装配，如果想按照名称来装配注入，则需要和@Qualifier 注解一起使用。

@Resource 注解有两个属性，分别是 name 和 type。name 属性指定 Bean 的实例名称，即按照名称来装配注入;type 属性指定 Bean 类型,即按照 Bean 的类型进行装配。

3)@Qualifier

该注解与@Autowired 注解配合使用。当@Autowired 注解需要按照名称来装配注入时,需要和该注解一起使用,Bean 的实例名称由@Qualifier 注解的参数指定。

▶1.3.3 基于注解的依赖注入

Spring IoC 容器(ApplicationContext)负责创建和注入 Bean。Spring 提供了使用 XML 配置、注解、Java 配置以及 groovy 配置实现 Bean 的创建和注入。本书尽量使用注解(@Component、@Repository、@Service 以及@Controller 等业务 Bean 的配置)和 Java 配置(全局配置如数据库、MVC 等相关配置)完全代替 XML 配置,这也是 Spring Boot 推荐的配置方式。

下面通过一个简单实例向读者演示基于注解的依赖注入的使用过程。

【例 1-2】 基于注解的依赖注入的使用过程。

其具体实现步骤如下。

❶ 使用 IntelliJ IDEA 创建模块 ch1_2 并导入 JAR 包

参考 1.2.3 节,在项目 ch1 中创建一个名为 ch1_2 的模块,同时给模块 ch1_2 添加 Web Application,并将 Spring 的 4 个基础包、Spring Commons Logging Bridge 对应的 JAR 包 spring-jcl-6.1.5.jar、jakarta.annotation.Resource 注解类对应的 JAR 包 annotations-api.jar 以及 spring-aop-6.1.5.jar(本节扫描注解,需要事先导入 Spring AOP 的 JAR 包)复制到 ch1_2 的 WEB-INF/lib 目录中,添加为模块依赖。

❷ 创建 DAO 层

在模块 ch1_2 的 src 中创建 annotation.dao 包,在该包下创建 TestDao 接口和 TestDaoImpl 实现类,并将实现类 TestDaoImpl 使用@Repository 注解标注为数据访问层。

TestDao 的代码如下:

```
package annotation.dao;
public interface TestDao {
    public void save();
}
```

TestDaoImpl 的代码如下:

```
package annotation.dao;
import org.springframework.stereotype.Repository;
@Repository("testDaoImpl")
/**相当于@Repository,但如果在 Service 层使用@Resource(name="testDaoImpl"),testDaoImpl 不能省略。**/
public class TestDaoImpl implements TestDao{
    @Override
    public void save() {
        System.out.println("testDao save");
    }
}
```

❸ 创建 Service 层

在模块 ch1_2 的 src 中创建 annotation.service 包,在该包下创建 TestService 接口和 TestServiceImpl 实现类,并将实现类 TestServiceImpl 使用@Service 注解标注为业务逻辑层。

TestService 的代码如下:

```
package annotation.service;
```

```java
public interface TestService {
    public void save();
}
```

TestServiceImpl 的代码如下：

```java
package annotation.service;
import jakarta.annotation.Resource;
import org.springframework.stereotype.Service;
import annotation.dao.TestDao;
@Service("testServiceImpl")                              //相当于@Service
public class TestServiceImpl implements TestService{
    @Resource(name="testDaoImpl")
    /**相当于@Autowired,@Autowired 默认按照 Bean 类型注入**/
    private TestDao testDao;
    @Override
    public void save() {
        testDao.save();
        System.out.println("testService save");
    }
}
```

❹ 创建 Controller 层

在模块 ch1_2 的 src 中创建 annotation.controller 包，在该包下创建 TestController 类，并将 TestController 类使用@Controller 注解标注为控制器层。

TestController 的代码如下：

```java
package annotation.controller;
import org.springframework.beans.factory.annotation.Autowired;
import org.springframework.stereotype.Controller;
import annotation.service.TestService;
@Controller
public class TestController {
    @Autowired
    private TestService testService;
    public void save() {
        testService.save();
        System.out.println("testController save");
    }
}
```

❺ 创建配置文件

在使用注解时，在 Spring 的配置文件中需要使用"＜context:component-scan base-package="Bean 所在的包路径"/＞"语句扫描使用注解的包，Spring IoC 容器根据 XML 配置文件的扫描信息，将包以及子包中使用注解的类的实例提供给应用程序使用。

在模块 ch1_2 的 src 中创建名为 config 的包，并在该包中创建名为 applicationContext.xml 的配置文件。

applicationContext.xml 的代码如下：

```xml
<?xml version="1.0" encoding="UTF-8"?>
<beans xmlns="http://www.springframework.org/schema/beans"
    xmlns:xsi="http://www.w3.org/2001/XMLSchema-instance"
    xmlns:context="http://www.springframework.org/schema/context"
    xsi:schemaLocation="http://www.springframework.org/schema/beans
        http://www.springframework.org/schema/beans/spring-beans.xsd
        http://www.springframework.org/schema/context
```

```
          http://www.springframework.org/schema/context/spring-context.xsd">
    <!-- 扫描annotation包及其子包中的注解 -->
    <context:component-scan base-package="annotation"/>
</beans>
```

❻ 创建测试类

在模块 ch1_2 的 src 目录中创建名为 annotation.test 的包,并在该包中创建测试类 TestAnnotation,具体代码如下:

```
package annotation.test;
import org.springframework.context.ApplicationContext;
import annotation.controller.TestController;
import org.springframework.context.support.ClassPathXmlApplicationContext;
public class TestAnnotation {
    public static void main(String[] args) {
        //初始化Spring容器ApplicationContext
        ApplicationContext appCon =
                new ClassPathXmlApplicationContext("config/applicationContext.xml");
        TestController tt = (TestController)appCon.getBean("testController");
        tt.save();
    }
}
```

❼ 运行结果

运行测试类 TestAnnotation 的 main 方法,运行结果如图 1.9 所示。

图 1.9 ch1_2 应用的运行结果

1.4 Spring AOP

Spring AOP 是 Spring 框架体系结构中非常重要的功能模块之一,该模块提供了面向切面编程的实现。面向切面编程在事务处理、日志记录、安全控制等操作中被广泛使用。

▶ 1.4.1 Spring AOP 的基本概念

❶ AOP 的概念

AOP(Aspect-Oriented Programming)即面向切面编程。它与 OOP(Object-Oriented Programming,面向对象编程)相辅相成,提供了与 OOP 不同的抽象软件结构的视角。在 OOP 中,以类作为程序的基本单元,而 AOP 中的基本单元是 Aspect(切面)。Struts 2 的拦截器设计就是基于 AOP 的思想,是一个比较经典的应用。

在业务处理代码中,通常都有日志记录、性能统计、安全控制、事务处理、异常处理等操作。尽管使用 OOP 可以通过封装或继承的方式达到代码的重用,但仍然存在同样的代码分散到各个方法中。因此,采用 OOP 处理日志记录等操作,不仅增加了开发者的工作量,而且加大了升级和维护的困难。为了解决此类问题,AOP 思想应运而生。AOP 采取横向抽取机制,即将分散在各个方法中的重复代码提取出来,然后在程序编译或运行阶段将这些代码应用到需要执行的地方。这种横向抽取机制,采用传统的 OOP 是无法办到的,因为 OOP 实现的是父子关系的纵向重用。AOP 不是 OOP 的替代品,而是 OOP 的补充,它们相辅相成。

在 AOP 中,横向抽取机制的类与切面的关系如图 1.10 所示。

从图 1.10 可以看出,通过切面 Aspect 分别在业务类 1 和业务类 2 中加入了日志记录、性能统计、安全控制、事务处理、异常处理等操作。

图 1.10　AOP 中类与切面的关系

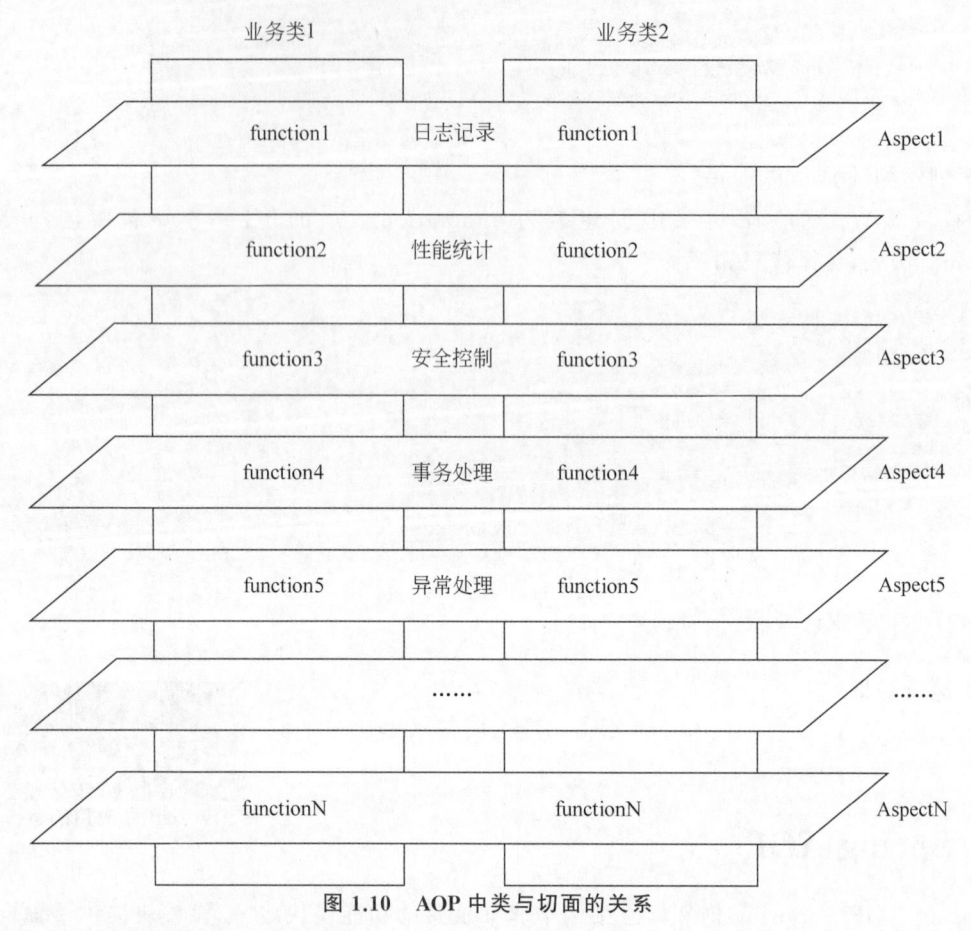

❷ **AOP 的术语**

在 Spring AOP 框架中涉及以下常用术语。

1）切面

切面（Aspect）是指封装横切到系统功能（如事务处理）的类。

2）连接点

连接点（Joinpoint）是指程序运行中的一些时间点，如方法的调用或异常的抛出。

3）切入点

切入点（Pointcut）是指那些需要处理的连接点。在 Spring AOP 中，所有的方法执行都是连接点，而切入点是一个描述信息，它修饰的是连接点，通过切入点确定哪些连接点需要被处理。切面、连接点和切入点的关系如图 1.11 所示。

4）通知（增强处理）

通知是由切面添加到特定的连接点（满足切入点规则）的一段代码，即在定义好的切入点处所要执行的程序代码。可以将其理解为切面开启后切面的方法。因此，通知是切面的具体实现。

5）引入

引入（Introduction）允许在现有的实现类中添加自定义的方法和属性。

6）目标对象

目标对象（Target Object）是指所有被通知的对象。如果 AOP 框架使用运行时代理的方式（动态的 AOP）来实现切面，那么通知对象总是一个代理对象。

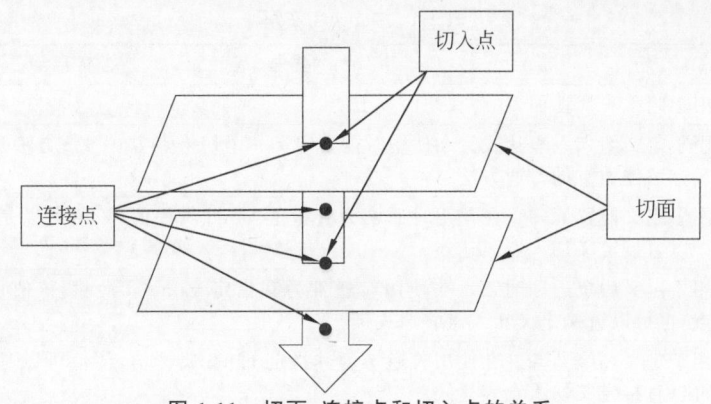

图 1.11　切面、连接点和切入点的关系

7）代理

代理（Proxy）是通知应用到目标对象之后被动态创建的对象。

8）组入

组入（Weaving）是将切面代码插入目标对象上，从而生成代理对象的过程。根据不同的实现技术，AOP 组入有三种方式：编译器组入需要有特殊的 Java 编译器；类装载器组入需要有特殊的类装载器；动态代理组入是在运行期为目标类添加通知生成子类的方式。Spring AOP 框架默认采用动态代理组入，而 AspectJ（基于 Java 语言的 AOP 框架）采用编译器组入和类装载器组入。

▶1.4.2　基于注解开发 AspectJ

基于注解开发 AspectJ 要比基于 XML 配置开发 AspectJ 便捷许多，所以在实际开发中推荐使用注解方式。在讲解 AspectJ 之前，先了解一下 Spring 的通知类型。根据 Spring 中通知在目标类方法中的连接点位置，通知可以分为如下 6 种类型：

❶ 环绕通知

环绕通知在目标方法执行前和执行后实施增强，可以应用于日志记录、事务处理等。

❷ 前置通知

前置通知在目标方法执行前实施增强，可以应用于权限管理等。

❸ 后置返回通知

后置返回通知在目标方法成功执行后实施增强，可以应用于关闭流、删除临时文件等。

❹ 后置（最终）通知

后置通知在目标方法执行后实施增强，与后置返回通知不同的是，不管是否发生异常都要执行该通知，该通知可以应用于释放资源。

❺ 异常通知

异常通知在方法抛出异常后实施增强，可以应用于处理异常、记录日志等。

❻ 引入通知

引入通知在目标类中添加一些新的方法和属性，可以应用于修改目标类（增强类）。

有关 AspectJ 注解的内容如表 1.1 所示。

表 1.1　AspectJ 注解

注 解 名 称	描　　述
@Aspect	用于定义一个切面，注解在切面类上
@Pointcut	用于定义切入点表达式。在使用时需要定义一个切入点方法。该方法是一个返回值 void，且方法体为空的普通方法
@Before	用于定义前置通知。在使用时通常为其指定 value 属性值，该值可以是已有的切入点，也可以直接定义切入点表达式
@AfterReturning	用于定义后置返回通知。在使用时通常为其指定 value 属性值，该值可以是已有的切入点，也可以直接定义切入点表达式
@Around	用于定义环绕通知。在使用时通常为其指定 value 属性值，该值可以是已有的切入点，也可以直接定义切入点表达式
@AfterThrowing	用于定义异常通知。在使用时通常为其指定 value 属性值，该值可以是已有的切入点，也可以直接定义切入点表达式。另外还有一个 throwing 属性，用于访问目标方法抛出的异常，该属性值与异常通知方法中同名的形参一致
@After	用于定义后置(最终)通知。在使用时通常为其指定 value 属性值，该值可以是已有的切入点，也可以直接定义切入点表达式

下面通过一个实例讲解基于注解开发 AspectJ 的过程。

【例 1-3】 基于注解开发 AspectJ 的过程。该实例的具体要求：首先在 DAO 层的实现类中定义 save、modify 和 delete 三个待增强的方法；然后使用@Aspect 注解定义一个切面，在该切面中定义各类型通知，增强 DAO 层中的 save、modify 和 delete 方法。

其具体实现步骤如下。

❶ 使用 IntelliJ IDEA 创建模块 ch1_3 并导入 JAR 包

在项目 ch1 中创建一个名为 ch1_3 的模块，同时给模块 ch1_3 添加 Web Application，并将 Spring 的 4 个基础包、Spring Commons Logging Bridge 对应的 JAR 包 spring-jcl-6.1.5.jar、spring-aop-6.1.5.jar、spring-aspects-6.1.5.jar(Spring 为 AspectJ 提供的实现)以及 AspectJ 框架所提供的规范包 aspectjweaver-1.9.22.jar（https://mvnrepository.com/artifact/org.aspectj/aspectjweaver）复制到 ch1_3 的 WEB-INF/lib 目录中，添加为模块依赖。

❷ 创建接口及实现类

在 src 目录中创建一个名为 aspectj.dao 的包，并在该包中创建接口 TestDao 和接口实现类 TestDaoImpl。该实现类作为目标类，在切面类中对其方法进行增强处理。使用@Repository 注解将目标类 aspectj.dao.TestDaoImpl 注解为目标对象。

TestDao 的代码如下：

```java
package aspectj.dao;
public interface TestDao {
    public void save();
    public void modify();
    public void delete();
}
```

TestDaoImpl 的代码如下：

```java
package aspectj.dao;
import org.springframework.stereotype.Repository;
@Repository("testDao")
public class TestDaoImpl implements TestDao{
```

第 1 章　Spring

```
    @Override
    public void save() {
        System.out.println("保存");
    }
    @Override
    public void modify() {
        System.out.println("修改");
    }
    @Override
    public void delete() {
        System.out.println("删除");
    }
}
```

❸ 创建切面类

在 src 目录中创建一个名为 aspectj.annotation 的包，并在该包中创建切面类 MyAspect。在该类中首先使用@Aspect 注解定义一个切面类，由于该类在 Spring 中是作为组件使用的，所以还需要使用@Component 注解；然后使用@Pointcut 注解定义切入点表达式，并通过定义方法来表示切入点名称；最后在每个通知方法上添加相应的注解，并将切入点名称作为参数传递给需要执行增强的通知方法。

MyAspect 的核心代码如下：

```
/**
 * 切面类,在此类中编写各种类型通知
 */
@Aspect                               //@Aspect 声明一个切面
@Component                            //@Component 让此切面成为 Spring 容器管理的 Bean
public class MyAspect {
    /**
     * 定义切入点,通知增强哪些方法。
     "execution(* aspectj.dao.*.*(..))"是定义切入点表达式,
     该切入点表达式的意思是匹配 aspectj.dao 包中任意类的任意方法的执行。
     execution()是表达式的主体,第一个 * 表示的是返回类型,使用 * 代表所有类型；
     aspectj.dao 表示的是需要匹配的包名,后面第二个 * 表示的是类名,使用 * 代表匹配包中所有的类；
     第三个 * 表示的是方法名,使用 * 表示所有方法；后面的(..)表示方法的参数,其中".."表示任意参数。
     另外,注意第一个 * 与包名之间有一个空格。
     */
    @Pointcut("execution(* aspectj.dao.*.*(..))")
    private void myPointCut() {
    }
    /**
     * 前置通知,使用 JoinPoint 接口作为参数获得目标对象信息
     */
    @Before("myPointCut()")            //myPointCut()是切入点的定义方法
    public void before(JoinPoint jp) {
        System.out.print("前置通知:模拟权限控制");
        System.out.println(",目标类对象:" + jp.getTarget()
            + ",被增强处理的方法:" + jp.getSignature().getName());
    }
    /**
     * 后置返回通知
     */
    @AfterReturning("myPointCut()")
    public void afterReturning(JoinPoint jp) {
        System.out.print("后置返回通知:" + "模拟删除临时文件");
        System.out.println(",被增强处理的方法:" + jp.getSignature().getName());
    }
    /**
```

```java
 * 环绕通知
 * ProceedingJoinPoint 是 JoinPoint 子接口,代表可以执行的目标方法
 * 返回值的类型必须是 Object
 * 必须一个参数是 ProceedingJoinPoint 类型
 * 必须 throws Throwable
 */
@Around("myPointCut()")
public Object around(ProceedingJoinPoint pjp) throws Throwable{
    //开始
    System.out.println("环绕开始:执行目标方法前,模拟开启事务");
    //执行当前目标方法
    Object obj = pjp.proceed();
    //结束
    System.out.println("环绕结束:执行目标方法后,模拟关闭事务");
    return obj;
}
/**
 * 异常通知
 */
@AfterThrowing(value="myPointCut()",throwing="e")
public void except(Throwable e) {
    System.out.println("异常通知:" + "程序执行异常" + e.getMessage());
}
/**
 * 后置(最终)通知
 */
@After("myPointCut()")
public void after() {
    System.out.println("最终通知:模拟释放资源");
}
}
```

❹ 创建配置文件

在 src 目录中创建一个名为 aspectj.config 的包,并在该包中创建配置文件 applicationContext.xml,在配置文件中指定需要扫描的包,使注解生效,同时需要启动基于注解的 AspectJ 支持。applicationContext.xml 的代码如下:

```xml
<?xml version="1.0" encoding="UTF-8"?>
<beans xmlns="http://www.springframework.org/schema/beans"
    xmlns:xsi="http://www.w3.org/2001/XMLSchema-instance"
    xmlns:aop="http://www.springframework.org/schema/aop"
    xmlns:context="http://www.springframework.org/schema/context"
    xsi:schemaLocation="http://www.springframework.org/schema/beans
        http://www.springframework.org/schema/beans/spring-beans.xsd
        http://www.springframework.org/schema/aop
        http://www.springframework.org/schema/aop/spring-aop.xsd
        http://www.springframework.org/schema/context
        http://www.springframework.org/schema/context/spring-context.xsd">
    <!-- 指定需要扫描的包,使注解生效 -->
    <context:component-scan base-package="aspectj"/>
    <!-- 启动基于注解的 AspectJ 支持 -->
    <aop:aspectj-autoproxy/>
</beans>
```

❺ 创建测试类

在 src 目录中创建一个名为 aspectj.test 的包,并在该包中创建测试类 AspectjAOPTest。AspectjAOPTest 的核心代码如下:

```java
public class AspectjAOPTest {
    public static void main(String[] args) {
        ApplicationContext appCon =
            new ClassPathXmlApplicationContext("aspectj/config/applicationContext.xml");
        //从容器中获取增强后的目标对象
        TestDao testDaoAdvice = (TestDao) appCon.getBean("testDao");
        //执行方法
        testDaoAdvice.save();
        System.out.println("==============");
        testDaoAdvice.modify();
        System.out.println("==============");
        testDaoAdvice.delete();
    }
}
```

❻ 运行测试类

运行测试类 AspectjAOPTest 的 main 方法，运行结果如图 1.12 所示。

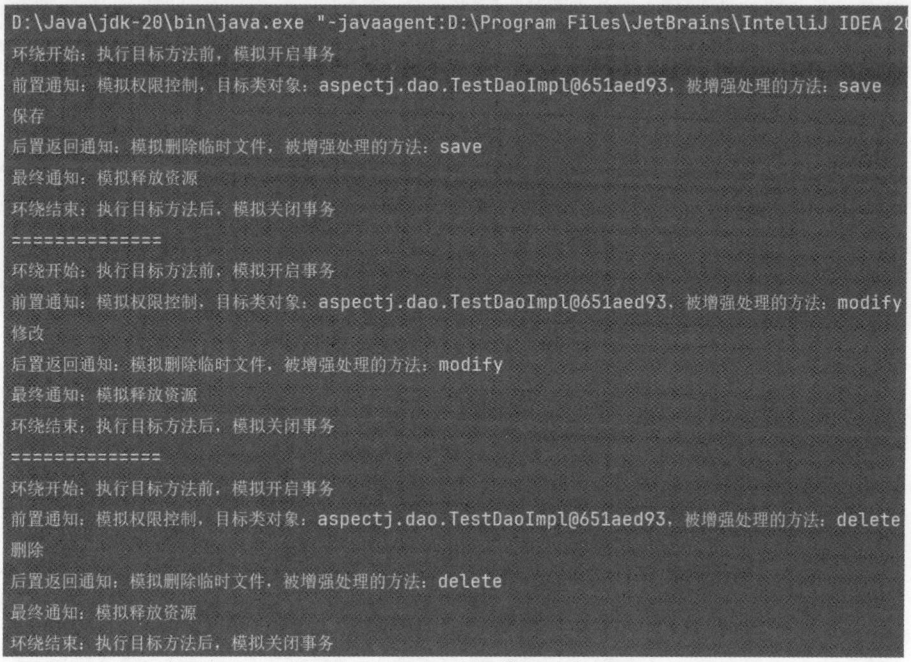

图 1.12　ch1_3 应用的运行结果

1.5　Spring Bean

在 Spring 应用中，Spring IoC 容器可以创建、装配和配置应用组件对象，这里的组件对象称为 Bean。

▶ 1.5.1　Bean 的实例化

在面向对象编程中，如果想使用某个对象，需要事先实例化该对象。同样，在 Spring 框架中，如果想使用 Spring 容器中的 Bean，也需要实例化 Bean。Spring 框架实例化 Bean 有三种方式，分别为构造方法实例化、静态工厂实例化和实例工厂实例化（其中，最常用的实例化方法是构造方法实例化）。使用实例工厂实例化具有解耦、降低代码复用等优点。

下面通过一个实例来演示 Bean 的实例化过程。

【例 1-4】 Bean 的实例化过程。

其具体实现步骤如下。

❶ **使用 IntelliJ IDEA 创建模块 ch1_4 并导入 JAR 包**

在项目 ch1 中创建一个名为 ch1_4 的模块，同时给模块 ch1_4 添加 Web Application，并将 Spring 的 4 个基础包、Spring Commons Logging Bridge 对应的 JAR 包 spring-jcl-6.1.5.jar 以及 spring-aop-6.1.5.jar 复制到 ch1_4 的 WEB-INF/lib 目录中，添加为模块依赖。

❷ **创建实例化 Bean 的类**

在 src 目录中创建一个名为 instance 的包，并在该包中创建 BeanClass、BeanInstanceFactory 以及 BeanStaticFactory 等实例化 Bean 的类。

BeanClass 的代码如下：

```java
package instance;
public class BeanClass {
    public String message;
    public BeanClass() {
        message = "构造方法实例化 Bean";
    }
    public BeanClass(String s) {
        message = s;
    }
}
```

BeanInstanceFactory 的代码如下：

```java
package instance;
public class BeanInstanceFactory {
    public BeanClass createBeanClassInstance() {
        return new BeanClass("调用实例工厂方法实例化 Bean");
    }
}
```

BeanStaticFactory 的代码如下：

```java
package instance;
public class BeanStaticFactory {
    private static BeanClass beanInstance = new BeanClass("调用静态工厂方法实例化 Bean");
    public static BeanClass createInstance() {
        return beanInstance;
    }
}
```

❸ **创建配置文件**

在 src 目录中创建名为 config 的包，并在该包中创建名为 applicationContext.xml 的配置文件，在配置文件中分别使用 BeanClass、BeanInstanceFactory 以及 BeanStaticFactory 实例化 Bean。

applicationContext.xml 的代码如下：

```xml
<?xml version="1.0" encoding="UTF-8"?>
<beans xmlns="http://www.springframework.org/schema/beans"
    xmlns:xsi="http://www.w3.org/2001/XMLSchema-instance"
    xsi:schemaLocation="http://www.springframework.org/schema/beans
        http://www.springframework.org/schema/beans/spring-beans.xsd">
    <!-- 构造方法实例化 Bean -->
    <bean id="constructorInstance" class="instance.BeanClass"/>
```

```xml
    <!-- 使用静态工厂方法实例化 Bean,createInstance 为静态工厂类 BeanStaticFactory 中的静态方
法-->
    <bean id="staticFactoryInstance"
        class="instance.BeanStaticFactory" factory-method="createInstance"/>
    <!-- 配置工厂 -->
    <bean id="myFactory" class="instance.BeanInstanceFactory"/>
    <!-- 使用 factory-bean 属性指定配置工厂,使用 factory-method 属性指定使用工厂中的哪个方法
实例化 Bean-->
    <bean id="instanceFactoryInstance"
        factory-bean="myFactory" factory-method="createBeanClassInstance"/>
</beans>
```

❹ 创建测试类

在 src 目录中创建 test 包,并在该包中创建测试类 TestInstance,核心代码如下:

```java
public class TestInstance {
    public static void main(String[] args) {
        ApplicationContext appCon =
            new ClassPathXmlApplicationContext("config/applicationContext.xml");
        //测试构造方法实例化 Bean
        BeanClass b1 = (BeanClass)appCon.getBean("constructorInstance");
        System.out.println(b1+ b1.message);
        //测试静态工厂方法实例化 Bean
        BeanClass b2 = (BeanClass)appCon.getBean("staticFactoryInstance");
        System.out.println(b2 + b2.message);
        //测试实例工厂方法实例化 Bean
        BeanClass b3 = (BeanClass)appCon.getBean("instanceFactoryInstance");
        System.out.println(b3 + b3.message);
    }
}
```

❺ 运行测试类

运行测试类 TestInstance 的 main 方法,运行结果如图 1.13 所示。

```
D:\Java\jdk-20\bin\java.exe "-javaagent:D:\Progra
instance.BeanClass@3b2da18f构造方法实例化Bean
instance.BeanClass@5906ebcb调用静态工厂方法实例化Bean
instance.BeanClass@258e2e41调用实例工厂方法实例化Bean
```

图 1.13 ch1_4 应用的运行结果

▶1.5.2 Bean 的作用域

在 Spring 中不仅可以完成 Bean 的实例化,还可以为 Bean 指定作用域。在 Spring 中为 Bean 的实例定义了如表 1.2 所示的作用域,通过 scope 属性来设定。

表 1.2 Bean 的作用域

作用域名称	描 述
singleton	默认的作用域,使用 singleton 定义的 Bean,在 Spring 容器中只有一个 Bean 实例
prototype	Spring 容器每次获取 prototype 定义的 Bean,容器都将创建一个新的 Bean 实例
request	在一次 HTTP 请求中容器将返回一个 Bean 实例,不同的 HTTP 请求返回不同的 Bean 实例。仅在 Web Spring 应用程序上下文中使用
session	在一个 HTTP Session 中,容器将返回同一个 Bean 实例。仅在 Web Spring 应用程序上下文中使用
application	为每个 ServletContext 对象创建一个实例,即同一个应用共享一个 Bean 实例。仅在 Web Spring 应用程序上下文中使用
websocket	为每个 WebSocket 对象创建一个 Bean 实例。仅在 Web Spring 应用程序上下文中使用

在表 1.2 所示的 6 种作用域中，singleton 和 prototype 是最常用的两种，后面 4 种作用域仅用在 Web Spring 应用程序上下文中。下面通过一个实例来演示 Bean 的作用域。

【例 1-5】 Bean 的作用域。该实例的具体要求：在应用 ch1_4 中分别定义作用域为 singleton 和 prototype 的两个 Bean。

其具体实现步骤如下。

❶ 添加配置文件内容

在应用 ch1_4 的配置文件 applicationContext.xml 中定义两个 Bean，一个 Bean 的作用域为 singleton，另一个 Bean 的作用域为 prototype。具体添加的配置内容如下：

```xml
<bean id="scope1" class="instance.BeanClass" scope="singleton"/>
<bean id="scope2" class="instance.BeanClass" scope="prototype"/>
```

❷ 创建测试类

在应用 ch1_4 的 test 包中创建测试类 TestScope，在该测试类中分别获得 id 为 scope1 和 scope2 的 Bean 实例，核心代码如下：

```java
public class TestScope {
    public static void main(String[] args) {
        ApplicationContext appCon =
            new ClassPathXmlApplicationContext("config/applicationContext.xml");
        BeanClass b1 = (BeanClass)appCon.getBean("scope1");
        System.out.println(b1);
        BeanClass b2 = (BeanClass)appCon.getBean("scope1");
        System.out.println(b2);
        System.out.println("==========");
        BeanClass b3 = (BeanClass)appCon.getBean("scope2");
        System.out.println(b3);
        BeanClass b4 = (BeanClass)appCon.getBean("scope2");
        System.out.println(b4);
    }
}
```

❸ 运行测试类

运行测试类 TestScope 的 main 方法，运行结果如图 1.14 所示。

从图 1.14 所示的运行结果可知，两次获取 id 为 scope1 的 Bean 实例时，IoC 容器返回两个相同的 Bean 实例；而两次获取 id 为 scope2 的 Bean 实例时，IoC 容器返回两个不同的 Bean 实例。这说明 singleton 定义的 Bean，在 Spring 容器中只有一个 Bean 实例。

图 1.14 TestScope 的运行结果

▶ 1.5.3 Bean 的初始化和销毁

在实际工程应用中，经常需要在 Bean 使用之前或之后做一些必要的操作，Spring 为 Bean 生命周期的操作提供了支持。在配置文件中定义 Bean 时，可以使用 init-method 和 destroy-method 属性对 Bean 进行初始化和销毁。下面通过一个实例来演示 Bean 的初始化和销毁。

【例 1-6】 Bean 的初始化和销毁。该实例的具体要求：在应用 ch1_4 中首先定义一个 MyService 类，在该类中定义初始化方法和销毁方法；然后在配置文件中使用 init-method 和 destroy-method 属性对 MyService 的 Bean 对象进行初始化和销毁。

具体实现步骤如下。

❶ 创建 Bean 的类

在应用 ch1_4 的 src 目录中创建一个名为 service 的包,并在该包中创建 MyService 类,具体代码如下:

```
package service;
public class MyService {
    public MyService() {
        System.out.println("执行构造方法");
    }
    public void initService() {
        System.out.println("initMethod");
    }
    public void destroyService() {
        System.out.println("destroyMethod");
    }
}
```

❷ 添加配置文件内容

在应用 ch1_4 的配置文件 applicationContext.xml 中配置一个 id 为 beanLife 的 Bean,并使用 init-method 属性指定初始化方法,使用 destroy-method 属性指定销毁方法。具体配置内容如下。

```
<!-- 配置 Bean,使用 init-method 属性指定初始化方法,使用 destroy-method 属性指定销毁方法-->
<bean id="beanLife" class="service.MyService"
        init-method="initService" destroy-method="destroyService"/>
```

❸ 创建测试类

在应用 ch1_4 的 test 包中创建测试类 TestInitAndDestroy,核心代码如下:

```
public class TestInitAndDestroy {
    public static void main(String[] args) {
        //为了方便演示销毁方法的执行,这里使用 ClassPathXmlApplicationContext
        ClassPathXmlApplicationContext appCon =
            new ClassPathXmlApplicationContext("config/applicationContext.xml");
        System.out.println("获得对象前");
        MyService blife = (MyService)appCon.getBean("beanLife");
        System.out.println("获得对象后" + blife);
        appCon.close();                              //关闭容器,销毁 Bean 对象
    }
}
```

❹ 运行测试类

运行测试类 TestInitAndDestroy 的 main 方法,运行结果如图 1.15 所示。

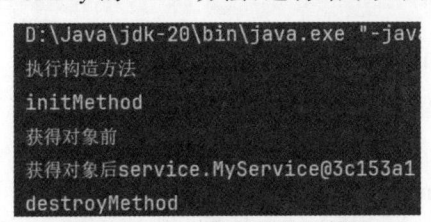

图 1.15 Bean 的初始化和销毁

从图 1.15 可以看出,在加载配置文件时,创建 Bean 对象,执行了 Bean 的构造方法和初始化方法 initService();在获得对象后,关闭容器时,执行了 Bean 的销毁方法 destroyService()。

1.6　Spring 的数据库编程

数据库编程是互联网编程的基础，Spring 框架为开发者提供了 JDBC 模板模式，即 jdbcTemplate，它可以简化许多代码，但在实际应用中 jdbcTemplate 并不常用。在工作中更多的时候用的是 Hibernate 框架和 MyBatis 框架进行数据库编程。本节仅简要介绍 Spring jdbcTemplate 的使用方法，Hibernate 框架和 MyBatis 框架的相关内容不属于本节的内容。

▶ 1.6.1　Spring JDBC 的 XML 配置

本节 Spring 数据库编程主要使用 Spring JDBC 模块的 core 和 dataSource 包。core 包是 JDBC 的核心功能包，包括常用的 JdbcTemplate 类；dataSource 包是访问数据源的工具类包。使用 Spring JDBC 操作数据库，需要对其进行配置。XML 配置文件的示例代码如下：

```xml
<!-- 配置数据源 -->
<bean id="dataSource" class="org.springframework.jdbc.datasource.DriverManagerDataSource">
    <!-- MySQL 数据库驱动 -->
    <property name="driverClassName" value="com.mysql.cj.jdbc.Driver"/>
    <!-- 连接数据库的 URL -->
    <property name="url" value="jdbc:mysql://127.0.0.1:3306/springtest?useUnicode=true&amp;characterEncoding=UTF-8&amp;allowMultiQueries=true&amp;serverTimezone=GMT%2B8"/>
    <!-- 连接数据库的用户名 -->
    <property name="username" value="root"/>
    <!-- 连接数据库的密码 -->
    <property name="password" value="root"/>
</bean>
<!-- 配置 JDBC 模板 -->
<bean id="jdbcTemplate" class="org.springframework.jdbc.core.JdbcTemplate">
    <property name="dataSource" ref="dataSource"/>
</bean>
```

在上述示例代码中，配置 JDBC 模板时，需要将 dataSource 注入 jdbcTemplate，而在数据访问层（如 Dao 类）使用 jdbcTemplate 时，也需要将 jdbcTemplate 注入对应的 Bean 中。示例代码如下：

```java
@Repository
public class TestDaoImpl implements TestDao{
    @Autowired
    //使用配置文件中的 JDBC 模板
    private JdbcTemplate jdbcTemplate;
    ……
}
```

▶ 1.6.2　Spring JdbcTemplate 的常用方法

在获取 JDBC 模板后，如何使用它是本节将要讲述的内容。首先需要了解 JdbcTemplate 类的常用方法，该类的常用方法是 update() 和 query()。

1）public int update(String sql,Object args[])

该方法可以对数据表进行增加、修改、删除等操作。使用 args[] 设置 SQL 语句中的参数，并返回更新的行数。示例代码如下：

```java
String insertSql = "insert into user values(null,?,?)";
Object param1[] = {"chenheng1", "男"};
jdbcTemplate.update(insertSql, param1);
```

2）public List<T> query(String sql，RowMapper<T> rowMapper，Object args[])

该方法可以对数据表进行查询操作。rowMapper 将结果集映射到用户自定义的类中（前提是自定义类中的属性与数据表的字段对应）。示例代码如下：

```
String selectSql ="select * from user";
RowMapper<MyUser> rowMapper = new BeanPropertyRowMapper<MyUser>(MyUser.class);
List<MyUser> list = jdbcTemplate.query(selectSql, rowMapper, null);
```

下面通过一个实例演示 Spring JDBC 的使用过程。

【例 1-7】 Spring JDBC 的使用过程。该实例的具体要求：首先在 MySQL 数据库中创建数据表 user，然后使用 Spring JDBC 对数据表 user 进行增、删、改、查。

具体实现步骤如下：

❶ 使用 IntelliJ IDEA 创建 Web 应用并导入相关 JAR 包

在项目 ch1 中创建一个名为 ch1_5 的模块，同时给模块 ch1_5 添加 Web Application，并将 Spring 的 4 个基础包、Spring Commons Logging Bridge 对应的 JAR 包 spring-jcl-6.1.5.jar、spring-aop-6.1.5.jar、MySQL 数据库的驱动 JAR 包（mysql-connector-java-8.0.29.jar）、Spring JDBC 的 JAR 包（spring-jdbc-6.1.5.jar）、Java 增强库（lombok-1.18.24.jar）以及 Spring 事务处理的 JAR 包（spring-tx-6.1.5.jar）复制到 ch1_5 的 WEB-INF/lib 目录中，并添加为模块依赖。

❷ 创建配置文件

在 src 目录中创建名为 config 的包，并在该包中创建名为 applicationContext.xml 的配置文件，在配置文件中配置数据源和 JDBC 模板，具体内容如下：

```xml
<?xml version="1.0" encoding="UTF-8"?>
<beans xmlns="http://www.springframework.org/schema/beans"
    xmlns:xsi="http://www.w3.org/2001/XMLSchema-instance"
    xmlns:context="http://www.springframework.org/schema/context"
    xsi:schemaLocation="http://www.springframework.org/schema/beans
        http://www.springframework.org/schema/beans/spring-beans.xsd
        http://www.springframework.org/schema/context
        http://www.springframework.org/schema/context/spring-context.xsd">
    <!-- 指定需要扫描的包(包括子包)，使注解生效 -->
    <context:component-scan base-package="dao"/>
    <context:component-scan base-package="service"/>
    <!-- 配置数据源 -->
    <bean id="dataSource" class="org.springframework.jdbc.datasource.DriverManagerDataSource">
        <!-- MySQL 数据库驱动 -->
        <property name="driverClassName" value="com.mysql.cj.jdbc.Driver"/>
        <!-- 连接数据库的 URL -->
        <property name="url" value="jdbc:mysql://127.0.0.1:3306/springtest?useUnicode=true&amp;characterEncoding=UTF-8&amp;allowMultiQueries=true&amp;serverTimezone=GMT%2B8"/>
        <!-- 连接数据库的用户名 -->
        <property name="username" value="root"/>
        <!-- 连接数据库的密码 -->
        <property name="password" value="root"/>
    </bean>
    <!-- 配置 JDBC 模板 -->
    <bean id="jdbcTemplate" class="org.springframework.jdbc.core.JdbcTemplate">
        <property name="dataSource" ref="dataSource"/>
    </bean>
</beans>
```

❸ 创建数据表与实体类

使用 Navicat for MySQL 创建数据库 springtest，并在该数据库中创建数据表 user，数据表 user 的结构如图 1.16 所示。

名	类型	长度	小数点	不是 null	虚拟	键
uid	int			☑	☐	🔑1
uname	varchar	20		☐	☐	
usex	varchar	10		☐	☐	

图 1.16 数据表 user 的结构

在模块 ch1_5 的 src 目录中创建一个名为 entity 的包，并在该包中创建实体类 MyUser，具体代码如下：

```
package entity;
import lombok.Data;
@Data
public class MyUser {
    private Integer uid;
    private String uname;
    private String usex;
    public String toString() {
        return "myUser [uid=" + uid +", uname=" + uname + ", usex=" + usex + "]";
    }
}
```

❹ 创建数据访问层

在模块 ch1_5 的 src 目录中创建一个名为 dao 的包，在该包中创建数据访问接口 TestDao 和接口实现类 TestDaoImpl。在实现类 TestDaoImpl 中使用@Repository 注解标注此类为数据访问层，并使用@Autowired 注解依赖注入 JdbcTemplate。

TestDao 的核心代码如下：

```
public interface TestDao{
    public int update(String sql, Object[] param);
    public List<MyUser> query(String sql, Object[] param);
}
```

TestDaoImpl 的核心代码如下：

```
@Repository
public class TestDaoImpl implements TestDao{
    @Autowired
    //使用配置类中的 JDBC 模板
    private JdbcTemplate jdbcTemplate;
    /**
     * 更新方法，包括添加、修改、删除
     * param 为 sql 中的参数，如通配符？
     */
    @Override
    public int update(String sql, Object[] param) {
        return jdbcTemplate.update(sql, param);
    }
    /**
     * 查询方法
     * param 为 sql 中的参数，如通配符？
     */
    @Override
    public List<MyUser> query(String sql, Object[] param) {
```

```
        RowMapper<MyUser> rowMapper = new BeanPropertyRowMapper<MyUser>(MyUser.class);
        return jdbcTemplate.query(sql, rowMapper);
    }
}
```

❺ 创建业务逻辑层

在模块 ch1_5 的 src 目录中创建一个名为 service 的包,在该包中创建数据访问接口 TestService 和接口实现类 TestServiceImpl。在实现类 TestServiceImpl 中使用@Service 注解标注此类为业务逻辑层,并使用@Autowired 注解依赖注入 TestDao。

TestService 的代码如下:

```
package service;
public interface TestService {
    public void testJDBC();
}
```

TestServiceImpl 的核心代码如下:

```
@Service
public class TestServiceImpl implements TestService{
    @Autowired
    public TestDao testDao;
    @Override
    public void testJDBC() {
        String insertSql = "insert into user values(null,?,?)";
        //数组 param 的值与 insertSql 语句中的?一一对应
        Object[] param1 = {"chenheng1", "男"};
        Object[] param2 = {"chenheng2", "女"};
        Object[] param3 = {"chenheng3", "男"};
        Object[] param4 = {"chenheng4", "女"};
        //添加用户
        testDao.update(insertSql, param1);
        testDao.update(insertSql, param2);
        testDao.update(insertSql, param3);
        testDao.update(insertSql, param4);
        //查询用户
        String selectSql ="select * from user";
        List<MyUser> list = testDao.query(selectSql, null);
        for(MyUser mu: list) {
            System.out.println(mu);
        }
    }
}
```

❻ 创建测试类

在模块 ch1_5 的 src 目录中创建一个名为 test 的包,并在该包中创建测试类 TestJDBC,核心代码如下:

```
public class TestJDBC {
    public static void main(String[] args) {
        ApplicationContext appCon =
            new ClassPathXmlApplicationContext("config/applicationContext.xml");
        TestService ts = (TestService)appCon.getBean("testServiceImpl");
        ts.testJDBC();
    }
}
```

❼ 运行测试类

运行测试类 TestJDBC 的 main 方法，运行结果如图 1.17 所示。

```
D:\Java\jdk-20\bin\java.exe "-javaagent:
myUser [uid=1, uname=chenheng1, usex=男]
myUser [uid=2, uname=chenheng2, usex=女]
myUser [uid=3, uname=chenheng3, usex=男]
myUser [uid=4, uname=chenheng4, usex=女]
```

图 1.17 ch1_5 应用的运行结果

▶1.6.3 基于@Transactional 注解的声明式事务管理

Spring 的声明式事务管理是通过 AOP 技术实现的事务管理，其本质是对方法前后进行拦截，然后在目标方法开始之前创建或者加入一个事务，在执行完目标方法之后根据执行情况提交或者回滚事务。

声明式事务管理最大的优点是不需要通过编程的方式管理事务，因此不需要在业务逻辑代码中掺杂事务处理的代码，只需相关的事务规则声明，便可以将事务规则应用到业务逻辑中。通常情况下，在开发中使用声明式事务处理，不仅因为其简单，更主要是因为这样使得纯业务代码不被污染，极大地方便后期的代码维护。

和编程式事务管理相比，声明式事务管理唯一不足的地方是最细粒度只能作用到方法级别，无法做到像编程式事务管理那样可以作用到代码块级别。但即便有这样的需求，也可以通过变通的方法解决，例如可以将需要进行事务处理的代码块独立为方法等。

Spring 的声明式事务管理可以通过两种方式来实现，一种是基于 XML 的方式，另一种是基于@Transactional 注解的方式。

@Transactional 注解可以作用于接口、接口方法、类以及类方法上。当作用于类上时，该类的所有 public 方法都将具有该类型的事务属性，同时也可以在方法级别使用该注解来覆盖类级别的定义。虽然@Transactional 注解可以作用于接口、接口方法、类以及类方法上，但是 Spring 小组不建议在接口或者接口方法上使用该注解，因为只有在使用基于接口的代理时它才会生效。可以使用@Transactional 注解的属性定制事务行为，具体属性如表 1.3 所示。

表 1.3 @Transactional 的属性

属　　性	属性值的含义	默　认　值
propagation	propagation 定义了事务的生命周期，主要有以下选项。 ① Propagation.REQUIRED：当需要事务支持的方法 A 被调用时，没有事务新建一个事务。当在方法 A 中调用另一个方法 B 时，方法 B 将使用相同的事务。如果方法 B 发生异常需要数据回滚，则整个事务数据回滚。 ② Propagation.REQUIRES_NEW：对于方法 A 和 B，在方法调用时，无论是否有事务都开启一个新的事务；方法 B 有异常不会导致方法 A 的数据回滚。 ③ Propagation.NESTED：和 Propagation.REQUIRES_NEW 类似，仅支持 JDBC，不支持 JPA 或 Hibernate。 ④ Propagation.SUPPORTS：方法调用时，有事务就使用事务，没有事务就不创建事务。 ⑤ Propagation.NOT_SUPPORTED：强制方法在事务中执行，若有事务，在方法调用到结束阶段时事务都将会被挂起。 ⑥ Propagation.NEVER：强制方法不在事务中执行，若有事务则抛出异常。 ⑦ Propagation.MANDATORY：强制方法在事务中执行，若无事务则抛出异常	Propagation.REQUIRED

续表

属　性	属性值的含义	默　认　值
isolation	isolation（隔离）决定了事务的完整性，是一种相同数据下的多事务处理机制，主要包含以下隔离级别（前提是当前数据库是否支持）： ① Isolation.READ_UNCOMMITTED：在 A 事务中修改了一条记录但没有提交事务，在 B 事务中可以读取到修改后的记录。可导致脏读、不可重复读以及幻读。 ② Isolation.READ_COMMITTED：只有当在 A 事务中修改了一条记录且提交事务后，B 事务才可以读取到提交后的记录。防止脏读，但可能导致不可重复读和幻读。 ③ Isolation.REPEATABLE_READ：不仅能实现 Isolation.READ_COMMITTED 的功能，还能保证当 A 事务读取了一条记录，B 事务将不允许修改该条记录。阻止脏读和不可重复读，但可出现幻读。 ④ Isolation.SERIALIZABLE：在此级别下事务是顺序执行的，可以避免上述级别的缺陷，但开销较大。 ⑤ Isolation.DEFAULT：使用当前数据库的默认隔离级别。例如 Oracle 和 SQL Server 是 READ_COMMITTED；MySQL 是 REPEATABLE_READ	Isolation.DEFAULT
timeout	timeout 指定事务过期时间，默认为当前数据库的事务过期时间	
readOnly	指定当前事务是否为只读事务	false
rollbackFor	指定哪个或哪些异常可以引起事务回滚（Class 对象数组，必须继承自 Throwable）	Throwable 的子类
rollbackForClassName	指定哪个或哪些异常可以引起事务回滚（类名数组，必须继承自 Throwable）	Throwable 的子类
noRollbackFor	指定哪个或哪些异常不可以引起事务回滚（Class 对象数组，必须继承自 Throwable）	Throwable 的子类
noRollbackForClassName	指定哪个或哪些异常不可以引起事务回滚（类名数组，必须继承自 Throwable）	Throwable 的子类

下面通过一个实例来演示基于 @Transactional 注解的声明式事务管理。

【例 1-8】 基于 @Transactional 注解的声明式事务管理。该例是通过修改例 1-7 中的代码实现的。

具体实现步骤如下。

❶ **添加配置文件内容**

在配置文件中使用＜tx:annotation-driven＞元素为事务管理器注册注解驱动器，同时为数据源添加事务管理器。添加的配置文件内容如下：

```xml
<!-- 为数据源添加事务管理器 -->
<bean id="txManager"
    class="org.springframework.jdbc.datasource.DataSourceTransactionManager">
    <property name="dataSource" ref="dataSource" />
</bean>
<!-- 为事务管理器注册注解驱动器 -->
<tx:annotation-driven transaction-manager="txManager" />
```

❷ **修改业务逻辑层**

在实际开发中通常通过 Service 层进行事务管理，因此需要为 Service 层添加 @Transactional 注解。

添加@Transactional 注解后的 TestServiceImpl 类的核心代码如下：

```
@Service
@Transactional
public class TestServiceImpl implements TestService{
    @Autowired
    public TestDao testDao;
    @Override
    public void testJDBC() {
        String insertSql = "insert into user values(null,?,?)";
        //数组 param 的值与 insertSql 语句中的?一一对应
        Object[] param1 = {"chenheng1", "男"};
        Object[] param2 = {"chenheng2", "女"};
        Object[] param3 = {"chenheng3", "男"};
        Object[] param4 = {"chenheng4", "女"};
        String insertSql1 = "insert into user values(?,?,?)";
        Object[] param5 = {1,"chenheng5", "女"};
        Object[] param6 = {1,"chenheng6", "女"};
        //添加用户
        testDao.update(insertSql, param1);
        testDao.update(insertSql, param2);
        testDao.update(insertSql, param3);
        testDao.update(insertSql, param4);
        //添加两个 ID 相同的用户,出现唯一性约束异常,使事务回滚
        testDao.update(insertSql1, param5);
        testDao.update(insertSql1, param6);
    }
}
```

❸ 测试事务处理

首先清空数据表 user 中的数据，然后运行测试类 TestJDBC，发现数据表 user 中并没有数据，这是因为最后执行添加数据时主键重复，事务回滚，即回到程序运行的初始状态。

▶ 1.6.4 如何在事务处理中捕获异常

声明式事务处理的流程是：①Spring 根据配置完成事务定义，设置事务属性。②执行开发者的代码逻辑。③如果开发者的代码产生异常（如主键重复）并且满足事务回滚的配置条件，则事务回滚，否则事务提交。④事务资源释放。

现在的问题是，如果开发者在代码逻辑中加入了 try...catch...语句，Spring 还能不能在声明式事务处理中正常得到事务回滚的异常信息？答案是不能。例如将 1.6.3 节中 TestServiceImpl 实现类的 testJDBC 方法的代码修改如下：

```
@Override
public void testJDBC() {
    String insertSql = "insert into user values(null,?,?)";
    //数组 param 的值与 insertSql 语句中的?一一对应
    Object[] param1 = {"chenheng1", "男"};
    Object[] param2 = {"chenheng2", "女"};
    Object[] param3 = {"chenheng3", "男"};
    Object[] param4 = {"chenheng4", "女"};
    String insertSql1 = "insert into user values(?,?,?)";
    Object[] param5 = {1,"chenheng5", "女"};
    Object[] param6 = {1,"chenheng6", "女"};
    try {
        //添加用户
        testDao.update(insertSql, param1);
        testDao.update(insertSql, param2);
        testDao.update(insertSql, param3);
```

```
        testDao.update(insertSql, param4);
        //添加两个ID相同的用户,出现唯一性约束异常,使事务回滚
        testDao.update(insertSql1, param5);
        testDao.update(insertSql1, param6);
    } catch (Exception e) {
        System.out.println("主键重复,事务回滚。");
    }
}
```

这时再运行测试类,发现主键重复但事务并没有回滚。这是因为在默认情况下,Spring只在发生未被捕获的 RuntimeException 时才事务回滚。现在如何在事务处理中捕获异常呢? 具体修改如下:

(1) 修改@Transactional 注解。

将 TestServiceImpl 类中的@Transactional 注解修改为:

```
@Transactional(rollbackFor= {Exception.class})
//rollbackFor 指定回滚生效的异常类,多个异常类用逗号分隔
//noRollbackFor 指定回滚失效的异常类
```

(2) 在 catch 语句中添加 "throw new RuntimeException();" 语句。

注意: 在实际工程应用中,经常仅需要在 catch 语句中添加 "TransactionAspectSupport.currentTransactionStatus().setRollbackOnly();" 语句。也就是说,不需要修改@Transactional 注解和在 catch 语句中添加 "throw new RuntimeException();" 语句。

本章小结

本章讲解了 Spring IoC、AOP、Bean 以及事务管理等基础知识,目的是让读者在学习 Spring MVC 之前对 Spring 有一个基本的了解。

习题 1

1. Spring 的核心容器由哪些模块组成?
2. 如何找到 Spring 框架的官方 API?
3. 什么是 Spring IoC? 什么是依赖注入?
4. 在配置文件中如何开启 Spring 对 AspectJ 的支持? 又如何开启 Spring 对声明式事务的支持?
5. 什么是 Spring AOP? 它与 OOP 是什么关系?

第 2 章　Spring MVC

学习目的与要求

本章重点讲解 Spring MVC 的工作原理、控制器以及数据绑定。通过本章的学习，了解 Spring MVC 的工作原理，掌握 Spring MVC 控制器接收请求参数的方式，掌握 Spring MVC 应用的开发步骤。

本章主要内容

- Spring MVC 的工作原理
- Spring MVC 的工作环境
- 基于注解的控制器
- 表单标签库与数据绑定
- 拦截器
- 文件的上传

MVC 思想将一个应用分成三个基本部分，即 Model（模型）、View（视图）和 Controller（控制器），让这三部分以最低的耦合进行协同工作，从而提高应用的可扩展性及可维护性。Spring MVC 是一款优秀的基于 MVC 思想的应用框架，它是 Spring 提供的一个实现了 Web MVC 设计模式的轻量级 Web 框架。

2.1　Spring MVC 的工作原理

Spring MVC 框架是高度可配置的，包含多种视图技术，如 JSP 技术、Velocity、Tiles、iText 和 POI。Spring MVC 框架并不关心使用的视图技术，也不会强迫开发者只使用 JSP 技术，但本章使用的视图是 JSP。

Spring MVC 框架主要由 DispatcherServlet、处理器映射、控制器、视图解析器、视图组成，其工作原理如图 2.1 所示。

从图 2.1 可总结出 Spring MVC 的工作流程如下：

（1）客户端请求提交到 DispatcherServlet。

（2）由 DispatcherServlet 控制器寻找一个或多个 HandlerMapping，找到处理请求的 Controller。

（3）DispatcherServlet 将请求提交到 Controller。

（4）Controller 调用业务逻辑处理后，返回 ModelAndView。

（5）DispatcherServlet 寻找一个或多个 ViewResolver 视图解析器，找到 ModelAndView 指定的视图。

（6）视图负责将结果发送到客户端。

在图 2.1 中包含 4 个 Spring MVC 接口，分别是 DispatcherServlet、HandlerMapping、Controller 和 ViewResolver。

Spring MVC 所有的请求都经过 DispatcherServlet 统一分发。DispatcherServlet 将请求分发给 Controller 之前，需要借助于 Spring MVC 提供的 HandlerMapping 定位到具体的 Controller。

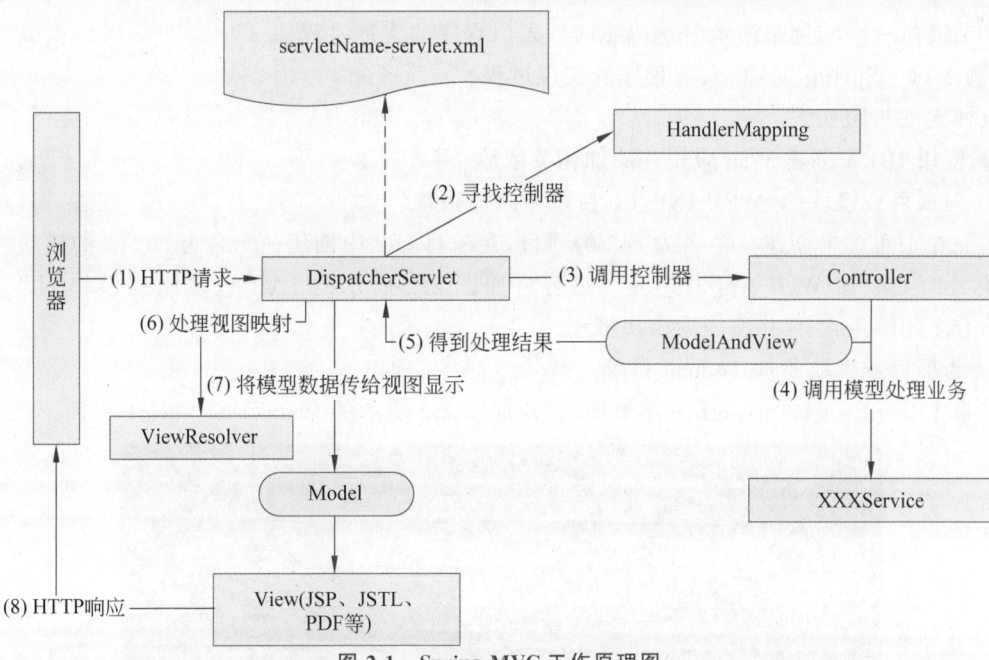

HandlerMapping 接口负责完成客户请求到 Controller 的映射。

图 2.1　Spring MVC 工作原理图

Controller 接口将处理用户请求，这和 Java Servlet 扮演的角色是一致的。一旦 Controller 处理完用户请求，则返回 ModelAndView 对象给 DispatcherServlet 前端控制器，ModelAndView 中包含了模型（Model）和视图（View）。从宏观角度考虑，DispatcherServlet 是整个 Web 应用的控制器；从微观角度考虑，Controller 是单个 HTTP 请求处理过程中的控制器，而 ModelAndView 是 HTTP 请求过程中返回的模型（Model）和视图（View）。

ViewResolver 接口（视图解析器）在 Web 应用中负责查找 View 对象，从而将相应结果渲染给客户。

2.2　Spring MVC 的工作环境

▶ 2.2.1　Spring MVC 所需要的 JAR 包

在第 1 章 Spring 开发环境的基础上导入 Spring MVC 的相关 JAR 包，即可开发 Spring MVC 应用。

对于 Spring MVC 框架的初学者，在开发 Spring MVC 应用时，只需要将 Spring 的 4 个基础 JAR 包、Spring Commons Logging Bridge 对应的 JAR 包 spring-jcl-6.1.5.jar、AOP 实现 JAR 包 spring-aop-6.1.5.jar 以及两个 Web 相关的 JAR 包（spring-web-6.1.5.jar 和 spring-webmvc-6.1.5.jar）复制到 Web 应用的 WEB-INF/lib 目录下即可。

Tomcat 10 运行 Spring MVC 应用时，DispatcherServlet 接口依赖性能监控 micrometer-observation 和 micrometer-commons 两个包进行请求分发。因此，Spring MVC 应用所添加的 JAR 包如图 2.2 所示。

图 2.2　Spring MVC 应用所添加的 JAR 包

2.2.2 使用 IntelliJ IDEA 开发 Spring MVC 应用

本节通过一个实例来演示 Spring MVC 入门程序的实现过程。

【例 2-1】 Spring MVC 入门程序的实现过程。

具体实现步骤如下。

❶ 使用 IDEA 创建 Web 应用并添加相关依赖

1）向项目或模块的/WEB-INF/lib 目录中添加依赖

首先在 IDEA 中创建一个名为 ch2 的项目,在项目 ch2 中创建一个名为 ch2_1 的模块,同时为模块 ch2_1 添加 Web Application。然后将如图 2.2 所示的 JAR 包复制到模块 ch2_1 的/WEB-INF/lib 目录中,并添加为模块依赖。

2）为项目或模块添加 Tomcat 依赖

选择 File→Project Structure 菜单项,打开如图 2.3 所示的 Project Structure 界面。

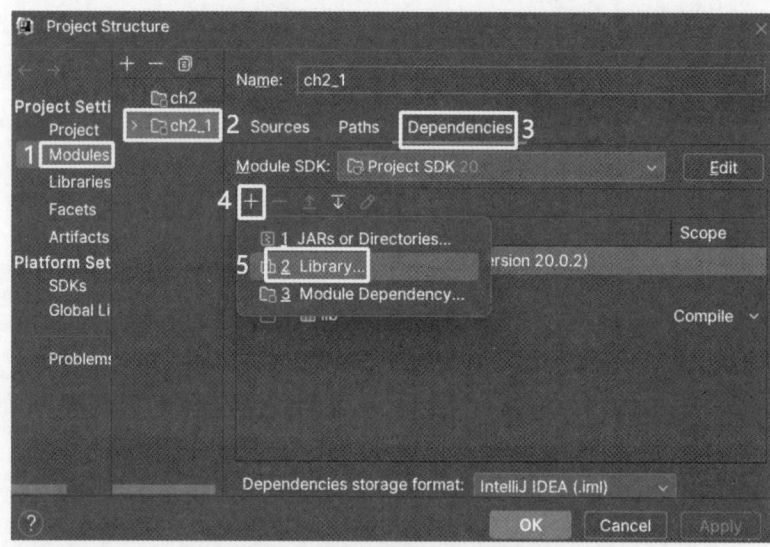

图 2.3　Project Structure 界面

按照图 2.3 中的操作顺序,打开如图 2.4 所示的 Choose Libraries 界面。

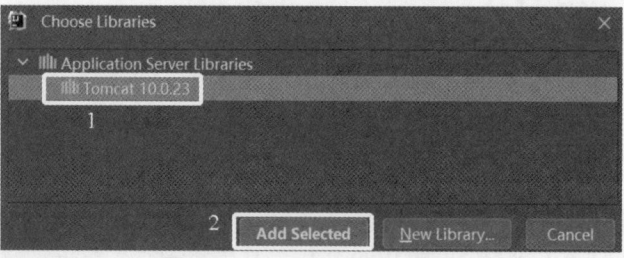

图 2.4　Choose Libraries 界面

按照图 2.4 中的操作顺序,返回如图 2.5 所示的 Project Structure 界面。

按照图 2.5 中的操作顺序,即可将 Tomcat 的相关 JAR 包添加到模块 ch2_1 中,为后续 Web 开发奠定基础。

❷ 在 web.xml 文件中部署 DispatcherServlet

在开发 Spring MVC 应用时,需要在 web.xml 中部署 DispatcherServlet,代码如下:

第 2 章 Spring MVC

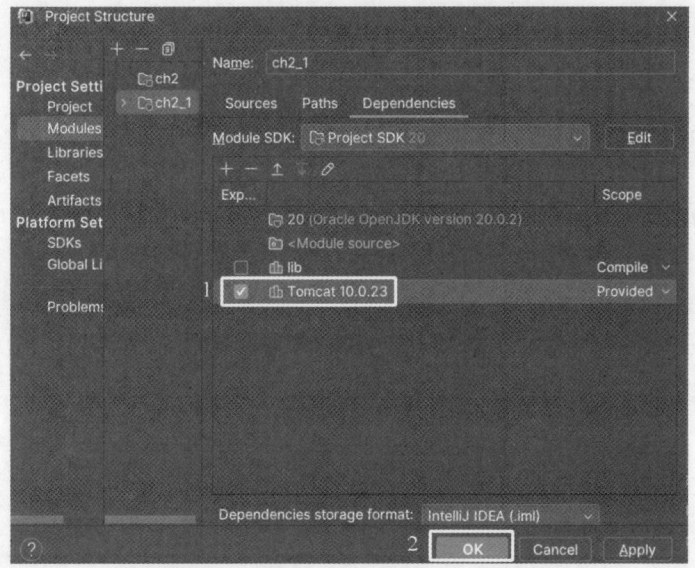

图 2.5 返回的 Project Structure 界面

```xml
<?xml version="1.0" encoding="UTF-8"?>
<web-app xmlns:xsi="http://www.w3.org/2001/XMLSchema-instance" xmlns="https://jakarta.ee/xml/ns/jakartaee" xmlns:web="http://xmlns.jcp.org/xml/ns/javaee" xsi:schemaLocation="https://jakarta.ee/xml/ns/jakartaee https://jakarta.ee/xml/ns/jakartaee/web-app_5_0.xsd" id="WebApp_ID" version="5.0">
    <display-name>ch2_1</display-name>
     <!--部署 DispatcherServlet -->
   <servlet>
        <servlet-name>springmvc</servlet-name>
        <servlet-class>org.springframework.web.servlet.DispatcherServlet</servlet-class>
        <!-- 表示容器在启动时立即加载 servlet -->
        <load-on-startup>1</load-on-startup>
   </servlet>
   <servlet-mapping>
        <servlet-name>springmvc</servlet-name>
        <!-- 处理所有 URL -->
        <url-pattern>/</url-pattern>
   </servlet-mapping>
</web-app>
```

上述 DispatcherServlet 的 servlet 对象 springmvc 初始化时，将在应用程序的 WEB-INF 目录下查找一个配置文件，该配置文件的命名规则是"servletName-servlet.xml"，如 springmvc-servlet.xml。

❸ 创建 Web 应用首页

在模块 ch2_1 的 web 目录下有一个应用首页 index.jsp。index.jsp 的代码如下：

```jsp
<%@ page language="java" contentType="text/html; charset=UTF-8" pageEncoding="UTF-8"%>
<!DOCTYPE html>
<html>
<head>
<meta charset="UTF-8">
<title>Insert title here</title>
</head>
<body>
    没注册的用户,请<a href="index/register">注册</a>!<br>
```

```
        已注册的用户,去<a href="index/login">登录</a>!
</body>
</html>
```

❹ **创建 Controller 类**

在模块 ch2_1 的 src 目录下创建 controller 包,并在该包中创建基于注解的名为 IndexController 的控制器类,在该类中有两个处理请求方法,分别处理首页中的"注册"和"登录"超链接请求。

```
package controller;
import org.springframework.stereotype.Controller;
import org.springframework.web.bind.annotation.RequestMapping;
/**"@Controller"表示 IndexController 的实例是一个控制器
 * @Controller 相当于@Controller("indexController")
 * 或@Controller(value = "indexController")
 */
@Controller
@RequestMapping("/index")
public class IndexController {
    @RequestMapping("/login")
    public String login() {
        /**login 代表逻辑视图名称,需要根据 Spring MVC 配置文件中
         * internalResourceViewResolver 的前缀和后缀找到对应的物理视图
         */
        return "login";
    }
    @RequestMapping("/register")
    public String register() {
        return "register";
    }
}
```

❺ **创建 Spring MVC 的配置文件**

在 Spring MVC 中,使用扫描机制找到应用中所有基于注解的控制器类。为了让控制器类被 Spring MVC 框架扫描到,需要在配置文件中声明 spring-context,并使用<context: component-scan/>元素指定控制器类的基本包(请确保所有控制器类都在基本包及其子包下)。另外,需要在配置文件中定义 Spring MVC 的视图解析器(ViewResolver),示例代码如下:

```
<bean class="org.springframework.web.servlet.view.InternalResourceViewResolver"
            id="internalResourceViewResolver">
    <!-- 前缀 -->
    <property name="prefix" value="/WEB-INF/jsp/" />
    <!-- 后缀 -->
    <property name="suffix" value=".jsp" />
</bean>
```

上述视图解析器配置了前缀和后缀两个属性。因此,控制器类中的视图路径仅需提供 register 和 login,视图解析器将会自动添加前缀和后缀。

在模块 ch2_1 的 WEB-INF 目录下创建名为 springmvc-servlet.xml 的配置文件,其代码如下:

```
<?xml version="1.0" encoding="UTF-8"?>
<beans xmlns="http://www.springframework.org/schema/beans"
    xmlns:xsi="http://www.w3.org/2001/XMLSchema-instance"
    xmlns:context="http://www.springframework.org/schema/context"
```

```xml
    xmlns:mvc="http://www.springframework.org/schema/mvc"
    xsi:schemaLocation="
        http://www.springframework.org/schema/beans
        http://www.springframework.org/schema/beans/spring-beans.xsd
        http://www.springframework.org/schema/context
        http://www.springframework.org/schema/context/spring-context.xsd">
    <!-- 使用扫描机制,扫描控制器类 -->
    <context:component-scan base-package="controller"/>
    <!-- 配置视图解析器 -->
    <bean class="org.springframework.web.servlet.view.InternalResourceViewResolver"
          id="internalResourceViewResolver">
        <!-- 前缀 -->
        <property name="prefix" value="/WEB-INF/jsp/" />
        <!-- 后缀 -->
        <property name="suffix" value=".jsp" />
    </bean>
</beans>
```

❻ **应用的其他页面**

IndexController 控制器的 register 方法处理成功后,跳转到"/WEB-INF/jsp/register.jsp"视图;IndexController 控制器的 login 方法处理成功后,跳转到"/WEB-INF/jsp/login.jsp"视图。因此,在模块 ch2_1 的"/WEB-INF/jsp"目录下应该有"register.jsp"和"login.jsp"页面(这两个 JSP 页面的代码略)。

❼ **发布并运行 Spring MVC 应用**

在 IDEA 中第一次运行 Spring MVC 应用时,需要将应用发布到 Tomcat,具体步骤如下。

在 IDEA 主界面中单击如图 2.6 所示的白色向下三角符号,选择 Edit Configurations 选项,打开如图 2.7 所示的页面。

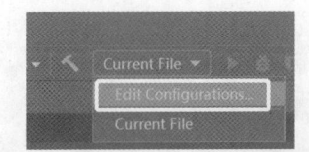

图 2.6 选择 Edit Configurations 选项

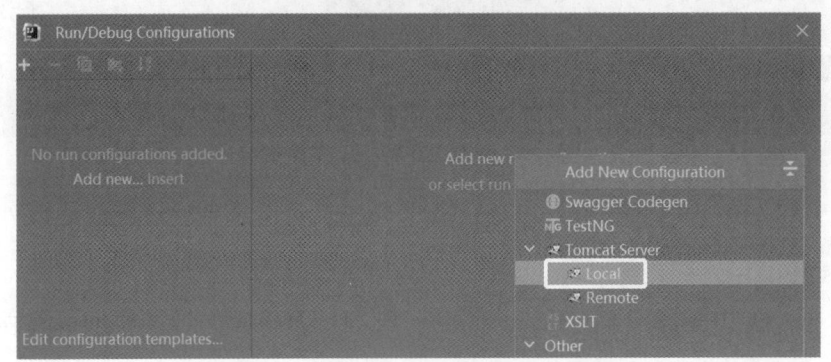

图 2.7 Run/Debug Configurations 页面

在图 2.7 中选择 Tomcat Server 下的 Local 选项,打开如图 2.8 所示的页面。

按照图 2.8 中的操作顺序,打开如图 2.9 所示的页面,并修改应用的上下文路径。

单击图 2.9 中的 OK 按钮,即可将模块 ch2_1 发布到 Tomcat。在发布成功后,回到如图 2.10 所示的 IDEA 主页面。

单击图 2.10 中的 ▶ 启动 Tomcat,即可运行 Web 应用 ch2_1,如图 2.11 所示。

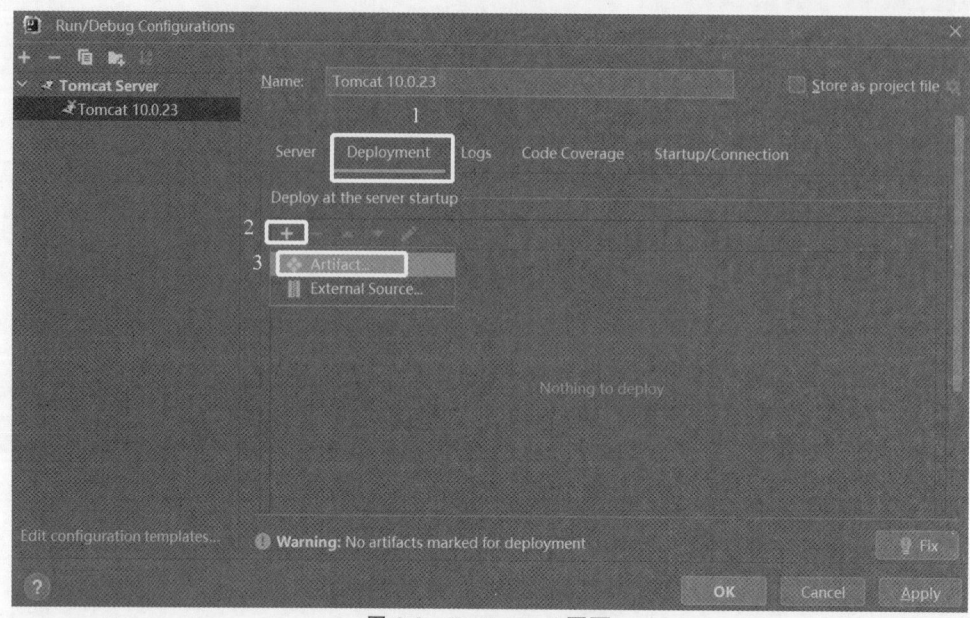

图 2.8　Deployment 页面

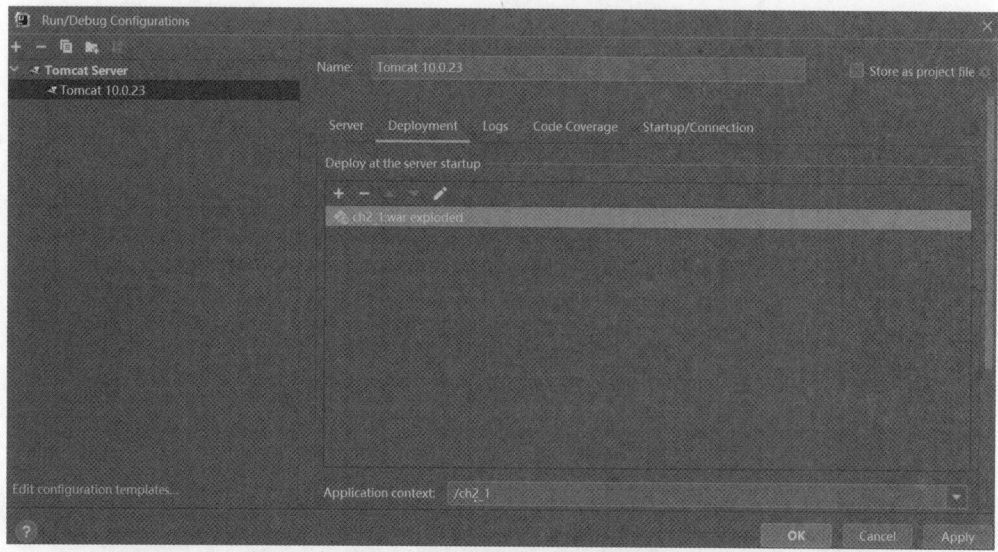

图 2.9　修改应用的上下文路径

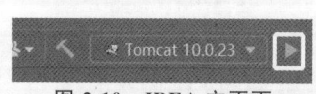

图 2.10　IDEA 主页面　　　　图 2.11　Web 应用 ch2_1 的 index.jsp 页面

在如图 2.11 所示的页面中，用户单击"注册"超链接时，根据 springmvc-servlet.xml 文件中的映射，将请求转发给 IndexController 控制器处理，处理后跳转到"/WEB-INF/jsp/register.jsp"视图。同理，单击"登录"超链接时，控制器处理后跳转到"/WEB-INF/jsp/login.jsp"视图。

2.3 基于注解的控制器

在使用 Spring MVC 进行 Web 应用开发时,Controller 是 Web 应用的核心。Controller 实现类包含对用户请求的处理逻辑,是用户请求和业务逻辑之间的"桥梁",是 Spring MVC 框架的核心部分,负责具体的业务逻辑处理。

2.3.1 Controller 注解类型

在 Spring MVC 中,使用 org.springframework.stereotype.Controller 注解类型声明某类的实例是一个控制器。例如,2.2.2 节中的 IndexController 控制器类。注意,在 Spring MVC 的配置文件中使用<context:component-scan/>元素(见例 2-1)指定控制器类的基本包,进而扫描所有注解的控制器类。

2.3.2 RequestMapping 注解类型

在基于注解的控制器类中,可以为每个请求编写对应的处理方法。如何将请求与处理方法一一对应呢?需要使用 org.springframework.web.bind.annotation.RequestMapping 注解类型将请求与处理方法一一对应。

❶ 方法级别注解

方法级别注解的示例代码如下:

```
@Controller
public class IndexController {
    @RequestMapping(value = "/index/login")
    public String login() {
        /**login 代表逻辑视图名称,需要根据 Spring MVC 配置中
         * internalResourceViewResolver 的前缀和后缀找到对应的物理视图
         */
        return "login";
    }
    @RequestMapping(value = "/index/register")
    public String register() {
        return "register";
    }
}
```

在上述示例中有两个 RequestMapping 注解语句,它们都作用在处理方法上。注解的 value 属性将请求 URI 映射到方法,value 属性是 RequestMapping 注解的默认属性,如果只有一个 value 属性,则可以省略该属性。用户可以使用如下 URL 访问 login 方法(请求处理方法)。

```
http://localhost:xxx/yyyy/index/login
```

❷ 类级别注解

类级别注解的示例代码如下:

```
@Controller
@RequestMapping("/index")
public class IndexController {
    @RequestMapping("/login")
    public String login() {
        return "login";
    }
    @RequestMapping("/register")
    public String register() {
```

```
            return "register";
    }
}
```

在类级别注解的情况下,控制器类中的所有方法都将映射为类级别的请求。用户可以使用如下URL访问login方法。

```
http://localhost:xxx/yyy/index/login
```

为了方便程序维护,建议开发者采用类级别注解,将相关处理放在同一个控制器类中。例如,对商品的增、删、改、查处理方法都可以放在一个名为GoodsOperate的控制器类中。

@RequestMapping注解的value属性表示请求路径;method属性表示请求方式。如果方法上的@RequestMapping注解没有设置method属性,则get和post请求都可以访问;如果方法上的@RequestMapping注解设置了method属性,则只能是相应的请求方式可以访问。

@RequestMapping还有特定于HTTP请求方式的组合注解,具体如下。

(1) @GetMapping：相当于@RequestMapping(method=RequestMethod.GET),处理get请求。使用@RequestMapping编写是@RequestMapping(value="requestpath",method=RequestMethod.GET);使用@GetMapping可简写为@GetMapping("requestpath")。其通常在查询数据时使用。

(2) @PostMapping：相当于@RequestMapping(method=HttpMethod.POST),处理post请求。使用@RequestMapping编写是@RequestMapping(value="requestpath",method=RequestMethod.POST);使用@PostMapping可简写为@PostMapping("requestpath")。其通常在新增数据时使用。

(3) @PutMapping、@PatchMapping：相当于@RequestMapping(method=RequestMethod.PUT/PATCH),处理put和patch请求。使用@RequestMapping编写是@RequestMapping(value="requestpath",method=RequestMethod.PUT/PATCH);使用@PutMapping可简写为@PutMapping("requestpath")。两者都是更新,@PutMapping为全局更新,@PatchMapping是对put方式的一种补充,put是对整体的更新,patch是对局部的更新。它们通常在更新数据时使用。

(4) @DeleteMapping：相当于@RequestMapping(method=RequestMethod.DELETE),处理delete请求。使用@RequestMapping编写是@RequestMapping(value="requestpath",method=RequestMethod.DELETE);使用@DeleteMapping可简写为@DeleteMapping("requestpath")。其通常在删除数据时使用。

▶2.3.3 编写请求处理方法

在控制器类中每个请求处理方法可以有多个不同类型的参数,以及一个多种类型的返回结果。

❶ 请求处理方法中常出现的参数类型

如果需要在请求处理方法中使用Servlet API类型,那么可以将这些类型作为请求处理方法的参数类型。Servlet API参数类型的示例代码如下:

```
@Controller
@RequestMapping("/index")
public class IndexController {
    @RequestMapping("/login")
    public String login(HttpSession session, HttpServletRequest request) {
```

```
        session.setAttribute("skey", "session 范围的值");
        request.setAttribute("rkey", "request 范围的值");
        return "login";
    }
}
```

除了 Servlet API 参数类型外,还有输入输出流、表单实体类、注解类型、与 Spring 框架相关的类型等。特别重要的类型是 org.springframework.ui.Model 类型,该类型是一个包含 Map 的 Spring 框架类型。每次调用请求处理方法时,Spring MVC 都将创建 org.springframework.ui.Model 对象。Model 参数类型的示例代码如下:

```
@Controller
@RequestMapping("/index")
public class IndexController {
    @RequestMapping("/register")
    public String register(Model model) {
        /*在视图中可以使用 EL 表达式${success}取出 model 中的值,有关 EL 的知识请读者参考相关内容,它们不属于本书的范畴。*/
        model.addAttribute("success", "注册成功");
        return "register";
    }
}
```

❷ 请求处理方法常见的返回类型

最常见的返回类型是代表逻辑视图名称的 String 类型。除了 String 类型外,还有 Model、View 以及其他任意的 Java 类型。

▶2.3.4　Controller 接收请求参数的常见方式

Controller 接收请求参数的方式有很多种,有的适合 get 请求,有的适合 post 请求,有的两者都适合。下面介绍几种常用的方式,读者可根据实际情况选择合适的接收方式。

❶ 通过实体 Bean 接收请求参数

通过一个实体 Bean 来接收请求参数,适用于 get 和 post 提交请求方式。需要注意的是,Bean 的属性名称必须与请求参数名称相同。Bean 的属性类型根据实际情况而定,例如接收表单输入的年龄,属性类型应该为 int 或 Integer。Spring MVC 框架将自动把表单输入的字符串转换为 Bean 属性对应的数据类型。

下面通过具体实例讲解如何使用实体 Bean 接收请求参数。

【例 2-2】　使用实体 Bean 接收请求参数。该实例的具体要求:单击如图 2.12 所示的"注册"超链接,打开如图 2.13 所示的注册页面;注册成功,打开如图 2.14 所示的登录页面;登录成功,打开如图 2.15 所示的主页面。

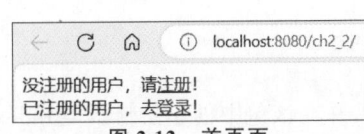

图 2.12　首页面

图 2.13　注册页面

图2.14 登录页面

图2.15 主页面

具体实现步骤如下。

1）创建 Web 应用并引入 JAR 包

在项目 ch2 中创建一个名为 ch2_2 的模块，同时为模块 ch2_2 添加 Web Application。在模块 ch2_2 的/WEB-INF/lib 目录中添加 Spring MVC 程序所需要的 JAR 包，包括 Spring 的 4 个基础 JAR 包、Spring Commons Logging Bridge 对应的 JAR 包 spring-jcl-6.1.5.jar、AOP 实现 JAR 包 spring-aop-6.1.5.jar、DispatcherServlet 接口所依赖的性能监控包（micrometer-observation.jar 和 micrometer-commons.jar）、Java 增强库（lombok-1.18.24.jar）以及两个与 Web 相关的 JAR 包（spring-web-6.1.5.jar 和 spring-webmvc-6.1.5.jar）。

2）为模块添加 Tomcat 依赖

参考 2.2.2 节，为模块 ch2_2 添加 Tomcat 依赖。

3）创建首页面

在模块 ch2_2 的 web 目录下创建 index.jsp 页面，具体代码如下。

```
<%@ page language="java" contentType="text/html; charset=UTF-8" pageEncoding="UTF-8"%>
<!DOCTYPE html>
<html>
<head>
<meta charset="UTF-8">
<title>Insert title here</title>
</head>
<body>
    没注册的用户，请<a href="index/register">注册</a>!<br>
    已注册的用户，去<a href="index/login">登录</a>!
</body>
</html>
```

4）创建实体 Bean 类

在模块 ch2_2 的 src 目录下创建 pojo 包，并在该包中创建实体类 UserForm，具体代码如下。

```
package pojo;
import lombok.Data;
/**
 * 使用@Data 注解实体属性无须 Setter 和 Getter 方法，需要给 IDEA 安装 Lombok 插件
 */
@Data
public class UserForm {
    private String uname;                    //与请求参数名称相同
    private String upass;
    private String reupass;
}
```

5）创建控制器类

在模块 ch2_2 的 src 目录下创建名为 controller 的包，并在该包中创建控制器类 IndexController 和 UserController。

IndexController 的核心代码如下。

```java
@Controller
@RequestMapping("/index")
public class IndexController {
    @GetMapping("/login")
    public String login() {
        return "login";                          //跳转到"/WEB-INF/jsp/login.jsp"
    }
    @GetMapping("/register")
    public String register() {
        return "register";
    }
}
```

UserController 的核心代码如下。

```java
package controller;
@Controller
@RequestMapping("/user")
public class UserController {
    //得到一个用来记录日志的对象
    private static final Log logger = LogFactory.getLog(UserController.class);
    /**
     * 处理登录
     * 使用 UserForm 对象(实体 Bean)user 接收注册页面提交的请求参数
     */
    @PostMapping("/login")
    public String login(UserForm user, HttpSession session, Model model) {
        if("zhangsan".equals(user.getUname())
                && "123456".equals(user.getUpass())) {
            session.setAttribute("u", user);
            logger.info("成功");
            return "main";                      //登录成功,跳转到 main.jsp
        }else{
            logger.info("失败");
            model.addAttribute("messageError", "用户名或密码错误");
            return "login";
        }
    }
    /**
     * 处理注册
     * 使用 UserForm 对象(实体 Bean)user 接收注册页面提交的请求参数
     */
    @PostMapping("/register")
    public String register(UserForm user, Model model) {
        if("zhangsan".equals(user.getUname())
                && "123456".equals(user.getUpass())) {
            logger.info("成功");
            return "login";                     //注册成功,跳转到 login.jsp
        }else{
            logger.info("失败");
            //在 register.jsp 页面上可以使用 EL 表达式取出 model 的 uname 值
            model.addAttribute("uname", user.getUname());
            return "register";                  //返回 register.jsp
        }
    }
}
```

6）创建配置文件

在模块 ch2_2 的 web/WEB-INF 目录下创建配置文件 springmvc-servlet.xml 和 web.xml。web.xml 的配置代码与 ch2_1 应用一样,这里不再赘述。springmvc-servlet.xml 的配置代码具

体如下。

```xml
<?xml version="1.0" encoding="UTF-8"?>
<beans xmlns="http://www.springframework.org/schema/beans"
    xmlns:xsi="http://www.w3.org/2001/XMLSchema-instance"
    xmlns:mvc="http://www.springframework.org/schema/mvc"
    xmlns:context="http://www.springframework.org/schema/context"
    xsi:schemaLocation="
        http://www.springframework.org/schema/beans
        http://www.springframework.org/schema/beans/spring-beans.xsd
        http://www.springframework.org/schema/mvc
        http://www.springframework.org/schema/mvc/spring-mvc.xsd
        http://www.springframework.org/schema/context
        http://www.springframework.org/schema/context/spring-context.xsd">
    <!-- 使用扫描机制,扫描控制器类 -->
    <context:component-scan base-package="controller" />
    <mvc:annotation-driven />
    <!-- annotation-driven用于简化开发的配置,替代注解处理器和适配器的配置 -->
    <!-- 使用 resources 过滤掉不需要 dispatcher servlet 的资源(即静态资源,如 CSS、JS、HTML、images)。
    在使用 resources 时,必须使用 annotation-driven,否则 resources 元素将阻止任意控制器被调用。
    -->
    <!-- 允许 static 目录下的所有文件可见 -->
    <mvc:resources location="/static/" mapping="/static/**"></mvc:resources>
    <bean class="org.springframework.web.servlet.view.InternalResourceViewResolver"
        id="internalResourceViewResolver">
        <!-- 前缀 -->
        <property name="prefix" value="/WEB-INF/jsp/" />
        <!-- 后缀 -->
        <property name="suffix" value=".jsp" />
    </bean>
</beans>
```

7）创建页面视图

在模块 ch2_2 的 web/WEB-INF 目录下创建 jsp 文件夹,并在该文件夹中创建 register.jsp（注册页面）、login.jsp（登录页面）和 main.jsp（主页面）。

register.jsp 的核心代码如下。

```html
<form class="form-horizontal" action="user/register" method="post">
    <br><br>
    <div class="form-group">
        <label class="col-sm-4 control-label">姓名</label>
        <div class="col-sm-4">
            <input name="uname" class="form-control" type="text"
            value="${uname}" placeholder="姓名" />
        </div>
    </div>
    <div class="form-group">
        <label class="col-sm-4 control-label">密码</label>
        <div class="col-sm-4">
            <input name="upass" class="form-control" type="password"
                placeholder="密码" />
        </div>
    </div>
    <div class="form-group">
        <label class="col-sm-4 control-label">确认密码</label>
        <div class="col-sm-4">
            <input name="reupass" class="form-control" type="password"
                placeholder="确认密码" />
```

```
                </div>
            </div>
            <div class="form-group">
                <div class="col-sm-offset-5 col-sm-6">
                    <button type="submit" class="btn btn-success">注册</button>
                    <button type="reset" class="btn btn-primary">重置</button>
                </div>
            </div>
</form>
```

当注册失败时,回到注册页面,并在 register.jsp 的代码中使用 EL 表达式语句"${uname}"取出"model.addAttribute("uname",user.getUname())"中的值。

login.jsp 的核心代码如下。

```
<form class="form-horizontal" action="user/login" method="post">
    <br><br>
    <div class="form-group">
        <label class="col-sm-4 control-label">姓名</label>
        <div class="col-sm-4">
            <input name="uname" class="form-control" type="text" placeholder="姓名" />
        </div>
    </div>
    <div class="form-group">
        <label class="col-sm-4 control-label">密码</label>
        <div class="col-sm-4">
            <input name="upass" class="form-control" type="password" placeholder="密码" />
        </div>
    </div>
    <div class="form-group">
        <div class="col-sm-offset-5 col-sm-6">
            <button type="submit" class="btn btn-success">登录</button>
            <button type="reset" class="btn btn-primary">重置</button>
        </div>
    </div>
    ${messageError}
</form>
```

main.jsp 的核心代码如下。

```
<body>
    <!-- 使用 EL 表达式从 session 中取出用户信息 -->
    欢迎${u.uname}登录成功
</body>
```

8) 测试应用

参考 2.2.2 节,发布并运行模块 ch2_2。

❷ **通过@RequestParam 接收请求参数**

通过@RequestParam("请求参数名")接收请求参数并注入处理方法的形参上,此方式适用于 get 请求和 post 请求。

在默认情况下,使用@RequestParam 注解的方法参数是必须提供的请求参数(即请求参数名与此注解的 value 相同,否则会出现 400 错误),但可以通过将@RequestParam 注解的required 属性设置为 false 来指定一个方法参数是可选的,例如@RequestParam(required = false, value = "uname")。另外,当@RequestParam 注解只有 value 属性时,value 属性名可以省略,例如@RequestParam("upass")。

可以将例 2-2 的控制器类 UserController 中 register 方法的代码修改如下:

```java
/**
 * 通过@RequestParam 接收请求参数
 */
@PostMapping("/register")
public String register(@RequestParam(required = false, value = "uname") String uname,
                       @RequestParam("upass") String upass, Model model) {
    if("zhangsan".equals(uname)
            && "123456".equals(upass)) {
        return "login";                          //注册成功,跳转到 login.jsp
    }else{
        //在 register.jsp 页面上可以使用 EL 表达式取出 model 的 uname 值
        model.addAttribute("uname", uname);
        return "register";                       //返回 register.jsp
    }
}
```

在上述代码中,uname 参数是可选的,upass 是必须提供的请求参数。如果将@RequestParam ("upass")修改为@RequestParam("upass11"),在运行程序时将出现 400 错误。

❸ 通过@ModelAttribute 接收请求参数

@ModelAttribute 注解放在处理方法的形参上,用于将多个请求参数封装到一个实体对象中,从而简化数据绑定流程,而且自动暴露为模型数据,用于视图页面展示。"通过实体 Bean 接收请求参数"只是将多个请求参数封装到一个实体对象,并不能暴露为模型数据(需要使用 model.addAttribute 语句才能暴露为模型数据,数据绑定与模型数据展示可参考 2.4 节的内容)。

通过@ModelAttribute 注解接收请求参数,适用于 get 和 post 请求方式。可以将例 2-2 的控制器类 UserController 中 register 方法的代码修改如下:

```java
@PostMapping("/register")
public String register(@ModelAttribute("user") UserForm user) {
    if("zhangsan".equals(user.getUname())
            && "123456".equals(user.getUpass())){
        return "login";                          //注册成功,跳转到 login.jsp
    }else{
        //使用@ModelAttribute("user")与 model.addAttribute("user", user)的功能相同
        //在 register.jsp 页面上可以使用 EL 表达式${user.uname}取出 ModelAttribute 的 uname 值
        return "register";                       //返回 register.jsp
    }
}
```

▶2.3.5 重定向与转发

重定向是将用户从当前处理请求定向到另一个视图(如 JSP)或处理请求,以前的请求(request)中存放的信息全部失效,并进入一个新的 request 作用域;转发是将用户对当前处理的请求转发给另一个视图或处理请求,以前的 request 中存放的信息不会失效。

转发是服务器行为,重定向是客户端行为。

转发过程:客户浏览器发送 HTTP 请求,Web 服务器接受此请求,调用内部的一个方法在容器内部完成请求处理和转发动作,将目标资源发送给客户。在这里转发的路径必须是同一个 Web 容器下的 URL,不能转到其他的 Web 路径上去,中间传递的是自己容器内的 request。在客户浏览器的地址栏中显示的仍然是其第一次访问的路径,也就是说客户是感觉不到服务器做了转发的。转发行为是浏览器只做了一次访问请求。

重定向过程:客户浏览器发送 HTTP 请求,Web 服务器接受后发送 302 状态码响应及对应新的 location 给客户浏览器,客户浏览器发现是 302 响应,则自动再发送一个新的 HTTP 请求,

请求 URL 是新的 location 地址，服务器根据此请求寻找资源并发送给客户。在这里 location 可以重定向到任意 URL，既然是浏览器重新发出了请求，则就没有什么 request 传递的概念了。在客户浏览器的地址栏中显示的是其重定向的路径，客户可以观察到地址的变化。重定向行为是浏览器做了至少两次的访问请求。

在 Spring MVC 框架中，控制器类中处理方法的 return 语句默认就是转发实现，只不过实现的是转发到视图。示例代码如下：

```
@RequestMapping("/register")
public String register() {
    return "register";                    //转发到 register.jsp
}
```

在 Spring MVC 框架中，重定向与转发的示例代码如下：

```
@Controller
@RequestMapping("/index")
public class IndexController {
    @RequestMapping("/login")
    public String login() {
        //转发到一个请求方法
        return "forward:/index/isLogin";
    }
    @RequestMapping("/isLogin")
    public String isLogin() {
        //重定向到一个请求方法
        return "redirect:/index/isRegister";
    }
    @RequestMapping("/isRegister")
    public String isRegister() {
        //转发到一个视图
        return "register";
    }
}
```

▶2.3.6 应用@Autowired 进行依赖注入

在前面学习的控制器中并没有体现 MVC 的 M 层，这是因为控制器既充当 C 层，又充当 M 层。这样设计程序的系统结构很不合理，应该将 M 层从控制器中分离出来。Spring MVC 框架本身就是一个非常优秀的 MVC 框架，它具有依赖注入的优点。可以通过 org.springframework.beans.factory.annotation.Autowired 注解类型将依赖注入一个属性（成员变量）或方法，例如：

```
@Autowired
public UserService userService;
```

在 Spring MVC 中，为了能被作为依赖注入，服务层的类必须使用 org.springframework.stereotype.Service 注解类型注明为 @Service（一个服务）。另外还需要在配置文件中使用 ＜context:component-scan base-package="基本包"/＞元素来扫描依赖基本包。下面将例 2-2 中 ch2_2 应用的"登录"和"注册"的业务逻辑处理分离出来，使用 Service 层实现。

首先创建 service 包，在该包中创建 UserService 接口和 UserServiceImpl 实现类。
UserService 接口的具体代码如下：

```
package service;
import pojo.UserForm;
public interface UserService {
```

```
    String login(UserForm user, HttpSession session, Model model);
    String register(UserForm user);
}
```

UserServiceImpl 实现类的具体代码如下：

```
package service;
import org.springframework.stereotype.Service;
import pojo.UserForm;
//注解为一个服务
@Service
public class UserServiceImpl implements UserService{
    @Override
    public String login(UserForm user, HttpSession session, Model model) {
        if("zhangsan".equals(user.getUname())
                && "123456".equals(user.getUpass())){
            session.setAttribute("u", user);
            return "main";                          //登录成功,跳转到 main.jsp
        } else{
            model.addAttribute("messageError", "用户名或密码错误");
            return "login";
        }
    }
    @Override
    public String register(UserForm user) {
        if("zhangsan".equals(user.getUname())
                && "123456".equals(user.getUpass()))
            return "login";                         //注册成功,跳转到 login.jsp
        return "register";                          //返回 register.jsp
    }
}
```

然后在配置文件中追加如下内容：

```
<context:component-scan base-package="service"/>
```

最后修改控制器类 UserController，核心代码如下：

```
@Controller
@RequestMapping("/user")
public class UserController {
    //将服务层依赖注入属性 userService
    @Autowired
    public UserService userService;
    /**
     * 处理登录
     */
    @PostMapping("/login")
    public String login(UserForm user, HttpSession session, Model model) {
        return userService.login(user, session, model);
    }
    /**
     * 处理注册
     */
    @PostMapping("/register")
    public String register(@ModelAttribute("user") UserForm user) {
        return userService.register(user);
    }
}
```

2.3.7 @ModelAttribute

通过 org.springframework.web.bind.annotation.ModelAttribute 注解类型可以实现如下两个功能。

❶ 绑定请求参数到实体对象（表单的命令对象）

该用法类同于 2.3.4 节中的"通过@ModelAttribute 接收请求参数"，示例代码如下：

```
@PostMapping("/register")
public String register(@ModelAttribute("user") UserForm user) {
    if("zhangsan".equals(user.getUname())
            && "123456".equals(user.getUpass())){
        return "login";
    }else{
        return "register";
    }
}
```

在上述代码中"@ModelAttribute("user") UserForm user"语句有两个功能，一是将请求参数的输入封装到 user 对象中；二是创建 UserForm 实例，以"user"为键值存储在 Model 对象中，与"model.addAttribute("user"，user)"语句的功能一样。如果没有指定键值，即"@ModelAttribute UserForm user"，那么创建 UserForm 实例时，以"userForm"为键值存储在 Model 对象中，与"model.addAttribute("userForm"，user)"语句的功能一样。

❷ 注解一个非请求处理方法

在控制器类中，被@ModelAttribute 注解的一个非请求处理方法，将在每次调用该控制器类的请求处理方法前被调用。这种特性可以用来控制登录权限，当然控制登录权限的方法有很多，例如用拦截器、过滤器等。

使用该特性控制登录权限的示例代码如下：

```
public class BaseController {
    @ModelAttribute
    public void isLogin(HttpSession session) throws Exception {
        if(session.getAttribute("user") == null){
            throw new Exception("没有权限");
        }
    }
}
@Controller
@RequestMapping("/admin")
public class ModelAttributeController extends BaseController{
    @RequestMapping("/add")
    public String add(){
        return "addSuccess";
    }
    @RequestMapping("/update")
    public String update(){
        return "updateSuccess";
    }
    @RequestMapping("/delete")
    public String delete(){
        return "deleteSuccess";
    }
}
```

上述 ModelAttributeController 类中的 add、update、delete 请求处理方法执行时，首先执行父类 BaseController 中的 isLogin 方法判断登录权限。

2.4 表单标签库与数据绑定

数据绑定是将用户参数输入值绑定到领域模型的一种特性，在 Spring MVC 的 Controller 和 View 参数数据传递中，所有 HTTP 请求参数的类型均为字符串，如果模型需要绑定的类型为 double 或 int，则需要手动进行类型转换，而有了数据绑定之后，就不再需要手动将 HTTP 请求中的 String 类型转换为模型需要的类型。数据绑定的另一个好处是，当输入验证失败时会重新生成一个 HTML 表单，无须重新填写输入字段。在 Spring MVC 中，为了方便、高效地使用数据绑定，还需要学习表单标签库。

▶ 2.4.1 表单标签库

表单标签库中包含了可以用在 JSP 页面中渲染 HTML 元素的标签。JSP 页面使用 Spring 表单标签库时，必须在 JSP 页面开头处声明 taglib 指令，指令代码如下：

```
<%@ taglib prefix="form" uri="http://www.springframework.org/tags/form" %>
```

在表单标签库中有 form、input、password、hidden、textarea、checkbox、checkboxes、radiobutton、radiobuttons、select、option、options、errors 等标签。

form：渲染表单元素。

input：渲染<input type="text"/>元素。

password：渲染<input type="password"/>元素。

hidden：渲染<input type="hidden"/>元素。

textarea：渲染 textarea 元素。

checkbox：渲染一个<input type="checkbox"/>元素。

checkboxes：渲染多个<input type="checkbox"/>元素。

radiobutton：渲染一个<input type="radio"/>元素。

radiobuttons：渲染多个<input type="radio"/>元素。

select：渲染一个选择元素。

option：渲染一个选项元素。

options：渲染多个选项元素。

errors：在 span 元素中渲染字段错误。

❶ 表单标签

表单标签的语法格式如下：

```
<form:form modelAttribute="xxx" method="post" action="xxx">
    ……
</form:form>
```

表单标签除了具有 HTML 表单元素的属性外，还具有 acceptCharset、commandName、cssClass、cssStyle、htmlEscape 和 modelAttribute 等属性。各属性的含义如下：

acceptCharset：定义服务器接受的字符编码列表。

commandName：暴露表单对象的模型属性名称，默认为 command。

cssClass：定义应用到 form 元素的 CSS 类。

cssStyle：定义应用到 form 元素的 CSS 样式。

htmlEscape：true 或 false，表示是否进行 HTML 转义。

modelAttribute：暴露 form backing object 的模型属性名称，默认为 command。
其中，commandName 和 modelAttribute 属性的功能基本一致，属性值绑定一个 JavaBean 对象。
假设控制器类 UserController 的方法 inputUser() 是返回 userAdd.jsp 的请求处理方法。
inputUser() 方法的代码如下：

```
@RequestMapping(value = "/input")
public String inputUser(Model model) {
    ……
    model.addAttribute("user", new User());
    return "userAdd";
}
```

userAdd.jsp 的表单标签代码如下：

```
<form:form modelAttribute="user" method="post" action="user/save">
    ……
</form:form>
```

注意：在 inputUser() 方法中，如果没有 Model 属性 user，userAdd.jsp 页面会抛出异常，因为表单标签无法找到在其 modelAttribute 属性中指定的 form backing object。

❷ **input 标签**

input 标签的语法格式如下：

```
<form:input path="xxx"/>
```

该标签除了具有 cssClass、cssStyle、htmlEscape 属性外，还具有一个最重要的属性——path。path 属性将文本框输入值绑定到 form backing object 的一个属性。其示例代码如下：

```
<form:form modelAttribute="user" method="post" action="user/save">
    <form:input path="userName"/>
</form:form>
```

上述代码将输入值绑定到 user 对象的 userName 属性。

❸ **password 标签**

password 标签的语法格式如下：

```
<form:password path="xxx"/>
```

该标签与 input 标签的用法完全一致，为了节省篇幅，不再赘述。

❹ **hidden 标签**

hidden 标签的语法格式如下：

```
<form:hidden path="xxx"/>
```

该标签与 input 标签的用法基本一致，只不过它不可显示，不支持 cssClass 和 cssStyle 属性。

❺ **textarea 标签**

textarea 是一个支持多行输入的 input 元素，其语法格式如下：

```
<form:textarea path="xxx"/>
```

该标签与 input 标签的用法完全一致，为了节省篇幅，不再赘述。

❻ **checkbox 标签**

checkbox 标签的语法格式如下：

```
<form:checkbox path="xxx" value="yyy"/>
```

多个 path 相同的 checkbox 标签是一个选项组,允许多选,选项值绑定到一个数组属性。其示例代码如下:

```
<form:checkbox path="friends" value="张三"/>张三
<form:checkbox path="friends" value="李四"/>李四
<form:checkbox path="friends" value="王五"/>王五
<form:checkbox path="friends" value="赵六"/>赵六
```

在上述示例代码中复选框的值绑定到一个字符串数组属性 friends(String[] friends)。该标签的其他用法与 input 标签基本一致,为了节省篇幅,不再赘述。

❼ checkboxes 标签

checkboxes 标签渲染多个复选框,是一个选项组,等价于多个 path 相同的 checkbox 标签。它有 3 个非常重要的属性,分别为 items、itemLabel 和 itemValue。

items:用于生成 input 元素的 Collection、Map 或 Array。

itemLabel:items 属性中指定的集合对象的属性,为每个 input 元素提供 label。

itemValue:items 属性中指定的集合对象的属性,为每个 input 元素提供 value。

checkboxes 标签的语法格式如下:

```
<form:checkboxes items="xxx" path="yyy"/>
```

其示例代码如下:

```
<form:checkboxes items="${hobbys}" path="hobby" />
```

上述示例代码将 model 属性 hobbys 的内容(集合元素)渲染为复选框。在 itemLabel 和 itemValue 默认情况下,如果集合是数组,复选框的 label 和 value 相同;如果是 Map 集合,复选框的 label 是 Map 的值(value),复选框的 value 是 Map 的关键字(key)。

❽ radiobutton 标签

radiobutton 标签的语法格式如下:

```
<form:radiobutton path="xxx" value="yyy"/>
```

多个 path 相同的 radiobutton 标签是一个选项组,只允许单选。

❾ radiobuttons 标签

radiobuttons 标签渲染多个 radio,是一个选项组,等价于多个 path 相同的 radiobutton 标签。radiobuttons 标签的语法格式如下:

```
<form:radiobuttons path="xxx" items="yyy"/>
```

该标签的 itemLabel 和 itemValue 属性与 checkboxes 标签的 itemLabel 和 itemValue 属性完全一样,但只允许单选。

❿ select 标签

select 标签的选项可能来自其属性 items 指定的集合,或者来自一个嵌套的 option 标签或 options 标签。其语法格式如下:

```
<form:select path="xxx" items="yyy" />
```

或

```
<form:select path="xxx" items="yyy" >
    <option value="xxx">xxx</option>
</ form:select>
```

或

```
<form:select path="xxx">
    <form:options items="yyy"/>
</form:select>
```

该标签的 itemLabel 和 itemValue 属性与 checkboxes 标签的 itemLabel 和 itemValue 属性完全一样。

⑪ options 标签

options 标签生成一个 select 标签的选项列表,因此需要与 select 标签一起使用,具体用法参见 select 标签。

⑫ errors 标签

errors 标签渲染一个或者多个 span 元素,每个 span 元素包含一个错误消息。它可以用于显示一个特定的错误消息,也可以显示所有错误消息。其语法格式如下:

```
<form:errors path="*"/>
```

或

```
<form:errors path="xxx"/>
```

其中,"*"表示显示所有错误消息;"xxx"表示显示由"xxx"指定的特定错误消息。

▶ 2.4.2 数据绑定

为了让读者进一步学习数据绑定和表单标签,本节给出了一个应用实例 ch2_3。在 ch2_3 应用中实现了 User 类属性和 JSP 页面中表单参数的绑定,同时在 JSP 页面中分别展示了 input、textarea、checkbox、checkboxs、select 等标签。

【例 2-3】 数据绑定和表单标签。

具体实现步骤如下:

❶ 创建应用并导入相关的 JAR 包

在项目 ch2 中创建一个名为 ch2_3 的模块,同时为模块 ch2_3 添加 Web Application。在模块 ch2_3 的/WEB-INF/lib 目录中除添加与模块 ch2_2 相同的 JAR 包外,还需要添加与 JSTL (本实例使用 JSTL 标签展示页面)相关的 JAR 包 taglibs-standard-impl-1.2.5-migrated-0.0.1.jar 和 taglibs-standard-spec-1.2.5-migrated-0.0.1.jar,它们位于 Tomcat 的 webapps\examples\WEB-INF\lib 目录下。

❷ 为模块添加 Tomcat 依赖

参考 2.2.2 节,为模块 ch2_3 添加 Tomcat 依赖。

❸ 创建领域模型

在模块 ch2_3 中实现了 User 类属性和 JSP 页面中表单参数的绑定,User 类包含了和表单参数名对应的属性。

在模块 ch2_3 的 src 目录下创建 pojo 包,并在该包中创建 User 类,具体代码如下:

```
package pojo;
import lombok.Data;
```

```
@Data
public class User {
    private String userName;
    private String[] hobby;
    private String[] friends;
    private String carrer;
    private String houseRegister;
    private String remark;
}
```

❹ 创建业务层

在模块 ch2_3 的 Service 层中使用静态集合变量 users 模拟数据库存储用户信息，包括添加用户和查询用户两个功能。

在模块 ch2_3 的 src 目录下创建 service 包，并在该包中创建 UserService 接口和 UserServiceImpl 实现类。

UserService 接口的核心代码如下。

```
public interface UserService {
    boolean addUser(User u);
    ArrayList<User> getUsers();
}
```

UserServiceImpl 实现类的核心代码如下。

```
@Service
public class UserServiceImpl implements UserService{
    //使用静态集合变量 users 模拟数据库
    private static ArrayList<User> users = new ArrayList<User>();
    @Override
    public boolean addUser(User u) {
        if(!"IT民工".equals(u.getCarrer())){              //不允许添加 IT 民工
            users.add(u);
            return true;
        }
        return false;
    }
    @Override
    public ArrayList<User> getUsers() {
        return users;
    }
}
```

❺ 创建控制层

在模块 ch2_3 的 Controller 类 UserController 中定义了请求处理方法，包括处理 user/input 请求的 inputUser 方法，以及处理 user/save 请求的 addUser 方法，其中在 addUser 方法中用到了重定向。在 UserController 类中，通过 @Autowired 注解在 UserController 对象中主动注入 UserService 对象，实现对 user 对象的添加和查询等操作；通过 model 的 addAttribute 方法将 User 类对象、HashMap 类型的 hobbys 对象、String[] 类型的 carrers 对象以及 String[] 类型的 houseRegisters 对象传递给 View(userAdd.jsp)。

在模块 ch2_3 的 src 目录下创建 controller 包，并在该包中创建 UserController 控制器类，核心代码如下。

```
@Controller
@RequestMapping("/user")
```

```java
public class UserController {
    private static final Log logger = LogFactory.getLog(UserController.class);
    @Autowired
    private UserService userService;
    private void initData(Model model) {
        HashMap<String, String> hobbys = new HashMap<String, String>();
        hobbys.put("篮球", "篮球");
        hobbys.put("乒乓球", "乒乓球");
        hobbys.put("电玩", "电玩");
        hobbys.put("游泳", "游泳");
        model.addAttribute("hobbys", hobbys);
        model.addAttribute("carrers", new String[] { "教师", "学生", "coding 搬运工", "IT民工", "其他" });
        model.addAttribute("houseRegisters", new String[] { "北京", "上海", "广州", "深圳", "其他" });
    }
    @GetMapping("/input")
    public String inputUser(Model model) {
        model.addAttribute("user", new User());
        initData(model);
        return "userAdd";
    }
    @PostMapping("/save")
    public String addUser(@ModelAttribute User user, Model model) {
        if (userService.addUser(user)) {
            logger.info("成功");
            return "redirect:/user/list";
        } else {
            logger.info("失败");
            initData(model);
            return "userAdd";
        }
    }
    @GetMapping("/list")
    public String listUsers(Model model) {
        List<User> users = userService.getUsers();
        model.addAttribute("users", users);
        return "userList";
    }
}
```

❻ 创建配置文件

在模块 ch2_3 的 web/WEB-INF 目录下创建配置文件 springmvc-servlet.xml 和 web.xml。web.xml 的配置代码与模块 ch2_2 一样，不再赘述。

springmvc-servlet.xml 的配置代码具体如下。

```xml
<?xml version="1.0" encoding="UTF-8"?>
<beans xmlns="http://www.springframework.org/schema/beans"
    xmlns:xsi="http://www.w3.org/2001/XMLSchema-instance"
    xmlns:mvc="http://www.springframework.org/schema/mvc"
    xmlns:context="http://www.springframework.org/schema/context"
    xsi:schemaLocation="
        http://www.springframework.org/schema/beans
        http://www.springframework.org/schema/beans/spring-beans.xsd
        http://www.springframework.org/schema/mvc
        http://www.springframework.org/schema/mvc/spring-mvc.xsd
        http://www.springframework.org/schema/context
        http://www.springframework.org/schema/context/spring-context.xsd">
    <!-- 使用扫描机制,扫描控制器类 -->
```

```xml
    <context:component-scan base-package="controller" />
    <context:component-scan base-package="service" />
    <mvc:annotation-driven />
    <!-- 允许static目录下的所有文件可见 -->
    <mvc:resources location="/static/" mapping="/static/**"></mvc:resources>
    <bean class="org.springframework.web.servlet.view.InternalResourceViewResolver"
        id="internalResourceViewResolver">
        <!-- 前缀 -->
        <property name="prefix" value="/WEB-INF/jsp/" />
        <!-- 后缀 -->
        <property name="suffix" value=".jsp" />
    </bean>
</beans>
```

❼ 创建视图层

View 层中包含两个 JSP 页面，一个是信息输入页面 userAdd.jsp，另一个是信息显示页面 userList.jsp。在模块 ch2_3 的 web/WEB-INF/jsp/ 目录下创建这两个 JSP 页面。

在 userAdd.jsp 页面中将 Map 类型的 hobbys 绑定到 checkboxes 上，将 String[] 类型的 carrers 和 houseRegisters 绑定到 select 上，实现通过 option 标签对 select 添加选项，同时表单的 method 方法需指定为 post 来避免中文乱码问题。

在 userList.jsp 页面中使用 JSTL 标签遍历集合中的用户信息。

userAdd.jsp 的核心代码具体如下。

```jsp
<form:form cssClass="form-horizontal" modelAttribute="user" method="post" action="user/save">
    <div class="form-group">
        <label class="col-sm-4 control-label"></label>
        <div class="col-sm-4">
            <h2>添加一个用户</h2>
        </div>
    </div>
    <div class="form-group">
        <label class="col-sm-4 control-label">用户名:</label>
        <div class="col-sm-4">
            <form:input path="userName" cssClass="form-control"/>
        </div>
    </div>
    <div class="form-group">
        <label class="col-sm-4 control-label">爱好:</label>
        <div class="col-sm-4">
            <form:checkboxes items="${hobbys}" path="hobby" />
        </div>
    </div>
    <div class="form-group">
        <label class="col-sm-4 control-label">朋友:</label>
        <div class="col-sm-4">
            <form:checkbox path="friends" value="张三"/>张三
            <form:checkbox path="friends" value="李四"/>李四
            <form:checkbox path="friends" value="王五"/>王五
            <form:checkbox path="friends" value="赵六"/>赵六
        </div>
    </div>
    <div class="form-group">
        <label class="col-sm-4 control-label">职业:</label>
        <div class="col-sm-4">
            <form:select path="carrer" cssClass="form-control">
                <form:option value="" label="请选择职业"/>
```

```html
                <form:options items="${carrers}"/>
            </form:select>
        </div>
    </div>
    <div class="form-group">
        <label class="col-sm-4 control-label">户籍:</label>
        <div class="col-sm-4">
            <form:select path="houseRegister" cssClass="form-control">
                <form:option value="" label="请选择户籍"/>
                <form:options items="${houseRegisters}"/>
            </form:select>
        </div>
    </div>
    <div class="form-group">
        <label class="col-sm-4 control-label">个人描述:</label>
        <div class="col-sm-4">
            <form:textarea path="remark" rows="5" cssClass="form-control"/>
        </div>
    </div>
    <div class="form-group">
        <div class="col-sm-offset-5 col-sm-6">
            <button type="submit" class="btn btn-success">添加</button>
            <button type="reset" class="btn btn-primary">重置</button>
        </div>
    </div>
</form:form>
```

userList.jsp 的核心代码具体如下。

```html
<div class="abox">
    <div class="box">
    <h3>用户列表</h3>
        <a href="<c:url value="user/input"/>">继续添加</a>
        <table class="table table-bordered table-hover">
            <tbody class="text-center">
            <tr>
                <th>用户名</th>
                <th>兴趣爱好</th>
                <th>朋友</th>
                <th>职业</th>
                <th>户籍</th>
                <th>个人描述</th>
            </tr>
            <c:forEach items="${users}" var="user">
                <tr>
                    <td>${user.userName}</td>
                    <td>
                        <c:forEach items="${user.hobby}" var="hobby">
                            ${hobby} 
                        </c:forEach>
                    </td>
                    <td>
                        <c:forEach items="${user.friends}" var="friend">
                            ${friend} 
                        </c:forEach>
                    </td>
                    <td>${user.carrer}</td>
                    <td>${user.houseRegister}</td>
                    <td>${user.remark}</td>
                </tr>
            </c:forEach>
```

```
            </tbody>
        </table>
    </div>
</div>
```

❽ 测试应用

参考 2.2.2 节,发布并运行模块 ch2_3。通过地址 http://localhost:8080/ch2_3/user/input 测试模块 ch2_3。如果在图 2.16 中职业选择"IT 民工",则添加失败。失败后回到添加页面,输入过的信息不再输入,自动回填(必须结合 form 标签)。自动回填是数据绑定的一个优点。失败页面如图 2.16 所示,添加成功页面如图 2.17 所示。

图 2.16 添加用户信息失败页面

图 2.17 添加成功页面

2.5 拦截器

Spring MVC 的拦截器(Interceptor)与 Java Servlet 的过滤器(Filter)类似,主要用于拦截用户的请求并做相应的处理,通常应用在权限验证、记录请求信息的日志、判断用户是否登录等功能上。

▶2.5.1 拦截器的定义

在 Spring MVC 框架中定义一个拦截器,需要对拦截器进行配置。拦截器通过实现 HandlerInterceptor 接口或继承 HandlerInterceptor 接口的实现类来定义,示例代码如下。

```
package interceptor;
import org.springframework.web.servlet.HandlerInterceptor;
```

```
import org.springframework.web.servlet.ModelAndView;
import jakarta.servlet.http.HttpServletRequest;
import jakarta.servlet.http.HttpServletResponse;
public class TestInterceptor implements HandlerInterceptor {
    @Override
    public boolean preHandle(HttpServletRequest request, HttpServletResponse response, Object handler)
            throws Exception {
        System.out.println("preHandle方法在控制器的处理请求方法调用前执行");
        /**返回true表示继续向下执行,返回false表示中断后续操作*/
        return true;
    }
    @Override
    public void postHandle(HttpServletRequest request, HttpServletResponse response, Object handler,
            ModelAndView modelAndView) throws Exception {
        System.out.println("postHandle方法在控制器的处理请求方法调用后,解析视图前执行");
    }
    @Override
    public void afterCompletion(HttpServletRequest request, HttpServletResponse response, Object handler, Exception ex)
            throws Exception {
        System.out.println("afterCompletion方法在控制器执行后执行,即视图渲染结束后执行");
    }
}
```

在上述拦截器的定义中实现了 HandlerInterceptor 接口,并实现了接口中的三个方法。有关这三个方法的描述如下。

preHandle()方法:该方法在控制器的处理请求方法前执行,其返回值表示是否中断后续操作,返回 true 表示继续向下执行,返回 false 表示中断后续操作。

postHandle()方法:该方法在控制器的处理请求方法调用之后,解析视图之前执行。可以通过此方法对请求域中的模型和视图做进一步修改。

afterCompletion()方法:该方法在控制器的处理请求方法执行完成后执行,即视图渲染结束后执行,可以通过此方法实现一些资源清理、记录日志信息等工作。

▶2.5.2 拦截器的配置

让自定义的拦截器生效,需要在 Spring MVC 的配置文件中进行配置,示例代码如下:

```
<!-- 配置拦截器 -->
<mvc:interceptors>
    <!-- 配置一个全局拦截器,拦截所有请求 -->
    <bean class="interceptor.TestInterceptor"/>
    <mvc:interceptor>
        <!-- 配置拦截器作用的路径,可以使用通配符*,也可以配置多个<mvc:mapping> -->
        <mvc:mapping path="/**"/>
        <!-- 配置不需要拦截作用的路径,可以使用通配符*,
        也可以配置多个<mvc:exclude-mapping> -->
        <mvc:exclude-mapping path=""/>
        <!-- 拦截器的实现类-->
        <bean class="interceptor.Interceptor1"/>
    </mvc:interceptor>
    <mvc:interceptor>
        <!-- 配置拦截器作用的路径 -->
        <mvc:mapping path="/gotoTest"/>
        <!-- 拦截器的实现类-->
        <bean class="interceptor.Interceptor2"/>
```

```
        </mvc:interceptor>
    </mvc:interceptors>
```

在上述配置示例代码中，<mvc:interceptors>元素用于配置一组拦截器，其子元素<bean>定义的是全局拦截器，即拦截所有的请求。<mvc:interceptor>元素中定义的是指定路径的拦截器，其子元素<mvc:mapping>用于配置拦截器作用的路径，该路径在其path属性中定义。如上述示例代码中，path的属性值"/**"表示拦截所有路径，"/gotoTest"表示拦截所有以"/gotoTest"结尾的路径。如果在请求路径中包含不需要拦截的内容，可以通过<mvc:exclude-mapping>子元素进行配置。

需要注意的是，<mvc:interceptor>元素的子元素必须按照<mvc:mapping.../>、<mvc:exclude-mapping.../>、<bean.../>的顺序配置。

▶2.5.3 拦截器的执行流程

在配置文件中，如果只定义了一个拦截器，程序将首先执行拦截器类中的preHandle()方法，如果该方法返回true，程序将继续执行控制器中处理请求的方法，否则中断执行。如果preHandle()方法返回true，并且控制器中处理请求的方法执行后返回视图前将执行postHandle()方法，返回视图后才执行afterCompletion()方法。

在实际Web应用中，通常有多个拦截器同时工作，这时它们的preHandle()方法将按照配置文件中拦截器的配置顺序执行，而它们的postHandle()方法和afterCompletion()方法则按照配置顺序的相反顺序执行。下面通过例2-4演示拦截器的执行流程。

【例2-4】 多个拦截器的执行过程。

具体实现步骤如下。

❶ 创建Web应用并导入相关的JAR包

首先在项目ch2中创建一个名为ch2_4的模块，同时为模块ch2_4添加Web Application，然后为模块ch2_4添加Tomcat依赖，最后将例2-1中模块ch2_1的JAR包及web.xml配置文件复制到模块ch2_4的对应位置。

❷ 创建控制器类

在src目录下创建名为controller的包，并在该包中创建控制器类InterceptorController，核心代码如下：

```
@Controller
public class InterceptorController {
    @RequestMapping("/gotoTest")
    public String gotoTest() {
        System.out.println("正在测试拦截器,执行控制器的处理请求方法中");
        return "test";
    }
}
```

❸ 创建拦截器类

在src目录下创建一个名为interceptor的包，并在该包中创建拦截器类TestInterceptor、Interceptor1和Interceptor2。TestInterceptor的代码与2.5.1节中的示例代码相同，为了节省篇幅，不再赘述。

Interceptor1类的核心代码如下：

```
public class Interceptor1 implements HandlerInterceptor{
```

```java
    @Override
    public boolean preHandle(HttpServletRequest request, HttpServletResponse
            response, Object handler) throws Exception {
        System.out.println("Interceptor1 preHandle 方法执行");
        /**返回 true 表示继续向下执行,返回 false 表示中断后续的操作 */
        return true;
    }
    @Override
    public void postHandle(HttpServletRequest request, HttpServletResponse
            response, Object handler, ModelAndView modelAndView) throws Exception {
        System.out.println("Interceptor1 postHandle 方法执行");
    }
    @Override
    public void afterCompletion(HttpServletRequest request, HttpServletResponse
            response, Object handler, Exception ex) throws Exception {
        System.out.println("Interceptor1 afterCompletion 方法执行");
    }
}
```

Interceptor2 类的代码如下:

```java
public class Interceptor2 implements HandlerInterceptor{
    @Override
    public boolean preHandle(HttpServletRequest request, HttpServletResponse
            response, Object handler) throws Exception {
        System.out.println("Interceptor2 preHandle 方法执行");
        /**返回 true 表示继续向下执行,返回 false 表示中断后续的操作 */
        return true;
    }
    @Override
    public void postHandle(HttpServletRequest request, HttpServletResponse
            response, Object handler, ModelAndView modelAndView) throws Exception {
        System.out.println("Interceptor2 postHandle 方法执行");
    }
    @Override
    public void afterCompletion(HttpServletRequest request, HttpServletResponse
            response, Object handler, Exception ex) throws Exception {
        System.out.println("Interceptor2 afterCompletion 方法执行");
    }
}
```

❹ 创建配置文件 springmvc.xml

在模块 ch2_4 的 web/WEB-INF 目录下创建配置文件 springmvc-servlet.xml。在配置文件中配置一个全局拦截器和两个局部拦截器,具体代码如下:

```xml
<?xml version="1.0" encoding="UTF-8"?>
<beans xmlns="http://www.springframework.org/schema/beans"
    xmlns:xsi="http://www.w3.org/2001/XMLSchema-instance"
    xmlns:context="http://www.springframework.org/schema/context"
    xmlns:mvc="http://www.springframework.org/schema/mvc"
    xsi:schemaLocation="
        http://www.springframework.org/schema/beans
        http://www.springframework.org/schema/beans/spring-beans.xsd
        http://www.springframework.org/schema/context
        http://www.springframework.org/schema/context/spring-context.xsd
        http://www.springframework.org/schema/mvc
        http://www.springframework.org/schema/mvc/spring-mvc.xsd">
    <!-- 使用扫描机制,扫描控制器类 -->
    <context:component-scan base-package="controller"/>
        <!-- 配置视图解析器 -->
```

```xml
<bean class="org.springframework.web.servlet.view.InternalResourceViewResolver"
    id="internalResourceViewResolver">
    <!-- 前缀 -->
    <property name="prefix" value="/WEB-INF/jsp/" />
    <!-- 后缀 -->
    <property name="suffix" value=".jsp" />
</bean>
<mvc:interceptors>
    <!-- 配置一个全局拦截器,拦截所有请求 -->
    <bean class="interceptor.TestInterceptor"/>
    <mvc:interceptor>
        <!-- 配置拦截器作用的路径 -->
        <mvc:mapping path="/**"/>
        <!-- 定义在<mvc:interceptor>元素中,表示匹配指定路径的请求才进行拦截 -->
        <bean class="interceptor.Interceptor1"/>
    </mvc:interceptor>
    <mvc:interceptor>
        <!-- 配置拦截器作用的路径 -->
        <mvc:mapping path="/gotoTest"/>
        <!-- 定义在<mvc:interceptor>元素中,表示匹配指定路径的请求才进行拦截 -->
        <bean class="interceptor.Interceptor2"/>
    </mvc:interceptor>
</mvc:interceptors>
</beans>
```

❺ **创建视图 JSP 文件**

在 WEB-INF 目录下创建一个名为 jsp 的文件夹,并在该文件夹中创建一个 JSP 文件 test.jsp。test.jsp 的核心代码如下:

```jsp
<body>
    视图
    <%System.out.println("视图渲染结束。"); %>
</body>
```

❻ **测试拦截器**

首先将应用 ch2_4 发布到 Tomcat 服务器,并启动 Tomcat 服务器,然后通过地址 http://localhost:8080/ch2_4/gotoTest 测试拦截器。程序正确执行后,控制台的输出结果如图 2.18 所示。

```
03-Apr-2024 15:37:37.485 其它 [Catalina-utility-2
preHandle方法在控制器的处理请求方法调用前执行
Interceptor1 preHandle方法执行
Interceptor2 preHandle方法执行
正在测试拦截器,执行控制器的处理请求方法中
Interceptor2 postHandle方法执行
Interceptor1 postHandle方法执行
postHandle方法在控制器的处理请求方法调用后,解析视图前执行
视图渲染结束。
Interceptor2 afterCompletion方法执行
Interceptor1 afterCompletion方法执行
afterCompletion方法在控制器执行后执行,即视图渲染结束后执行
```

图 2.18 多个拦截器的执行过程

2.6 文件的上传

org.springframework.web.multipart.MultipartResolver 是一个解析文件上传的 multipart 请求接口。从 Spring 6.0 和 Servlet 5.0＋开始，基于 Apache Commons FileUpload 组件的 MultipartResolver 接口实现类 CommonsMultipartResolver 被弃用，改用基于 Servlet 容器的 MultipartResolver 接口实现类 StandardServletMultipartResolver 进行 multipart 请求解析。

在 Spring MVC 框架中，上传文件时，将文件的相关信息及操作封装到 MultipartFile 接口对象中。因此，开发者只需要使用 MultipartFile 类型声明模型类的一个属性，即可对上传文件进行操作。该接口具有如下方法。

- byte[] getBytes()：以字节数组的形式返回文件的内容。
- String getContentType()：返回文件的内容类型。
- InputStream getInputStream()：返回一个 InputStream，从中读取文件的内容。
- String getName()：返回请求参数的名称。
- String getOriginalFilename()：返回客户端提交的原始文件名称。
- long getSize()：返回文件的大小，单位为字节。
- boolean isEmpty()：判断被上传文件是否为空。
- void transferTo(File destination)：将上传文件保存到目标目录下。

在 Spring MVC 应用中，上传文件时，需要完成以下三点配置或设置：

（1）在 Spring MVC 配置文件中，使用 Spring 的 org.springframework.web.multipart.support. StandardServletMultipartResolver 类配置一个名为 multipartResolver（MultipartResolver 接口）的 Bean，用于文件上传。

（2）必须将文件上传表单的 method 设置为 post，并将 enctype 设置为 multipart/form-data，只有这样设置，浏览器才能将所选文件的二进制数据发送给服务器。

（3）需要通过 Servlet 容器配置启用 Servlet multipart 请求解析，可以在 web.xml 配置文件中的 Servlet 声明部分添加<multipart-config>子元素进行配置，示例代码如下。

```xml
<servlet>
    <servlet-name>springmvc</servlet-name>
    <servlet-class>org.springframework.web.servlet.DispatcherServlet</servlet-class>
    <load-on-startup>1</load-on-startup>
    <multipart-config>
        <!-- 允许上传文件的最大值,默认为-1(不限制) -->
        <max-file-size>20848820</max-file-size>
        <!-- multipart/form-data 请求允许的最大值,默认为-1(不限制) -->
        <max-request-size>418018841</max-request-size>
        <!-- 文件将写入磁盘的阈值大小,默认为 0 -->
        <file-size-threshold>1048576</file-size-threshold>
    </multipart-config>
</servlet>
```

本节通过一个应用案例讲解 Spring MVC 框架如何实现文件上传。

【例 2-5】 Spring MVC 框架实现文件上传。

具体实现步骤如下。

❶ 创建应用并导入 JAR 包

首先在项目 ch2 中创建一个名为 ch2_5 的模块，同时为模块 ch2_5 添加 Web Application，然后在模块 ch2_5 的/WEB-INF/lib 目录中添加与模块 ch2_3 相同的 JAR 包,最后为模块 ch2_5 添加 Tomcat 依赖。

❷ 创建 web.xml 文件

在模块 ch2_5 的 web/WEB-INF 目录下创建 web.xml 文件,并在 web.xml 文件中添加 <multipart-config>子元素进行配置,启用 Servlet multipart 请求解析,具体配置代码如下。

```xml
<?xml version="1.0" encoding="UTF-8"?>
<web-app xmlns:xsi="http://www.w3.org/2001/XMLSchema-instance"
    xmlns="https://jakarta.ee/xml/ns/jakartaee"
    xmlns:web="http://xmlns.jcp.org/xml/ns/javaee"
    xsi:schemaLocation="https://jakarta.ee/xml/ns/jakartaee
    https://jakarta.ee/xml/ns/jakartaee/web-app_5_0.xsd"
    id="WebApp_ID" version="5.0">
    <display-name>ch2_5</display-name>
    <!--部署 DispatcherServlet -->
    <servlet>
        <servlet-name>springmvc</servlet-name>
        <servlet-class>org.springframework.web.servlet.DispatcherServlet</servlet-class>
        <load-on-startup>1</load-on-startup>
        <multipart-config>
            <!-- 允许上传文件的最大值,默认为-1(不限制) -->
            <max-file-size>20848820</max-file-size>
            <!-- multipart/form-data 请求允许的最大值,默认为-1(不限制) -->
            <max-request-size>418018841</max-request-size>
            <!-- 文件将写入磁盘的阈值大小,默认为 0 -->
            <file-size-threshold>1048576</file-size-threshold>
        </multipart-config>
    </servlet>
    <servlet-mapping>
        <servlet-name>springmvc</servlet-name>
        <!-- 处理所有 URL -->
        <url-pattern>/</url-pattern>
    </servlet-mapping>
</web-app>
```

❸ 创建文件选择页面

在模块 ch2_5 的 web 目录下创建 JSP 页面 uploadFile.jsp,在该页面中使用表单上传多个文件,具体代码如下。

```jsp
<%@ page language="java" contentType="text/html; charset=UTF-8" pageEncoding="UTF-8"%>
<%
String path = request.getContextPath();
String basePath = request.getScheme()+"://"+request.getServerName()+":"+request.getServerPort()+path+"/";
%>
<!DOCTYPE html>
<html>
<head>
<base href="<%=basePath%>">
<meta charset="UTF-8">
<title>文件上传</title>
</head>
<body>
<form action="multifile" method="post" enctype="multipart/form-data">
    选择文件 1:<input type="file" name="myfile"> <br>
    文件描述 1:<input type="text" name="description"> <br>
    选择文件 2:<input type="file" name="myfile"> <br>
```

```html
            文件描述 2:<input type="text" name="description"> <br>
            选择文件 3:<input type="file" name="myfile"> <br>
            文件描述 3:<input type="text" name="description"> <br>
            <button type="submit">提交</button>
        </form>
    </body>
</html>
```

❹ 创建 POJO 类

在模块 ch2_5 的 src 目录下创建 pojo 包,并在该包中创建 POJO 类 MultiFileDomain。在该 POJO 类中声明一个 List<MultipartFile>类型的属性,封装被上传的文件信息,属性名与文件选择页面 uploadFile.jsp 中 file 类型的表单参数名 myfile 相同。MultiFileDomain 的具体代码如下。

```java
package pojo;
import java.util.List;
import org.springframework.web.multipart.MultipartFile;
import lombok.Data;
@Data
public class MultiFileDomain {
    private List<String> description;
    private List<MultipartFile> myfile;
}
```

❺ 创建控制器类

在模块 ch2_5 的 src 目录下创建 controller 包,并在该包中创建 FileUploadController 控制器类,核心代码如下。

```java
@Controller
public class FileUploadController {
    @PostMapping("/multifile")
    public String multiFileUpload(@ModelAttribute MultiFileDomain multiFileDomain, HttpServletRequest request){
    //D:\idea-workspace\ch2\out\artifacts\ch2_5\uploadfiles
        String realpath = request.getServletContext().getRealPath("uploadfiles");
        File targetDir = new File(realpath);
        if(!targetDir.exists()){
            targetDir.mkdirs();
        }
        List<MultipartFile> files = multiFileDomain.getMyfile();
        for (int i = 0; i < files.size(); i++) {
            MultipartFile file = files.get(i);
            String fileName = file.getOriginalFilename();
            File targetFile = new File(realpath,fileName);
            //上传
            try {
                file.transferTo(targetFile);
            } catch (Exception e) {
                e.printStackTrace();
            }
        }
        return "showMulti";
    }
}
```

❻ 创建 Spring MVC 的配置文件

在模块 ch2_5 的 web/WEB-INF 目录下创建 springmvc-servlet.xml 配置文件。在上传文件时,

需要在配置文件中使用 Spring 的 StandardServletMultipartResolver 类配置 MultipartResolver,配置文件 springmvc-servlet.xml 的代码如下。

```xml
<?xml version="1.0" encoding="UTF-8"?>
<beans xmlns="http://www.springframework.org/schema/beans"
xmlns:xsi="http://www.w3.org/2001/XMLSchema-instance"
xmlns:context="http://www.springframework.org/schema/context"
xsi:schemaLocation="
    http://www.springframework.org/schema/beans
    http://www.springframework.org/schema/beans/spring-beans.xsd
    http://www.springframework.org/schema/context
    http://www.springframework.org/schema/context/spring-context.xsd">
<!-- 使用扫描机制,扫描控制器类 -->
<context:component-scan base-package="controller" />
<bean class="org.springframework.web.servlet.view.InternalResourceViewResolver"
    id="internalResourceViewResolver">
    <!-- 前缀 -->
    <property name="prefix" value="/WEB-INF/jsp/" />
    <!-- 后缀 -->
    <property name="suffix" value=".jsp" />
</bean>
<!-- 配置一个名为 multipartResolver 的 Bean,用于文件上传 -->
<bean id="multipartResolver" class="org.springframework.web.multipart.support.StandardServletMultipartResolver"/>
</beans>
```

❼ 创建成功显示页面

在模块 ch2_5 的 web/WEB-INF 目录下创建名为 jsp 的文件夹,并在该文件夹中创建文件上传成功的显示页面 showMulti.jsp,具体代码如下。

```jsp
<%@ page language="java" contentType="text/html; charset=UTF-8" pageEncoding="UTF-8"%>
<%@ taglib uri="http://java.sun.com/jsp/jstl/core" prefix="c" %>
<%
String path = request.getContextPath();
String basePath = request.getScheme()+"://"+ request.getServerName()+":"+ request.getServerPort()+path+"/";
%>
<!DOCTYPE html>
<html>
<head>
<base href="<%=basePath%>">
<meta charset="UTF-8">
<title>Insert title here</title>
</head>
<body>
    <table>
        <tr>
            <td>详情</td><td>文件名</td>
        </tr>
        <!-- 同时取两个数组的元素 -->
        <c:forEach items="${multiFileDomain.description}" var="description" varStatus="loop">
            <tr>
                <td>${description}</td>
                <td>${multiFileDomain.myfile[loop.count-1].originalFilename}</td>
            </tr>
        </c:forEach>
        <!-- fileDomain.getMyfile().getOriginalFilename() -->
    </table>
```

```
        </body>
</html>
```

❽ 测试文件上传

发布模块 ch2_5 到 Tomcat 服务器，并启动 Tomcat 服务器，然后通过地址 http://localhost:8080/ch2_5/uploadFile.jsp 运行文件选择页面，运行结果如图 2.19 所示。

图 2.19 文件选择页面

在图 2.19 中选择文件，并输入文件描述，然后单击"提交"按钮上传多个文件，文件上传成功显示如图 2.20 所示的界面。

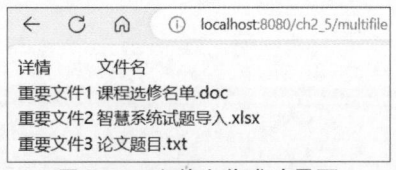

图 2.20 文件上传成功界面

本章小结

本章简单介绍了 Spring MVC 框架基础，包括 Spring MVC 的工作流程、控制器、表单标签与数据绑定、拦截器以及文件上传等内容。

习题 2

1. 在开发 Spring MVC 应用时如何配置 DispatcherServlet？
2. 简述 Spring MVC 的工作流程。
3. 举例说明数据绑定的优点。
4. Spring MVC 有哪些表单标签？其中，可以绑定集合数据的标签有哪些？
5. @ModelAttribute 可实现哪些功能？

第 3 章　MyBatis

学习目的与要求

本章讲解 MyBatis 的环境构建、工作原理、SQL 映射文件以及 SSM 框架的整合开发。通过本章的学习，了解 MyBatis 的工作原理，掌握 MyBatis 的环境构建以及 SSM 框架的整合开发，了解 MyBatis 的核心配置文件的配置信息，掌握 MyBatis 的 SQL 映射文件的编写，熟悉级联查询的 MyBatis 实现，掌握 MyBatis 的动态 SQL 的编写。

本章主要内容

- MyBatis 的环境构建
- MyBatis 的工作原理
- SSM 框架整合开发
- 核心配置文件
- SQL 映射文件
- 级联查询
- 动态 SQL
- MyBatis 的缓存机制

MyBatis 是主流的 Java 持久层框架之一，它与 Hibernate 一样，也是一种 ORM（Object-Relational Mapping，即对象-关系映射）框架。其因性能优异，且具有高度的灵活性、可优化性、易维护以及简单易学等特点，受到了广大互联网企业和编程爱好者的青睐。

3.1　MyBatis 简介

MyBatis 本是 Apache Software Foundation 的一个开源项目 iBatis，在 2010 年这个项目由 Apache Software Foundation 迁移到 Google Code，并改名为 MyBatis。

MyBatis 是一个基于 Java 的持久层框架。MyBatis 提供的持久层框架包括 SQL Maps 和 Data Access Objects（DAO），它消除了几乎所有的 JDBC 代码和参数的手工设置以及结果集的检索。MyBatis 使用简单的 XML 或注解来配置和映射原始类型，将接口和 Java 的 POJOs（Plain Old Java Objects，普通的 Java 对象）映射成数据库中的记录。

目前，Java 的持久层框架产品有许多，常见的有 Hibernate 和 MyBatis。MyBatis 是一个半自动映射的框架，因为 MyBatis 需要手动匹配 POJO、SQL 和映射关系；而 Hibernate 是一个全表映射的框架，只需提供 POJO 和映射关系即可。MyBatis 是一个小巧、方便、高效、简单、直接、半自动化的持久层框架；Hibernate 是一个强大、方便、高效、复杂、间接、全自动化的持久化框架。两个持久层框架各有优缺点，开发者应根据实际应用选择它们。

3.2　MyBatis 的环境构建

在编写本书时，MyBatis 的最新版本是 3.5.14，因此编者选择该版本作为本书的实践环境，也希望读者下载该版本，以便于学习。

第 3 章 MyBatis

如果读者不使用 Maven 或 Gradle 下载 MyBatis，可通过网址 https://github.com/mybatis/mybatis-3/releases/tag/mybatis-3.5.14 下载。解压后得到如图 3.1 所示的目录。

在图 3.1 中 mybatis-3.5.14.jar 是 MyBatis 的核心包，mybatis-3.5.14.pdf 是 MyBatis 的使用手册，lib 文件夹下的 JAR 文件是 MyBatis 的依赖包。

在使用 MyBatis 框架时，需要将它的核心包和依赖包引入应用程序中。如果是 Web 应用，只需将核心包和依赖包复制到/WEB-INF/lib 目录中。

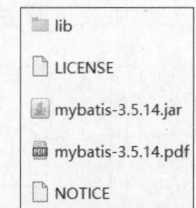

图 3.1　MyBatis 解压后的目录

3.3　MyBatis 的工作原理

在学习 MyBatis 程序之前，读者需要了解一下 MyBatis 的工作原理，以便于理解程序。MyBatis 的工作原理如图 3.2 所示。

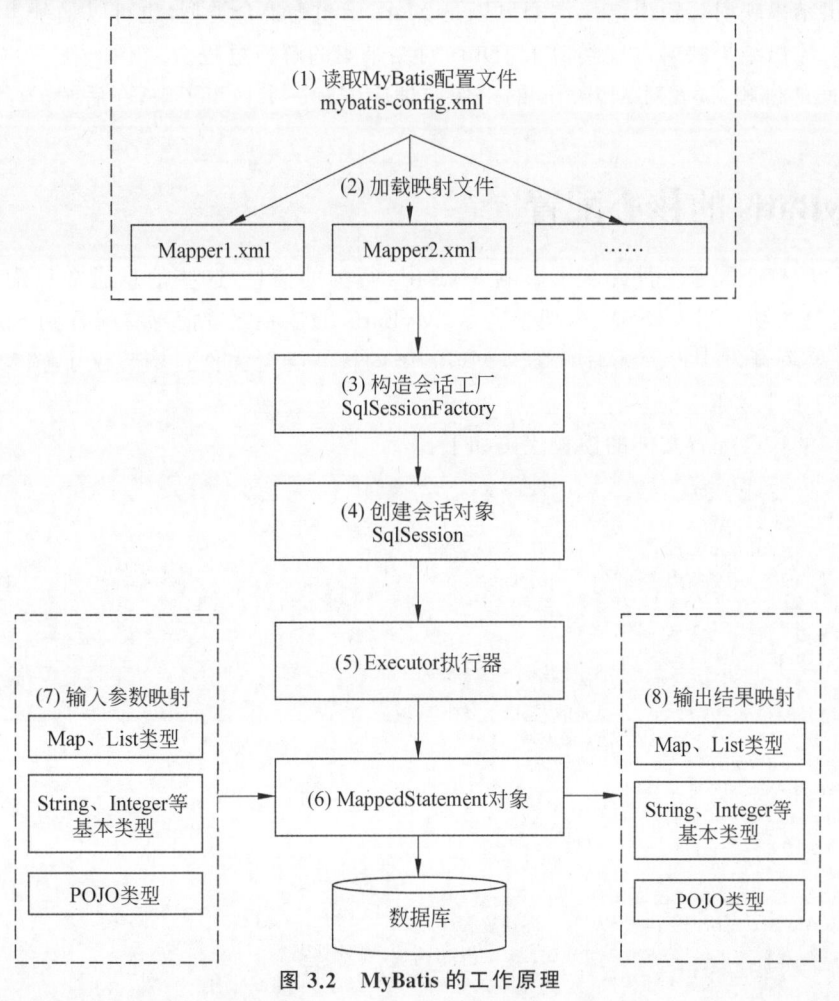

图 3.2　MyBatis 的工作原理

下面对图 3.2 中的每一步进行说明。

（1）读取 MyBatis 配置文件 mybatis-config.xml。mybatis-config.xml 为 MyBatis 的全局配

69

置文件，配置了 MyBatis 的运行环境等信息，如数据库连接信息。

（2）加载映射文件。映射文件即 SQL 映射文件，在该文件中配置了操作数据库的 SQL 语句，需要在 MyBatis 配置文件 mybatis-config.xml 中加载。mybatis-config.xml 文件可以加载多个映射文件。

（3）构造会话工厂。通过 MyBatis 的环境等配置信息，构建会话工厂 SqlSessionFactory。

（4）创建会话对象（SqlSession）。由会话工厂创建 SqlSession 对象，在该对象中包含执行 SQL 语句的所有方法。

（5）MyBatis 底层定义了一个 Executor 接口来操作数据库，它将根据 SqlSession 传递的参数动态地生成需要执行的 SQL 语句，同时负责查询缓存的维护。

（6）在 Executor 接口的执行方法中有一个 MappedStatement 类型的参数，该参数是对映射信息的封装，用于存储要映射的 SQL 语句的 ID、参数等信息。

（7）输入参数映射。输入参数类型可以是 Map、List 等集合类型，也可以是基本数据类型和 POJO 类型。输入参数映射过程类似于 JDBC 对 preparedStatement 对象设置参数的过程。

（8）输出结果映射。输出结果类型可以是 Map、List 等集合类型，也可以是基本数据类型和 POJO 类型。输出结果映射过程类似于 JDBC 对结果集的解析过程。

通过上面的讲解，读者对 MyBatis 的工作原理应该有一个初步了解，在后续的学习中将慢慢加深理解。

3.4 MyBatis 的核心配置

MyBatis 的核心配置文件配置了影响 MyBatis 行为的信息，这些信息通常只配置在一个文件中，并不轻易改动。另外，SSM 框架整合后，MyBatis 的核心配置信息将配置到 Spring 配置文件中，因此在实际开发中很少编写或修改 MyBatis 的核心配置文件。本节仅了解 MyBatis 的核心配置文件的主要元素。

MyBatis 的核心配置文件的模板代码如下：

```xml
<?xml version="1.0" encoding="UTF-8"?>
<!DOCTYPE configuration
PUBLIC "-//mybatis.org//DTD Config 3.0//EN"
"http://mybatis.org/dtd/mybatis-3-config.dtd">
<configuration>
    <properties/><!-- 属性 -->
    <settings><!-- 设置 -->
        <setting name="" value=""/>
    </settings>
    <typeAliases/><!-- 类型命名(别名) -->
    <typeHandlers/><!-- 类型处理器 -->
    <objectFactory type=""/><!-- 对象工厂 -->
    <plugins><!-- 插件 -->
        <plugin interceptor=""></plugin>
    </plugins>
    <environments default=""><!-- 配置环境 -->
        <environment id=""><!-- 环境变量 -->
            <transactionManager type=""/><!-- 事务管理器 -->
            <dataSource type=""/><!-- 数据源 -->
        </environment>
    </environments>
    <databaseIdProvider type=""/><!-- 数据库厂商标识 -->
    <mappers><!-- 映射器,告诉 MyBatis 到哪里去找映射文件-->
```

```xml
        <mapper resource="com/mybatis/UserMapper.xml"/>
    </mappers>
</configuration>
```

MyBatis 的核心配置文件中的元素配置顺序不能颠倒，否则在 MyBatis 启动阶段将发生异常。

3.5 使用 IntelliJ IDEA 开发 MyBatis 入门程序

本节使用 1.6.2 节 MySQL 数据库 springtest 的 user 数据表进行讲解。下面通过一个实例讲解如何使用 IntelliJ IDEA 开发 MyBatis 入门程序。

【例 3-1】 使用 IntelliJ IDEA 开发 MyBatis 入门程序。

具体实现步骤如下。

❶ **创建 Web 应用并导入相关 JAR 包**

首先在 IDEA 中创建一个名为 ch3 的项目，在项目 ch3 中创建一个名为 ch3_1 的模块，同时为模块 ch3_1 添加 Web Application，然后在模块 ch3_1 的 /WEB-INF/lib 目录中添加 MyBatis 的核心 JAR 包、MyBatis 的依赖 JAR 包、Java 增强库（lombok-1.18.24.jar）以及 MySQL 的驱动连接 JAR 包。

❷ **创建 Log4j 的日志配置文件**

MyBatis 可以使用 Log4j 输出日志信息，如果开发者需要查看控制台输出的 SQL 语句，那么需要在 classpath 路径下配置其日志文件。在模块 ch3_1 的 src 目录下创建 log4j.properties 文件，其内容如下。

```properties
#Global logging configuration
log4j.rootLogger=ERROR, stdout
#MyBatis logging configuration...
log4j.logger.com.mybatis.mapper=DEBUG
#Console output...
log4j.appender.stdout=org.apache.log4j.ConsoleAppender
log4j.appender.stdout.layout=org.apache.log4j.PatternLayout
log4j.appender.stdout.layout.ConversionPattern=%5p [%t] - %m%n
```

在上述日志文件中配置了全局的日志配置、MyBatis 的日志配置和控制台输出，其中 MyBatis 的日志配置用于将 com.mybatis.mapper 包下所有类的日志记录级别设置为 DEBUG。该配置文件的内容不需要开发者全部手写，可以从 MyBatis 使用手册中的 Logging 小节复制，然后进行简单修改。

Log4j 是 Apache 的一个开源代码项目，通过使用 Log4j，可以控制日志信息输送的目的地是控制台、文件或 GUI 组件等；也可以控制每一条日志的输出格式；通过定义每一条日志信息的级别，能够更加详细地控制日志的生成过程。这些都可以通过一个配置文件来灵活地进行配置，而不需要修改应用的代码。有关 Log4j 的使用方法，读者可参考相关资料学习。

❸ **创建持久化类**

在模块 ch3_1 的 src 目录下创建一个名为 com.mybatis.po 的包，并在该包中创建持久化类 MyUser。在该类中声明的属性与数据表 user（创建表的代码请参见源代码 user.sql）中的字段一致。

MyUser 类的代码如下。

```
package com.mybatis.po;
import lombok.Data;
@Data
public class MyUser {
    private Integer uid;                            //主键
    private String uname;
    private String usex;
    @Override
    public String toString() {                      //为了方便查看结果,重写了 toString 方法
        return "User [uid=" + uid +",uname=" + uname + ",usex=" + usex +"]";
    }
}
```

❹ **创建 MyBatis 的核心配置文件**

在模块 ch3_1 的 src 目录下创建 MyBatis 的核心配置文件 mybatis-config.xml。在该文件中配置了数据库环境和映射文件的位置,具体内容如下。

```
<?xml version="1.0" encoding="UTF-8" ?>
<!DOCTYPE configuration
PUBLIC "-//mybatis.org//DTD Config 3.0//EN"
"http://mybatis.org/dtd/mybatis-3-config.dtd">
<configuration>
    <!-- 数据库连接信息 -->
    <properties>
        <property name="username" value="root" />
        <property name="password" value="root" />
        <property name="driver" value="com.mysql.cj.jdbc.Driver" />
        <property name="url"
            value="jdbc:mysql://127.0.0.1:3306/springtest? useUnicode=
true&amp;characterEncoding=UTF-8&amp;allowMultiQueries=true&amp;serverTimezone=GMT%2B8" />
    </properties>
    <settings>
        <setting name="logImpl" value="LOG4J" />
    </settings>
    <!--为实体类 com.mybatis.po.MyUser 配置一个别名 MyUser -->
    <!--    <typeAliases>
        <typeAlias type="com.mybatis.po.MyUser" alias="MyUser" />
    </typeAliases>-->
    <!-- 为 com.mybatis.po 包下的所有实体类配置别名,MyBatis 默认的设置别名的方式就是去除类所在
的包后简单的类名,例如 com.mybatis.po.MyUser 这个实体类的别名会被设置成 MyUser -->
    <typeAliases>
        <package name="com.mybatis.po" />
    </typeAliases>
    <!-- SSM 整合后 environments 配置将废除 -->
    <environments default="development">
        <environment id="development">
            <!-- 使用 JDBC 事务管理 -->
            <transactionManager type="JDBC" />
            <!-- 数据库连接池 -->
            <dataSource type="POOLED">
                <property name="driver" value="${driver}" />
                <property name="url" value="${url}" />
                <property name="username" value="${username}" />
                <property name="password" value="${password}" />
            </dataSource>
        </environment>
    </environments>
    <!-- 加载映射文件 -->
    <mappers>
        <mapper resource="com/mybatis/mapper/UserMapper.xml" />
```

```
        </mappers>
</configuration>
```

上述映射文件和配置文件不需要读者完全手动编写，从 MyBatis 使用手册中复制，然后做简单修改即可。

❺ **创建 SQL 映射文件**

在模块 ch3_1 的 src 目录下创建一个名为 com.mybatis.mapper 的包，并在该包中创建 SQL 映射文件 UserMapper.xml。

SQL 映射文件 UserMapper.xml 的内容如下。

```xml
<?xml version="1.0" encoding="UTF-8" ?>
<!DOCTYPE mapper
PUBLIC "-//mybatis.org//DTD Mapper 3.0//EN"
"http://mybatis.org/dtd/mybatis-3-mapper.dtd">
<mapper namespace="com.mybatis.mapper.UserMapper">
    <!-- 根据 uid 查询一个用户信息 -->
    <select id="selectUserById" parameterType="Integer"
        resultType="com.mybatis.po.MyUser">
        select * from user where uid = #{uid}
    </select>
    <!-- 查询所有用户信息 -->
    <select id="selectAllUser" resultType="MyUser">
        select * from user
    </select>
    <!-- 添加一个用户,#{uname}为 MyUser 的属性值-->
    <insert id="addUser" parameterType="MyUser">
        insert into user (uname,usex) values(#{uname},#{usex})
    </insert>
    <!-- 修改一个用户 -->
    <update id="updateUser" parameterType="MyUser">
        update user set uname = #{uname},usex = #{usex} where uid = #{uid}
    </update>
    <!-- 删除一个用户 -->
    <delete id="deleteUser" parameterType="Integer">
        delete from user where uid = #{uid}
    </delete>
</mapper>
```

在上述映射文件中，<mapper>元素是配置文件的根元素，它包含了一个 namespace 属性，该属性值通常设置为"包名+SQL 映射文件名"，指定了唯一的命名空间。子元素<select>、<insert>、<update>以及<delete>中的信息是用于执行查询、添加、修改以及删除操作的配置。在定义的 SQL 语句中，"#{}"表示一个占位符，相当于"?"，而"#{uid}"表示该占位符待接收参数的名称为 uid。

❻ **创建测试类**

在模块 ch3_1 的 src 目录下创建一个名为 com.mybatis.test 的包，并在该包中创建 MyBatisTest 测试类。在该测试类中，首先使用输入流读取配置文件，然后根据配置信息构建 SqlSessionFactory 对象。接下来通过 SqlSessionFactory 对象创建 SqlSession 对象，并使用 SqlSession 对象执行数据库操作。MyBatisTest 测试类的核心代码如下。

```java
public class MyBatisTest {
    public static void main(String[] args) {
        try {
            //读取配置文件 mybatis-config.xml
            InputStream config = Resources.getResourceAsStream("mybatis-config.xml");
```

```java
            //根据配置文件构建 SqlSessionFactory
            SqlSessionFactory ssf = new SqlSessionFactoryBuilder().build(config);
            //通过 SqlSessionFactory 创建 SqlSession
            SqlSession ss = ssf.openSession();
            //SqlSession 执行映射文件中定义的 SQL,并返回映射结果
            /* com.mybatis.mapper.UserMapper.selectUserById 为
            UserMapper.xml 中的命名空间+select 的 id */
            //查询一个用户
            MyUser mu = ss.selectOne("com.mybatis.mapper.UserMapper.selectUserById", 1);
            System.out.println(mu);
            //添加一个用户
            MyUser addmu = new MyUser();
            addmu.setUname("陈恒");
            addmu.setUsex("男");
            ss.insert("com.mybatis.mapper.UserMapper.addUser",addmu);
            //修改一个用户
            MyUser updatemu = new MyUser();
            updatemu.setUid(1);
            updatemu.setUname("张三");
            updatemu.setUsex("女");
            ss.update("com.mybatis.mapper.UserMapper.updateUser", updatemu);
            //删除一个用户
            ss.delete("com.mybatis.mapper.UserMapper.deleteUser", 2);
            //查询所有用户
            List<MyUser> listMu = ss.selectList("com.mybatis.mapper.UserMapper.selectAllUser");
            for (MyUser myUser: listMu) {
                System.out.println(myUser);
            }
            //提交事务
            ss.commit();
            //关闭 SqlSession
            ss.close();
        } catch (IOException e) {
            e.printStackTrace();
        }
    }
}
```

上述测试类的运行结果如图 3.3 所示。

```
DEBUG [main] - ==>  Preparing: select * from user where uid = ?
DEBUG [main] - ==> Parameters: 1(Integer)
DEBUG [main] - <==      Total: 1
User [uid=1,uname=陈恒1,usex=女]
DEBUG [main] - ==>  Preparing: insert into user (uname,usex) values(?,?)
DEBUG [main] - ==> Parameters: 陈恒(String), 男(String)
DEBUG [main] - <==    Updates: 1
DEBUG [main] - ==>  Preparing: update user set uname = ?,usex = ? where uid = ?
DEBUG [main] - ==> Parameters: 张三(String), 女(String), 13(Integer)
DEBUG [main] - <==    Updates: 0
DEBUG [main] - ==>  Preparing: select * from user
DEBUG [main] - ==> Parameters:
DEBUG [main] - <==      Total: 5
User [uid=1,uname=陈恒1,usex=女]
User [uid=2,uname=陈恒2,usex=男]
User [uid=3,uname=陈恒3,usex=女]
User [uid=4,uname=陈恒4,usex=男]
User [uid=36,uname=陈恒,usex=男]
```

图 3.3　MyBatis 入门程序的运行结果

3.6 SSM 框架整合开发

从 3.5 节的测试类的代码中可以看出,直接使用 MyBatis 框架的 SqlSession 访问数据库并不简便。MyBatis 框架的重点是 SQL 映射文件,为了方便后续学习,从本节开始讲解 SSM 框架整合开发。在本书 MyBatis 的后续学习中将使用整合后的框架进行演示。

▶ 3.6.1 相关 JAR 包

实现 SSM 框架整合开发需要导入相关 JAR 包,包括 MyBatis、Spring、Spring MVC、MySQL 连接器、MyBatis 与 Spring 桥接器、Log4j 以及 DBCP 等 JAR 包。

❶ **MyBatis 框架所需的 JAR 包**

MyBatis 框架所需的 JAR 包包括核心包和依赖包,详情见 3.2 节。

❷ **Spring 框架所需的 JAR 包**

Spring 框架所需的 JAR 包包括它的核心模块 JAR、AOP 开发使用的 JAR、JDBC 和事务的 JAR 包、Spring MVC 所需要的 JAR 包、DispatcherServlet 接口所依赖的性能监控包以及 Java 增强库(lombok)。

❸ **MyBatis 与 Spring 整合的中间 JAR 包**

在编写本书时,该中间 JAR 包的最新版本为 mybatis-spring-3.0.3.jar。此版本可以从地址 http://mvnrepository.com/artifact/org.mybatis/mybatis-spring 下载。

❹ **数据库驱动 JAR 包**

本书所使用的 MySQL 数据库驱动包为 mysql-connector-java-8.0.29.jar。

❺ **数据源所需的 JAR 包**

在整合时使用的是 DBCP 数据源,需要准备 DBCP 和连接池的 JAR 包。在编写本书时,最新版本的 DBCP 的 JAR 包为 commons-dbcp2-2.12.0.jar,可以从地址 http://commons.apache.org/proper/commons-dbcp/download_dbcp.cgi 下载;最新版本的连接池的 JAR 包为 commons-pool2-2.12.0.jar,可以从地址 http://commons.apache.org/proper/commons-pool/download_pool.cgi 下载。

▶ 3.6.2 MapperScannerConfigurer 方式

在一般情况下,将数据源及 MyBatis 工厂配置在 Spring 的配置文件中,实现 MyBatis 与 Spring 的无缝整合。在 Spring 的配置文件中,首先使用 org.apache.commons.dbcp2.BasicDataSource 配置数据源,然后使用 org.springframework.jdbc.datasource.DataSourceTransactionManager 为数据源添加事务管理器,最后使用 org.mybatis.spring.SqlSessionFactoryBean 配置 MyBatis 工厂,同时指定数据源,并与 MyBatis 完美整合。

使用 Spring 管理 MyBatis 的数据操作接口的方式有多种。其中,最常用、最简洁的一种是基于 org.mybatis.spring.mapper.MapperScannerConfigurer 的整合实现 Mapper 代理开发。MapperScannerConfigurer 将包(<property name="basePackage" value="xxx" />)中的所有接口自动装配为 MyBatis 映射接口 Mapper 的实现类的实例(映射器),所有映射器都被自动注入 SqlSessionFactory 实例,同时扫描包中的 SQL 映射文件,MyBatis 核心配置文件不再加载 SQL 映射文件(但要保证接口与 SQL 映射文件名相同)。配置文件的示例代码如下:

```
<!-- 配置数据源 -->
<bean id="dataSource" class="org.apache.commons.dbcp2.BasicDataSource">
```

```xml
        <property name="driverClassName" value="${jdbc.driver}" />
        <property name="url" value="${jdbc.url}" />
        <property name="username" value="${jdbc.username}" />
        <property name="password" value="${jdbc.password}" />
        <!-- 最大连接数 -->
        <property name="maxTotal" value="${jdbc.maxTotal}" />
        <!-- 最大空闲连接数 -->
        <property name="maxIdle" value="${jdbc.maxIdle}" />
        <!-- 初始化连接数 -->
        <property name="initialSize" value="${jdbc.initialSize}" />
</bean>
<!-- 添加事务支持 -->
<bean id="txManager" class="org.springframework.jdbc.datasource.
DataSourceTransactionManager">
        <property name="dataSource" ref="dataSource" />
</bean>
<!-- 开启事务注解 -->
<tx:annotation-driven transaction-manager="txManager" />
<!-- 配置MyBatis工厂,同时指定数据源,并与MyBatis完美整合 -->
<bean id="sqlSessionFactory" class="org.mybatis.spring.SqlSessionFactoryBean">
        <property name="dataSource" ref="dataSource" />
        <!-- configLocation的属性值为MyBatis的核心配置文件 -->
        <property name="configLocation" value="classpath:config/mybatis-config.xml" />
</bean>
<!--Mapper代理开发,MapperScannerConfigurer将包中的所有接口自动装配为MyBatis映射接口
Mapper的实现类的实例(映射器),所有映射器都被自动注入SqlSessionFactory实例,同时扫描包中的SQL
映射文件,MyBatis核心配置文件不再加载SQL映射文件 -->
<bean class="org.mybatis.spring.mapper.MapperScannerConfigurer">
        <!-- mybatis-spring组件的扫描器,basePackage属性可以包含多个包名,多个包名之间用逗号或分
号隔开 -->
        <property name="basePackage" value="dao" />
        <property name="sqlSessionFactoryBeanName" value="sqlSessionFactory" />
</bean>
```

▶3.6.3 整合示例

下面通过 SSM 框架整合实现例 3-1 的功能。

【例 3-2】 SSM 框架整合开发。

具体实现步骤如下。

❶ 创建 Web 应用并导入相关 JAR 包

首先在项目 ch3 中创建一个名为 ch3_2 的模块,同时为模块 ch3_2 添加 Web Application,然后参考 3.6.1 节将相关 JAR 包复制到模块 ch3_2 的 WEB-INF/lib 目录中,最后为模块 ch3_2 添加 Tomcat 依赖。

❷ 创建数据库连接信息属性文件及 Log4j 的日志配置文件

在模块 ch3_2 的 src 目录下创建名为 config 的包,并在该包中创建数据库连接信息属性文件 jdbc.properties,具体内容如下。

```
jdbc.driver=com.mysql.cj.jdbc.Driver
jdbc.url = jdbc: mysql://localhost: 3306/springtest? useUnicode = true&characterEncoding = UTF-
8&allowMultiQueries=true&serverTimezone=GMT%2B8
jdbc.username=root
jdbc.password=root
jdbc.maxTotal=30
jdbc.maxIdle=10
jdbc.initialSize=5
```

在模块 ch3_2 的 src 目录下创建 Log4j 的日志配置文件 log4j.properties，其内容与 3.5 节中的例 3-1 相同，为了节省篇幅，不再赘述。

❸ 创建持久化类

在模块 ch3_2 的 src 目录下创建一个名为 com.mybatis.po 的包，并在该包中创建持久化类 MyUser。该类与 3.5 节中的例 3-1 相同，为了节省篇幅，不再赘述。

❹ 创建 SQL 映射文件

在模块 ch3_2 的 src 目录下创建一个名为 com.mybatis.mapper 的包，并在该包中创建 SQL 映射文件 UserMapper.xml。该文件与 3.5 节中的例 3-1 相同，为了节省篇幅，不再赘述。

❺ 创建 MyBatis 的核心配置文件

在模块 ch3_2 的 config 包中创建 MyBatis 的核心配置文件 mybatis-config.xml，在该文件中配置实体类的别名、日志输出等，具体内容如下。

```xml
<?xml version="1.0" encoding="UTF-8" ?>
<!DOCTYPE configuration
PUBLIC "-//mybatis.org//DTD Config 3.0//EN"
"http://mybatis.org/dtd/mybatis-3-config.dtd">
<configuration>
    <settings>
        <setting name="logImpl" value="LOG4J" />
    </settings>
    <typeAliases>
        <package name="com.mybatis.po" />
    </typeAliases>
</configuration>
```

❻ 创建 Mapper 接口

在模块 ch3_2 的 com.mybatis.mapper 包中创建接口 UserMapper，并使用@Repository 注解标注该接口是数据访问层。该接口中的方法与 SQL 映射文件 UserMapper.xml 中的 id 一致。UserMapper 接口的核心代码如下。

```java
@Repository
public interface UserMapper {
    public MyUser selectUserById(Integer id);
    public List<MyUser> selectAllUser();
    public int addUser(MyUser myUser);
    public int updateUser(MyUser myUser);
    public int deleteUser(Integer id);
}
```

❼ 创建控制器类

在模块 ch3_2 的 src 目录下创建一个名为 controller 的包，并在该包中创建控制器类 TestController。在该控制器类中调用 Mapper 接口中的方法操作数据库，核心代码如下。

```java
@Controller
public class TestController {
    @Autowired
    private UserMapper userMapper;
    @GetMapping("/test")
    public String test() {
        //查询一个用户
        MyUser mu = userMapper.selectUserById(1);
        System.out.println(mu);
        //添加一个用户
```

```java
            MyUser addmu = new MyUser();
            addmu.setUname("陈恒");
            addmu.setUsex("男");
            userMapper.addUser(addmu);
            //修改一个用户
            MyUser updatemu = new MyUser();
            updatemu.setUid(1);
            updatemu.setUname("张三");
            updatemu.setUsex("女");
            userMapper.updateUser(updatemu);
            //删除一个用户
            userMapper.deleteUser(3);
            //查询所有用户
            List<MyUser> listMu = userMapper.selectAllUser();
            for (MyUser myUser: listMu) {
                System.out.println(myUser);
            }
            return "test";
        }
    }
```

❽ 创建测试页面

在模块 ch3_2 的 web/WEB-INF/目录下创建一个名为 jsp 的文件夹，并在该文件夹中创建 test.jsp 文件，test.jsp 的代码略。

❾ 创建 Web、Spring、Spring MVC 的配置文件

在模块 ch3_2 的 config 包中创建 Spring 配置文件 applicationContext.xml 和 Spring MVC 配置文件 springmvc.xml，在模块 ch3_2 的 web/WEB-INF/目录中创建 Web 配置文件 web.xml。

在 Spring 配置文件 applicationContext.xml 中，首先使用＜context：property-placeholder/＞加载数据库连接信息属性文件，然后使用 org.apache.commons.dbcp2.BasicDataSource 配置数据源，并使用 org.springframework.jdbc.datasource.DataSourceTransactionManager 为数据源添加事务管理器，接着使用 org.mybatis.spring.SqlSessionFactoryBean 配置 MyBatis 工厂，同时指定数据源，并与 MyBatis 完美整合，最后使用 org.mybatis.spring.mapper.MapperScannerConfigurer 实现 Mapper 代理开发，用 basePackage 属性指定包中的所有接口自动装配为 MyBatis 映射接口 Mapper 的实现类的实例（映射器），所有映射器都被自动注入 SqlSessionFactory 实例，同时扫描包中的 SQL 映射文件，MyBatis 核心配置文件不再加载 SQL 映射文件。Spring 配置文件 applicationContext.xml 的具体代码如下。

```xml
<?xml version="1.0" encoding="UTF-8"?>
<beans xmlns="http://www.springframework.org/schema/beans"
xmlns:xsi="http://www.w3.org/2001/XMLSchema-instance"
xmlns:tx="http://www.springframework.org/schema/tx"
xmlns:context="http://www.springframework.org/schema/context"
xsi:schemaLocation="http://www.springframework.org/schema/beans
        http://www.springframework.org/schema/beans/spring-beans.xsd
        http://www.springframework.org/schema/tx
        http://www.springframework.org/schema/tx/spring-tx.xsd
        http://www.springframework.org/schema/context
        http://www.springframework.org/schema/context/spring-context.xsd">
    <!-- 加载数据库配置文件 -->
    <context:property-placeholder location="classpath:config/jdbc.properties" />
    <!-- 配置数据源 -->
```

```xml
<bean id="dataSource" class="org.apache.commons.dbcp2.BasicDataSource">
    <property name="driverClassName" value="${jdbc.driver}" />
    <property name="url" value="${jdbc.url}" />
    <property name="username" value="${jdbc.username}" />
    <property name="password" value="${jdbc.password}" />
    <!-- 最大连接数 -->
    <property name="maxTotal" value="${jdbc.maxTotal}" />
    <!-- 最大空闲连接数 -->
    <property name="maxIdle" value="${jdbc.maxIdle}" />
    <!-- 初始化连接数 -->
    <property name="initialSize" value="${jdbc.initialSize}" />
</bean>
<!-- 添加事务支持 -->
<bean id="txManager" class="org.springframework.jdbc.datasource.DataSourceTransactionManager">
    <property name="dataSource" ref="dataSource" />
</bean>
<!-- 开启事务注解 -->
<tx:annotation-driven transaction-manager="txManager" />
<!-- 配置MyBatis工厂,同时指定数据源,并与MyBatis完美整合 -->
<bean id="sqlSessionFactory" class="org.mybatis.spring.SqlSessionFactoryBean">
    <property name="dataSource" ref="dataSource" />
    <property name="configLocation" value="classpath:config/mybatis-config.xml" />
</bean>
<bean class="org.mybatis.spring.mapper.MapperScannerConfigurer">
    <property name="basePackage" value="com.mybatis.mapper" />
    <property name="sqlSessionFactoryBeanName" value="sqlSessionFactory" />
</bean>
</beans>
```

在 Spring MVC 配置文件 springmvc.xml 中使用＜context:component-scan/＞扫描控制器包,并使用 org.springframework.web.servlet.view.InternalResourceViewResolver 配置视图解析器,具体代码如下。

```xml
<?xml version="1.0" encoding="UTF-8"?>
<beans xmlns="http://www.springframework.org/schema/beans"
    xmlns:xsi="http://www.w3.org/2001/XMLSchema-instance"
    xmlns:context="http://www.springframework.org/schema/context"
    xsi:schemaLocation="
        http://www.springframework.org/schema/beans
        http://www.springframework.org/schema/beans/spring-beans.xsd
        http://www.springframework.org/schema/context
        http://www.springframework.org/schema/context/spring-context.xsd">
    <context:component-scan base-package="controller"/>
    <bean class="org.springframework.web.servlet.view.InternalResourceViewResolver"
        id="internalResourceViewResolver">
        <property name="prefix" value="/WEB-INF/jsp/" />
        <property name="suffix" value=".jsp" />
    </bean>
</beans>
```

在 Web 配置文件 web.xml 中,首先通过＜context-param＞加载 Spring 配置文件 applicationContext.xml,并通过 org.springframework.web.context.ContextLoaderListener 启动 Spring 容器,然后配置 Spring MVC DispatcherServlet,并加载 Spring MVC 配置文件 springmvc.xml。Web 配置文件 web.xml 的代码如下。

```xml
<?xml version="1.0" encoding="UTF-8"?>
<web-app xmlns:xsi="http://www.w3.org/2001/XMLSchema-instance"
```

```xml
xmlns="http://xmlns.jcp.org/xml/ns/javaee"
xsi:schemaLocation="http://xmlns.jcp.org/xml/ns/javaee
http://xmlns.jcp.org/xml/ns/javaee/web-app_4_0.xsd"
id="WebApp_ID" version="4.0">
<!-- 实例化 ApplicationContext 容器 -->
<context-param>
    <!-- 加载 applicationContext.xml 文件 -->
    <param-name>contextConfigLocation</param-name>
    <param-value>
        classpath:config/applicationContext.xml
    </param-value>
</context-param>
<!-- 指定以 ContextLoaderListener 方式启动 Spring 容器 -->
<listener>
    <listener-class>org.springframework.web.context.ContextLoaderListener</listener-class>
</listener>
<!--配置 Spring MVC DispatcherServlet -->
<servlet>
    <servlet-name>springmvc</servlet-name>
    <servlet-class>org.springframework.web.servlet.DispatcherServlet</servlet-class>
    <init-param>
        <param-name>contextConfigLocation</param-name>
        <!-- classpath是指到 src 目录查找配置文件 -->
        <param-value>classpath:config/springmvc.xml</param-value>
    </init-param>
    <load-on-startup>1</load-on-startup>
</servlet>
<servlet-mapping>
    <servlet-name>springmvc</servlet-name>
    <url-pattern>/</url-pattern>
</servlet-mapping>
</web-app>
```

❿ 测试应用

首先将模块 ch3_2 发布到 Tomcat 服务器，然后启动 Tomcat 服务器，通过地址 http://localhost:8080/ch3_2/test 测试应用。成功运行后，控制台信息输出结果如图 3.4 所示。

```
DEBUG [http-nio-8080-exec-9] - ==>  Preparing: select * from user where uid = ?
DEBUG [http-nio-8080-exec-9] - ==> Parameters: 1(Integer)
DEBUG [http-nio-8080-exec-9] - <==      Total: 1
User [uid=1,uname=陈恒1,usex=女]
DEBUG [http-nio-8080-exec-9] - ==>  Preparing: insert into user (uname,usex) values(?,?)
DEBUG [http-nio-8080-exec-9] - ==> Parameters: 陈恒(String), 男(String)
DEBUG [http-nio-8080-exec-9] - <==    Updates: 1
DEBUG [http-nio-8080-exec-9] - ==>  Preparing: update user set uname = ?,usex = ? where uid = ?
DEBUG [http-nio-8080-exec-9] - ==> Parameters: 张三(String), 女(String), 1(Integer)
DEBUG [http-nio-8080-exec-9] - <==    Updates: 1
DEBUG [http-nio-8080-exec-9] - ==>  Preparing: delete from user where uid = ?
DEBUG [http-nio-8080-exec-9] - ==> Parameters: 3(Integer)
DEBUG [http-nio-8080-exec-9] - <==    Updates: 1
DEBUG [http-nio-8080-exec-9] - ==>  Preparing: select * from user
DEBUG [http-nio-8080-exec-9] - ==> Parameters:
DEBUG [http-nio-8080-exec-9] - <==      Total: 5
User [uid=1,uname=张三,usex=女]
User [uid=2,uname=陈恒2,usex=男]
User [uid=4,uname=陈恒4,usex=男]
User [uid=36,uname=陈恒,usex=男]
User [uid=37,uname=陈恒,usex=男]
```

图 3.4 模块 ch3_2 的控制台信息输出结果

3.6.4 SqlSessionDaoSupport 方式

从 3.6.3 节的示例可知,在 MyBatis 中,当编写好访问数据库的映射器接口后,MapperScannerConfigurer 就能自动根据这些接口生成 DAO 对象,然后使用@Autowired 把这些 DAO 对象注入业务逻辑层或控制层。在这种情况下的 DAO 层中几乎不用编写代码,而且也没有地方编写,因为只有接口。这固然方便,不过当需要在 DAO 层编写代码时,这种方式就无能为力。幸运的是,MyBatis-Spring 提供了以继承 SqlSessionDaoSupport 类的方式访问数据库。

org.mybatis.spring.support.SqlSessionDaoSupport 类继承了 org.springframework.dao.support.DaoSupport 类,是一个抽象类,作为 DAO 的基类使用,需要一个 SqlSessionFactory。在继承 SqlSessionDaoSupport 类的子类中通过调用 SqlSessionDaoSupport 类的 getSqlSession()方法来获取这个 SqlSessionFactory 提供的 SqlSessionTemplate 对象。SqlSessionTemplate 类实现了 SqlSession 接口,即可以进行数据库访问。所以,需要 Spring 框架给 SqlSessionDaoSupport 类的子类的对象(多个 DAO 对象)注入一个 SqlSessionFactory。

自 mybatis-spring-1.2.0 以来,SqlSessionDaoSupport 的 setSqlSessionTemplate 和 setSqlSessionFactory 两个方法上的 @Autowired 注解被删除,这意味着继承于 SqlSessionDaoSupport 的 DAO 类,它们的对象不能被自动注入 SqlSessionFactory 或 SqlSessionTemplate 对象。如果在 Spring 的配置文件中一个一个地配置,显然太麻烦。比较好的解决办法是在 DAO 类中覆盖这两个方法之一,并加上@Autowired 或@Resource 注解。那么如果在每个 DAO 类中都这么做,显然很低效。更合理的做法是编写一个继承于 SqlSessionDaoSupport 的 BaseDao,在 BaseDao 中完成这个工作,然后其他的 DAO 类都继承 BaseDao。BaseDao 的示例代码如下。

```
package dao;
import javax.annotation.Resource;
import org.apache.ibatis.session.SqlSessionFactory;
import org.mybatis.spring.support.SqlSessionDaoSupport;
public class BaseDao extends SqlSessionDaoSupport{
    //依赖注入 sqlSession 工厂
    @Resource(name = "sqlSessionFactory")
    public void setSqlSessionFactory(SqlSessionFactory sqlSessionFactory) {
        super.setSqlSessionFactory(sqlSessionFactory);
    }
}
```

下面通过实例讲解以继承 SqlSessionDaoSupport 类的方式访问数据库。

【例 3-3】 在 3.6.3 节中例 3-2 的基础上实现以继承 SqlSessionDaoSupport 类的方式访问数据库。

为了节省篇幅,相同的实现不再赘述,其他的具体实现如下。

❶ 创建 Web 应用并导入相关 JAR 包

首先在项目 ch3 中创建一个名为 ch3_3 的模块,同时为模块 ch3_3 添加 Web Application,然后参考 3.6.1 节,除将相关 JAR 包复制到模块 ch3_3 的 WEB-INF/lib 目录中外,还需要复制@Resource 注解所在的 JAR 包 annotations-api.jar(可以从 Tomcat 的 lib 目录中复制),最后为模块 ch3_3 添加 Tomcat 依赖。

❷ 复制数据库连接信息属性文件及 Log4j 的日志配置文件

在模块 ch3_3 的 src 目录下创建名为 config 的包,并将模块 ch3_2 的数据库连接信息属性文件 jdbc.properties 复制到该包中。

将模块 ch3_2 的 Log4j 日志配置文件 log4j.properties 复制到模块 ch3_3 的 src 目录中，并将其中的"log4j.logger.com.mybatis.mapper=DEBUG"修改为"log4j.logger.dao=DEBUG"。

❸ 创建持久化类

在模块 ch3_3 的 src 目录下创建一个名为 po 的包，并在该包中创建持久化类 MyUser。该类与 3.5 节中的例 3-1 相同，为了节省篇幅，不再赘述。

❹ 创建 SQL 映射文件

在模块 ch3_3 的 src 目录下创建一个名为 dao 的包，并在该包中创建 SQL 映射文件 UserMapper.xml，文件内容如下：

```xml
<?xml version="1.0" encoding="UTF-8" ?>
<!DOCTYPE mapper
PUBLIC "-//mybatis.org//DTD Mapper 3.0//EN"
"http://mybatis.org/dtd/mybatis-3-mapper.dtd">
<mapper namespace="dao.UserMapper">
    <!-- 根据 uid 查询一个用户信息 -->
    <select id="selectUserById" parameterType="Integer" resultType="MyUser">
        select * from user where uid = #{uid}
    </select>
    <!-- 查询所有用户信息 -->
    <select id="selectAllUser" resultType="MyUser">
        select * from user
    </select>
</mapper>
```

❺ 创建 MyBatis 的核心配置文件

在模块 ch3_3 的 config 包中创建 MyBatis 的核心配置文件 mybatis-config.xml。在该文件中配置实体类的别名、日志输出、指定映射文件位置等，具体内容如下：

```xml
<?xml version="1.0" encoding="UTF-8" ?>
<!DOCTYPE configuration
PUBLIC "-//mybatis.org//DTD Config 3.0//EN"
"http://mybatis.org/dtd/mybatis-3-config.dtd">
<configuration>
    <settings>
        <setting name="logImpl" value="LOG4J" />
    </settings>
    <typeAliases>
        <package name="po" />
    </typeAliases>
    <!-- 告诉 MyBatis 到哪里去找映射文件 -->
    <mappers>
        <mapper resource="dao/UserMapper.xml"/>
    </mappers>
</configuration>
```

❻ 创建 DAO 接口和接口实现类

在模块 ch3_3 的 dao 包中创建 SQL 映射文件 UserMapper.xml 对应的接口 UserMapper。UserMapper 接口的代码如下：

```java
package dao;
import java.util.List;
import po.MyUser;
public interface UserMapper {
    public MyUser selectUserById(int id);
    public List<MyUser> selectAllUser();
}
```

在模块 ch3_3 的 dao 包中创建 BaseMapper 类，在该类中使用 @Resource（name = "sqlSessionFactory"）注解依赖注入 sqlSession 工厂。BaseMapper 类的代码如下。

```java
package dao;
import org.apache.ibatis.session.SqlSessionFactory;
import org.mybatis.spring.support.SqlSessionDaoSupport;
import jakarta.annotation.Resource;
public class BaseMapper extends SqlSessionDaoSupport{
    //依赖注入 sqlSession 工厂
    @Resource(name = "sqlSessionFactory")
    public void setSqlSessionFactory(SqlSessionFactory sqlSessionFactory) {
        super.setSqlSessionFactory(sqlSessionFactory);
    }
}
```

在模块 ch3_3 的 dao 包中创建接口 UserMapper 的实现类 UserMapperImpl，在该类中使用 @Repository 注解标注该类的实例是数据访问对象。UserMapperImpl 类的核心代码如下。

```java
@Repository
public class UserMapperImpl extends BaseMapper implements UserMapper {
    public MyUser selectUserById(int id) {
        //获取 SqlSessionFactory 提供的 SqlSessionTemplate 对象
        SqlSession session = getSqlSession();
        //dao.UserMapper.selectUserById 为数据接口中的访问方法
        return session.selectOne("dao.UserMapper.selectUserById", id);
    }
    public List<MyUser> selectAllUser() {
        SqlSession session = getSqlSession();
        return session.selectList("dao.UserMapper.selectAllUser");
    }
}
```

❼ 创建控制器类

在模块 ch3_3 的 src 目录下创建一个名为 controller 的包，并在该包中创建控制器类 MyController。在该控制器类中调用 UserMapper 接口中的方法操作数据库，具体核心代码如下。

```java
@Controller
public class MyController {
    @Autowired
    private UserMapper userMapper;
    @GetMapping("/test")
    public String test() {
        //查询一个用户
        MyUser mu = userMapper.selectUserById(1);
        System.out.println(mu);
        //查询所有用户
        List<MyUser> listMu = userMapper.selectAllUser();
        for (MyUser myUser: listMu) {
            System.out.println(myUser);
        }
        return "test";
    }
}
```

❽ 创建测试页面

在模块 ch3_3 的 web/WEB-INF/目录下创建一个名为 jsp 的文件夹，并在该文件夹中创建 test.jsp 文件，test.jsp 的代码略。

❾ 创建 Web、Spring、Spring MVC 的配置文件

在模块 ch3_3 的 config 包中创建 Spring 配置文件 applicationContext.xml 和 Spring MVC 配置文件 springmvc.xml，在模块 ch3_3 的 web/WEB-INF/目录中创建 Web 配置文件 web.xml。

在 Spring 配置文件 applicationContext.xml 中，首先使用＜context:property-placeholder/＞加载数据库连接信息属性文件，然后使用 org.apache.commons.dbcp2.BasicDataSource 配置数据源，并使用 org.springframework.jdbc.datasource.DataSourceTransactionManager 为数据源添加事务管理器，最后使用 org.mybatis.spring.SqlSessionFactoryBean 配置 MyBatis 工厂，同时指定数据源，并与 MyBatis 完美整合。Spring 配置文件 applicationContext.xml 的具体内容如下。

```xml
<?xml version="1.0" encoding="UTF-8"?>
<beans xmlns="http://www.springframework.org/schema/beans"
    xmlns:xsi="http://www.w3.org/2001/XMLSchema-instance"
    xmlns:tx="http://www.springframework.org/schema/tx"
    xmlns:context="http://www.springframework.org/schema/context"
    xsi:schemaLocation="http://www.springframework.org/schema/beans
        http://www.springframework.org/schema/beans/spring-beans.xsd
        http://www.springframework.org/schema/tx
        http://www.springframework.org/schema/tx/spring-tx.xsd
        http://www.springframework.org/schema/context
        http://www.springframework.org/schema/context/spring-context.xsd">
    <!-- 加载数据库配置文件 -->
    <context:property-placeholder location="classpath:config/jdbc.properties" />
    <!-- 配置数据源 -->
    <bean id="dataSource" class="org.apache.commons.dbcp2.BasicDataSource">
        <property name="driverClassName" value="${jdbc.driver}" />
        <property name="url" value="${jdbc.url}" />
        <property name="username" value="${jdbc.username}" />
        <property name="password" value="${jdbc.password}" />
        <!-- 最大连接数 -->
        <property name="maxTotal" value="${jdbc.maxTotal}" />
        <!-- 最大空闲连接数 -->
        <property name="maxIdle" value="${jdbc.maxIdle}" />
        <!-- 初始化连接数 -->
        <property name="initialSize" value="${jdbc.initialSize}" />
    </bean>
    <!-- 添加事务支持 -->
    <bean id="txManager" class="org.springframework.jdbc.datasource.DataSourceTransactionManager">
        <property name="dataSource" ref="dataSource" />
    </bean>
    <!-- 开启事务注解 -->
    <tx:annotation-driven transaction-manager="txManager" />
    <bean id="sqlSessionFactory" class="org.mybatis.spring.SqlSessionFactoryBean">
        <property name="dataSource" ref="dataSource" />
        <property name="configLocation" value="classpath:config/mybatis-config.xml">
        </property>
    </bean>
</beans>
```

在 Spring MVC 配置文件 springmvc.xml 中，使用＜context:component-scan/＞扫描包，并使用 org.springframework.web.servlet.view.InternalResourceViewResolver 配置视图解析器，具体代码如下。

```xml
<?xml version="1.0" encoding="UTF-8"?>
<beans xmlns="http://www.springframework.org/schema/beans"
    xmlns:xsi="http://www.w3.org/2001/XMLSchema-instance"
    xmlns:context="http://www.springframework.org/schema/context"
    xsi:schemaLocation="
        http://www.springframework.org/schema/beans
        http://www.springframework.org/schema/beans/spring-beans.xsd
        http://www.springframework.org/schema/context
        http://www.springframework.org/schema/context/spring-context.xsd">
    <context:component-scan base-package="controller"/>
    <context:component-scan base-package="dao"/>
    <bean class="org.springframework.web.servlet.view.InternalResourceViewResolver"
          id="internalResourceViewResolver">
        <property name="prefix" value="/WEB-INF/jsp/" />
        <property name="suffix" value=".jsp" />
    </bean>
</beans>
```

在 Web 配置文件 web.xml 中，首先通过 ＜context-param＞ 加载 Spring 配置文件 applicationContext.xml，并通过 org.springframework.web.context.ContextLoaderListener 启动 Spring 容器，然后配置 Spring MVC DispatcherServlet，并加载 Spring MVC 配置文件 springmvc.xml。Web 配置文件 web.xml 的代码与模块 ch3_2 的相同，不再赘述。

❿ 测试应用

发布模块 ch3_3 到 Web 服务器 Tomcat，然后通过地址 http://localhost:8080/ch3_3/test 测试应用。

3.7 使用 MyBatis Generator 插件自动生成映射文件

使用 MyBatis Generator 插件自动生成 MyBatis 的 DAO 接口、实体模型类、Mapper 映射文件，将生成的代码复制到项目工程中，以便把更多精力放在业务逻辑上。

MyBatis Generator 有三种常用方法自动生成代码，分别为命令行、IDEA 插件和 Maven 插件。本节使用比较简单的方法（命令行）自动生成相关代码，具体步骤如下。

❶ 准备相关 JAR 包

需要准备的 JAR 包为 mysql-connector-java-8.0.29.jar 和 mybatis-generator-core-1.4.1.jar（https://mvnrepository.com/artifact/org.mybatis.generator/mybatis-generator-core/1.4.1）。

❷ 创建文件目录

在某磁盘的根目录下新建一个文件目录，例如 C:\generator，并将 mysql-connector-java-8.0.29.jar 和 mybatis-generator-core-1.4.1.jar 文件复制到 generator 目录下。另外，在 generator 目录下创建 src 子目录存放生成的相关代码文件。

❸ 创建配置文件

在第 2 步创建的文件目录（C:\generator）下创建配置文件 generator.xml 与 src 文件夹，文件目录如图 3.5 所示。

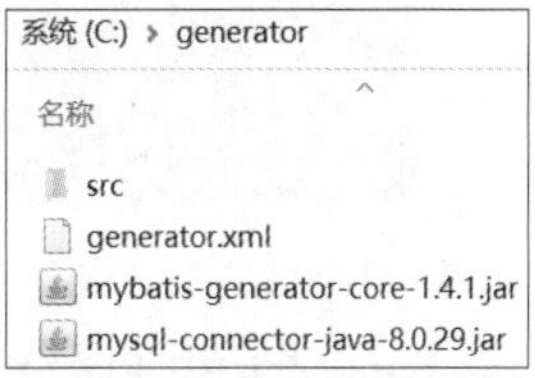

图 3.5 generator 目录

generator.xml 配置文件的内容如下（具体含义见注释）。

```xml
<?xml version="1.0" encoding="UTF-8"?>
<!DOCTYPE generatorConfiguration PUBLIC " -//mybatis.org//DTD MyBatis Generator Configuration 1.0//EN" "http://mybatis.org/dtd/mybatis-generator-config_1_0.dtd">
```

```xml
<generatorConfiguration>
    <!-- 数据库驱动包的位置 -->
    <classPathEntry location="C:\generator\mysql-connector-java-8.0.29.jar" />
    <context id="mysqlTables" targetRuntime="MyBatis3">
        <commentGenerator>
            <property name="suppressAllComments" value="true" />
        </commentGenerator>
        <!-- 数据库链接URL、用户名、密码(前提是数据库springtest存在) -->
        <jdbcConnection
                driverClass="com.mysql.cj.jdbc.Driver"
                connectionURL="jdbc:mysql://127.0.0.1:3306/springtest?useUnicode=true&amp;characterEncoding=UTF-8&amp;allowMultiQueries=true&amp;serverTimezone=GMT%2B8"
                userId="root" password="root">
        </jdbcConnection>
        <javaTypeResolver>
            <property name="forceBigDecimals" value="false" />
        </javaTypeResolver>
        <!-- 生成模型(MyBatis里面用到实体类)的包名和位置-->
        <javaModelGenerator targetPackage="com.po" targetProject="C:\generator\src">
            <property name="enableSubPackages" value="true" />
            <property name="trimStrings" value="true" />
        </javaModelGenerator>
        <!-- 生成映射文件(MyBatis 的 SQL 语句 XML 文件)的包名和位置-->
        <sqlMapGenerator targetPackage="mybatis" targetProject="C:\generator\src">
            <property name="enableSubPackages" value="true" />
        </sqlMapGenerator>
        <!-- 生成DAO的包名和位置 -->
        <javaClientGenerator type="XMLMAPPER" targetPackage="com.dao"
targetProject="C:\generator\src">
            <property name="enableSubPackages" value="true" />
        </javaClientGenerator>
        <!-- 要生成哪些表(更改tableName和domainObjectName就可以,数据库springtest中的user表已创建)-->
        <table tableName="user" domainObjectName="User" enableCountByExample="false" enableUpdateByExample="false" enableDeleteByExample="false" enableSelectByExample="false" selectByExampleQueryId="false" />
    </context>
</generatorConfiguration>
```

❹ 使用命令生成代码

打开命令提示符,进入C:\generator目录,输入命令java -jar mybatis-generator-core-1.4.1.jar -configfile generator.xml -overwrite,如图3.6所示。该命令成功执行的前提是配置了Java的系统环境变量classpath。

```
C:\>cd generator

C:\generator>java -jar mybatis-generator-core-1.4.0.jar -configfile generator.xml -overwrite
MyBatis Generator finished successfully.

C:\generator>
```

图3.6 使用命令行生成映射文件

3.8 映射器概述

映射器是MyBatis最复杂且最重要的组件,由一个接口和一个XML文件(SQL映射文件)组成。MyBatis的映射器也可以使用注解完成,但在实际应用中使用不多,原因主要来自三个方面:其一,面对复杂的SQL会显得无力;其二,注解的可读性较差;其三,注解丢失了XML上下文相互引用的功能。因此,推荐大家使用XML文件开发MySQL的映射器。

SQL 映射文件的常用配置元素如表 3.1 所示。

表 3.1　SQL 映射文件的常用配置元素

元素名称	描　　述	备　　注
select	查询语句，最常用、最复杂的元素之一	可以自定义参数，返回结果集等
insert	插入语句	执行后返回一个整数，代表插入的行数
update	更新语句	执行后返回一个整数，代表更新的行数
delete	删除语句	执行后返回一个整数，代表删除的行数
sql	定义一部分 SQL，在多个位置被引用	例如，一张表的列名，一次定义，可以在多个 SQL 语句中使用
resultMap	用来描述从数据库结果集中加载对象，是最复杂、最强大的元素之一	提供映射规则

3.9　<select>元素

在 SQL 映射文件中，<select>元素用于映射 SQL 的 select 语句，其示例代码如下。

```
<!-- 根据 uid 查询一个用户信息 -->
<select id="selectUserById" parameterType="Integer" resultType="MyUser">
    select * from user where uid = #{uid}
</select>
```

在上述示例代码中，id 的值是唯一标识符（对应 Mapper 接口的某个方法），它接收一个 Integer 类型的参数，返回一个 MyUser 类型的对象，结果集自动映射到 MyUser 的属性。需要注意的是，MyUser 的属性名称一定与查询结果的列名相同。

<select>元素除了有上述示例代码中的几个属性外，还有一些常用的属性，如表 3.2 所示。

表 3.2　<select>元素的常用属性

属性名称	描　　述
id	和 Mapper 的命名空间组合起来使用（对应 Mapper 接口的某个方法），是唯一标识符，供 MyBatis 调用
parameterType	表示传入 SQL 语句的参数类型的全限定名或别名。是一个可选属性，MyBatis 能推断出具体传入语句的参数
resultType	SQL 语句执行后返回的类型（全限定名或者别名）。如果是集合类型，返回的是集合元素的类型。在返回时可以使用 resultType 或 resultMap 之一
resultMap	映射集的引用，与<resultMap>元素一起使用。在返回时可以使用 resultType 或 resultMap 之一
flushCache	作用是在调用 SQL 语句后是否要求 MyBatis 清空之前查询的本地缓存和二级缓存。其默认值为 false，如果设置为 true，则任何时候只要 SQL 语句被调用，都将清空本地缓存和二级缓存
useCache	启动二级缓存的开关。其默认值为 true，表示将查询结果存入二级缓存中
timeout	用于设置超时参数，单位是秒。若超时将抛出异常
fetchSize	获取记录的总条数设定
statementType	告诉 MyBatis 使用哪个 JDBC 的 Statement 工作，取值为 STATEMENT（Statement）、PREPARED（PreparedStatement）、CALLABLE（CallableStatement），默认值为 PREPARED

续表

属性名称	描述
resultSetType	这是针对 JDBC 的 ResultSet 接口而言，其值可设置为 FORWARD_ONLY（只允许向前访问）、SCROLL_SENSITIVE（双向滚动，但不及时更新）、SCROLL_INSENSITIVE（双向滚动，及时更新）

▶3.9.1 使用 Map 接口传递参数

在实际开发中，查询 SQL 语句经常需要多个参数，例如多条件查询。当多个参数传递时，＜select＞元素的 parameterType 属性值的类型是什么呢？在 MyBatis 中，允许 Map 接口通过键-值对传递多个参数。

假设数据操作接口中有一个实现查询陈姓男性用户信息功能的方法：

```
public List<MyUser> testMapSelect(Map<String, Object> param);
```

此时传递给 MyBatis 映射器的是一个 Map 对象，使用该 Map 对象在 SQL 中设置对应的参数，对应 SQL 映射文件的代码如下：

```xml
<!-- 查询陈姓男性用户信息 -->
<select id="testMapSelect" resultType="MyUser" parameterType="map">
    select * from user
    where uname like concat(#{u_name},'%')
    and usex = #{u_sex}
</select>
```

在上述 SQL 映射文件中，参数名 u_name 和 u_sex 是 Map 中的 key。

【例 3-4】 在 3.6.3 节中例 3-2 的基础上实现使用 Map 接口传递参数。

为了节省篇幅，相同的实现不再赘述，其他的具体实现如下。

❶ 添加接口方法

在模块 ch3_2 的 com.mybatis.mapper.UserMapper 接口中添加接口方法（见上述），实现查询陈姓男性用户信息。

❷ 添加 SQL 映射

在模块 ch3_2 的 SQL 映射文件 UserMapper.xml 中添加 SQL 映射（见上述），实现查询陈姓男性用户信息。

❸ 添加请求处理方法

在模块 ch3_2 的 TestController 控制器类中添加测试方法 testMapSelect，具体代码如下。

```java
@GetMapping("/testMapSelect")
public String testMapSelect(Model model) {
    //查询所有陈姓男性用户
    Map<String, Object> map = new HashMap<>();
    map.put("u_name", "陈");
    map.put("u_sex", "男");
    List<MyUser> unameAndUsexList = userMapper.testMapSelect(map);
    model.addAttribute("unameAndUsexList", unameAndUsexList);
    return "showUnameAndUsexUser";
}
```

❹ 创建查询结果显示页面

在模块 ch3_2 的 web/WEB-INF/jsp 目录下创建查询结果显示页面 showUnameAndUsexUser.

jsp。在该页面中使用 JSTL 标签，所以需要将 taglibs-standard-impl-1.2.5-migrated-0.0.1.jar 和 taglibs-standard-spec-1.2.5-migrated-0.0.1.jar 复制到 WEB-INF/lib 目录中。另外，使用 BootStrap 美化页面，所以需要将相关的 CSS 及 JS 复制到/web/static 目录中，同时在应用 ch3_2 的 springmvc.xml 文件中使用<mvc:resources location="/static/" mapping="/static/**"></mvc:resources>允许 web/static 目录下的所有静态资源可见。showUnameAndUsexUser.jsp 的代码如下。

```jsp
<%@ page language="java" contentType="text/html; charset=UTF-8" pageEncoding="UTF-8"%>
<%@ taglib uri="http://java.sun.com/jsp/jstl/core" prefix="c" %>
<%
String path = request.getContextPath();
String basePath = request.getScheme() +"://" + request.getServerName() + ":" + request.getServerPort()+path+"/";
%>
<!DOCTYPE html>
<html>
<head>
<base href="<%=basePath%>">
<meta charset="UTF-8">
<title>Insert title here</title>
<link href="static/css/bootstrap.min.css" rel="stylesheet">
</head>
<body>
    <div class="container">
        <div class="panel panel-primary">
            <div class="panel-heading">
                <h3 class="panel-title">陈姓男性用户列表</h3>
            </div>
            <div class="panel-body">
                <div class="table table-responsive">
                    <table class="table table-bordered table-hover">
                        <tbody class="text-center">
                            <tr>
                                <th>用户 ID</th>
                                <th>姓名</th>
                                <th>性别</th>
                            </tr>
                            <c:forEach items="${unameAndUsexList}" var="user">
                                <tr>
                                    <td>${user.uid}</td>
                                    <td>${user.uname}</td>
                                    <td>${user.usex}</td>
                                </tr>
                            </c:forEach>
                        </tbody>
                    </table>
                </div>
            </div>
        </div>
    </div>
</body>
</html>
```

❺ 测试应用

发布模块 ch3_2 到 Web 服务器 Tomcat，然后通过地址 http://localhost:8080/ch3_2/testMapSelect 测试应用，成功运行后如图 3.7 所示。

Map 是一个键值对应的集合，使用者通过阅读它的键才能了解其作用。另外，使用 Map 不能限定其传递的数据类型，所以业务性不强，可读性差。如果 SQL 语句很复杂，参数很多，使用

图3.7 查询所有陈姓男性用户

Map 就很不方便。幸运的是，MyBatis 还提供了使用 Java Bean 传递参数。

▶3.9.2 使用 Java Bean 传递参数

在 MyBatis 中，当需要将多个参数传递给映射器时，可以将它们封装在一个 Java Bean 中。下面通过具体实例讲解如何使用 Java Bean 传递参数。

【例 3-5】 在 3.6.3 节中例 3-2 的基础上实现使用 Java Bean 传递参数。

为了节省篇幅，相同的实现不再赘述，其他的具体实现如下。

❶ 添加接口方法

在模块 ch3_2 的 com.mybatis.mapper.UserMapper 接口中添加接口方法 selectAllUserByJavaBean()，在该方法中使用 MyUser 类的对象将参数信息封装。接口方法 selectAllUserByJavaBean() 的定义如下：

```java
public List<MyUser> selectAllUserByJavaBean(MyUser user);
```

❷ 添加 SQL 映射

在模块 ch3_2 的 SQL 映射文件 UserMapper.xml 中添加接口方法对应的 SQL 映射，具体代码如下。

```xml
<!-- 通过 Java Bean 传递参数查询陈姓男性用户信息,#{uname}中的 uname 为参数 MyUser 的属性-->
<select id="selectAllUserByJavaBean" resultType="MyUser" parameterType="MyUser">
    select * from user
    where uname like concat(#{uname},'%')
    and usex = #{usex}
</select>
```

❸ 添加请求处理方法

在模块 ch3_2 的控制器类 TestController 中添加请求处理方法 selectAllUserByJavaBean()，具体代码如下。

```java
@GetMapping("/selectAllUserByJavaBean")
public String selectAllUserByJavaBean(Model model) {
    //通过 MyUser 封装参数,查询所有陈姓男性用户
    MyUser mu = new MyUser();
    mu.setUname("陈");
    mu.setUsex("男");
    List<MyUser> unameAndUsexList = userMapper.selectAllUserByJavaBean(mu);
    model.addAttribute("unameAndUsexList", unameAndUsexList);
    return "showUnameAndUsexUser";
}
```

❹ 测试应用

重启 Web 服务器 Tomcat，通过地址 http://localhost:8080/ch3_2/selectAllUserByJavaBean 测试应用。

▶3.9.3 使用@Param 注解传递参数

不管是 Map 传参还是 Java Bean 传参，它们都是将多个参数封装在一个对象中，实际上传递的还是一个参数，而使用@Param 注解可以将多个参数依次传递给 MyBatis 映射器，示例代码如下。

```java
public List<MyUser> selectByParam(@Param("puname") String uname, @Param("pusex") String usex);
```

在上述示例代码中，puname 和 pusex 是传递给 MyBatis 映射器的参数名。

下面通过实例讲解如何使用@Param 注解传递参数。

【例 3-6】 在 3.6.3 节中例 3-2 的基础上实现使用@Param 注解传递参数。

为了节省篇幅，相同的实现不再赘述，其他的具体实现如下。

❶ 添加接口方法

在模块 ch3_2 的 UserMapper 接口中添加数据操作接口方法 selectAllUserByParam()，在该方法中使用@Param 注解传递两个参数。接口方法 selectAllUserByParam()的定义如下。

```java
public List<MyUser> selectAllUserByParam(@Param("puname") String uname, @Param("pusex") String usex);
```

❷ 添加 SQL 映射

在模块 ch3_2 的 SQL 映射文件 UserMapper.xml 中添加接口方法 selectAllUserByParam()对应的 SQL 映射，具体代码如下。

```xml
<!-- 通过@Param注解传递参数查询陈姓男性用户信息，这里不需要定义参数类型-->
<select id="selectAllUserByParam" resultType="MyUser">
    select * from user
    where uname like concat(#{puname},'%')
    and usex = #{pusex}
</select>
```

❸ 添加请求处理方法

在模块 ch3_2 的控制器类 TestController 中添加请求处理方法 selectAllUserByParam()，具体代码如下。

```java
@GetMapping("/selectAllUserByParam")
public String selectAllUserByParam(Model model) {
    //通过@Param注解传递参数，查询所有陈姓男性用户
    List<MyUser> unameAndUsexList = userMapper.selectAllUserByParam("陈", "男");
    model.addAttribute("unameAndUsexList", unameAndUsexList);
    return "showUnameAndUsexUser";
}
```

❹ 测试应用

重启 Web 服务器 Tomcat，通过地址 http://localhost:8080/ch3_2/selectAllUserByParam 测试应用。

在实际应用中是选择 Map 接口、Java Bean 还是选择@Param 传递多个参数，应根据实际情况而定。如果参数较少，建议选择@Param；如果参数较多，建议选择 Java Bean。

3.9.4 <resultMap> 元素

<resultMap>元素表示结果映射集,是 MyBatis 中最重要、最强大的元素之一,其主要用来定义映射规则、级联的更新以及定义类型转化器等。<resultMap>元素中包含了一些子元素,结构如下:

```
<resultMap type="" id="">
    <constructor>              <!-- 类在实例化时,用来注入结果到构造方法 -->
        <idArg/>               <!-- ID参数,结果为ID -->
        <arg/>                 <!-- 注入构造方法的一个普通结果 -->
    </constructor>
    <id/>                      <!-- 用于表示哪个列是主键 -->
    <result/>                  <!-- 注入字段或POJO属性的普通结果 -->
    <association property=""/> <!-- 用于一对一级联 -->
    <collection property=""/>  <!-- 用于一对多、多对多级联 -->
</resultMap>
```

<resultMap>元素的 type 属性表示需要的 POJO,id 属性是 resultMap 的唯一标识。子元素<constructor>用于配置构造方法(当 POJO 未定义无参数的构造方法时使用)。子元素<id>用于表示哪个列是主键。子元素<result>用于表示 POJO 和数据表普通列的映射关系。子元素<association>和<collection>用在级联的情况下。关于级联的问题比较复杂,将在 3.11 节学习。

一条查询 SQL 语句执行后,结果可以使用 Map 存储,也可以使用 POJO(Java Bean)存储。

3.9.5 使用 POJO 存储结果集

在 3.9.1~3.9.3 节中都是直接使用 Java Bean(MyUser)存储结果集,这是因为 MyUser 的属性名与查询结果集的列名相同。如果查询结果集的列名与 Java Bean 的属性名不同,那么可以结合<resultMap>元素将 Java Bean 的属性名与查询结果集的列名一一对应。

下面通过一个实例讲解如何使用<resultMap>元素将 Java Bean 的属性名与查询结果集的列名一一对应。

【例 3-7】 在 3.6.3 节中例 3-2 的基础上使用<resultMap>元素将 Java Bean 的属性名与查询结果集的列名一一对应。

为了节省篇幅,相同的实现不再赘述,其他的具体实现如下。

❶ 创建 POJO 类

在模块 ch3_2 的 com.mybatis.po 包中创建一个名为 MapUser 的 POJO(Plain Ordinary Java Object,普通的 Java 类)类,具体代码如下。

```
package com.mybatis.po;
import lombok.Data;
@Data
public class MapUser {
    private Integer m_uid;
    private String m_uname;
    private String m_usex;
    @Override
    public String toString() {
        return "User [uid=" + m_uid +",uname=" + m_uname + ",usex=" + m_usex +"]";
    }
}
```

❷ 添加接口方法

在模块 ch3_2 的 UserMapper 接口中添加数据操作接口方法 selectAllUserPOJO(),该方法的返回值的类型是List＜MapUser＞。接口方法selectAllUserPOJO()的定义如下。

```
public List<MapUser> selectAllUserPOJO();
```

❸ 添加SQL映射

在模块ch3_2的SQL映射文件UserMapper.xml中，首先使用＜resultMap＞元素将MapUser类的属性名与查询结果的列名一一对应，然后添加接口方法selectAllUserPOJO()对应的SQL映射，具体代码如下。

```xml
<!-- 使用自定义结果集类型 -->
<resultMap type="com.mybatis.po.MapUser" id="myResult">
    <!-- property 是 MapUser 类中的属性-->
    <!-- column 是查询结果的列名,可以来自不同的表 -->
    <id property="m_uid" column="uid"/>
    <result property="m_uname" column="uname"/>
    <result property="m_usex" column="usex"/>
</resultMap>
<!-- 使用自定义结果集类型查询所有用户 -->
<select id="selectAllUserPOJO" resultMap="myResult">
    select * from user
</select>
```

❹ 添加请求处理方法

在模块ch3_2的控制器类TestController中添加请求处理方法selectAllUserPOJO()，具体代码如下。

```java
@GetMapping("/selectAllUserPOJO")
public String selectAllUserPOJO(Model model) {
    List<MapUser> unameAndUsexList = userMapper.selectAllUserPOJO();
    model.addAttribute("unameAndUsexList", unameAndUsexList);
    return "showUnameAndUsexUserPOJO";
}
```

❺ 创建显示查询结果的页面

在模块ch3_2的WEB-INF/jsp目录下创建showUnameAndUsexUserPOJO.jsp文件显示查询结果，核心代码如下。

```jsp
<c:forEach items="${unameAndUsexList}" var="user">
    <tr>
        <td>${user.m_uid}</td>
        <td>${user.m_uname}</td>
        <td>${user.m_usex}</td>
    </tr>
</c:forEach>
```

❻ 测试应用

重启Web服务器Tomcat，通过地址http://localhost:8080/ch3_2/selectAllUserPOJO测试应用。

▶3.9.6 使用Map存储结果集

在MyBatis中，任何查询结果都可以使用Map存储。下面通过一个实例讲解如何使用Map存储查询结果。

【例3-8】 在3.6.3节中例3-2的基础上使用Map存储查询结果。

为了节省篇幅，相同的实现不再赘述，其他的具体实现如下。

❶ 添加接口方法

在模块 ch3_2 的 UserMapper 接口中添加数据操作接口方法 selectAllUserMap()，该方法的返回值的类型是 List＜Map＜String，Object＞＞。接口方法 selectAllUserMap()的定义如下。

```
public List<Map<String, Object>> selectAllUserMap();
```

❷ 添加 SQL 映射

在模块 ch3_2 的 SQL 映射文件 UserMapper.xml 中添加接口方法 selectAllUserMap()对应的 SQL 映射，具体代码如下。

```
<!-- 使用 Map 存储查询结果,查询结果的列名作为 Map 的 key,列值为 Map 的 value -->
<select id="selectAllUserMap" resultType="map">
    select * from user
</select>
```

❸ 添加请求处理方法

在模块 ch3_2 的控制器类 TestController 中添加请求处理方法 selectAllUserMap()，具体代码如下。

```
@GetMapping("/selectAllUserMap")
public String selectAllUserMap(Model model) {
    //使用 Map 存储查询结果
    List<Map<String, Object>> unameAndUsexList = userMapper.selectAllUserMap();
    model.addAttribute("unameAndUsexList", unameAndUsexList);
    //在 showUnameAndUsexUser.jsp 页面中遍历时,属性名与查询结果的列名(Map 的 key)相同
    return "showUnameAndUsexUser";
}
```

❹ 测试应用

重启 Web 服务器 Tomcat，通过地址 http://localhost:8080/ch3_2/selectAllUserMap 测试应用。

3.10 ＜insert＞、＜update＞、＜delete＞以及＜sql＞元素

3.10.1 ＜insert＞元素

＜insert＞元素用于映射添加语句，MyBatis 在执行完一条添加语句后，将返回一个整数，表示其影响的行数。它的属性与＜select＞元素的属性大部分相同，在本节讲解它的几个特有属性。

keyProperty：添加时将自动生成的主键值回填给 PO(Persistant Object)类的某个属性，通常会设置为主键对应的属性。如果是联合主键，可以在多个值之间用逗号隔开。

keyColumn：设置第几列是主键，当主键列不是表中的第一列时需要设置。如果是联合主键，可以在多个值之间用逗号隔开。

useGeneratedKeys：该属性将使 MyBatis 用 JDBC 的 getGeneratedKeys()方法获取由数据库内部产生的主键，如 MySQL、SQL Server 等自动递增的字段，其默认值为 false。

❶ 主键(自动递增)回填

MySQL、SQL Server 等数据库的表格可以使用自动递增的字段作为主键。有时可能需要使用这个刚产生的主键，用于关联其他业务。因为本书使用的数据库是 MySQL 数据库，所以可

以直接使用 ch3_2 应用讲解自动递增主键回填的使用方法。

【例 3-9】 在 3.6.3 节中例 3-2 的基础上实现自动递增主键回填。

为了节省篇幅，相同的实现不再赘述，其他的具体实现如下。

1）添加接口方法

在模块 ch3_2 的 UserMapper 接口中添加数据操作接口方法 addUserBack()，该方法的返回值的类型是 int。接口方法 addUserBack() 的定义如下。

```
public int addUserBack(MyUser mu);
```

2）添加 SQL 映射

在模块 ch3_2 的 SQL 映射文件 UserMapper.xml 中添加接口方法 addUserBack() 对应的 SQL 映射，具体代码如下。

```xml
<!-- 添加一个用户,成功后将主键值回填给 uid(PO 类的属性)-->
<insert id="addUserBack" parameterType="MyUser" keyProperty="uid" useGeneratedKeys="true">
    insert into user (uname,usex) values(#{uname},#{usex})
</insert>
```

3）添加请求处理方法

在模块 ch3_2 的控制器类 TestController 中添加请求处理方法 addUserBack()，具体代码如下。

```java
@GetMapping("/addUserBack")
public String addUserBack(Model model) {
    //添加一个用户
    MyUser addmu = new MyUser();
    addmu.setUname("陈恒主键回填");
    addmu.setUsex("男");
    userMapper.addUserBack(addmu);
    model.addAttribute("addmu", addmu);
    return "showAddUser";
}
```

4）创建显示被添加的用户信息的页面

在模块 ch3_2 的 /WEB-INF/jsp 目录下创建 showAddUser.jsp 文件显示添加的用户信息，核心代码如下。

```html
<div class="container">
    <div class="panel panel-primary">
        <div class="panel-heading">
            <h3 class="panel-title">添加的用户信息</h3>
        </div>
        <div class="panel-body">
            <div class="table table-responsive">
                <table class="table table-bordered table-hover">
                    <tbody class="text-center">
                        <tr>
                            <th>用户 ID(回填的主键)</th>
                            <th>姓名</th>
                            <th>性别</th>
                        </tr>
                        <tr>
                            <td>${addmu.uid}</td>
                            <td>${addmu.uname}</td>
                            <td>${addmu.usex}</td>
                        </tr>
```

```
                </tbody>
            </table>
        </div>
    </div>
</div>
```

5）测试应用

重启 Web 服务器 Tomcat，通过地址 http://localhost:8080/ch3_2/addUserBack 测试应用，运行结果如图 3.8 所示。

用户ID（回填的主键）	姓名	性别
38	陈恒主键回填	男

图 3.8 回填主键值

❷ 自定义主键

如果在实际工程中使用的数据库不支持主键自动递增（如 Oracle），或者取消了主键自动递增的规则，可以使用 MyBatis 的＜selectKey＞元素来自定义生成主键，具体配置示例代码如下：

```
<insert id="insertUser" parameterType="MyUser">
    <!-- 先使用 selectKey 元素定义主键，再定义 SQL 语句 -->
    <selectKey keyProperty="uid" resultType="Integer" order="BEFORE">
        select decode(max(uid), null, 1, max(uid)+1) as newUid from user
    </selectKey>
    insert into user (uid,uname,usex) values(#{uid},#{uname},#{usex})
</insert>
```

在执行上述示例代码时，＜selectKey＞元素首先被执行，该元素通过自定义的语句设置数据表的主键，然后执行添加语句。

＜selectKey＞元素的 keyProperty 属性指定了新生主键值返回给 PO 类（MyUser）的哪个属性。order 属性可以设置为 BEFORE 或 AFTER，BEFORE 表示先执行＜selectKey＞元素再执行插入语句；AFTER 表示先执行插入语句再执行＜selectKey＞元素。

3.10.2 ＜update＞与＜delete＞元素

＜update＞和＜delete＞元素比较简单，它们的属性和＜insert＞元素的属性基本一样，在执行后也返回一个整数，表示影响数据库的记录行数。配置示例代码如下。

```
<!-- 修改一个用户 -->
<update id="updateUser" parameterType="MyUser">
    update user set uname = #{uname},usex = #{usex} where uid = #{uid}
</update>
<!-- 删除一个用户 -->
<delete id="deleteUser" parameterType="Integer">
    delete from user where uid = #{uid}
</delete>
```

3.10.3 ＜sql＞元素

＜sql＞元素的作用是定义 SQL 语句的一部分（代码片段），以方便后续的 SQL 语句引用它，例如反复使用的列名。在 MyBatis 中只需使用＜sql＞元素编写一次便能在其他元素中引用它。配置示例代码如下。

```
<sql id="comColumns">id,uname,usex</sql>
<select id="selectUser" resultType="MyUser">
    select <include refid="comColumns"/> from user
</select>
```

在上述代码中，使用<include>元素的 refid 属性引用了自定义的代码片段。

3.11 级联查询

级联关系是一个数据库实体的概念，有三种级联关系，分别是一对一级联、一对多级联以及多对多级联。级联的优点是获取关联数据十分方便，但是级联过多会增加数据库系统的复杂度，同时降低系统的性能。在实际开发中要根据实际情况判断是否需要使用级联。更新和删除的级联关系很简单，由数据库内在机制即可完成。本节仅讲述级联查询的相关实现。

如果表 A 中有一个外键引用了表 B 的主键，A 表就是子表，B 表就是父表。当查询表 A 的数据时，通过表 A 的外键，也将表 B 的相关记录返回，这就是级联查询。例如，查询一个人的信息时，同时根据外键（身份证号）将他的身份证信息返回。

▶ 3.11.1 一对一级联查询

一对一级联关系在现实生活中是十分常见的。例如一个大学生只有一张一卡通，一张一卡通只属于一个学生。再如人与身份证的关系也是一对一级联关系。

MyBatis 如何处理一对一级联查询呢？在 MyBatis 中，通过<resultMap>元素的子元素<association>处理这种一对一级联关系。在<association>元素中通常使用以下属性。

- property：指定映射到实体类的对象属性。
- column：指定表中对应的字段（即查询返回的列名）。
- javaType：指定映射到实体对象属性的类型。
- select：指定引入嵌套查询的子 SQL 语句，该属性用于关联映射中的嵌套查询。

下面以个人与身份证之间的关系为例，讲解一对一级联查询的处理过程，读者只需要参考该实例即可学会一对一级联查询的 MyBatis 实现。

【例 3-10】 在 3.6.3 节中例 3-2 的基础上进行一对一级联查询的 MyBatis 实现。

为了节省篇幅，相同的实现不再赘述，其他的具体实现如下。

❶ 创建数据表

本实例需要在数据库 springtest 中创建两张数据表，一张是身份证表 idcard，另一张是个人信息表 person。这两张表具有一对一级联关系，它们的创建代码如下：

```
CREATE TABLE `idcard` (
  `id` int(11) NOT NULL AUTO_INCREMENT,
  `code` varchar(18) COLLATE utf8_unicode_ci DEFAULT NULL,
  PRIMARY KEY (`id`)
);
CREATE TABLE `person` (
  `id` int(11) NOT NULL AUTO_INCREMENT,
  `name` varchar(20) COLLATE utf8_unicode_ci DEFAULT NULL,
  `age` int(11) DEFAULT NULL,
  `idcard_id` int(11) DEFAULT NULL,
  PRIMARY KEY (`id`),
  KEY `idcard_id` (`idcard_id`),
```

```
    CONSTRAINT 'idcard_id' FOREIGN KEY ('idcard_id') REFERENCES 'idcard' ('id')
);
```

❷ 创建实体类

在模块 ch3_2 的 com.mybatis.po 包中创建数据表对应的持久化类 Idcard 和 Person。Idcard 的代码如下。

```
package com.mybatis.po;
import lombok.Data;
@Data
public class Idcard {
    private Integer id;
    private String code;
    @Override
    public String toString() {
        return "Idcard [id=" + id + ",code="+ code + "]";
    }
}
```

Person 的代码如下。

```
package com.mybatis.po;
import lombok.Data;
@Data
public class Person {
    private Integer id;
    private String name;
    private Integer age;
    //个人身份证关联
    private Idcard card;
    @Override
    public String toString() {
        return "Person [id=" + id + ",name=" + name + ",age=" + age +",card=" + card +"]";
    }
}
```

❸ 创建 SQL 映射文件

在模块 ch3_2 的 com.mybatis.mapper 包中创建两张表对应的映射文件 IdCardMapper.xml 和 PersonMapper.xml。在 PersonMapper.xml 文件中以三种方式实现"根据个人 id 查询个人信息"的功能，详情请看代码备注。

IdCardMapper.xml 的代码如下。

```
<?xml version="1.0" encoding="UTF-8"?>
<!DOCTYPE mapper
PUBLIC "-//mybatis.org//DTD Mapper 3.0//EN"
"http://mybatis.org/dtd/mybatis-3-mapper.dtd">
<mapper namespace="com.mybatis.mapper.IdCardMapper">
    <select id="selectCodeById" parameterType="Integer" resultType="Idcard">
        select * from idcard where id=#{id}
    </select>
</mapper>
```

PersonMapper.xml 的代码如下。

```
<?xml version="1.0" encoding="UTF-8"?>
<!DOCTYPE mapper
PUBLIC "-//mybatis.org//DTD Mapper 3.0//EN"
"http://mybatis.org/dtd/mybatis-3-mapper.dtd">
```

```xml
<mapper namespace="com.mybatis.mapper.PersonMapper">
    <!-- 一对一 根据id查询个人信息:级联查询的第一种方法(嵌套查询,执行两个SQL语句) -->
    <resultMap type="Person" id="cardAndPerson1">
        <id property="id" column="id"/>
        <result property="name" column="name"/>
        <result property="age" column="age"/>
        <!-- 一对一级联查询 -->
        <association property="card" column="idcard_id" javaType="Idcard"
        select="com.mybatis.mapper.IdCardMapper.selectCodeById"/>
    </resultMap>
    <select id="selectPersonById1" parameterType="Integer" resultMap="cardAndPerson1">
        select * from person where id=#{id}
    </select>
    <!-- 一对一 根据id查询个人信息:级联查询的第二种方法(嵌套结果,执行一个SQL语句) -->
    <resultMap type="Person" id="cardAndPerson2">
        <id property="id" column="id"/>
        <result property="name" column="name"/>
        <result property="age" column="age"/>
        <!-- 一对一级联查询 -->
        <association property="card" javaType="Idcard">
            <id property="id" column="idcard_id"/>
            <result property="code" column="code"/>
        </association>
    </resultMap>
    <select id="selectPersonById2" parameterType="Integer" resultMap="cardAndPerson2">
        select p.*,ic.code
        from person p, idcard ic
        where p.idcard_id = ic.id and p.id=#{id}
    </select>
    <!-- 一对一 根据id查询个人信息:连接查询(使用POJO存储结果) -->
    <select id="selectPersonById3" parameterType="Integer" resultType=
"SelectPersonById">
        select p.*,ic.code
        from person p, idcard ic
        where p.idcard_id = ic.id and p.id=#{id}
    </select>
</mapper>
```

❹ **创建 POJO 类**

在模块 ch3_2 的 com.mybatis.po 包中创建 POJO 类 SelectPersonById(第 3 步使用的 POJO 类)。

SelectPersonById 的代码如下。

```java
package com.mybatis.po;
import lombok.Data;
@Data
public class SelectPersonById {
    private Integer id;
    private String name;
    private Integer age;
    private String code;
    @Override
    public String toString() {
        return "Person [id=" + id + ",name=" + name + ",age="
        + age + ",code=" + code + "]";
    }
}
```

❺ **创建 Mapper 接口**

在模块 ch3_2 的 com.mybatis.mapper 包中创建第 3 步映射文件对应的数据操作接口 IdCardMapper 和 PersonMapper。

IdCardMapper 的核心代码如下。

```
@Repository
public interface IdCardMapper{
    public Idcard selectCodeById(Integer i);
}
```

PersonMapper 的核心代码如下。

```
@Repository
public interface PersonMapper{
    public Person selectPersonById1(Integer id);
    public Person selectPersonById2(Integer id);
    public SelectPersonById selectPersonById3(Integer id);
}
```

❻ **创建控制器类**

在模块 ch3_2 的 controller 包中创建控制器类 OneToOneController，并在该类中调用第 5 步的接口方法，核心代码如下。

```
@Controller
public class OneToOneController{
    @Autowired
    private PersonMapper personMapper;
    @GetMapping("/oneToOneTest")
    public String oneToOneTest() {
        System.out.println("级联查询的第一种方法(嵌套查询,执行两个 SQL 语句)");
        Person p1 = personMapper.selectPersonById1(1);
        System.out.println(p1);
        System.out.println("========================");
        System.out.println("级联查询的第二种方法(嵌套结果,执行一个 SQL 语句)");
        Person p2 = personMapper.selectPersonById2(1);
        System.out.println(p2);
        System.out.println("========================");
        System.out.println("连接查询(使用 POJO 存储结果)");
        SelectPersonById p3 = personMapper.selectPersonById3(1);
        System.out.println(p3);
        return "test";
    }
}
```

❼ **测试应用**

发布应用到 Web 服务器 Tomcat，通过地址 http://localhost:8080/ch3_2/oneToOneTest 测试应用。在测试时需要事先为数据表手动添加数据。运行结果如图 3.9 所示。

▶ 3.11.2 一对多级联查询

在 3.11.1 节中学习了 MyBatis 如何处理一对一级联查询，那么 MyBatis 如何处理一对多级联查询呢？在实际生活中有许多一对多的关系，例如一个用户可以有多个订单，而一个订单只属于一个用户。

下面以用户和订单之间的关系为例，讲解一对多级联查询(实现"根据用户 id 查询用户及其关联的订单信息"的功能)的处理过程，读者只需要参考该实例即可学会一对多级联查询的 MyBatis 实现。

第 3 章　MyBatis

```
级联查询的第一种方法（嵌套查询，执行两个SQL语句）
DEBUG [http-nio-8080-exec-7] - ==>  Preparing: select * from person where id=?
DEBUG [http-nio-8080-exec-7] - ==> Parameters: 1(Integer)
DEBUG [http-nio-8080-exec-7] - ====>  Preparing: select * from idcard where id=?
DEBUG [http-nio-8080-exec-7] - ====> Parameters: 1(Integer)
DEBUG [http-nio-8080-exec-7] - <====      Total: 1
DEBUG [http-nio-8080-exec-7] - <==       Total: 1
Person [id=1,name=陈恒,age=88,card=Idcard [id=1,code=123456789]]
==========================
级联查询的第二种方法（嵌套结果，执行一个SQL语句）
DEBUG [http-nio-8080-exec-7] - ==>  Preparing: select p.*,ic.code from person p,
DEBUG [http-nio-8080-exec-7] - ==> Parameters: 1(Integer)
DEBUG [http-nio-8080-exec-7] - <==       Total: 1
Person [id=1,name=陈恒,age=88,card=Idcard [id=1,code=123456789]]
==========================
连接查询（使用POJO存储结果）
DEBUG [http-nio-8080-exec-7] - ==>  Preparing: select p.*,ic.code from person p,
DEBUG [http-nio-8080-exec-7] - ==> Parameters: 1(Integer)
DEBUG [http-nio-8080-exec-7] - <==       Total: 1
Person [id=1,name=陈恒,age=88,code=123456789]
```

图 3.9　一对一级联查询结果

【例 3-11】 在 3.6.3 节中例 3-2 的基础上进行一对多级联查询的 MyBatis 实现。为了节省篇幅，相同的实现不再赘述，其他的具体实现如下。

❶ 创建数据表

本实例需要两张数据表，一张是用户表 user，另一张是订单表 orders。这两张表具有一对多级联关系。user 表在前面已创建，orders 的创建代码如下。

```sql
CREATE TABLE 'orders' (
  'id' int(11) NOT NULL AUTO_INCREMENT,
  'ordersn' varchar(10) COLLATE utf8_unicode_ci DEFAULT NULL,
  'user_id' int(11) DEFAULT NULL,
  PRIMARY KEY ('id'),
  KEY 'user_id' ('user_id'),
  CONSTRAINT 'user_id' FOREIGN KEY ('user_id') REFERENCES 'user' ('uid')
);
```

❷ 创建持久化类

在模块 ch3_2 的 com.mybatis.po 包中创建数据表 orders 对应的持久化类 Orders 以及数据表 user 对应的持久化类 MyUserOrder。

MyUserOrder 类的代码如下。

```java
package com.mybatis.po;
import lombok.Data;
import java.util.List;
@Data
public class MyUserOrder{
    private Integer uid;                            //主键
    private String uname;
    private String usex;
    private List<Orders> ordersList;
    @Override
    public String toString() {
    return "User [uid=" + uid +",uname=" + uname + ",usex=" + usex + ",ordersList=
" + ordersList +"]";
    }
}
```

Orders 类的代码如下。

```
package com.mybatis.po;
import lombok.Data;
@Data
public class Orders {
    private Integer id;
    private String ordersn;
    @Override
    public String toString() {
        return "Orders [id=" + id + ",ordersn=" + ordersn + "]";
    }
}
```

❸ 创建并修改映射文件

在模块 ch3_2 的 com.mybatis.mapper 中创建 orders 表对应的映射文件 OrdersMapper.xml。在映射文件 UserMapper.xml 中添加实现一对多级联查询（根据用户 id 查询用户及其关联的订单信息）的 SQL 映射。

在 UserMapper.xml 文件中添加的 SQL 映射如下。

```xml
<!-- 一对多 根据 uid 查询用户及其关联的订单信息:级联查询的第一种方法(嵌套查询) -->
<resultMap type="MyUserOrder" id="userAndOrders1">
    <id property="uid" column="uid"/>
    <result property="uname" column="uname"/>
    <result property="usex" column="usex"/>
    <!-- 一对多级联查询,ofType 表示集合中元素的类型,将 uid 传递给 selectOrdersById-->
    <collection property="ordersList" ofType="Orders" column="uid"  select="com.mybatis.mapper.OrdersMapper.selectOrdersById"/>
</resultMap>
<select id="selectUserOrdersById1" parameterType="Integer" resultMap="userAndOrders1">
    select * from user where uid = #{id}
</select>
<!-- 一对多 根据 uid 查询用户及其关联的订单信息:级联查询的第二种方法(嵌套结果) -->
<resultMap type="MyUserOrder" id="userAndOrders2">
    <id property="uid" column="uid"/>
    <result property="uname" column="uname"/>
    <result property="usex" column="usex"/>
    <!-- 一对多级联查询,ofType 表示集合中元素的类型 -->
    <collection property="ordersList" ofType="Orders" >
        <id property="id" column="id"/>
        <result property="ordersn" column="ordersn"/>
    </collection>
</resultMap>
<select id="selectUserOrdersById2" parameterType="Integer" resultMap="userAndOrders2">
    select u.*,o.id,o.ordersn from user u, orders o where u.uid = o.user_id and u.uid=#{id}
</select>
<!-- 一对多 根据 uid 查询用户及其关联的订单信息:连接查询(使用 map 存储结果) -->
<select id="selectUserOrdersById3" parameterType="Integer" resultType="map">
    select u.*,o.id,o.ordersn from user u, orders o where u.uid = o.user_id and u.uid=#{id}
</select>
```

OrdersMapper.xml 的代码如下。

```xml
<?xml version="1.0" encoding="UTF-8" ?>
<!DOCTYPE mapper
PUBLIC "-//mybatis.org//DTD Mapper 3.0//EN"
"http://mybatis.org/dtd/mybatis-3-mapper.dtd">
<mapper namespace="com.mybatis.mapper.OrdersMapper">
    <!-- 根据用户 uid 查询订单信息 -->
```

```xml
    <select id="selectOrdersById" parameterType="Integer" resultType="Orders">
        select * from orders where user_id=#{id}
    </select>
</mapper>
```

❹ 创建并修改 Mapper 接口

在模块 ch3_2 的 com.mybatis.mapper 包中创建第 3 步映射文件对应的数据操作接口 OrdersMapper，并在 UserMapper 接口中添加接口方法。

OrdersMapper 的核心代码如下。

```java
@Repository
public interface OrdersMapper{
    public List<Orders> selectOrdersById(Integer uid);
}
```

在 UserMapper 接口中添加如下接口方法。

```java
public MyUserOrder selectUserOrdersById1(Integer uid);
public MyUserOrder selectUserOrdersById2(Integer uid);
public List<Map<String, Object>> selectUserOrdersById3(Integer uid);
```

❺ 创建控制器类

在模块 ch3_2 的 controller 包中创建控制器类 OneToMoreController，并在该类中调用第 4 步的接口方法，核心代码如下。

```java
@Controller
public class OneToMoreController {
    @Autowired
    private UserMapper userMapper;
    @GetMapping("/oneToMoreTest")
    public String oneToMoreTest() {
        //查询一个用户及订单信息
        System.out.println("级联查询的第一种方法(嵌套查询,执行两个 SQL 语句)");
        MyUserOrder auser1 = userMapper.selectUserOrdersById1(1);
        System.out.println(auser1);
        System.out.println("==================================");
        System.out.println("级联查询的第二种方法(嵌套结果,执行一个 SQL 语句)");
        MyUserOrder auser2 = userMapper.selectUserOrdersById2(1);
        System.out.println(auser2);
        System.out.println("==================================");
        System.out.println("连接查询(使用 map 存储结果)");
        List<Map<String, Object>> auser3 = userMapper.selectUserOrdersById3(1);
        System.out.println(auser3);
        return "test";
    }
}
```

❻ 测试应用

重启 Web 服务器 Tomcat，通过地址 http://localhost:8080/ch3_2/oneToMoreTest 测试应用。在测试时需要事先为数据表手动添加数据。运行结果如图 3.10 所示。

▶3.11.3 多对多级联查询

其实，MyBatis 没有实现多对多级联，这是因为多对多级联可以用两个一对多级联进行替换。例如，一个订单可以有多种商品，一种商品可以对应多个订单，订单和商品就是多对多级联关系。这里使用一个中间表——订单记录表，可以将多对多级联转换成两个一对多级联。下面

```
级联查询的第一种方法（嵌套查询，执行两个SQL语句）
DEBUG [http-nio-8080-exec-7] - ==>  Preparing: select * from user where uid = ?
DEBUG [http-nio-8080-exec-7] - ==> Parameters: 1(Integer)
DEBUG [http-nio-8080-exec-7] - ====>  Preparing: select * from orders where user_id=?
DEBUG [http-nio-8080-exec-7] - ====> Parameters: 1(Integer)
DEBUG [http-nio-8080-exec-7] - <====      Total: 2
DEBUG [http-nio-8080-exec-7] - <==      Total: 1
User [uid=1,uname=张三,usex=女,ordersList=[Orders [id=1,ordersn=123456], Orders [id=2,ordersn=67890]]
======================================
级联查询的第二种方法（嵌套结果，执行一个SQL语句）
DEBUG [http-nio-8080-exec-7] - ==>  Preparing: select u.*,o.id,o.ordersn from user u, orders o where
DEBUG [http-nio-8080-exec-7] - ==> Parameters: 1(Integer)
DEBUG [http-nio-8080-exec-7] - <==      Total: 2
User [uid=1,uname=张三,usex=女,ordersList=[Orders [id=1,ordersn=123456], Orders [id=2,ordersn=67890]]
======================================
连接查询（使用map存储结果）
DEBUG [http-nio-8080-exec-7] - ==>  Preparing: select u.*,o.id,o.ordersn from user u, orders o where
DEBUG [http-nio-8080-exec-7] - ==> Parameters: 1(Integer)
DEBUG [http-nio-8080-exec-7] - <==      Total: 2
[{uid=1, uname=张三, ordersn=123456, usex=女, id=1}, {uid=1, uname=张三, ordersn=67890, usex=女, id=2}]
```

图 3.10 一对多级联查询结果

以订单和商品（实现"查询所有订单以及每个订单对应的商品信息"的功能）为例，讲解多对多级联查询。

【例 3-12】 在 3.11.2 节中例 3-11 的基础上进行多对多级联查询的 MyBatis 实现。

为了节省篇幅，相同的实现不再赘述，其他的具体实现如下。

❶ 创建数据表

订单表在前面已创建，这里需要创建商品表 product 和订单记录表 orders_detail，创建代码如下。

```
CREATE TABLE `product` (
  `id` int(11) NOT NULL AUTO_INCREMENT,
  `name` varchar(50) COLLATE utf8_unicode_ci DEFAULT NULL,
  `price` double DEFAULT NULL,
  PRIMARY KEY (`id`)
);
CREATE TABLE `orders_detail` (
  `id` int(11) NOT NULL AUTO_INCREMENT,
  `orders_id` int(11) DEFAULT NULL,
  `product_id` int(11) DEFAULT NULL,
  PRIMARY KEY (`id`),
  KEY `orders_id` (`orders_id`),
  KEY `product_id` (`product_id`),
  CONSTRAINT `orders_id` FOREIGN KEY (`orders_id`) REFERENCES `orders` (`id`),
  CONSTRAINT `product_id` FOREIGN KEY (`product_id`) REFERENCES `product` (`id`)
);
```

❷ 创建持久化类

在模块 ch3_2 的 com.mybatis.po 包中创建数据表 product 对应的持久化类 Product，中间表 orders_detail 不需要持久化类，但需要在订单表 orders 对应的持久化类 Orders 中添加关联属性。

Product 的代码如下。

```
package com.mybatis.po;
import java.util.List;
```

```java
import lombok.Data;
@Data
public class Product {
    private Integer id;
    private String name;
    private Double price;
    //多对多中的一个一对多
    private List<Orders> orders;
    @Override
    public String toString() {
        return "Product [id=" + id + ",name=" + name + ",price=" + price +"]";
    }
}
```

修改后 Orders 的代码如下。

```java
package com.mybatis.po;
import java.util.List;
import lombok.Data;
@Data
public class Orders {
    private Integer id;
    private String ordersn;
    //多对多中的另一个一对多
    private List<Product> products;
    @Override
    public String toString() {
        return "Orders [id=" + id + ",ordersn=" + ordersn + ",products=" + products + "]";
    }
}
```

❸ 创建映射文件

本实例只需要在 com.mybatis.mapper 的 OrdersMapper.xml 文件中追加以下配置,即可实现多对多级联查询。

```xml
<!-- 多对多级联 查询所有订单以及每个订单对应的商品信息(嵌套结果) -->
<resultMap type="Orders" id="allOrdersAndProducts">
    <id property="id" column="id"/>
    <result property="ordersn" column="ordersn"/>
    <!-- 多对多级联 -->
    <collection property="products" ofType="Product">
        <id property="id" column="pid"/>
        <result property="name" column="name"/>
        <result property="price" column="price"/>
    </collection>
</resultMap>
<select id="selectallOrdersAndProducts" resultMap="allOrdersAndProducts">
    select o.*,p.id as pid,p.name,p.price
    from orders o,orders_detail od,product p
    where od.orders_id = o.id
    and od.product_id = p.id
</select>
```

❹ 添加 Mapper 接口方法

在 OrdersMapper 接口中添加以下接口方法。

```java
public List<Orders> selectallOrdersAndProducts();
```

❺ 创建控制器类

在模块 ch3_2 的 controller 包中创建控制器类 MoreToMoreController,并在该类中调用第

4 步的接口方法。MoreToMoreController 的核心代码如下。

```
@Controller
public class MoreToMoreController {
    @Autowired
    private OrdersMapper ordersMapper;
    @GetMapping("/moreToMoreTest")
    public String test() {
        List<Orders> os = ordersMapper.selectallOrdersAndProducts();
        for (Orders orders: os) {
            System.out.println(orders);
        }
        return "test";
    }
}
```

❻ 测试应用

重启 Web 服务器 Tomcat，通过地址 http://localhost:8080/ch3_2/moreToMoreTest 测试应用。在测试时需要事先为数据表手动添加数据。运行结果如图 3.11 所示。

```
DEBUG [http-nio-8080-exec-4] - ==>  Preparing: select o.*,p.id as pid,p.name,p.price from orders o,orders_det
DEBUG [http-nio-8080-exec-4] - ==> Parameters:
DEBUG [http-nio-8080-exec-4] - <==      Total: 3
Orders [id=1,ordersn=123456,products=[Product [id=1,name=苹果,price=10.0], Product [id=2,name=桔子,price=8.0]]]
Orders [id=2,ordersn=67890,products=[Product [id=2,name=桔子,price=8.0]]]
```

图 3.11 多对多级联查询结果

3.12 动态 SQL

开发人员通常根据需求手动拼接 SQL 语句，这是一个极其麻烦的工作，而 MyBatis 提供了对 SQL 语句动态组装的功能，恰好能解决这一问题。MyBatis 的动态 SQL 元素和 JSTL 或其他类似基于 XML 的文本处理器相似，常用元素有＜if＞、＜choose＞、＜when＞、＜otherwise＞、＜trim＞、＜where＞、＜set＞、＜foreach＞和＜bind＞等。

▶3.12.1 <if> 元素

动态 SQL 通常要做的事情是有条件地包含 where 子句的一部分，所以在 MyBatis 中＜if＞元素是最常用的元素。它类似于 Java 中的 if 语句。下面通过一个实例讲解＜if＞元素的使用过程。

【例 3-13】 在 3.6.3 节中例 3-2 的基础上讲解＜if＞元素的使用过程。

为了节省篇幅，相同的实现不再赘述，其他实现具体如下。

❶ 添加 Mapper 接口方法

在模块 ch3_2 的 UserMapper 接口中添加数据操作接口方法 selectAllUserByIf()，并在该方法中使用 MyUser 类的对象将参数信息封装。接口方法 selectAllUserByIf() 的定义如下。

```
public List<MyUser> selectAllUserByIf(MyUser user);
```

❷ 添加 SQL 映射

在模块 ch3_2 的 SQL 映射文件 UserMapper.xml 中添加接口方法 selectAllUserByIf() 对应的 SQL 映射，具体代码如下。

```
<!-- 使用 if 元素,根据条件动态查询用户信息 -->
<select id="selectAllUserByIf" resultType="MyUser" parameterType="MyUser">
```

```
    select * from user where 1=1
    <if test="uname !=null and uname!=''">
        and uname like concat(#{uname},'%')
    </if>
    <if test="usex !=null and usex!=''">
        and usex = #{usex}
    </if>
</select>
```

❸ 添加请求处理方法

在模块 ch3_2 的控制器类 TestController 中添加请求处理方法 selectAllUserByIf()，具体代码如下。

```
@GetMapping("/selectAllUserByIf")
public String selectAllUserByIf(Model model) {
    MyUser mu = new MyUser();
    mu.setUname("陈");
    mu.setUsex("男");
    List<MyUser> unameAndUsexList = userMapper.selectAllUserByIf(mu);
    model.addAttribute("unameAndUsexList", unameAndUsexList);
    return "showUnameAndUsexUser";
}
```

❹ 测试应用

重启 Web 服务器 Tomcat，通过地址 http://localhost:8080/ch3_2/selectAllUserByIf 测试应用。

▶3.12.2 <choose>、<when>、<otherwise> 元素

有时不需要用到所有的条件语句，而只需从中选择一二。针对这种情况，MyBatis 提供了<choose>元素，它有点像 Java 中的 switch 语句。下面通过一个实例讲解<choose>元素的使用过程。

【例 3-14】 在 3.6.3 节中例 3-2 的基础上讲解<choose>元素的使用过程。

为了节省篇幅，相同的实现不再赘述，其他实现具体如下。

❶ 添加 Mapper 接口方法

在模块 ch3_2 的 UserMapper 接口中添加数据操作接口方法 selectUserByChoose()，并在该方法中使用 MyUser 类的对象将参数信息封装。接口方法 selectUserByChoose() 的定义如下。

```
public List<MyUser> selectUserByChoose(MyUser user);
```

❷ 添加 SQL 映射

在模块 ch3_2 的 SQL 映射文件 UserMapper.xml 中添加接口方法 selectUserByChoose()对应的 SQL 映射，具体代码如下。

```
<!-- 使用 choose、when、otherwise 元素,根据条件动态查询用户信息 -->
<select id="selectUserByChoose" resultType="MyUser" parameterType="MyUser">
    select * from user where 1=1
    <choose>
        <when test="uname !=null and uname!=''">
            and uname like concat(#{uname},'%')
        </when>
        <when test="usex !=null and usex!=''">
```

```
                and usex = #{usex}
            </when>
            <otherwise>
                and uid > 3
            </otherwise>
        </choose>
</select>
```

❸ 添加请求处理方法

在模块 ch3_2 的控制器类 TestController 中添加请求处理方法 selectUserByChoose()，具体代码如下。

```
@GetMapping("/selectUserByChoose")
public String selectUserByChoose(Model model) {
    MyUser mu = new MyUser();
    mu.setUname("");
    mu.setUsex("");
    List<MyUser> unameAndUsexList = userMapper.selectUserByChoose(mu);
    model.addAttribute("unameAndUsexList", unameAndUsexList);
    return "showUnameAndUsexUser";
}
```

❹ 测试应用

重启 Web 服务器 Tomcat，通过地址 http://localhost:8080/ch3_2/selectUserByChoose 测试应用。

▶ 3.12.3 <trim> 元素

＜trim＞元素可以在自己包含的内容前加上某些前缀，也可以在其后加上某些后缀，与之对应的属性是 prefix 和 suffix；可以把包含的首部的某些内容覆盖，即忽略，也可以把包含的尾部的某些内容覆盖，对应的属性是 prefixOverrides 和 suffixOverrides。因为＜trim＞元素有这样的功能，所以可以非常简单地使用它来代替＜where＞元素。下面通过一个实例讲解＜trim＞元素的使用过程。

【例 3-15】 在 3.6.3 节中例 3-2 的基础上讲解＜trim＞元素的使用过程。

为了节省篇幅，相同的实现不再赘述，其他实现具体如下。

❶ 添加 Mapper 接口方法

在模块 ch3_2 的 UserMapper 接口中添加数据操作接口方法 selectUserByTrim()，并在该方法中使用 MyUser 类的对象将参数信息封装。接口方法 selectUserByTrim() 的定义如下。

```
public List<MyUser> selectUserByTrim(MyUser user);
```

❷ 添加 SQL 映射

在模块 ch3_2 的 SQL 映射文件 UserMapper.xml 中添加接口方法 selectUserByTrim() 对应的 SQL 映射，具体代码如下。

```
<!-- 使用 trim 元素, 根据条件动态查询用户信息 -->
<select id="selectUserByTrim" resultType="MyUser" parameterType="MyUser">
    select * from user
    <trim prefix="where" prefixOverrides="and|or">
        <if test="uname !=null and uname!=''">
            and uname like concat(#{uname},'%')
        </if>
        <if test="usex !=null and usex!=''">
```

```xml
            and usex = #{usex}
        </if>
    </trim>
</select>
```

❸ 添加请求处理方法

在模块 ch3_2 的控制器类 TestController 中添加请求处理方法 selectUserByTrim()，具体代码如下：

```java
@GetMapping("/selectUserByTrim")
public String selectUserByTrim(Model model) {
    MyUser mu = new MyUser();
    mu.setUname("陈");
    mu.setUsex("男");
    List<MyUser> unameAndUsexList = userMapper.selectUserByTrim(mu);
    model.addAttribute("unameAndUsexList", unameAndUsexList);
    return "showUnameAndUsexUser";
}
```

❹ 测试应用

重启 Web 服务器 Tomcat，通过地址 http://localhost:8080/ch3_2/selectUserByTrim 测试应用。

▶3.12.4 <where> 元素

<where>元素的作用是输出一个 where 语句，优点是不考虑<where>元素的条件输出，MyBatis 将智能处理。如果所有的条件都不满足，那么 MyBatis 将会查出所有记录，如果输出是以 and 开头，MyBatis 将把第一个 and 忽略，如果是以 or 开头，MyBatis 也将把它忽略。此外，在<where>元素中不考虑空格的问题，MyBatis 将智能加上。下面通过一个实例讲解<where>元素的使用过程。

【例 3-16】 在 3.6.3 节中例 3-2 的基础上讲解<where>元素的使用过程。

为了节省篇幅，相同的实现不再赘述，其他实现具体如下。

❶ 添加 Mapper 接口方法

在模块 ch3_2 的 UserMapper 接口中添加数据操作接口方法 selectUserByWhere()，并在该方法中使用 MyUser 类的对象将参数信息封装。接口方法 selectUserByWhere()的定义如下。

```java
public List<MyUser> selectUserByWhere(MyUser user);
```

❷ 添加 SQL 映射

在模块 ch3_2 的 SQL 映射文件 UserMapper.xml 中添加接口方法 selectUserByWhere()对应的 SQL 映射，具体代码如下。

```xml
<!-- 使用 where 元素,根据条件动态查询用户信息 -->
<select id="selectUserByWhere" resultType="MyUser" parameterType="MyUser">
    select * from user
    <where>
        <if test="uname !=null and uname!=''">
            and uname like concat(#{uname},'%')
        </if>
        <if test="usex !=null and usex!=''">
            and usex = #{usex}
        </if>
    </where>
</select>
```

❸ 添加请求处理方法

在模块 ch3_2 的控制器类 TestController 中添加请求处理方法 selectUserByWhere(),具体代码如下:

```java
@GetMapping("/selectUserByWhere")
public String selectUserByWhere(Model model) {
    MyUser mu = new MyUser();
    mu.setUname("陈");
    mu.setUsex("男");
    List<MyUser> unameAndUsexList = userMapper.selectUserByWhere(mu);
    model.addAttribute("unameAndUsexList", unameAndUsexList);
    return "showUnameAndUsexUser";
}
```

❹ 测试应用

重启 Web 服务器 Tomcat,通过地址 http://localhost:8080/ch3_2/selectUserByWhere 测试应用。

▶3.12.5 \<set\> 元素

在 update 语句中可以使用\<set\>元素动态更新列。下面通过一个实例讲解\<set\>元素的使用过程。

【例 3-17】 在 3.6.3 节中例 3-2 的基础上讲解\<set\>元素的使用过程。

为了节省篇幅,相同的实现不再赘述,其他实现具体如下。

❶ 添加 Mapper 接口方法

在模块 ch3_2 的 UserMapper 接口中添加数据操作接口方法 updateUserBySet(),并在该方法中使用 MyUser 类的对象将参数信息封装。接口方法 updateUserBySet()的定义如下。

```java
public int updateUserBySet(MyUser user);
```

❷ 添加 SQL 映射

在模块 ch3_2 的 SQL 映射文件 UserMapper.xml 中添加接口方法 updateUserBySet()对应的 SQL 映射,具体代码如下。

```xml
<!-- 使用 set 元素,动态修改一个用户 -->
<update id="updateUserBySet" parameterType="MyUser">
    update user
    <set>
        <if test="uname != null">uname=#{uname},</if>
        <if test="usex != null">usex=#{usex}</if>
    </set>
    where uid = #{uid}
</update>
```

❸ 添加请求处理方法

在模块 ch3_2 的控制器类 TestController 中添加请求处理方法 updateUserBySet(),具体代码如下:

```java
@GetMapping("/updateUserBySet")
public String updateUserBySet(Model model) {
    MyUser setmu = new MyUser();
    setmu.setUid(3);
    setmu.setUname("张九");
    userMapper.updateUserBySet(setmu);
    //查询出来看 id 为 3 的用户是否被修改
```

```
        List<Map<String, Object>> unameAndUsexList = userMapper.selectAllUserMap();
        model.addAttribute("unameAndUsexList", unameAndUsexList);
        return "showUnameAndUsexUser";
}
```

❹ 测试应用

重启 Web 服务器 Tomcat，通过地址 http://localhost:8080/ch3_2/updateUserBySet 测试应用。

▶3.12.6 <foreach> 元素

＜foreach＞元素主要用于构建 in 条件，它可以在 SQL 语句中迭代一个集合。＜foreach＞元素的属性主要有 item、index、collection、open、separator、close。item 表示集合中每一个元素进行迭代时的别名，index 指定一个名字，用于表示在迭代过程中每次迭代到的位置，open 表示该语句以什么开始，separator 表示在每次迭代之间以什么符号作为分隔符，close 表示以什么结束。在使用＜foreach＞时，最关键的也是最容易出错的是 collection 属性，该属性是必选的，但在不同情况下该属性的值是不一样的，主要有以下三种情况：

- 如果传入的是单参数且参数类型是一个 List，则 collection 的属性值为 list。
- 如果传入的是单参数且参数类型是一个 array，则 collection 的属性值为 array。
- 如果传入的参数是多个，则需要把它们封装成一个 Map，当然单参数也可以封装成 Map。Map 的 key 是参数名，所以 collection 属性值是传入的 List 或 array 对象在自己封装的 Map 中的 key。

下面通过一个实例讲解＜foreach＞元素的使用过程。

【例 3-18】 在 3.6.3 节中例 3-2 的基础上讲解＜foreach＞元素的使用过程。

为了节省篇幅，相同的实现不再赘述，其他的具体实现如下。

❶ 添加 Mapper 接口方法

在模块 ch3_2 的 UserMapper 接口中添加数据操作接口方法 selectUserByForeach()，并在该方法中使用 List 作为参数。接口方法 selectUserByForeach() 的定义如下。

```
public List<MyUser> selectUserByForeach(List<Integer> listId);
```

❷ 添加 SQL 映射

在模块 ch3_2 的 SQL 映射文件 UserMapper.xml 中添加接口方法 selectUserByForeach() 对应的 SQL 映射，具体代码如下。

```
<!-- 使用 foreach 元素,查询用户信息 -->
<select id="selectUserByForeach" resultType="MyUser" parameterType="List">
    select * from user where uid in
    <foreach item="item" index="index" collection="list"
        open="(" separator="," close=")">
        #{item}
    </foreach>
</select>
```

❸ 添加请求处理方法

在模块 ch3_2 的控制器类 TestController 中添加请求处理方法 selectUserByForeach()，具体代码如下。

```
@GetMapping("/selectUserByForeach")
public String selectUserByForeach(Model model) {
```

```
        List<Integer> listId = new ArrayList<Integer>();
        listId.add(4);
        listId.add(5);
        List<MyUser> unameAndUsexList = userMapper.selectUserByForeach(listId);
        model.addAttribute("unameAndUsexList", unameAndUsexList);
        return "showUnameAndUsexUser";
    }
```

❹ 测试应用

重启 Web 服务器 Tomcat,通过地址 http://localhost:8080/ch3_2/selectUserByForeach 测试应用。

▶3.12.7 <bind> 元素

在进行模糊查询时,如果使用"${}"拼接字符串,则无法防止 SQL 注入问题。如果使用字符串拼接函数或连接符号,但不同数据库的拼接函数或连接符号不同,如 MySQL 的 concat 函数、Oracle 的连接符号"||"。这样,SQL 映射文件就需要根据不同的数据库提供不同的实现,显然比较麻烦,且不利于代码的移植。幸运的是,MyBatis 提供了<bind>元素来解决这一问题。

下面通过一个实例讲解<bind>元素的使用过程。

【例 3-19】 在 3.6.3 节中例 3-2 的基础上讲解<bind>元素的使用过程。

为了节省篇幅,相同的实现不再赘述,其他实现具体如下。

❶ 添加 Mapper 接口方法

在模块 ch3_2 的 UserMapper 接口中添加数据操作接口方法 selectUserByBind(),接口方法 selectUserByBind()的定义如下。

```
public List<MyUser> selectUserByBind(MyUser user);
```

❷ 添加 SQL 映射

在模块 ch3_2 的 SQL 映射文件 UserMapper.xml 中添加接口方法 selectUserByBind()对应的 SQL 映射,具体代码如下。

```
<!-- 使用 bind 元素进行模糊查询 -->
<select id="selectUserByBind" resultType="MyUser" parameterType="MyUser">
    <!-- bind 中的 uname 是 com.po.MyUser 的属性名 -->
    <bind name="paran_uname" value="uname + '%'"/>
    select * from user where uname like #{paran_uname}
</select>
```

❸ 添加请求处理方法

在模块 ch3_2 的控制器类 TestController 中添加请求处理方法 selectUserByBind(),具体代码如下：

```
@GetMapping("/selectUserByBind")
public String selectUserByBind(Model model) {
    MyUser bindmu = new MyUser();
    bindmu.setUname("陈");
    List<MyUser> unameAndUsexList = userMapper.selectUserByBind(bindmu);
    model.addAttribute("unameAndUsexList", unameAndUsexList);
    return "showUnameAndUsexUser";
}
```

❹ 测试应用

重启 Web 服务器 Tomcat,通过地址 http://localhost:8080/ch3_2/selectUserByBind 测试

应用。

3.13 MyBatis 的缓存机制

我们知道内存的读取速度远快于硬盘的读取速度。当需要重复地获取相同数据时，一次一次地请求数据库或者远程服务，会导致大量的时间消耗在数据库查询或者远程方法的调用上，最终导致程序的性能降低，这就是数据缓存要解决的问题。

MyBatis 提供数据查询缓存，用于减轻数据库的压力，提高数据库的性能。MyBatis 提供了一级缓存和二级缓存。

▶3.13.1 一级缓存（SqlSession 级别的缓存）

在操作数据库时，需要构造 SqlSession 对象，在对象中有一个数据结构（HashMap）用于存储缓存数据，不同 SqlSession 之间的缓存区域是互不影响的。

❶ 一级缓存配置

MyBatis 的一级缓存不需要任何配置，在每一个 SqlSession 中都有一个一级缓存区域，作用范围是 SqlSession。

在 MyBatis 的一级缓存中，当第一次发起查询 ID 为 1 的用户信息，先去找缓存中是否有 ID 为 1 的用户信息，如果没有则从数据库中查询用户信息，将用户信息存储到一级缓存中；如果 sqlSession 去执行插入、更新、删除（执行 commit 操作），将会清空 SqlSession 中的一级缓存，这样做是为了让缓存存储的是最新的信息，避免脏读；当第二次发起查询 ID 为 1 的用户信息，先去找缓存中是否有 ID 为 1 的用户信息，如果缓存中有则直接从缓存中获取用户信息。这里涉及一个缓存命中率（Cache Hit Ratio），指的是在缓存中查询到的次数与总共在缓存中查询的次数之比。

❷ 一级缓存实验

【例 3-20】 在 3.5 节中例 3-1 的基础上讲解一级缓存实验。

为了节省篇幅，相同的实现不再赘述，其他实现具体如下。

1）修改测试类

在模块 ch3_1 的测试类 MyBatisTest 中多次发起查询用户 ID 为 1 的用户信息，修改后的代码如下。

```java
public class MyBatisTest {
    public static void main(String[] args) {
        try {
            //读取配置文件 mybatis-config.xml
            InputStream config = Resources.getResourceAsStream("mybatis-config.xml");
            //根据配置文件构建 SqlSessionFactory
            SqlSessionFactory ssf = new SqlSessionFactoryBuilder().build(config);
            //通过 SqlSessionFactory 创建 SqlSession
            SqlSession ss = ssf.openSession();
            //查询一个用户
            MyUser mu = ss.selectOne("com.mybatis.mapper.UserMapper.selectUserById", 1);
            System.out.println(mu);
            //测试一级缓存
            mu = ss.selectOne("com.mybatis.mapper.UserMapper.selectUserById", 1);
            System.out.println(mu);
            //添加一个用户
            MyUser addmu = new MyUser();
            addmu.setUname("陈恒");
```

```
            addmu.setUsex("男");
            ss.insert("com.mybatis.mapper.UserMapper.addUser",addmu);
            //修改一个用户
            MyUser updatemu = new MyUser();
            updatemu.setUid(2);
            updatemu.setUname("张三");
            updatemu.setUsex("女");
            ss.update("com.mybatis.mapper.UserMapper.updateUser", updatemu);
            //测试一级缓存
            mu = ss.selectOne("com.mybatis.mapper.UserMapper.selectUserById", 1);
            System.out.println(mu);
            //提交事务
            ss.commit();
            //关闭 SqlSession
            ss.close();
        } catch (IOException e) {
            //TODO Auto-generated catch block
            e.printStackTrace();
        }
    }
}
```

2）测试缓存，运行程序

运行修改后的测试类 MyBatisTest，运行结果如图 3.12 所示。

```
MyBatisTest
DEBUG [main] - ==>  Preparing: select * from user where uid = ?
DEBUG [main] - ==> Parameters: 1(Integer)
DEBUG [main] - <==      Total: 1
User [uid=1,uname=张三,usex=女]
User [uid=1,uname=张三,usex=女]
DEBUG [main] - ==>  Preparing: insert into user (uname,usex) values(?,?)
DEBUG [main] - ==> Parameters: 陈恒(String), 男(String)
DEBUG [main] - <==    Updates: 1
DEBUG [main] - ==>  Preparing: update user set uname = ?,usex = ? where uid = ?
DEBUG [main] - ==> Parameters: 张三(String), 女(String), 1(Integer)
DEBUG [main] - <==    Updates: 1
DEBUG [main] - ==>  Preparing: delete from user where uid = ?
DEBUG [main] - ==> Parameters: 2(Integer)
DEBUG [main] - <==    Updates: 0
DEBUG [main] - ==>  Preparing: select * from user where uid = ?
DEBUG [main] - ==> Parameters: 1(Integer)
DEBUG [main] - <==      Total: 1
User [uid=1,uname=张三,usex=女]
```

图 3.12　一级缓存测试结果

从图 3.12 所示的运行结果可知，前两次发起查询 ID 为 1 的用户信息时，只执行了一次 SQL 语句（即只查询一次数据库），第 2 次直接从缓存中返回数据。但经过添加和修改操作后，第 3 次发起查询 ID 为 1 的用户信息时，又重新查询了数据库（清空 SqlSession 中的一级缓存）。

▶3.13.2　二级缓存（Mapper 级别的缓存）

MyBatis 的二级缓存需要手动开启才能启动，与一级缓存的最大区别在于二级缓存的作用范围比一级缓存大，二级缓存是多个 SqlSession 可以共享一个 Mapper 的二级缓存区域，二级缓存的作用范围是 Mapper 中的同一个命名空间（namespace）的 statement。在 MyBatis 的核心配置文件中默认开启二级缓存。在默认开启二级缓存的情况下，如果每一个 namespace 都开启了二级缓存，则都对应一个二级缓存区域，同一个 namespace 共用一个二级缓存区域。

❶ 二级缓存配置

在 MyBatis 默认开启二级缓存的情况下，当 SqlSession 1 去查询 ID 为 1 的用户信息，查询到用户信息会将查询数据存储到二级缓存中；当 SqlSession 2 去执行相同 Mapper 下的 statement，执行 commit 提交，清空该 Mapper 下二级缓存区域中的数据；当 SqlSession 3 去查询 ID 为 1 的用户信息，先去缓存中找是否存在数据，如果存在则直接从缓存中取出数据，不存在则去数据库中查询读取。二级缓存配置具体如下：

1）开启二级缓存

第一个需要配置的地方是核心配置文件（此步可以省略，因为默认是开启的，配置的目的是方便维护）。配置示例代码如下。

```
<settings>
    <!-- 开启二级缓存 -->
    <setting name="cacheEnabled" value="true"/>
</settings>
```

2）开启 namespace 下的二级缓存

在需要开启二级缓存的 statement 的命名空间（namespace）中配置标签<cache></cache>。配置示例代码如下：

```
<mapper namespace="dao.UserMapper">
    <!-- 开启 namespace 下的二级缓存 -->
    <cache></cache>
</mapper>
```

<cache>标签有以下 6 个参数。

type：指定缓存（cache）接口的实现类型，当需要和 EhCache 整合时更改该参数值即可。

flushInterval：刷新间隔，可以被设置为任意正整数，单位为毫秒，默认不设置。

size：引用数目，可以被设置为任意正整数，缓存的对象数目等于运行环境的可用内存资源数目，默认为 1024。

readOnly：只读，取值为 true 或 false，只读的缓存会给所有的调用者返回缓存对象的相同实例。其默认为 false。

eviction：缓存收回策略，取值为 LRU（最近最少使用的）、FIFO（先进先出）、SOFT（软引用）、WEAK（弱引用），默认为 LRU。

在 Mapper 的 select 中可以设置 useCache＝"false"来禁用缓存，缓存默认是开启的；在 insert、update、delete 中设置 flushCache＝"true"来清空缓存（刷新缓存），默认会清空缓存。

3）POJO 类实现序列化

在使用二级缓存时，持久化类需要序列化，即 POJO 类实现 Serializable 接口。示例代码如下。

```
public class MyUser implements Serializable{}
```

❷ 二级缓存实验

【例 3-21】 在 3.6.4 节中例 3-3 的基础上讲解二级缓存实验。

为了节省篇幅，相同的实现不再赘述，其他实现具体如下。

1）开启 namespace 下的二级缓存

在模块 ch3_3 的 Mapper 映射文件 UserMapper.xml 中添加标签<cache></cache>。

2）POJO 类实现序列化

模块 ch3_3 的持久化类 MyUser 实现序列化接口 Serializable。

3）修改控制器类

将模块 ch3_3 的控制器类 MyController 修改如下。

```java
@Controller
public class MyController {
    @Autowired
    private UserMapper userMapper;
    @RequestMapping("/test")
    public String test() {
        //查询一个用户
        MyUser mu = userMapper.selectUserById(1);
        System.out.println(mu);
        //测试二级缓存
        mu = userMapper.selectUserById(1);
        System.out.println(mu);
        return "test";
    }
}
```

4）测试缓存，运行程序

发布应用 ch3_3 到 Web 服务器 Tomcat，通过地址 http://localhost:8080/ch3_3/test 测试应用，运行结果如图 3.13 所示。

图 3.13 二级缓存测试结果

从图 3.13 可以看出在第 1 次查询 ID 为 1 的用户信息时，缓存命中率为 0，说明先访问缓存，读取缓存中是否有 ID 为 1 的用户数据，然后发现缓存中没有就去数据库中查询用户信息；第 2 次查询 ID 为 1 的用户信息时，发现缓存中有对应数据，就直接从缓存中读取。

二级缓存一般应用在访问多的查询请求且对查询结果的实时性要求不高的情况，此时可采用 MyBatis 二级缓存技术降低数据库的访问量，提高访问速度，例如耗时比较高的统计分析的 SQL。

本章小结

本章重点讲述了 MyBatis 的 SQL 映射文件的编写以及 SSM 框架整合开发的流程。通过本章的学习，读者不仅要掌握 SSM 框架整合开发的流程，还应该熟悉 MyBatis 的基本应用。

习题 3

1. MyBatis Generator 有哪几种方法自动生成代码？
2. 简述 SSM 框架集成的步骤。
3. MyBatis 在实现查询时，返回的结果集有几种常见的存储方式？请举例说明。
4. 在 MyBatis 中针对不同的数据库软件，<insert>元素如何将主键回填？
5. 在 MyBatis 中如何给 SQL 语句传递参数？
6. 在动态 SQL 元素中类似分支语句的元素有哪些？如何使用它们？

第 4 章　名片管理系统的设计与实现 (SSM+JSP)

学习目的与要求

本章通过名片管理系统的设计与实现讲述如何使用 SSM 框架来实现一个 Web 应用。通过本章的学习，掌握 SSM 框架应用开发的流程、方法以及技术。

本章主要内容

- 系统设计
- 数据库设计
- 系统管理
- 组件设计
- 系统实现

本章系统使用 SSM 框架实现各个模块，Web 引擎为 Tomcat 10，数据库采用的是 MySQL 8，集成开发环境为 IntelliJ IDEA。

4.1　系统设计

▶ 4.1.1　系统功能需求

名片管理系统是针对注册用户使用的系统。该系统提供的功能如下：

(1) 非注册用户可以注册为注册用户。
(2) 成功注册的用户可以登录系统。
(3) 成功登录的用户可以添加、修改、删除以及浏览自己的客户名片信息。
(4) 成功登录的用户可以修改密码。

▶ 4.1.2　系统模块划分

用户登录成功后进入管理主页面（main.jsp），可以对自己的客户名片进行管理。系统模块划分如图 4.1 所示。

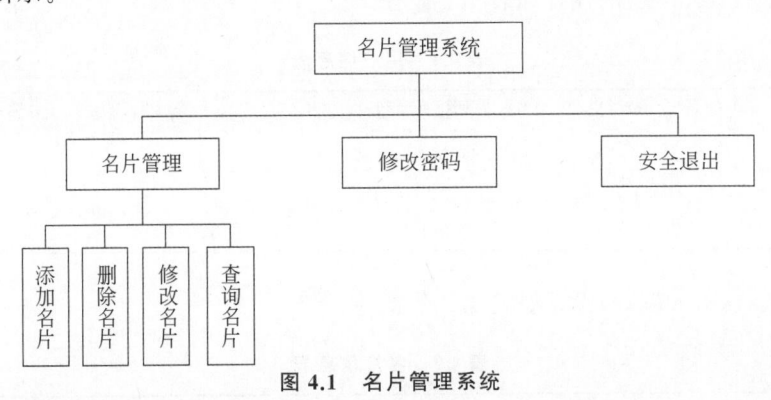

图 4.1　名片管理系统

4.2 数据库设计

该系统采用加载纯 Java 数据库驱动程序的方式连接 MySQL 8 数据库。在 MySQL 8 的数据库 ch4 中共创建两张与系统相关的数据表，分别为 usertable 和 cardtable。

▶4.2.1 数据库概念结构设计

根据系统设计与分析，可以设计出如下数据结构：

❶ 用户

用户包括 ID、用户名以及密码，注册用户名唯一。

❷ 名片

名片包括 ID、姓名、电话、E-Mail、单位、职位、地址、Logo 以及所属用户。其中，ID 唯一，"所属用户"与"1.用户"的用户 ID 关联。

根据以上数据结构，结合数据库设计的特点，可以画出如图 4.2 所示的数据库概念结构图。

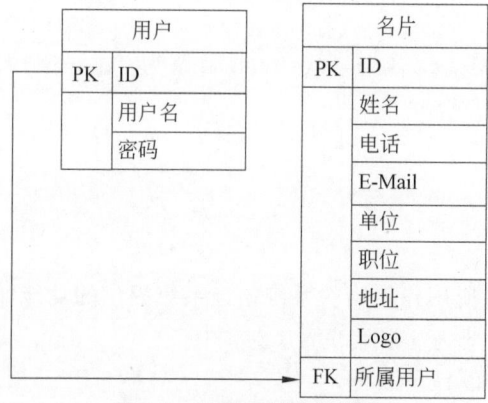

图 4.2　数据库概念结构图

图 4.2 中 ID 为正整数，值是从 1 开始递增的序列。

▶4.2.2 数据库逻辑结构设计

将数据库概念结构图转换为 MySQL 数据库所支持的实际数据模型，即数据库的逻辑结构。

用户信息表（usertable）的设计如表 4.1 所示。

表 4.1　用户信息表

字　　段	含　　义	类　　型	长　　度	是否为空
id	编号（PK）	int		no
uname	用户名	varchar	50	no
upwd	密码	varchar	32	no

名片信息表（cardtable）的设计如表 4.2 所示。

表 4.2　名片信息表

字　　段	含　　义	类　　型	长　　度	是否为空
id	编号（PK）	int		no
name	姓名	varchar	50	no

续表

字段	含义	类型	长度	是否为空
telephone	电话	varchar	20	no
E-Mail	邮箱	varchar	50	
company	单位	varchar	50	
post	职位	varchar	50	
address	地址	varchar	50	
logoName	照片	varchar	30	
user_id	所属用户	int		no

4.3 系统管理

▶ 4.3.1 所需 JAR 包

使用 IntelliJ IDEA 创建一个名为 ch4 的 Web 应用，并将所依赖的 JAR 包（包括 MyBatis、Spring、Spring MVC、MySQL 连接器、MyBatis 与 Spring 桥接器、Log4j、DBCP、Jackson 开源包、JSTL、DispatcherServlet 接口所依赖的性能监控包以及 Java 增强库）复制到应用的/WEB-INF/lib 目录中，具体参见本书提供的源代码。

▶ 4.3.2 JSP 页面管理

为了方便管理，在/web/static 目录下存放与系统相关的静态资源，如与 BootStrap 相关的 CSS 与 JS；在/WEB-INF/jsp 目录下存放与系统相关的 JSP 页面。由于篇幅受限，本章仅附上部分 JSP 和 Java 文件的核心代码，具体代码请读者参见本书提供的源代码 ch4。

❶ 首页面

在/web/目录下创建应用的首页面 index.jsp，首页面重定向到 user/toLogin 请求，打开登录页面。index.jsp 的核心代码如下：

```
<body>
    <%response.sendRedirect("user/toLogin");%>
</body>
```

❷ 异常信息显示页面

本系统使用 Spring 框架的统一异常处理机制处理未登录异常和程序错误异常。为了显示异常信息，需要在/WEB-INF/jsp 目录下创建一个名为 error.jsp 的页面。error.jsp 的核心代码如下：

```
<body>
    <c:if test="${mymessage == 'noLogin'}">
        <h2>没登录,您没有权限访问,<a href="user/toLogin">登录</a>!</h2>
    </c:if>
    <c:if test="${mymessage == 'noError'}">
        <h2>服务器内部错误或资源不存在!</h2>
    </c:if>
</body>
```

▶ 4.3.3 包管理

❶ config 包

该包中存放的配置文件是系统的配置，包括 Spring 配置、Spring MVC 配置以及 MyBatis 的核心配置。

❷ controller 包

该包中存放的类是系统的控制器类和异常处理类,包括与名片管理相关的控制器类、与用户相关的控制器类、验证码控制器类以及全局异常处理类。

❸ dao 包

该包中存放的 Java 程序是@Repository 注解的数据操作接口以及 SQL 映射文件,包括名片和用户相关的数据访问接口、SQL 映射文件。

❹ model 包

该包中存放的类是两个模型实体类,分别为 Card 封装名片信息和 MyUser 封装用户信息。

❺ service 包

该包中存放的类是业务处理类,它们是控制器和 dao 的"桥梁"。在该包下有 Service 接口和 Service 实现类。

❻ util 包

该包中存放的类是工具类,包括 MyUtil 类(文件重命名)和 MD5Util 类(MD5 加密)。

▶4.3.4 配置管理

名片管理系统共有 4 个配置,分别是 Web 的配置 web.xml、Spring 的配置 applicationContext.xml、Spring MVC 的配置 springmvc.xml 和 MyBatis 的核心配置 mybatis-config.xml,具体代码请读者参见本书提供的源代码 ch4。

在 web.xml 文件中实例化 ApplicationContext 容器、配置 Spring MVC DispatcherServlet 以及配置文件上传的限制信息;在 applicationContext.xml 文件中配置数据源、为数据源添加事务管理器、配置 MyBatis 工厂以及 Mapper 代理开发;在 springmvc.xml 文件中扫描注解的包、配置视图解析器、设置静态资源可见以及进行上传文件的相关设置;在 mybatis-config.xml 文件中配置实体类的别名以及日志实现 logImpl。

4.4 组件设计

名片管理系统的组件包括工具类、统一异常处理类和验证码类。

▶4.4.1 工具类

名片管理系统的工具类包括 MyUtil 和 MD5Util。在 MyUtil 类中定义一个文件重命名方法 getNewFileName,在 MD5Util 类中定义 MD5 加密方法,具体代码请读者参见本书提供的源代码 ch4。

▶4.4.2 统一异常处理

名片管理系统采用@ControllerAdvice 注解(控制器增强)实现异常的统一处理,统一处理了 NoLoginException 和 Exception 异常,核心代码如下:

```
/**
* 统一异常处理
*/
@ControllerAdvice
public class GlobalExceptionHandleController {
    @ExceptionHandler(value=Exception.class)
    public String exceptionHandler(Exception e, Model model) {
```

```
            String message = "";
            if (e instanceof NoLoginException) {
                message = "noLogin";
            } else {                                          //未知异常
                message = "noError";
            }
            model.addAttribute("mymessage",message);
            return "error";
        }
    }
```

未登录异常类 NoLoginException 的代码如下：

```
package controller;
public class NoLoginException extends Exception{
    private static final long serialVersionUID = 1L;
    public NoLoginException() {
        super();
    }
    public NoLoginException(String message) {
        super(message);
    }
}
```

▶4.4.3 验证码

本系统中验证码的使用步骤如下：

❶ 创建产生验证码的控制器类

在 controller 包中创建产生验证码的控制器类 ValidateCodeController，具体代码请读者参见本书提供的源代码 ch4。

❷ 使用验证码

在需要使用验证码的 JSP 页面中，调用产生验证码的控制器显示验证码，示例代码片段如下：

```
<td><img src="validateCode" id="mycode"></td>
```

4.5 名片管理

▶4.5.1 领域模型与持久化实体类

在本系统中使用实体类 Card（位于 model 包中）创建领域模型，它的作用是将某个指定页面的所有数据封装起来，与表单对应，数据传递方向为 View→Controller→Service→Dao。

同时，在本系统中实体类 Card 也作为持久化实体类，cardtable 数据表中的每个字段对应到实体类属性上，数据传递方向为 Dao→Service→Controller→View。

▶4.5.2 Controller 实现

在本系统中，与名片管理相关的功能包括添加、修改、删除、查询等，由控制器类 CardController 负责处理。由系统功能需求可知，用户必须成功登录才能管理自己的名片，所以在 CardController 处理添加、修改、删除、查询名片等功能前需要进行登录权限验证。在 CardController 中使用@ModelAttribute 注解的方法进行登录权限验证，CardController 的核心代码如下：

```java
@Controller
@RequestMapping("/card")
@SuppressWarnings("all")
public class CardController {
    @Autowired
    private CardService cardService;
    /**
     * 权限控制
     */
    @ModelAttribute
    public void checkLogin(HttpSession session) throws NoLoginException{
        if(session.getAttribute("userLogin") == null) {
            throw new NoLoginException();
        }
    }
    /**
     * 查询、修改查询、删除查询
     */
    @GetMapping("/selectAllCardsByPage")
    public String selectAllCardsByPage(Model model,
                @RequestParam("currentPage") int currentPage,
                HttpSession session) {
        return cardService.selectAllCardsByPage(model, currentPage,session);
    }
    /**
     * 打开添加页面
     */
    @GetMapping("/toAddCard")
    public String toAddCard(@ModelAttribute Card card) {
        return "addCard";
    }
    /**
     * 实现添加及修改功能
     */
    @PostMapping("/addCard")
    public String addCard(@ModelAttribute Card card,
                HttpServletRequest request, @RequestParam("act") String act,
                HttpSession session) throws IllegalStateException, IOException {
        return cardService.addCard(card, request, act, session);
    }
    /**
     * 打开详情及修改页面
     */
    @GetMapping("/detail")
    public String detail(Model model,
                @RequestParam("id") int id,
                @RequestParam("act") String act) {
        return cardService.detail(model, id, act);
    }
    /**
     * 删除
     */
    @PostMapping("/delete")
    @ResponseBody
    public String delete(@RequestParam("id") int id) {
        return cardService.delete(id);
    }
    /**
     * 安全退出
     */
    @GetMapping("/loginOut")
```

```java
public String loginOut(Model model, HttpSession session) {
    return cardService.loginOut(model, session);
}
/**
 * 打开修改密码页面
 */
@GetMapping("/toUpdatePwd")
public String toUpdatePwd(Model model, HttpSession session) {
    return cardService.toUpdatePwd(model, session);
}
/**
 * 修改密码
 */
@PostMapping("/updatePwd")
public String updatePwd(@ModelAttribute MyUser myuser) {
    return cardService.updatePwd(myuser);
}
}
```

▶ 4.5.3　Service 实现

与名片管理相关的 Service 接口和实现类分别为 CardService 和 CardServiceImpl。在控制器获取一个请求后，需要调用 Service 层中的业务处理方法，在 Service 层中需要调用 Dao 层，所以 Service 层是控制器层和 Dao 层的"桥梁"。CardService 接口的代码略。

CardServiceImpl 实现类的核心代码如下：

```java
@Service
public class CardServiceImpl implements CardService{
    @Autowired
    private CardMapper cardMapper;
    /**
     * 查询、修改查询、删除查询、分页查询
     */
    @Override
    public String selectAllCardsByPage(Model model, int currentPage, HttpSession session) {
        MyUserTable mut = (MyUserTable)session.getAttribute("userLogin");
        List<Map<String, Object>> allUser = cardMapper.selectAllCards(mut.getId());
        //共多少个用户
        int totalCount = allUser.size();
        //计算共多少页
        int pageSize = 5;
        int totalPage = (int)Math.ceil(totalCount * 1.0/pageSize);
        List<Map<String, Object>> cardsByPage =
            cardMapper.selectAllCardsByPage((currentPage-1) * pageSize, pageSize, mut.getId());
        model.addAttribute("allCards", cardsByPage);
        model.addAttribute("totalPage", totalPage);
        model.addAttribute("currentPage", currentPage);
        return "main";
    }
    /**
     * 添加与修改名片
     */
    @Override
    public String addCard(Card card, HttpServletRequest  request, String act, HttpSession session) throws IllegalStateException, IOException {
        MultipartFile myfile = card.getLogo();
        //如果选择了上传文件,将文件上传到指定的目录 static/images
        if(!myfile.isEmpty()) {
            //上传文件路径(生产环境)
```

```java
            String path = request.getServletContext().getRealPath("/static/images/");
            //获得上传文件的原名
            String fileName = myfile.getOriginalFilename();
            //对文件重命名
            String fileNewName = MyUtil.getNewFileName(fileName);
            File filePath = new File(path + File.separator + fileNewName);
            //如果文件目录不存在,创建目录
            if(!filePath.getParentFile().exists()) {
                filePath.getParentFile().mkdirs();
            }
            //将上传文件保存到一个目标文件中
            myfile.transferTo(filePath);
            //将重命名后的图片名保存到 card 对象中,在添加时使用
            card.setLogoName(fileNewName);
        }
        if("add".equals(act)) {
            MyUserTable mut = (MyUserTable)session.getAttribute("userLogin");
            card.setUser_id(mut.getId());
            int n = cardMapper.addCard(card);
            if(n > 0)                              //成功
                return "redirect:/card/selectAllCardsByPage?currentPage=1";
            //失败
            return "addCard";
        }else {                                    //修改
            int n = cardMapper.updateCard(card);
            if(n > 0)                              //成功
                return "redirect:/card/selectAllCardsByPage?currentPage=1";
            //失败
            return "updateCard";
        }
    }
    /**
     * 打开详情与修改页面
     */
    @Override
    public String detail(Model model, int id, String act) {
        CardTable ct = cardMapper.selectACard(id);
        model.addAttribute("card", ct);
        if("detail".equals(act)) {
            return "cardDetail";
        }else {
            return "updateCard";
        }
    }
    /**
     * 删除
     */
    @Override
    public String delete(int id) {
        cardMapper.deleteACard(id);
        return "/card/selectAllCardsByPage?currentPage=1";
    }
    /**
     * 安全退出
     */
    @Override
    public String loginOut(Model model, HttpSession session) {
        session.invalidate();
        model.addAttribute("myUser", new MyUser());
        return "login";
    }
```

```
/**
 * 打开修改密码页面
 */
@Override
public String toUpdatePwd(Model model, HttpSession session) {
    MyUserTable mut = (MyUserTable)session.getAttribute("userLogin");
    model.addAttribute("myuser", mut);
    return "updatePwd";
}
/**
 * 修改密码
 */
@Override
public String updatePwd(MyUser myuser) {
    //将明文变成密文
    myuser.setUpwd(MD5Util.MD5(myuser.getUpwd()));
    cardMapper.updatePwd(myuser);
    return "login";
}
```

▶4.5.4 Dao 实现

Dao 层是数据访问层，即@Repository 注解的数据操作接口（接口中的方法与 SQL 映射文件中元素的 id 对应），与名片管理相关的数据访问层为 CardMapper，CardMapper 接口的代码略。

▶4.5.5 SQL 映射文件

SQL 映射文件的 namespace 属性与数据操作接口对应。与名片管理功能相关的 SQL 映射文件是 CardMapper.xml（位于 dao 包中），具体代码如下：

```xml
<?xml version="1.0" encoding="UTF-8" ?>
<!DOCTYPE mapper
PUBLIC "-//mybatis.org//DTD Mapper 3.0//EN"
"http://mybatis.org/dtd/mybatis-3-mapper.dtd">
<mapper namespace="dao.CardMapper">
    <!-- 查询所有名片 -->
    <select id="selectAllCards" resultType="map">
        select * from cardtable where user_id = #{uid}
    </select>
    <!-- 分页查询名片 -->
    <select id="selectAllCardsByPage" resultType="map">
        select * from cardtable where user_id = #{uid} limit #{startIndex}, #{perPageSize}
    </select>
    <!-- 添加名片 -->
    <insert id="addCard" parameterType="Card">
        insert into cardtable (id, name, telephone, email, company, post, address, logoName, user_id)
        values (null, #{name}, #{telephone}, #{email}, #{company}, #{post}, #{address}, #{logoName}, #{user_id})
    </insert>
    <!-- 修改名片 -->
    <update id="updateCard" parameterType="Card">
        update cardtable set
            name = #{name},
            telephone =#{telephone},
            email = #{email},
            company = #{company},
```

```xml
                post = #{post},
                address = #{address},
                logoName = #{logoName}
            where id = #{id}
    </update>
    <!-- 查询一个名片,在修改及查询详情时使用 -->
    <select id="selectACard" parameterType="integer" resultType="CardTable">
        select * from cardtable where id = #{id}
    </select>
    <!-- 删除一个名片 -->
    <delete id="deleteACard" parameterType="integer">
        delete from cardtable where id = #{id}
    </delete>
    <!-- 修改密码 -->
    <update id="updatePwd" parameterType="myuser">
        update usertable set upwd = #{upwd} where id = #{id}
    </update>
</mapper>
```

▶4.5.6 添加名片

用户登录成功后,进入名片管理系统的主页面。在名片管理系统的主页面中单击"添加名片"超链接打开添加名片页面,输入客户的姓名、电话、E-Mail、单位、职务、地址、照片,然后单击"添加"按钮实现添加。如果添加成功,则跳转到名片管理系统的主页面;如果添加失败,则回到添加名片页面。

addCard.jsp 页面提供添加名片信息的输入界面,如图 4.3 所示。addCard.jsp 的代码请读者参见本书提供的源代码 ch4。

图 4.3 添加名片页面

单击图 4.3 中的"添加"按钮,将添加请求通过"card/addCard? act=add"提交给控制器类 CardController(4.5.2 节)的 addCard 方法进行添加功能的处理。如果添加成功,跳转到名片管理系统的主页面;如果添加失败,回到添加名片页面。

4.5.7 名片管理主页面

用户登录成功后，进入名片管理系统的主页面（main.jsp），运行效果如图 4.4 所示。

图 4.4 名片管理系统的主页面

在主页面中单击"详情"超链接，打开名片详细信息页面。"详情"超链接的目标地址是一个 URL 请求，请求路径为"card/detail?id=＄{card.id}&act=detail"。根据请求路径找到对应控制器类 CardController 的 detail 方法处理查询一个名片的功能。根据动作类型（"修改"以及"详情"）将查询结果转发到不同视图。名片详细信息页面的运行效果如图 4.5 所示。

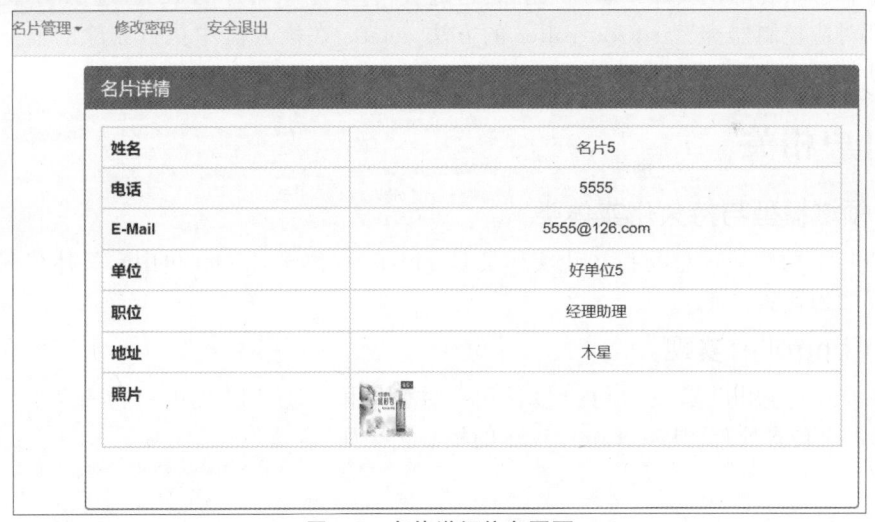

图 4.5 名片详细信息页面

4.5.8 修改名片

单击名片管理系统主页面中的"修改"超链接，打开修改名片信息页面 updateCard.jsp。"修改"超链接的目标地址是一个 URL 请求，请求路径为"card/detail?id=＄{card.id}&act=update"。找到对应控制器类 CardController 的方法 detail，在该方法中根据动作类型将查询结果转发给 updateCard.jsp 页面显示。

输入要修改的信息，然后单击"修改"按钮，将名片信息提交给控制器类。找到对应控制器类 CardController 的方法 addCard，在 addCard 方法中根据动作类型执行修改的业务处理。如果修改成功,进入名片管理系统的主页面;如果修改失败,回到 updateCard.jsp 页面。

updateCard.jsp 页面的运行效果如图 4.6 所示。

图 4.6　updateCard.jsp 页面

▶4.5.9　删除名片

在名片管理系统的主页面中单击"删除"超链接,将要删除名片的 ID 通过 Ajax 提交给控制器类。找到对应控制器类 CardController 的方法 delete,在该方法中执行删除的业务处理。如果删除成功,进入名片管理系统的主页面。

4.6　用户相关

▶4.6.1　领域模型与持久化实体类

与用户相关的领域模型与持久化实体类是 MyUser(位于 model 包中),具体代码请读者参见本书提供的源代码 ch4。

▶4.6.2　Controller 实现

在本系统中,与用户相关的功能包括用户注册、用户登录以及用户检查等,由控制器类 UserController 负责处理。UserController 的核心代码如下:

```
@Controller
@RequestMapping("/user")
@SuppressWarnings("all")
public class UserController {
    @Autowired
    private UserService userService;
    @GetMapping("/toLogin")
    public String toLogin(@ModelAttribute MyUser myUser) {
        return "login";
    }
    @GetMapping("/toRegister")
    public String toRegister(@ModelAttribute MyUser myUser) {
        return "register";
    }
```

```java
@PostMapping("/checkUname")
@ResponseBody
public String checkUname(@RequestBody MyUser myUser) {
    return userService.checkUname(myUser);
}
@PostMapping("/register")
public String register(@ModelAttribute MyUser myUser, Model model) {
    return userService.register(myUser);
}
@PostMapping("/login")
public String login(@ModelAttribute MyUser myUser, Model model, HttpSession session) {
    return userService.login(myUser, model, session);
}
}
```

4.6.3 Service 实现

与用户相关的 Service 接口和实现类分别为 UserService 和 UserServiceImpl。在控制器获取一个请求后，需要调用 Service 层中的业务处理方法，在 Service 层中需要调用 Dao 层，所以 Service 层是控制器层和 Dao 层的"桥梁"。UserService 接口的代码略。

UserServiceImpl 实现类的核心代码如下：

```java
@Service
public class UserServiceImpl implements UserService{
    @Autowired
    private UserMapper userMapper;
    /***
     * 检查用户名是否可用
     */
    @Override
    public String checkUname(MyUser myUser) {
        List<MyUserTable> userList = userMapper.selectByUname(myUser);
        if(userList.size() > 0)
            return "no";
        return "ok";
    }
    /**
     * 实现注册功能
     */
    @Override
    public String register(MyUser myUser) {
        //将明文变成密文
        myUser.setUpwd(MD5Util.MD5(myUser.getUpwd()));
        if(userMapper.register(myUser) > 0)
            return "login";
        return "register";
    }
    /**
     * 实现登录功能
     */
    @Override
    public String login(MyUser myUser, Model model, HttpSession session) {
        //ValidateCodeController 中的 rand
        String code = (String)session.getAttribute("rand");
        if(!code.equalsIgnoreCase(myUser.getCode())) {
            model.addAttribute("errorMessage", "验证码错误!");
            return "login";
        }else {
            //将明文变成密文
```

```
            myUser.setUpwd(MD5Util.MD5(myUser.getUpwd()));
            List<MyUserTable> list = userMapper.login(myUser);
            if(list.size() > 0){
                session.setAttribute("userLogin", list.get(0));
                return "redirect:/card/selectAllCardsByPage?currentPage=1";
            }else {
                model.addAttribute("errorMessage", "用户名或密码错误!");
                return "login";
            }
        }
    }
}
```

▶4.6.4　Dao 实现

Dao 层是数据访问层,即@Repository 注解的数据操作接口(接口中的方法与 SQL 映射文件中元素的 id 对应),与用户相关的数据访问层为 UserMapper,UserMapper 的代码略。

▶4.6.5　SQL 映射文件

SQL 映射文件的 namespace 属性与数据操作接口对应。与用户相关的 SQL 映射文件是 UserMapper.xml(位于 dao 包中),具体代码如下：

```
<?xml version="1.0" encoding="UTF-8" ?>
<!DOCTYPE mapper PUBLIC "-//mybatis.org//DTD Mapper 3.0//EN"
"http://mybatis.org/dtd/mybatis-3-mapper.dtd">
<mapper namespace="dao.UserMapper">
    <select id="selectByUname" resultType="MyUserTable" parameterType="MyUser">
        select * from usertable where uname = #{uname}
    </select>
    <insert id="register" parameterType="MyUser">
        insert into usertable (id,uname,upwd) values(null,#{uname},#{upwd})
    </insert>
    <select id="login" parameterType="MyUser" resultType="MyUserTable">
        select * from usertable where uname=#{uname} and upwd=#{upwd}
    </select>
</mapper>
```

▶4.6.6　注册

在登录页面 login.jsp 中单击"注册"链接,打开注册页面 register.jsp,效果如图 4.7 所示。

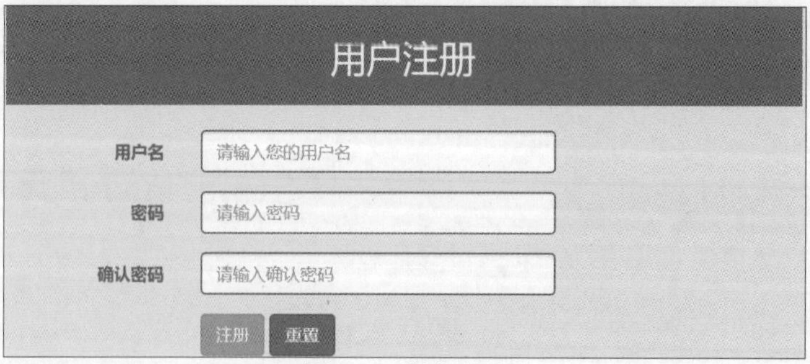

图 4.7　注册页面

在图 4.7 所示的注册页面中输入"用户名"后,系统将通过 Ajax 提交"user/checkUname"请求检测"用户名"是否可用。输入合法的用户信息后,单击"注册"按钮,实现注册功能。

▶ 4.6.7 登录

在浏览器中通过地址 http://localhost:8080/ch4 打开登录页面 login.jsp，效果如图 4.8 所示。

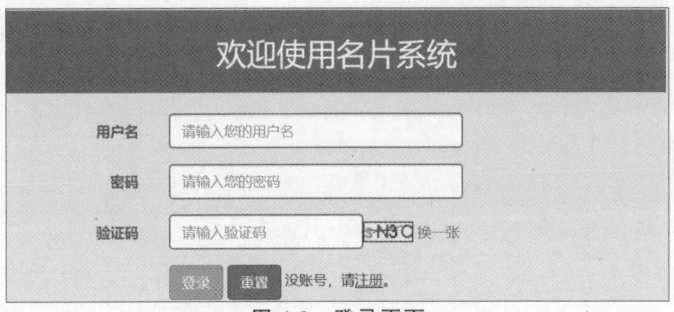

图 4.8 登录页面

用户输入用户名、密码和验证码后，系统将对用户名、密码和验证码进行验证。如果用户名、密码和验证码同时正确，则登录成功，将用户信息保存到 session 对象，并进入名片管理系统的主页面（main.jsp）；如果输入有误，则提示错误。

▶ 4.6.8 修改密码

单击名片管理系统主页面中的"修改密码"，打开密码修改页面 updatePwd.jsp，如图 4.9 所示。

图 4.9 密码修改页面

在图 4.9 中输入新密码后，单击"修改"按钮，将请求通过"card/updatePwd"提交给控制器类。根据请求路径找到对应控制器类 CardController（4.5.2 节）的 updatePwd 方法处理密码修改请求。这里找控制器类 CardController 处理密码修改，是因为用户必须登录成功后才能修改密码。

▶ 4.6.9 安全退出

在名片管理系统的主页面中单击"安全退出"，将返回登录页面。"安全退出"超链接的目标地址是一个请求（card/loginOut），根据请求路径找到控制器类 CardController（4.5.2 节）的 loginOut 方法处理安全退出请求。这里找控制器类 CardController 处理安全退出，是因为用户必须登录成功后才能安全退出。

本章小结

本章讲述了名片管理系统的设计与实现。通过本章的学习，读者不仅要掌握 SSM 框架整合开发的流程、方法和技术，还应该熟悉名片管理的业务需求、设计以及实现。

习题 4

1. 在名片管理系统中是如何控制登录权限的？
2. 在名片管理系统中安全退出功能的程序做了什么工作？

第二阶段　Spring Boot 框架开发

第 5 章　Spring Boot 入门

学习目的与要求

本章首先介绍什么是 Spring Boot，然后介绍 Spring Boot 应用的开发环境，最后介绍如何快速构建一个 Spring Boot 应用。通过本章的学习，掌握如何构建 Spring Boot 应用的开发环境以及 Spring Boot 应用。

本章主要内容

- SpringBoot 概述
- 快速构建 Spring Boot 应用

从前两章的学习可知，Spring 框架非常优秀，但问题在于"配置过多"，造成开发效率低、部署流程复杂以及集成难度大等问题。为了解决上述问题，Spring Boot 应运而生。在编写本书时，Spring Boot 的最新正式版是 3.2.4，建议读者在测试本书示例代码时使用相同的版本。

5.1　Spring Boot 概述

5.1.1　什么是 Spring Boot

Spring Boot 是由 Pivotal 团队提供的全新框架，其设计目的是用来简化新 Spring 应用的初始搭建以及开发过程。使用 Spring Boot 框架可以做到专注于 Spring 应用的开发，无须过多关注样板化的配置。

在 Spring Boot 框架中，使用"约定优于配置（Convention Over Configuration，COC）"的理念。针对企业应用开发，提供了符合各种场景的 spring-boot-starter 自动配置依赖模块，这些模块都是基于"开箱即用"的原则，进而使企业应用开发更加快捷和高效。可以说，Spring Boot 是开发者和 Spring 框架的中间层，目的是帮助开发者管理应用的配置，提供应用开发中常见配置的默认处理（即约定优于配置），简化 Spring 应用的开发和运维，降低开发人员对框架的关注度，使开发人员把更多精力放在业务逻辑代码上。通过"约定优于配置"的原则，Spring Boot 致力于在蓬勃发展的快速应用开发领域成为领导者。

5.1.2　Spring Boot 的优点

Spring Boot 具有如下优点。

（1）使编码变得简单：推荐使用注解。

（2）使配置变得快捷：自动配置、快速构建项目、快速集成第三方技术的能力。

（3）使部署变得简便：内嵌 Tomcat、Jetty 等 Web 容器。

（4）使监控变得容易：自带项目监控。

5.1.3　Spring Boot 的主要特性

Spring Boot 的主要特性如下。

❶ 约定优于配置

Spring Boot 遵循"约定优于配置"的原则，只需很少的配置，大多数情况直接使用默认配置即可。

❷ 独立运行的 Spring 应用

Spring Boot 可以以 JAR 包的形式独立运行，使用 java -jar 命令或者在项目的主程序中执行 main 方法运行 Spring Boot 应用（项目）。

❸ 内嵌 Web 容器

内嵌 Servlet 容器，Spring Boot 可以选择内嵌 Tomcat、Jetty 等 Web 容器，无须以 WAR 包的形式部署应用。

❹ 提供 starter 简化 Maven 配置

Spring Boot 提供了一系列的 starter pom 简化 Maven 的依赖加载，基本上可以做到自动化配置，高度封装，开箱即用。

❺ 自动配置 Spring

Spring Boot 根据项目依赖（在类路径中的 JAR 包、类）自动配置 Spring 框架，极大地减少了项目的配置。

❻ 提供准生产的应用监控

Spring Boot 提供基于 HTTP、SSH、Telnet 对运行的项目进行跟踪监控。

❼ 无代码生成和 XML 配置

Spring Boot 不是借助于代码生成来实现的，而是通过条件注解来实现的，提倡使用 Java 配置和注解配置相结合的配置方式，非常方便、快捷。

5.2 第一个 Spring Boot 应用

▶ 5.2.1 Maven 简介

Apache Maven 是一个软件项目管理工具，基于项目对象模型（Project Object Model，POM）的理念，通过一段核心描述信息来管理项目构建、报告和文档信息。在 Java 项目中，Maven 主要完成两件工作：①统一开发规范与工具；②统一管理 JAR 包。

Maven 统一管理项目开发所需要的 JAR 包，但这些 JAR 包不再包含在项目内（即不在 lib 目录下），而是存放于仓库中。仓库主要包括中央仓库和本地仓库。

❶ 中央仓库

中央仓库存放开发过程中的所有 JAR 包，例如 JUnit，可以通过互联网从中央仓库中下载，仓库的地址为 http://mvnrepository.com。

❷ 本地仓库

本地仓库指本地计算机中的仓库。官方下载 Maven 的本地仓库配置在"％MAVEN_HOME％\conf\settings.xml"文件中，找到"localRepository"即可。

Maven 项目首先从本地仓库中获取所需要的 JAR 包，当无法获取指定 JAR 包时，本地仓库将从远程仓库（中央仓库）中下载 JAR 包，并存入本地仓库以备将来使用。

▶ 5.2.2 Maven 的 pom.xml

Maven 是基于项目对象模型的理念管理项目的，所以 Maven 的项目都有一个 pom.xml 配置文件来管理项目的依赖以及项目的编译等功能。

在 Maven Web 项目中重点关注以下元素：

❶ properties 元素

在<properties></properties>之间可以定义变量，以便在<dependency></dependency>中

引用,示例代码如下:

```xml
<properties>
    <!-- Spring 版本号 -->
    <spring.version>6.0.0</spring.version>
</properties>
<dependencies>
    <dependency>
        <groupId>org.springframework</groupId>
        <artifactId>spring-core</artifactId>
        <version>${spring.version}</version>
    </dependency>
</dependencies>
```

❷ **dependencies 元素**

＜dependencies＞＜/dependencies＞元素包含多个项目依赖需要使用的＜dependency＞＜/dependency＞元素。

❸ **dependency 元素**

＜dependency＞＜/dependency＞元素内部通过＜groupId＞＜/groupId＞、＜artifactId＞＜/artifactId＞、＜version＞＜/version＞三个子元素确定唯一的依赖,也可以称为三个坐标。其示例代码如下:

```xml
<dependency>
    <!--groupId组织的唯一标识-->
    <groupId>org.springframework</groupId>
    <!--artifactId项目的唯一标识-->
    <artifactId>spring-core</artifactId>
    <!--version项目的版本号-->
    <version>${spring.version}</version>
</dependency>
```

❹ **scope 子元素**

在＜dependency＞＜/dependency＞元素中有时使用＜scope＞＜/scope＞子元素管理依赖的部署。＜scope＞＜/scope＞子元素可以使用以下 5 个值。

(1) compile(编译范围):compile 是默认值,即默认范围。如果依赖没有提供范围,那么该依赖的范围就是编译范围。编译范围的依赖在所有的 classpath 中可用,同时也会被打包发布。

(2) provided(已提供范围):provided 表示已提供范围,只有当 JDK 或者容器已提供该依赖时才可以使用。已提供范围的依赖不具有传递性,也不会被打包发布。

(3) runtime(运行时范围):runtime 范围依赖在运行和测试系统时需要,但在编译时不需要。

(4) test(测试范围):test 范围依赖在一般的编译和运行时都不需要,它们只有在测试编译和测试运行阶段可用,不会随项目发布。

(5) system(系统范围):system 范围与 provided 范围类似,但需要显式地提供包含依赖的 JAR 包,Maven 不会在 Repository 中查找它。

▶5.2.3 使用 IntelliJ IDEA 快速构建 Spring Boot 应用

下面详细讲解如何使用 IDEA 集成开发工具快速构建一个 Spring Boot Web 应用,具体步骤如下。

❶ **新建 Spring Project**

打开 IDEA,通过选择 File→New→Project 打开新建项目窗口。在新建项目窗口的左侧选中 Spring Initializr 选项,打开如图 5.1 所示的窗口输入项目信息。

第 5 章　Spring Boot 入门

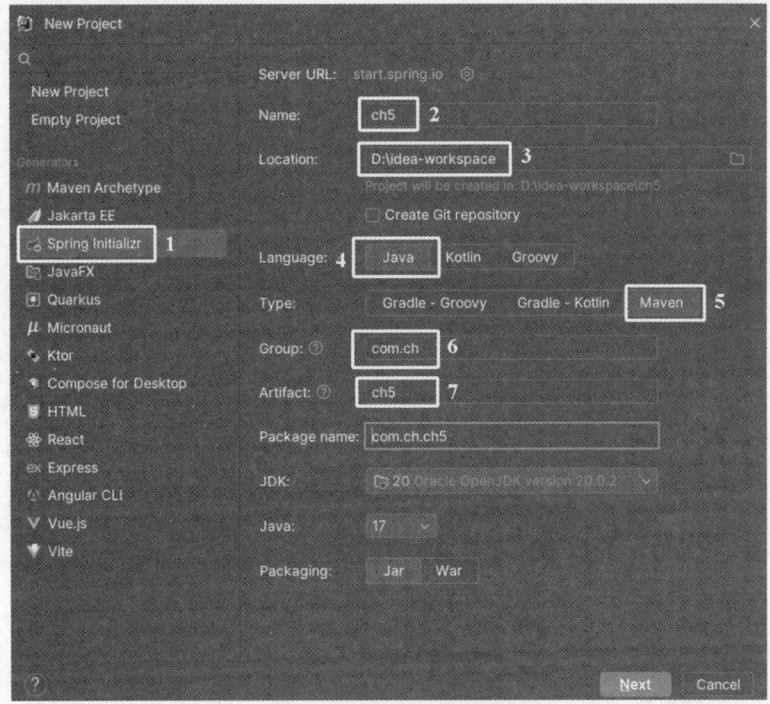

图 5.1　输入项目信息

❷ 选择项目依赖

在图 5.1 中输入项目信息后，单击 Next 按钮，打开如图 5.2 所示的选择项目依赖窗口。在图 5.2 中选择项目依赖（如 Spring Web）后，单击图 5.2 中的 Create 按钮，即可完成 Spring Boot Web 应用的创建。

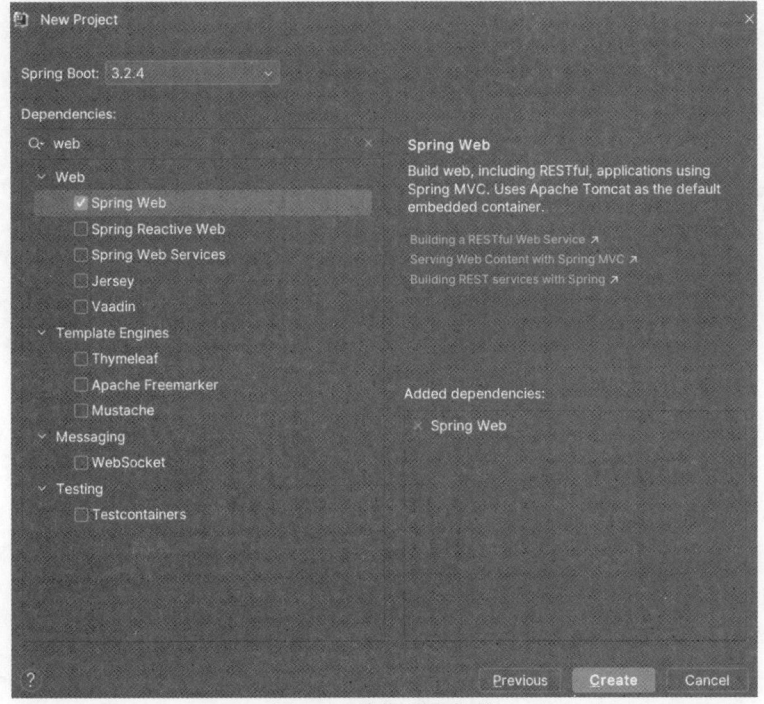

图 5.2　选择项目依赖

❸ 编写测试代码

在应用 ch5 的 src/main/java 目录下创建 com.ch.ch5.test 包,并在该包中创建 TestController 类,代码如下:

```java
package com.ch.ch5.test;
import org.springframework.web.bind.annotation.GetMapping;
import org.springframework.web.bind.annotation.RestController;
@RestController
public class TestController {
    @GetMapping("/hello")
    public String hello() {
        return "您好,Spring Boot!";
    }
}
```

在上述代码中使用的@RestController 注解是一个组合注解,相当于 Spring MVC 中的@Controller 和@ResponseBody 注解的组合,具体应用如下:

(1) 如果只是使用@RestController 注解 Controller,则 Controller 中的方法无法返回 JSP、HTML 等视图,返回的内容就是 return 的内容。

(2) 如果需要返回到指定页面,则需要用@Controller 注解。如果需要返回 JSON、XML 或自定义 mediaType 内容到页面,则需要在对应的方法上加@ResponseBody 注解。

❹ 应用程序的 App 类

在应用 ch5 的 com.ch.ch5 包中自动生成了应用程序的 App 类 Ch5Application。Ch5Application 的代码略。

❺ 运行 main 方法启动 Spring Boot 应用

运行 Ch5Application 类的 main 方法后,控制台信息如图 5.3 所示。

图 5.3 启动 Spring Boot 应用后的控制台信息

从控制台信息可以看到 Tomcat 的启动过程、Spring MVC 的加载过程。注意,Spring Boot 3 内嵌 Tomcat 10,因此 Spring Boot Web 应用不需要开发者配置与启动 Tomcat。

❻ 测试 Spring Boot 应用

启动 Spring Boot 应用后,默认访问地址为 http://localhost:8080/,将项目路径直接设置为根路径,这是 Spring Boot 的默认设置。因此,可以通过 http://localhost:8080/hello 测试应用(hello 与测试类 TestController 中的@GetMapping("/hello")对应),测试效果如图 5.4 所示。

图 5.4 访问 Spring Boot 应用

本章小结

本章首先简单地介绍了 Spring Boot 应运而生的缘由，然后演示了如何使用 IDEA 快速构建 Spring Boot 应用。IDEA 在业界被公认为较好的 Java 开发工具，开发者可根据实际工程需要选择合适的 IDE 构建 Spring Boot 应用。

习题 5

1. Spring、Spring MVC、Spring Boot 三者有什么联系？为什么要学习 Spring Boot？
2. 在 IDEA 中如何快速构建 Spring Boot 的 Web 应用？

第 6 章　Spring Boot 核心

学习目的与要求

本章将详细介绍 Spring Boot 的核心注解、基本配置、自动配置原理以及条件注解。通过本章的学习，掌握 Spring Boot 的核心注解与基本配置，理解 Spring Boot 的自动配置原理与条件注解。

本章主要内容

- SpringBoot 的基本配置
- 读取应用配置
- SpringBoot 的自动配置原理
- Spring Boot 的条件注解

在 Spring Boot 产生之前，Spring 项目会存在多个配置文件，例如 web.xml、application.xml，应用程序自身也需要多个配置文件，同时需要编写程序读取这些配置文件。现在 Spring Boot 简化了 Spring 项目配置的管理和读取，仅需要一个 application.properties 文件，并提供了多种读取配置文件的方式。本章将学习 Spring Boot 的基本配置与运行原理。

6.1　Spring Boot 的基本配置

6.1.1　启动类和核心注解@SpringBootApplication

Spring Boot 应用通常都有一个名为 *Application 的程序入口类，该入口类需要使用 Spring Boot 的核心注解@SpringBootApplication 标注为应用的启动类。另外，该入口类有一个标准的 Java 应用程序的 main 方法，在 main 方法中通过"SpringApplication.run(*Application.class, args);"启动 Spring Boot 应用。

Spring Boot 的核心注解 @SpringBootApplication 是一个组合注解，主要组合了 @SpringBootConfiguration、@EnableAutoConfiguration 和@ComponentScan 注解。其源代码可以从 spring-boot-autoconfigure-x.y.z.jar 依赖包中查看 org/springframework/boot/autoconfigure/SpringBootApplication.java。

❶ @SpringBootConfiguration 注解

@SpringBootConfiguration 是 Spring Boot 应用的配置注解，该注解也是一个组合注解，源代码可以从 spring-boot-x.y.z.jar 依赖包中查看 org/springframework/boot/SpringBootConfiguration.java。在 Spring Boot 应用中推荐使用@SpringBootConfiguration 注解代替@Configuration 注解。

❷ @EnableAutoConfiguration 注解

@EnableAutoConfiguration 注解可以让 Spring Boot 根据当前应用项目所依赖的 JAR 包自动配置项目的相关配置。例如，在 Spring Boot 项目的 pom.xml 文件中添加了 spring-boot-starter-web 依赖，Spring Boot 项目会自动添加 Tomcat 和 Spring MVC 的依赖，同时对 Tomcat 和 Spring MVC 进行自动配置。打开 pom.xml 文件，单击窗口右侧的 Maven 即可查看 spring-boot-starter-web 的相关依赖，如图 6.1 所示。

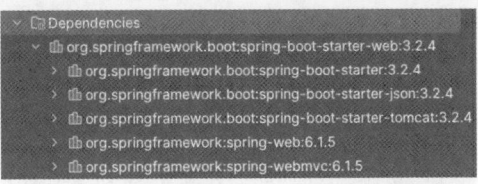

图 6.1　spring-boot-starter-web 的相关依赖

❸ @ComponentScan 注解

该注解的功能是让 Spring Boot 自动扫描 @SpringBootApplication 所在类的同级包以及其子包中的配置，所以建议将 @SpringBootApplication 注解的入口类放置在项目包（Group Id＋Artifact Id 组合的包名）中，并将用户自定义的程序放置在项目包及其子包中，这样可以保证 Spring Boot 自动扫描项目所有包中的配置。

6.1.2　Spring Boot 的全局配置文件

Spring Boot 的全局配置文件（application.properties 或 application.yml）位于 Spring Boot 应用的 src/main/resources 目录下。

❶ 设置端口号

全局配置文件主要用于修改项目的默认配置。例如，在 Spring Boot 应用 ch6 的 src/main/resources 目录下找到名为 application.properties 的全局配置文件，添加如下配置内容：

```
server.port=8888
```

可以将内嵌的 Tomcat 的默认端口改为 8888。

❷ 设置 Web 应用的上下文路径

如果开发者想设置一个 Web 应用程序的上下文路径，可以在 application.properties 文件中配置如下内容：

```
server.servlet.context-path=/XXX
```

这时应该通过"http://localhost:8080/XXX/testStarters"访问如下控制器类中的请求处理方法：

```
@GetMapping("/testStarters")
public String index() {
}
```

❸ 配置文档

在 Spring Boot 的全局配置文件中可以配置与修改多个参数，如果读者想了解参数的详细说明和描述，可以查看官方文档说明（https://docs.spring.io/spring-boot/docs/current/reference/htmlsingle/#appendix.application-properties）。

6.1.3　Spring Boot 的 Starters

Spring Boot 提供了许多简化企业级开发的"开箱即用"的 Starters。Spring Boot 项目只要使用所需要的 Starters，Spring Boot 即可自动关联项目开发所需要的相关依赖。例如，在应用的 pom.xml 文件中添加如下依赖配置：

```xml
<dependency>
    <groupId>org.springframework.boot</groupId>
    <artifactId>spring-boot-starter-web</artifactId>
</dependency>
```

Spring Boot 将自动关联 Web 开发的相关依赖，如 Tomcat、spring-webmvc 等，进而对 Web 开发支持，并将相关技术的配置实现自动配置。

通过访问"https://docs.spring.io/spring-boot/docs/current/reference/htmlsingle/#using.build-systems.starters"官网，可以查看 Spring Boot 官方提供的 Starters。

除了 Spring Boot 官方提供的 Starters 外，还可以通过访问"https://github.com/spring-projects/spring-boot/blob/master/spring-boot-project/spring-boot-starters/README.adoc"网站查看第三方为 Spring Boot 贡献的 Starters。

6.2 读取应用配置

Spring Boot 提供了三种方式读取项目的 application.properties 配置文件的内容。这三种方式分别为使用 Environment 类、@Value 注解以及 @ConfigurationProperties 注解。

▶ 6.2.1 Environment

Environment 是一个通用的读取应用程序运行时的环境变量的类，可以通过 key-value 方式读取 application.properties、命令行输入参数、系统属性、操作系统环境变量等。下面通过一个实例来演示如何使用 Environment 类读取 application.properties 配置文件的内容。

【例 6-1】 使用 Environment 类读取 application.properties 配置文件的内容。

其具体实现步骤如下。

❶ 创建 Spring Boot Web 应用 ch6

使用 IDEA 快速创建 Spring Boot Web 应用 ch6，同时给应用 ch6 添加如图 6.2 所示的依赖。在 6.2.3 节使用 @ConfigurationProperties 注解读取配置时需要 Spring Configuration Processor 依赖，在此一起添加。

❷ 添加配置文件内容

在应用 ch6 的 src/main/resources 目录下找到全局配置文件 application.properties，并添加如下内容：

图 6.2 应用 ch6 的依赖

```
test.msg=read config
```

❸ 创建控制器类 EnvReaderConfigController

在应用 ch6 的 src/main/java 目录下创建名为 com.ch.ch6.controller 的包（com.ch.ch6 包的子包，保障注解全部被扫描），并在该包下创建控制器类 EnvReaderConfigController。在控制器类 EnvReaderConfigController 中使用 @Autowired 注解依赖注入 Environment 类的对象，核心代码如下：

```
@RestController
public class EnvReaderConfigController{
    @Autowired
    private Environment env;
    @GetMapping("/testEnv")
    public String testEnv() {
        return "方法一:" + env.getProperty("test.msg") ;
        //test.msg 为配置文件 application.properties 中的 key
    }
}
```

❹ 启动 Spring Boot 应用

运行 Ch6Application 类的 main 方法,启动 Spring Boot 应用。

❺ 测试应用

在启动 Spring Boot 应用后,默认访问地址为 http://localhost:8080/,将项目路径直接设置为根路径,这是 Spring Boot 的默认设置。因此,可以通过 http://localhost:8080/testEnv 测试应用(testEnv 与控制器类 EnvReaderConfigController 中的@GetMapping("/testEnv")对应),测试效果如图 6.3 所示。

图 6.3　使用 Environment 类读取 application.properties 文件的内容

6.2.2　@Value

使用@Value 注解读取配置文件内容的示例代码如下:

```
@Value("${test.msg}")              //test.msg 为配置文件 application.properties 中的 key
private String msg;                //通过@Value 注解将配置文件中 key 对应的 value 赋值给变量 msg
```

下面通过实例讲解如何使用@Value 注解读取配置文件的内容。

【例 6-2】　使用@Value 注解读取配置文件的内容。

其具体实现步骤如下。

❶ 创建控制器类 ValueReaderConfigController

在 ch6 应用的 com.ch.ch6.controller 包中创建名为 ValueReaderConfigController 的控制器类,并在该控制器类中使用@Value 注解读取配置文件的内容,核心代码如下:

```
@RestController
public class ValueReaderConfigController {
    @Value("${test.msg}")
    private String msg;
    @GetMapping("/testValue")
    public String testValue() {
        return "方法二:" + msg;
    }
}
```

❷ 启动并测试应用

首先运行 Ch6Application 类的 main 方法,启动 Spring Boot 应用,然后通过 http://localhost:8080/testValue 测试应用。

6.2.3　@ConfigurationProperties

使用@ConfigurationProperties 首先建立配置文件与对象的映射关系,然后在控制器方法中使用@Autowired 注解将对象注入。

下面通过实例讲解如何使用@ConfigurationProperties 读取配置文件的内容。

【例 6-3】　使用@ConfigurationProperties 读取配置文件的内容。

其具体实现步骤如下。

❶ 添加配置文件内容

在 ch6 项目的 src/main/resources 目录下找到全局配置文件 application.properties,并添加

如下内容：

```
#nest Simple properties
obj.sname=chenheng
obj.sage=88
#List properties
obj.hobby[0]=running
obj.hobby[1]=basketball
#Map Properties
obj.city.cid=dl
obj.city.cname=dalian
```

❷ 建立配置文件与对象的映射关系

在 ch6 项目的 src/main/java 目录下创建名为 com.ch.ch6.model 的包，并在该包中创建实体类 StudentProperties，在该类中使用@ConfigurationProperties 注解建立配置文件与对象的映射关系，核心代码如下：

```
@Component                                          //使用 Component 注解声明一个组件被控制器依赖注入
@ConfigurationProperties(prefix = "obj")  //obj 为配置文件中 key 的前缀
@Data
public class StudentProperties {
    private String sname;
    private int sage;
    private List<String> hobby;
    private Map<String, String> city;
    @Override
    public String toString() {
        return "StudentProperties [sname=" + sname
            + ", sage=" + sage
            + ", hobby0=" + hobby.get(0)
            + ", hobby1=" + hobby.get(1)
            + ", city=" + city +  "]";
    }
}
```

❸ 创建控制器类 ConfigurationPropertiesController

在 ch6 项目的 com.ch.ch6.controller 包中创建名为 ConfigurationPropertiesController 的控制器类，并在该控制器类中使用@Autowired 注解依赖注入 StudentProperties 对象，核心代码如下：

```
@RestController
public class ConfigurationPropertiesController {
    @Autowired
    StudentProperties studentProperties;
    @GetMapping("/testConfigurationProperties")
    public String testConfigurationProperties() {
        return studentProperties.toString();
    }
}
```

❹ 启动并测试应用

首先运行 Ch6Application 类的 main 方法，启动 Spring Boot 应用，然后通过 http://localhost:8080/testConfigurationProperties 测试应用。

▶6.2.4 @PropertySource

开发者希望读取项目的其他配置文件，而不是全局配置文件 application.properties，该如何实现呢？使用@PropertySource注解找到项目的其他配置文件，然后结合6.2.1～6.2.3节中的任意一种方式读取即可。

下面通过实例讲解如何使用@PropertySource＋@Value读取其他配置文件的内容。

【例6-4】 使用@PropertySource＋@Value读取其他配置文件的内容。

其具体实现步骤如下。

❶ 创建配置文件

在ch6的src/main/resources目录下创建配置文件ok.properties和test.properties，并在ok.properties文件中添加如下内容：

```
your.msg=hello.
```

在test.properties文件中添加如下内容：

```
my.msg=test PropertySource
```

❷ 创建控制器类PropertySourceValueReaderOtherController

在ch6项目的com.ch.ch6.controller包中创建名为PropertySourceValueReaderOtherController的控制器类。在该控制器类中首先使用@PropertySource注解找到其他配置文件，然后使用@Value注解读取配置文件的内容，核心代码如下：

```
@RestController
@PropertySource({"test.properties","ok.properties"})
public class PropertySourceValueReaderOtherController {
    @Value("${my.msg}")
    private String mymsg;
    @Value("${your.msg}")
    private String yourmsg;
    @GetMapping("/testProperty")
    public String testProperty() {
        return "其他配置文件test.properties:" + mymsg + "<br>"
                + "其他配置文件ok.properties:" + yourmsg;
    }
}
```

❸ 启动并测试应用

首先运行Ch6Application类的main方法，启动Spring Boot应用，然后通过http://localhost:8080/testProperty测试应用，测试结果如图6.4所示。

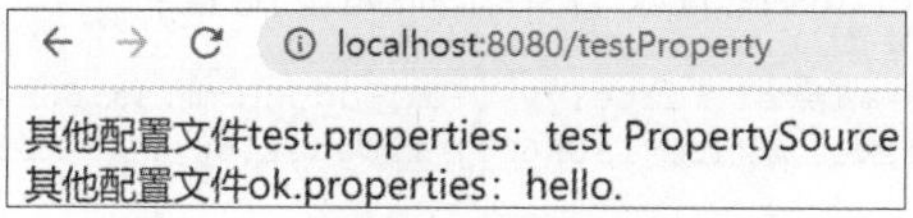

图6.4 读取其他配置文件的内容

6.3 日志配置

在默认情况下，Spring Boot项目使用LogBack实现日志，使用apache Commons Logging作为日志接口，因此在代码中通常如下使用日志：

```
@RestController
public class LogTestController {
    private Log log = LogFactory.getLog(LogTestController.class);
    @GetMapping("/testLog")
```

```
    public String testLog() {
        log.info("测试日志");
        return "测试日志";
    }
}
```

通过地址 http://localhost:8080/testLog 运行上述控制器类代码，可以在控制台输出如图 6.5 所示的日志。

```
2024-04-05T21:23:56.413+08:00  INFO 9356 --- [ch6] [           main] com.ch.ch6.Ch6Application                : Starte
2024-04-05T21:24:02.713+08:00  INFO 9356 --- [ch6] [nio-8080-exec-1] o.a.c.c.C.[Tomcat].[localhost].[/]       : Initia
2024-04-05T21:24:02.713+08:00  INFO 9356 --- [ch6] [nio-8080-exec-1] o.s.web.servlet.DispatcherServlet        : Initia
2024-04-05T21:24:02.714+08:00  INFO 9356 --- [ch6] [nio-8080-exec-1] o.s.web.servlet.DispatcherServlet        : Comple
2024-04-05T21:24:02.742+08:00  INFO 9356 --- [ch6] [nio-8080-exec-1] c.ch.ch6.controller.LogTestController    : 测试日志
```

图 6.5 默认日志

日志级别有 ERROR、WARN、INFO、DEBUG 和 TRACE。Spring Boot 默认的日志级别为 INFO，日志信息可以打印到控制台。开发者可以自己设定 Spring Boot 项目的日志输出级别，例如在 application.properties 配置文件中加入以下配置：

```
#设定日志的默认级别为 info
logging.level.root=info
#设定 org 包下的日志级别为 warn
logging.level.org=warn
#设定 com.ch.ch6 包下的日志级别为 debug
logging.level.com.ch.ch6=debug
```

Spring Boot 项目默认并没有输出日志到文件，但开发者可以在 application.properties 配置文件中指定日志输出到文件，配置示例如下：

```
logging.file=my.log
```

日志输出到 my.log 文件，该日志文件位于 Spring Boot 项目运行的当前目录（项目工程目录下）。开发者也可以指定日志文件目录，配置示例如下：

```
logging.file=C:/log/my.log
```

这样将在 C:/log 目录下生成一个名为 my.log 的日志文件。不管日志文件位于何处，当日志文件的大小达到 10MB 时将自动生成一个新日志文件。

Spring Boot 使用内置的 LogBack 支持对控制台日志输出和文件输出进行格式控制，例如开发者可以在 application.properties 配置文件中添加如下配置：

```
logging.pattern.console=%level %date{yyyy-MM-dd HH:mm:ss:SSS} %logger{50}.%M %L :%m%n
logging.pattern.file=%level %date{ISO8601} %logger{50}.%M %L :%m%n
```

logging.pattern.console：指定控制台日志格式。

logging.pattern.file：指定日志文件格式。

%level：指定输出日志级别。

%date：指定日志发生的时间。ISO8601 表示标准日期，相当于 yyyy-MM-dd HH:mm:ss:SSS。

%logger：指定输出 Logger 的名字，形式为包名+类名，{n}限定了输出长度。

%M：指定日志发生时的方法名。

%L：指定日志调用时所在的代码行，适用于开发调试，在线上运行时不建议使用此参数，因为获取代码行对性能有消耗。

%m：表示日志消息。

%n：表示日志换行。

6.4　Spring Boot 的自动配置原理

从 6.1.1 节可知 Spring Boot 使用核心注解@SpringBootApplication 将一个带有 main 方法的类标注为应用的启动类。@SpringBootApplication 注解最主要的功能之一是为 Spring Boot 开启了一个@EnableAutoConfiguration 注解的自动配置功能。

@EnableAutoConfiguration 注解主要使用了一个类名为 AutoConfigurationImportSelector 的选择器向 Spring 容器自动配置一些组件。@EnableAutoConfiguration 注解的源代码可以从 spring-boot-autoconfigure-x.y.z.jar(org.springframework.boot.autoconfigure)依赖包中查看，核心代码如下：

```
@Import(AutoConfigurationImportSelector.class)
public @interface EnableAutoConfiguration {
    String ENABLED_OVERRIDE_PROPERTY = "spring.boot.enableautoconfiguration";
    Class<?>[] exclude() default {};
    String[] excludeName() default {};
}
```

在 AutoConfigurationImportSelector(源代码位于 org.springframework.boot.autoconfigure 包中)类中有一个名为 selectImports 的方法，该方法规定了向 Spring 容器自动配置的组件。

selectImports 方法的代码如下：

```
@Override
public String[] selectImports(AnnotationMetadata annotationMetadata) {
    //判断@EnableAutoConfiguration 注解是否开启，默认开启
    if (!isEnabled(annotationMetadata)) {
        return NO_IMPORTS;
    }
    //获得自动配置
    AutoConfigurationEntry autoConfigurationEntry =
    getAutoConfigurationEntry(annotationMetadata);
    return StringUtils.toStringArray(autoConfigurationEntry.getConfigurations());
}
```

在 selectImports 方法中，调用 getAutoConfigurationEntry 方法获得自动配置。进入该方法，查看到的源代码如下：

```
protected AutoConfigurationEntry getAutoConfigurationEntry(AnnotationMetadata
annotationMetadata) {
    if (!isEnabled(annotationMetadata)) {
        return EMPTY_ENTRY;
    }
    AnnotationAttributes attributes = getAttributes(annotationMetadata);
    //获取自动配置数据
    List<String> configurations = getCandidateConfigurations(annotationMetadata,
            attributes);
    //去重
    configurations = removeDuplicates(configurations);
    //去除一些多余的类
    Set<String> exclusions = getExclusions(annotationMetadata, attributes);
    checkExcludedClasses(configurations, exclusions);
    configurations.removeAll(exclusions);
    //过滤掉一些条件没有满足的配置
```

```
        configurations = getConfigurationClassFilter().filter(configurations);
        fireAutoConfigurationImportEvents(configurations, exclusions);
        return new AutoConfigurationEntry(configurations, exclusions);
    }
```

在 getAutoConfigurationEntry 方法中，调用 getCandidateConfigurations 方法获取自动配置数据。进入该方法，查看到的源代码如下：

```
    protected List<String> getCandidateConfigurations(AnnotationMetadata metadata,
AnnotationAttributes attributes) {
        List<String> configurations = ImportCandidates.load(AutoConfiguration.class,
getBeanClassLoader()).getCandidates();
        Assert.notEmpty(configurations, "No auto configuration classes found in "
            + "META-INF/spring/org.springframework.boot.autoconfigure.
AutoConfiguration.imports. If you are using a custom packaging, make sure that file is
correct.");
        return configurations;
    }
```

在 getCandidateConfigurations 方法中，调用 ImportCandidates 类的静态方法 load。进入该方法，查看到的源代码如下：

```
    public static ImportCandidates load(Class<?> annotation, ClassLoader classLoader) {
        Assert.notNull(annotation, "'annotation' must not be null");
        ClassLoader classLoaderToUse = decideClassloader(classLoader);
        String location = String.format("META-INF/spring/%s.imports", annotation.getName());
        Enumeration<URL> urls = findUrlsInClasspath(classLoaderToUse, location);
        List<String> importCandidates = new ArrayList();
        while(urls.hasMoreElements()) {
            URL url = (URL)urls.nextElement();
            importCandidates.addAll(readCandidateConfigurations(url));
        }
        return new ImportCandidates(importCandidates);
    }
```

在 load 方法中可以看到加载了一个字符串常量"META-INF/spring/%s.imports"。

从上述源代码中可以看出，最终 Spring Boot 是通过加载所有（in multiple JAR files）META-INF/spring/XXX.imports 配置文件进行自动配置的。所以@SpringBootApplication 注解通过使用@EnableAutoConfiguration 注解自动配置的原理是从 classpath 中搜索所有 META-INF/spring/XXX.imports 配置文件，并将其中 org.springframework.boot.autoconfigure 对应的配置项通过 Java 反射机制进行实例化，然后汇总并加载到 Spring 的 IoC 容器。

在 Spring Boot 项目的 Maven Dependencies 的 spring-boot-autoconfigure-x.y.z.jar 目录下，可以找到 META-INF/spring/org.springframework.boot.autoconfigure.AutoConfiguration.imports 配置文件，该文件中定义了许多自动配置。

6.5 Spring Boot 的条件注解

打开 META-INF/spring/org.springframework.boot.autoconfigure.AutoConfiguration.imports 配置文件中的任意一个 AutoConfiguration，一般都可以找到条件注解。例如，打开 org.springframework.boot.autoconfigure.aop.AopAutoConfiguration 的源代码，可以看到@ConditionalOnClass 和@ConditionalOnProperty 等条件注解。

通过 org.springframework.boot.autoconfigure.aop.AopAutoConfiguration 的源代码可以看

出 Spring Boot 的自动配置是使用 Spring 的 @Conditional 注解实现的，因此本节将介绍相关的条件注解，并讲述如何自定义 Spring 的条件注解。

▶ 6.5.1 条件注解

所谓 Spring 的条件注解，就是应用程序的配置类在满足某些特定条件时才会被自动启用此配置类的配置项。Spring Boot 的条件注解位于 spring-boot-autoconfigure-x.y.z.jar 的 org.springframework.boot.autoconfigure.condition 包下，具体如表 6.1 所示。

表 6.1 Spring Boot 的条件注解

注 解 名	条件实现类	条 件
@ConditionalOnBean	OnBeanCondition	Spring 容器中存在指定的实例 Bean
@ConditionalOnClass	OnClassCondition	类加载器（类路径）中存在对应的类
@ConditionalOnCloudPlatform	OnCloudPlatformCondition	是否在云平台
@ConditionalOnExpression	OnExpressionCondition	判断 SpEL 表达式是否成立
@ConditionalOnJava	OnJavaCondition	指定 Java 版本是否符合要求
@ConditionalOnJndi	OnJndiCondition	在 JNDI（Java 命名和目录接口）存在的条件下查找指定的位置
@ConditionalOnMissingBean	OnBeanCondition	Spring 容器中不存在指定的实例 Bean
@ConditionalOnMissingClass	OnClassCondition	类加载器（类路径）中不存在对应的类
@ConditionalOnNotWebApplication	OnWebApplicationCondition	当前应用程序不是 Web 程序
@ConditionalOnProperty	OnPropertyCondition	应用环境中属性是否存在指定的值
@ConditionalOnResource	OnResourceCondition	是否存在指定的资源文件
@ConditionalOnSingleCandidate	OnBeanCondition	Spring 容器中是否存在且只存在一个对应的实例 Bean
@ConditionalOnWarDeployment	OnWarDeploymentCondition	当前应用程序是传统 WAR 部署
@ConditionalOnWebApplication	OnWebApplicationCondition	当前应用程序是 Web 程序

表 6.1 中的条件注解都是组合了 @Conditional 元注解，只是针对不同的条件去实现。

▶ 6.5.2 自定义条件

Spring 的 @Conditional 注解根据满足某特定条件创建一个特定的 Bean。例如，当某 JAR 包在类路径下时自动配置多个 Bean。这就是根据特定条件控制 Bean 的创建行为，这样开发者就可以使用这个特性进行一些自动配置。那么，开发者如何自己构造条件呢？在 Spring 框架中可以通过实现 Condition 接口并重写 matches 方法来构造条件。下面通过实例讲解条件的构造过程。

【例 6-5】 如果类路径 classpath（src/main/resources）下存在 test.properties 文件，输出 "test.properties 文件存在。"，否则输出 "test.properties 文件不存在！"。

其具体实现步骤如下。

❶ 构造条件

在 Spring Boot 应用 ch6 的 src/main/java 目录下创建 com.ch.ch6.conditional 包，并在该包中分别创建条件实现类 MyCondition（存在文件 test.properties）和 YourCondition（不存在文件 test.properties）。

MyCondition 的核心代码如下：

```
public class MyCondition implements Condition{
    @Override
    public boolean matches(ConditionContext context, AnnotatedTypeMetadata metadata) {
         return context.getResourceLoader().getResource("classpath:test.properties").exists();
    }
}
```

YourCondition 的核心代码如下：

```
public class YourCondition implements Condition{
    @Override
    public boolean matches(ConditionContext context, AnnotatedTypeMetadata metadata) {
         return !context.getResourceLoader().getResource("classpath:test.properties").exists();
    }
}
```

❷ 创建不同条件下 Bean 的类

在 com.ch.ch6.conditional 包中创建接口 MessagePrint，并分别创建该接口的实现类 MyMessagePrint 和 YourMessagePrint。

MessagePrint 的代码如下：

```
package com.ch.ch6.conditional;
public interface MessagePrint {
    public String showMessage();
}
```

MyMessagePrint 的代码如下：

```
package com.ch.ch6.conditional;
public class MyMessagePrint implements MessagePrint{
    @Override
    public String showMessage() {
        return "test.properties 文件存在。";
    }
}
```

YourMessagePrint 的代码如下：

```
package com.ch.ch6.conditional;
public class YourMessagePrint implements MessagePrint{
    @Override
    public String showMessage() {
        return "test.properties 文件不存在!";
    }
}
```

❸ 创建配置类

在 com.ch.ch6.conditional 包中创建配置类 ConditionConfig，并在该配置类中使用@Bean 和@Conditional 实例化符合条件的 Bean。

ConditionConfig 的核心代码如下：

```
@Configuration
public class ConditionConfig {
    @Bean
    @Conditional(MyCondition.class)
    public MessagePrint myMessage() {
```

```
        return new MyMessagePrint();
    }
    @Bean
    @Conditional(YourCondition.class)
    public MessagePrint yourMessage() {
        return new YourMessagePrint();
    }
}
```

❹ 创建测试类

在 com.ch.ch6.conditional 包中创建测试类 TestMain，具体代码如下：

```
package com.ch.ch6.conditional;
import org.springframework.context.annotation.AnnotationConfigApplicationContext;
public class TestMain {
    public static void main(String[] args) {
        AnnotationConfigApplicationContext context =
            new AnnotationConfigApplicationContext(ConditionConfig.class);
        MessagePrint mp = context.getBean(MessagePrint.class);
        System.out.println(mp.showMessage());
    }
}
```

❺ 运行

当 Spring Boot 应用 ch6 的 src/main/resources 目录下存在 test.properties 文件时，运行测试类，结果如图 6.6 所示；当 Spring Boot 应用 ch6 的 src/main/resources 目录下不存在 test.properties 文件时，运行测试类，结果如图 6.7 所示。

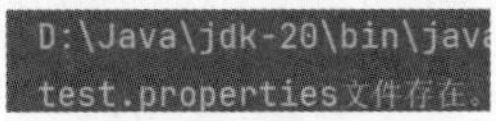

图 6.6　存在 test.properties 文件 　　　　图 6.7　不存在 test.properties 文件

▶6.5.3　自定义 Starters

从 6.1.3 节可知，第三方为 Spring Boot 贡献了许多 Starters。那么，开发者是否也可以贡献自己的 Starters？在学习 Spring Boot 的自动配置机制后，答案是肯定的。下面通过实例讲解如何自定义 Starters。

【例 6-6】　自定义一个 Starter（spring_boot_mystarters）。要求：当类路径中存在 MyService 类时自动配置该类的 Bean，并可以将相应 Bean 的属性在 application.properties 中配置。

其具体实现步骤如下。

❶ 新建 Spring Boot 项目 spring_boot_mystarters

首先在 IntelliJ IDEA 中通过选择 File→New→Project 打开 New Project 对话框，然后在该对话框中输入项目名称 spring_boot_mystarters，并单击 Next 与 Create 按钮。

❷ 修改 POM 文件

修改 Spring Boot 项目 spring_boot_mystarters 的 POM 文件，增加 Spring Boot 自身的自动配置作为依赖；使用 @ConfigurationProperties 注解读取配置，需要添加 Spring Configuration Processor 依赖；使用 spring-boot-maven-plugin 插件将其打包成普通 JAR 包，需要给插件配置 <skip>true</skip>。因此，POM 文件的核心内容具体如下：

```xml
<dependencies>
    <dependency>
        <groupId>org.springframework.boot</groupId>
        <artifactId>spring-boot-starter</artifactId>
    </dependency>
    <dependency>
        <groupId>org.springframework.boot</groupId>
        <artifactId>spring-boot-autoconfigure</artifactId>
    </dependency>
    <dependency>
        <groupId>org.springframework.boot</groupId>
        <artifactId>spring-boot-configuration-processor</artifactId>
        <optional>true</optional>
    </dependency>
    <dependency>
        <groupId>org.springframework.boot</groupId>
        <artifactId>spring-boot-starter-test</artifactId>
        <scope>test</scope>
    </dependency>
</dependencies>
<build>
    <plugins>
        <plugin>
            <groupId>org.springframework.boot</groupId>
            <artifactId>spring-boot-maven-plugin</artifactId>
            <configuration>
                <skip>true</skip>
            </configuration>
        </plugin>
    </plugins>
</build>
```

❸ 创建属性配置类 MyProperties

在项目 spring_boot_mystarters 的 com.ch.spring_boot_mystarters 包中创建属性配置类 MyProperties。在使用 spring_boot_mystarters 的 Spring Boot 项目的配置文件 application.properties 中，可以使用"my.msg="设置属性；若不设置，默认为 my.msg＝默认值。属性配置类 MyProperties 的代码如下：

```java
package com.ch.spring_boot_mystarters;
import org.springframework.boot.context.properties.ConfigurationProperties;
//在 application.properties 中通过"my.msg="设置属性
@ConfigurationProperties(prefix="my")
public class MyProperties {
    private String msg = "默认值";
    public String getMsg() {
        return msg;
    }
    public void setMsg(String msg) {
        this.msg = msg;
    }
}
```

❹ 创建判断依据类 MyService

在项目 spring_boot_mystarters 的 com.ch.spring_boot_mystarters 包中创建判断依据类 MyService。本例自定义的 Starters 将根据该类存在与否来创建该类的 Bean，该类可以是第三方类库的类。判断依据类 MyService 的代码如下：

```java
package com.ch.spring_boot_mystarters;
public class MyService {
    private String msg;
    public String sayMsg() {
        return "my " + msg;
    }
    public String getMsg() {
        return msg;
    }
    public void setMsg(String msg) {
        this.msg = msg;
    }
}
```

❺ 创建自动配置类 MyAutoConfiguration

在项目 spring_boot_mystarters 的 com.ch.spring_boot_mystarters 包中创建自动配置类 MyAutoConfiguration。在该类中使用@EnableConfigurationProperties 注解开启属性配置类 MyProperties 提供参数；使用@ConditionalOnClass 注解判断类加载器（类路径）中是否存在 MyService 类；使用@ConditionalOnMissingBean 注解判断容器中是否存在 MyService 的 Bean，若不存在，自动配置这个 Bean。自动配置类 MyAutoConfiguration 的核心代码如下：

```java
@Configuration
//开启属性配置类 MyProperties 提供参数
@EnableConfigurationProperties(MyProperties.class)
//类加载器(类路径)中是否存在对应的类
@ConditionalOnClass(MyService.class)
//应用环境中属性是否存在指定的值
@ConditionalOnProperty(prefix = "my", value = "enabled", matchIfMissing = true)
public class MyAutoConfiguration {
    @Autowired
    private MyProperties myProperties;
    @Bean
    //当容器中不存在 MyService 的 Bean 时创建 Bean 对象
    @ConditionalOnMissingBean(MyService.class)
    public MyService myService() {
        MyService myService = new MyService();
        myService.setMsg(myProperties.getMsg());
        return myService;
    }
}
```

❻ 注册配置

在项目 spring_boot_mystarters 的 src/main/resources 目录下新建文件夹 META-INF/spring，并在该文件夹下创建名为 com.ch.spring_boot_mystarters.MyAutoConfiguration.imports 的文件。在此文件中添加如下内容注册自动配置类 MyAutoConfiguration：

```
com.ch.spring_boot_mystarters.MyAutoConfiguration
```

在上述文件中，若有多个自动配置类，换行即可配置另一个自动配置类。

经过上述6个步骤后，自定义 Starters(spring_boot_mystarters)已经完成。可以将 spring_boot_mystarters 安装到 Maven 的本地库，或者将 JAR 包发布到 Maven 的私服上。

下面讲解如何将 spring_boot_mystarters 安装到 Maven 的本地库。具体做法：使用 IntelliJ IDEA 打开 spring_boot_mystarters 项目，进一步打开右侧的 Maven，双击 Lifecycle 中的 install 即可将 spring_boot_mystarters 安装到 Maven 的本地库，如图6.8所示。

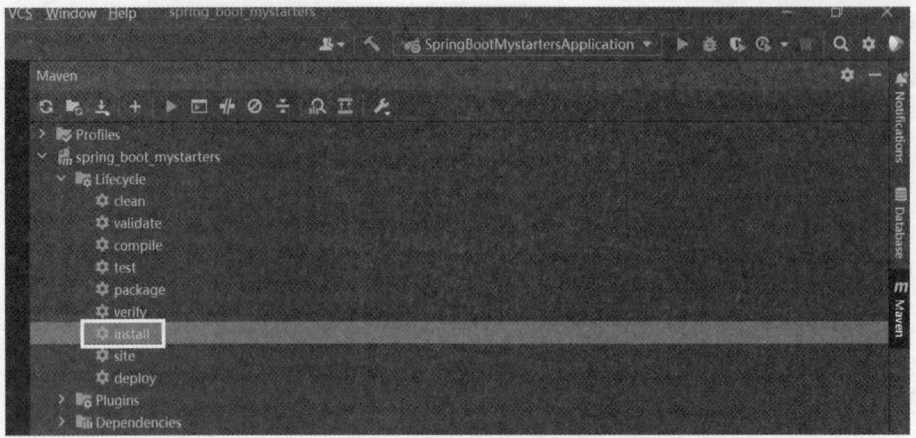

图 6.8　将 spring_boot_mystarters 安装到 Maven 的本地库

将 spring_boot_mystarters 成功安装到 Maven 的本地库后，在项目 spring_boot_mystarters 的 target 目录下可以看到生成了 spring_boot_mystarters-0.0.1-SNAPSHOT.jar 文件。

下面讲解如何在另一个 Spring Boot 应用中使用 spring_boot_mystarters-0.0.1-SNAPSHOT.jar 作为项目依赖。

【例 6-7】　创建 Spring Boot 的 Web 应用 ch6_1，并在 ch6_1 中使用 spring_boot_mystarters 作为项目依赖。

其具体实现步骤如下。

❶ 创建 Spring Boot 的 Web 应用 ch6_1

使用 IntelliJ IDEA 快速创建 Spring Boot 的 Web 应用 ch6_1。

❷ 添加 spring_boot_mystarters 的依赖

在 Web 应用 ch6_1 的 pom.xml 文件中添加 spring_boot_mystarters 的依赖，代码如下：

```xml
<dependency>
    <groupId>com.ch</groupId>
    <artifactId>spring_boot_mystarters</artifactId>
    <version>0.0.1-SNAPSHOT</version>
</dependency>
```

添加依赖后，可以在 Maven 的依赖中查看 spring_boot_mystarters 依赖，如图 6.9 所示。

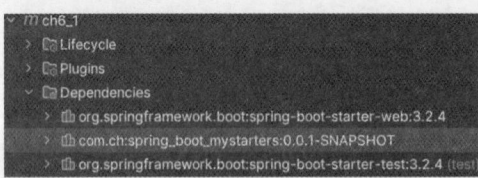

图 6.9　查看 spring_boot_mystarters 依赖

❸ 修改程序入口类 Ch61Application，测试 spring_boot_mystarters

类 Ch61Application 修改后的核心代码如下：

```java
@RestController
//扫描 com.ch.spring_boot_mystarters 包
@SpringBootApplication(scanBasePackages = {"com.ch.spring_boot_mystarters"})
public class Ch61Application {
    @Autowired
    private MyService myService;
```

```
    public static void main(String[] args) {
        SpringApplication.run(Ch61Application.class, args);
    }
    @GetMapping("/testStarters")
    public String index() {
        return myService.sayMsg();
    }
}
```

运行 Ch61Application 应用程序，启动 Web 应用。通过访问 http://localhost:8080/testStarters 测试 spring_boot_mystarters，运行效果如图 6.10 所示。

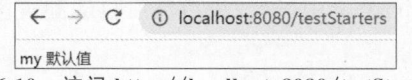

图 6.10 访问 http://localhost:8080/testStarters

这时在 Web 应用 ch6_1 的 application.properties 文件中配置 msg 的内容：

```
my.msg=starter pom
```

然后运行 Ch61Application 应用程序，重新启动 Web 应用。再次访问 http://localhost:8080/testStarters，运行效果如图 6.11 所示。

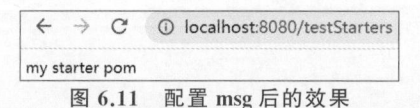

图 6.11 配置 msg 后的效果

另外，可以在 Web 应用 ch6_1 的 application.properties 文件中配置 debug 属性（debug＝true）查看自动配置报告。重新启动 Web 应用，可以在控制台中查看到如图 6.12 所示的自定义 Starters 的自动配置。

```
MyAutoConfiguration matched:
   - @ConditionalOnClass found required class 'com.ch.spring_boot_mystarters.MyService'
   - @ConditionalOnProperty (my.enabled) matched (OnPropertyCondition)

MyAutoConfiguration#myService matched:
   - @ConditionalOnMissingBean (types: com.ch.spring_boot_mystarters.MyService; SearchS
```

图 6.12 查看自动配置报告

本章小结

本章重点讲解了 Spring Boot 的基本配置、读取应用配置、日志配置、自动配置原理以及条件注解。通过本章的学习，读者应掌握 Spring Boot 的核心注解@SpringBootApplication 的基本用法，理解 Spring Boot 的自动配置原理，了解 Spring Boot 的条件注解。重要的是，开发者可以使用 Spring Boot 的自动配置与条件注解贡献自己的 Starters。

习题 6

1. 如何读取 Spring Boot 项目的应用配置？请举例说明。
2. 参考例 6-5，编写 Spring Boot 应用程序 practice6_2。要求：以不同的操作系统作为条件，若在 Windows 系统下运行程序，则输出列表的命令为 dir；若在 Linux 操作系统下运行程序，则输出列表的命令为 ls。

3. 参考例 6-6 与例 6-7，自定义一个 Starter（spring_boot_addstarters）和 Spring Boot 的 Web 应用 practice6_3。在 practice6_3 中，使用 spring_boot_addstarters 计算两个整数的和，通过访问 http://localhost:8080/testAddStarters 返回两个整数的和。在 spring_boot_addstarters 中，首先创建属性配置类 AddProperties（有 Integer 类型的 number1 与 number2 两个属性），在该属性配置类中使用@ConfigurationProperties（prefix="add"）注解设置属性的前缀为 add；然后创建判断依据类 AddService（有 Integer 类型的 number1 与 number2 两个属性），在 AddService 类中提供 add 方法（计算 number1 与 number2 的和）；接着创建自动配置类 AddAutoConfiguration，当类路径中存在 AddService 类时自动配置该类的 Bean，并可以将相应 Bean 的属性在 application.properties 中配置；最后注册自动配置类 AddAutoConfiguration。

第 7 章　Spring Boot 的 Web 开发

学习目的与要求

本章首先介绍 Spring Boot 的 Web 开发支持，然后介绍 Thymeleaf 视图模板引擎技术，最后介绍 Spring Boot 的 Web 开发技术（JSON 数据交互、文件的上传与下载、异常统一处理以及对 JSP 的支持）。通过本章的学习，掌握 Spring Boot 的 Web 开发技术。

本章主要内容

- Thymeleaf 模板引擎
- 用 Spring Boot 处理 JSON 数据
- Spring Boot 文件的上传与下载
- Spring Boot 的异常处理
- Spring Boot 对 JSP 的支持

Web 开发是一种基于 B/S 架构（即浏览器/服务器）的应用软件开发技术，分为前端（用户接口）和后端（业务逻辑和数据），前端的可视化及用户交互由浏览器实现，即通过浏览器作为客户端，实现客户与服务器远程的数据交互。Spring Boot 的 Web 开发内容主要包括内嵌 Servlet 容器和 Spring MVC。

7.1　Spring Boot 的 Web 开发支持

Spring Boot 提供了 spring-boot-starter-web 依赖模块，该依赖模块包含了 Spring Boot 预定义的 Web 开发常用依赖包，为 Web 开发者提供了内嵌的 Servlet 容器（Tomcat）以及 Spring MVC 的依赖。如果开发者希望开发 Spring Boot 的 Web 应用程序，可以在 Spring Boot 项目的 pom.xml 文件中添加如下依赖配置：

```
<dependency>
    <groupId>org.springframework.boot</groupId>
    <artifactId>spring-boot-starter-web</artifactId>
</dependency>
```

Spring Boot 将自动关联 Web 开发的相关依赖，如 Tomcat、spring-webmvc 等，进而对 Web 开发支持，并将相关技术的配置实现自动配置。

另外，开发者也可以使用 IDEA 集成开发工具快速创建 Spring Initializr，在 New Project 对话框中添加 Spring Boot 的 Web 依赖。

7.2　Thymeleaf 模板引擎

在 Spring Boot 的 Web 应用中，建议开发者使用 HTML 完成动态页面。Spring Boot 提供了许多模板引擎，主要包括 FreeMarker、Groovy、Thymeleaf、Velocity 和 Mustache。因为 Thymeleaf 提供了完美的 Spring MVC 支持，所以在 Spring Boot 的 Web 应用中推荐使用 Thymeleaf 作为模板引擎。

Thymeleaf 是一个 Java 类库，是一个 XML/XHTML/HTML5 的模板引擎，能够处理 HTML、XML、JavaScript 以及 CSS，可以作为 MVC Web 应用的 View 层显示数据。

7.2.1　Spring Boot 的 Thymeleaf 支持

在 Spring Boot 1.X 版本中，spring-boot-starter-thymeleaf 依赖包含了 spring-boot-starter-web 模块。但是，在 Spring 5 中 WebFlux 的出现使 Web 应用的解决方案不再唯一。所以，spring-boot-starter-thymeleaf 依赖不再包含 spring-boot-starter-web 模块，需要开发人员自己选择 spring-boot-starter-web 模块依赖。下面通过一个实例讲解如何创建基于 Thymeleaf 模板引擎的 Spring Boot Web 应用 ch7_1。

【例 7-1】　创建基于 Thymeleaf 模板引擎的 Spring Boot Web 应用 ch7_1。
其具体实现步骤如下。

❶ **创建基于 Thymeleaf 模板引擎的 Spring Boot Web 应用 ch7_1**

在 IDEA 中选择 File→New→Project，打开 New Project 对话框；在 New Project 对话框中选择和输入相关信息后，单击 Next 按钮，打开新的 New Project 对话框；在新的 New Project 对话框中选择 Thymeleaf、Lombok 和 Spring Web 依赖，单击 Create 按钮即可创建应用 ch7_1。

❷ **打开项目目录**

创建如图 7.1 所示的基于 Thymeleaf 模板引擎的 Spring Boot Web 应用 ch7_1。

Tymeleaf 模板默认将 JS 脚本、CSS 样式、图片等静态文件放置在 src/main/resources/static 目录下；将视图页面放在 src/main/resources/templates 目录下。

图 7.1　基于 Thymeleaf 模板引擎的 Spring Boot Web 应用 ch7_1

❸ **设置 Web 应用 ch7_1 的上下文路径**

在 ch7_1 的 application.properties 文件中配置如下内容：

```
server.servlet.context-path=/ch7_1
```

❹ **创建控制器类**

创建一个名为 com.ch.ch7_1.controller 的包，并在该包中创建控制器类 TestThymeleafController，核心代码如下：

```
@Controller
public class TestThymeleafController {
    @GetMapping("/")
    public String test(){
        //根据 Tymeleaf 模板，默认将返回 src/main/resources/templates/index.html
        return "index";
    }
}
```

❺ **新建 index.html 页面**

在 src/main/resources/templates 目录下新建 index.html 页面，代码略。

❻ **运行测试**

首先运行 Ch71Application 主类，然后访问 http://localhost:8080/ch7_1/，运行结果如图 7.2 所示。

图 7.2　例 7-1 的运行结果

7.2.2 Thymeleaf 基础语法

❶ 引入 Thymeleaf

首先将 View 层页面文件的 html 标签修改如下：

```html
<html xmlns:th="http://www.thymeleaf.org">
```

然后在 View 层页面文件的其他标签中使用 th:* 动态处理页面。示例代码如下：

```html
<img th:src="'images/' + ${aBook.picture}"/>
```

其中，${aBook.picture}获得数据对象 aBook 的 picture 属性值。

❷ 输出内容

使用 th:text 和 th:utext(不对文本转义，正常输出)将文本内容输出到所在标签的 body 中。假如在国际化资源文件 messages_en_US.properties 中有消息文本"test.myText=<strong>Test International Message</strong>"，那么在页面中可以使用以下两种方式获得消息文本：

```html
<p th:text="#{test.myText}"></p>
<!-- 对文本转义，即输出<strong>Test International Message</strong> -->
<p th:utext="#{test.myText}"></p>
<!-- 不对文本转义，即输出加粗的"Test International Message"-->
```

❸ 基本表达式

1) 变量表达式：${...}

变量表达式用于访问容器上下文环境中的变量，示例代码如下：

```html
<span th:text="${information}">
```

2) 选择变量表达式：*{...}

选择变量表达式计算的是选定的对象(th:object 属性绑定的对象)，示例代码如下：

```html
<div th:object="${session.user}">
    name: <span th: text="*{firstName}"></span><br>
    <!-- firstName 为 user 对象的属性-->
    surname: <span th: text="*{lastName}"></span><br>
    nationality: <span th: text="*{nationality}"></span><br>
</div>
```

3) 信息表达式：#{...}

信息表达式一般用于显示页面静态文本，将可能要根据需求整体变动的静态文本放在 properties 文件中以便维护(如国际化)。信息表达式通常与 th:text 属性一起使用，示例代码如下：

```html
<p th:text="#{test.myText}"></p>
```

❹ 引入 URL

Thymeleaf 模板通过@{...}表达式引入 URL，示例代码如下：

```html
<!-- 默认访问 src/main/resources/static 下的 css 文件夹-->
<link rel="stylesheet" th:href="@{css/bootstrap.min.css}" />
<!--访问相对路径-->
<a th:href="@{/}">去看看</a>
<!--访问绝对路径-->
<a th:href="@{http://www.tup.tsinghua.edu.cn/index.html(param1='传参')}">去清华大学出版社</a>
```

```
<!-- 默认访问 src/main/resources/static 下的'images 文件夹-->
<img th:src="'images/' + ${aBook.picture}"/>
```

❺ 访问 WebContext 对象中的属性

Thymeleaf 模板通过一些专门的表达式从模板的 WebContext 获取请求参数、请求、会话和应用程序中的属性，具体如下：

${xxx}将返回存储在 Thymeleaf 模板上下文中的变量 xxx 或请求 request 作用域中的属性 xxx。

${param.xxx}将返回一个名为 xxx 的请求参数（可能是多个值）。

${session.xxx}将返回一个名为 xxx 的 HttpSession 作用域中的属性。

${application.xxx}将返回一个名为 xxx 的全局 ServletContext 上下文作用域中的属性。

与 EL 表达式一样，使用 ${xxx}获得变量值，使用 ${对象变量名.属性名}获取 JavaBean 属性值。需要注意的是，${}表达式只能在 th 标签内部有效。

❻ 运算符

在 Thymeleaf 模板的表达式中可以使用＋、－、＊、/、％等各种算术运算符，也可以使用＞、＜、＜=、＞=、==、!=等各种逻辑运算符。示例代码如下：

```
<tr th:class="(${row}== 'even')? 'even' : 'odd'">...</tr>
```

❼ 条件判断

1）if 和 unless

设置标签只有在 th：if 条件成立时才显示，th：unless 与 th：if 相反，只有在条件不成立时才显示标签内容。示例代码如下：

```
<a href="success.html" th:if="${user != null}">成功</a>
<a href="success.html" th:unless="${user = null}">成功</a>
```

2）switch 语句

Thymeleaf 模板支持多路选择 switch 语句结构，默认属性 default 可用"＊"表示。示例代码如下：

```
<div th:switch="${user.role}">
    <p th:case="'admin'">User is an administrator</p>
    <p th:case="'teacher'">User is a teacher</p>
    <p th:case="*">User is a student </p>
</div>
```

❽ 循环

1）基本循环

Thymeleaf 模板使用 th：each＝"obj,iterStat：${objList}"标签进行迭代循环，迭代对象可以是 java.util.List、java.util.Map 或数组等。示例代码如下：

```
<!-- 循环取出集合数据 -->
<div class="col-md-4 col-sm-6" th:each="book:${books}">
    <a href="">
        <img th:src="'images/' + ${book.picture}" alt="图书封面" style="height: 180px; width: 40%;"/>
    </a>
    <div class="caption">
        <h4 th:text="${book.bname}"></h4>
```

```
        <p th:text="${book.author}"></p>
        <p th:text="${book.isbn}"></p>
        <p th:text="${book.price}"></p>
        <p th:text="${book.publishing}"></p>
    </div>
</div>
```

2）循环状态的使用

在 th：each 标签中可以使用循环状态变量，该变量具有如下属性。

index：当前迭代对象的索引值（从 0 开始计数）。

count：当前迭代对象的计数（从 1 开始计数）。

size：迭代对象的大小。

current：当前迭代变量。

even/odd：布尔值，当前循环是否为偶数/奇数（从 0 开始计数）。

first：布尔值，当前循环是否为第一个。

last：布尔值，当前循环是否为最后一个。

使用循环状态变量的示例代码如下：

```
<!-- 循环取出集合数据 -->
<div class="col-md-4 col-sm-6" th:each="book,bookStat:${books}">
    <a href="">
        <img th:src="'images/' + ${book.picture}" alt="图书封面" style="height: 180px; width: 40%;"/>
    </a>
    <div class="caption">
        <!--循环状态 bookStat-->
        <h3 th:text="${bookStat.count}"></h3>
        <h4 th:text="${book.bname}"></h4>
        <p th:text="${book.author}"></p>
        <p th:text="${book.isbn}"></p>
        <p th:text="${book.price}"></p>
        <p th:text="${book.publishing}"></p>
    </div>
</div>
```

▶7.2.3　Thymeleaf 的常用属性

通过 7.2.2 节的学习，发现 Thymeleaf 都是通过在 HTML 页面的标签中添加 th：xxx 关键字来实现模板套用，且其属性与 HTML 页面的标签基本类似。其常用属性如下：

❶ th：action

th：action 定义后台控制器路径，类似于＜form＞标签的 action 属性。示例代码如下：

```
<form th:action="@{/login}">...</form>
```

❷ th：each

th：each 用于集合对象的遍历，功能类似于 JSTL 标签＜c：forEach＞。示例代码如下：

```
<div class="col-md-4 col-sm-6" th:each="gtype:${gtypes}">
    <div class="caption">
        <p th:text="${gtype.id}"></p>
        <p th:text="${gtype.typename}"></p>
    </div>
</div>
```

❸ th:field

th:field 常用于表单参数的绑定,通常与 th:object 一起使用。示例代码如下:

```
<form th:action="@{/login}" th:object="${user}">
    <input type="text" value="" th:field="*{username}"></input>
    <input type="text" value="" th:field="*{role}"></input>
</form>
```

❹ th:href

th:href 定义超链接,类似于<a>标签的 href 属性。其 value 形式为@{/logout},示例代码如下:

```
<a th:href="@{/gogo}"></a>
```

❺ th:id

th:id 用于 div 的 id 声明,类似于 HTML 标签中的 id 属性,示例代码如下:

```
<div th:id ="stu+(${rowStat.index}+1) "></div>
```

❻ th:if

th:if 用于条件判断,如果为否,则标签不显示,示例代码如下:

```
<div th:if="${rowStat.index} == 0">... do something ...</div>
```

❼ th:fragment

th:fragment 定义被包含的模板片段,常用于头文件、页尾文件的引入。其常与 th:include、th:replace 一起使用。

假如在 ch7_1 的 src/main/resources/templates 目录下声明模板片段文件 footer.html,代码如下:

```html
<!DOCTYPE html>
<html xmlns:th="http://www.thymeleaf.org">
<head>
<meta charset="UTF-8">
<title>Insert title here</title>
</head>
<body>
  <!-- 声明片段 content -->
  <div th:fragment="content" >
    主体内容
  </div>
  <!-- 声明片段 copy -->
  <div th:fragment="copy" >
    ©清华大学出版社
  </div>
</body>
</html>
```

那么,可以在 ch7_1 的 src/main/resources/templates/index.html 文件中引入模板片段,代码如下:

```html
<!DOCTYPE html>
<html xmlns:th="http://www.thymeleaf.org">
<head>
<meta charset="UTF-8">
<title>Insert title here</title>
```

```
</head>
<body>
    测试 Spring Boot 的 Thymeleaf 支持<br>
    引入主体内容模板片段：
    <div th:include="footer::content"></div>
    引入版权所有模板片段：
    <div th:replace="footer::copy" ></div>
</body>
</html>
```

❽ th:object

th:object 用于表单数据对象的绑定，将表单绑定到后台 controller 的一个 JavaBean 参数。其常与 th:field 一起使用，进行表单数据的绑定。下面通过实例讲解表单提交及数据绑定的实现过程。

【例 7-2】 表单提交及数据绑定的实现过程。

其具体实现步骤如下。

1）创建实体类

在 Web 应用 ch7_1 的 src/main/java 目录下创建 com.ch.ch7_1.model 包，并在该包中创建实体类 LoginBean，代码如下：

```
package com.ch.ch7_1.model;
import lombok.Data;
@Data
public class LoginBean {
    String uname;
    String urole;
}
```

2）创建控制器类

在 Web 应用 ch7_1 的 com.ch.ch7_1.controller 包中创建控制器类 LoginController，核心代码如下：

```
@Controller
public class LoginController {
    @GetMapping("/toLogin")
    public String toLogin(Model model) {
        /* loginBean 与 login.html 页面中的 th:object="${loginBean}"相同 */
        model.addAttribute("loginBean", new LoginBean());
        return "login";
    }
    @PostMapping("/login")
    public String greetingSubmit(@ModelAttribute LoginBean loginBean) {
         /* @ModelAttribute LoginBean loginBean 接收 login.html 页面中的表单数据，并将
loginBean 对象保存到 model 中返回给 result.html 页面显示。 */
        System.out.println("测试提交的数据:" + loginBean.getUname());
        return "result";
    }
}
```

3）创建页面表示层

在 Web 应用 ch7_1 的 src/main/resources/templates 目录下创建页面 login.html 和 result.html。

login.html 页面的代码如下：

```html
<!DOCTYPE html>
<html lang="en" xmlns:th="http://www.thymeleaf.org">
<head>
<meta charset="UTF-8">
<title>Insert title here</title>
</head>
<body>
    <h1>Form</h1>
    <form action="#" th:action="@{/login}" th:object="${loginBean}" method="post">
    <!--th:field="*{uname}"的 uname 与实体类的属性相同，即绑定 loginBean 对象-->
        <p>Uname: <input type="text" th:field="*{uname}" th:placeholder="请输入用户名" /></p>
        <p>Urole: <input type="text" th:field="*{urole}" th:placeholder="请输入角色" /></p>
        <p><input type="submit" value="Submit" /> <input type="reset" value="Reset" /></p>
    </form>
</body>
</html>
```

result.html 页面的核心代码如下：

```html
<body>
    <h1>Result</h1>
    <p th:text="'Uname: ' + ${loginBean.uname}" />
    <p th:text="'Urole: ' + ${loginBean.urole}" />
    <a href="toLogin">继续提交</a>
</body>
```

4）运行

首先运行 Ch71Application 主类，然后访问 http://localhost:8080/ch7_1/toLogin，运行结果如图 7.3 所示。

在图 7.3 的文本框中输入信息后，单击 Submit 按钮，打开如图 7.4 所示的页面。

图 7.3　login.html 页面的运行结果　　　　图 7.4　result.html 页面的运行结果

❾ th:src

th:src 用于外部资源的引入，类似于＜script＞标签的 src 属性。示例代码如下：

```html
<img th:src="'images/' + ${aBook.picture}" />
```

❿ th:text

th:text 用于文本的显示，将文本内容显示到所在标签的 body 中。示例代码如下：

```html
<td th:text="${username}"></td>
```

⓫ th:value

th:value 用于标签的赋值，类似于标签的 value 属性。示例代码如下：

```
<option th:value="Adult">Adult</option>
<input type="hidden" th:value="${msg}" />
```

⑫ th:style

th:style 用于修改标签 style,示例代码如下:

```
<span th:style="'display:' + @{(${myVar} ? 'none' : 'inline-block')}"> myVar 是一个变量
</span>
```

⑬ th:onclick

th:onclick 用于修改点击事件,示例代码如下:

```
<button th:onclick="'getCollect()'"></button>
```

▶7.2.4 用 Spring Boot 与 Thymeleaf 实现页面信息的国际化

在 Spring Boot 的 Web 应用中实现页面信息的国际化非常简单,下面通过实例讲解国际化信息的实现过程。

【例 7-3】 国际化信息的实现过程。

其具体实现步骤如下。

❶ 编写国际化资源属性文件

1) 编写管理员模块的国际化信息

在 ch7_1 的 src/main/resources 目录下创建 i18n/admin 文件夹,并在该文件夹下创建 adminMessages.properties、adminMessages_en_US.properties 和 adminMessages_zh_CN.properties 资源属性文件。adminMessages.properties 表示默认加载的信息;adminMessages_en_US.properties 表示英文信息(en 代表语言代码,US 代表国家地区);adminMessages_zh_CN.properties 表示中文信息。

adminMessages.properties 的内容如下:

```
test.admin=\u6D4B\u8BD5\u540E\u53F0
admin=\u540E\u53F0\u9875\u9762
```

adminMessages_en_US.properties 的内容如下:

```
test.admin=test admin
admin=admin
```

adminMessages_zh_CN.properties 的内容如下:

```
test.admin=\u6D4B\u8BD5\u540E\u53F0
admin=\u540E\u53F0\u9875\u9762
```

2) 编写用户模块的国际化信息

在 ch7_1 的 src/main/resources 目录下创建 i18n/before 文件夹,并在该文件夹下创建 beforeMessages.properties、beforeMessages_en_US.properties 和 beforeMessages_zh_CN.properties 资源属性文件。

beforeMessages.properties 的内容如下:

```
test.before=\u6D4B\u8BD5\u524D\u53F0
before=\u524D\u53F0\u9875\u9762
```

beforeMessages_en_US.properties 的内容如下:

```
test.before=test before
before=before
```

beforeMessages_zh_CN.properties 的内容如下：

```
test.before=\u6D4B\u8BD5\u524D\u53F0
before=\u524D\u53F0\u9875\u9762
```

3）编写公共模块的国际化信息

在 ch7_1 的 src/main/resources 目录下创建 i18n/common 文件夹，并在该文件夹下创建 commonMessages.properties、commonMessages_en_US.properties 和 commonMessages_zh_CN.properties 资源属性文件。

commonMessages.properties 的内容如下：

```
chinese.key=\u4E2D\u6587\u7248
english.key=\u82F1\u6587\u7248
return=\u8FD4\u56DE\u9996\u9875
```

commonMessages_en_US.properties 的内容如下：

```
chinese.key=chinese
english.key=english
return=return
```

commonMessages_zh_CN.properties 的内容如下：

```
chinese.key=\u4E2D\u6587\u7248
english.key=\u82F1\u6587\u7248
return=\u8FD4\u56DE\u9996\u9875
```

❷ 添加配置文件内容，引入资源属性文件

在 ch7_1 应用的配置文件中添加如下内容，引入资源属性文件。

```
spring.messages.basename=i18n/admin/adminMessages,i18n/before/beforeMessages,
i18n/common/commonMessages
```

❸ 重写 localeResolver 方法配置语言区域

在 ch7_1 应用的 com.ch.ch7_1 包中创建配置类 LocaleConfig，该配置类实现 WebMvcConfigurer 接口，并配置语言区域。LocaleConfig 的核心代码如下：

```
@Configuration
@EnableAutoConfiguration
public class LocaleConfig implements WebMvcConfigurer {
    /**
     * 根据用户本次会话过程中的语义设定语言区域(如用户进入首页时选择的语言种类)
     */
    @Bean
    public LocaleResolver localeResolver() {
        SessionLocaleResolver slr = new SessionLocaleResolver();
        //默认语言
        slr.setDefaultLocale(Locale.CHINA);
        return slr;
    }
    /**
     * 在使用 SessionLocaleResolver 存储语言区域时，必须配置 localeChangeInterceptor 拦截器
     */
    @Bean
```

```
    public LocaleChangeInterceptor localeChangeInterceptor() {
        LocaleChangeInterceptor lci = new LocaleChangeInterceptor();
        //选择语言的参数名
        lci.setParamName("locale");
        return lci;
    }
    /**
     * 注册拦截器
     */
    @Override
    public void addInterceptors(InterceptorRegistry registry) {
        registry.addInterceptor(localeChangeInterceptor());
    }
}
```

❹ 创建控制器类 I18nTestController

在 ch7_1 应用的 com.ch.ch7_1.controller 包中创建控制器类 I18nTestController,核心代码如下:

```
@Controller
@RequestMapping("/i18n")
public class I18nTestController {
    @GetMapping("/first")
    public String testI18n(){
        return "/i18n/first";
    }
    @GetMapping("/admin")
    public String admin(){
        return "/i18n/admin";
    }
    @GetMapping("/before")
    public String before(){
        return "/i18n/before";
    }
}
```

❺ 创建视图页面,并获得国际化信息

在 ch7_1 应用的 src/main/resources/templates 目录下创建文件夹 i18n,在该文件夹中创建 admin.html、before.html 和 first.html 视图页面,并在这些视图页面中使用 th:text="#{xxx}" 获得国际化信息。

admin.html 的核心代码如下:

```
<body>
    <span th:text="#{admin}"></span><br>
    <a th:href="@{/i18n/first}" th:text="#{return}"></a>
</body>
```

before.html 的核心代码如下:

```
<body>
    <span th:text="#{before}"></span><br>
    <a th:href="@{/i18n/first}" th:text="#{return}"></a>
</body>
```

first.html 的核心代码如下:

```
<body>
    <a th:href="@{/i18n/first(locale='zh_CN')}" th:text="#{chinese.key}"></a>
```

```html
        <a th:href="@{/i18n/first(locale='en_US')}" th:text="#{english.key}"></a>
        <br>
        <a th:href="@{/i18n/admin}" th:text="#{test.admin}"></a><br>
        <a th:href="@{/i18n/before}" th:text="#{test.before}"></a><br>
</body>
```

❻ 运行

首先运行 Ch71Application 主类，然后访问 http://localhost:8080/ch7_1/i18n/first，运行效果如图 7.5 所示。

单击图 7.5 中的"英文版"，打开如图 7.6 所示的页面效果。

图 7.5　程序入口页面 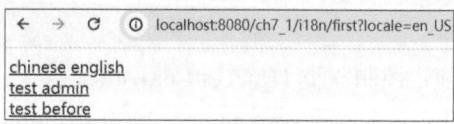　　　　　图 7.6　英文版效果

7.2.5　Spring Boot 与 Thymeleaf 的表单验证

JSR 是 Java Specification Requests 的缩写，意思是 Java 规范提案。关于数据校验，最新的是 JSR380，也就是 Bean Validation 2.0。

Bean Validation 是一个通过配置注解来验证数据的框架，它包含 Bean Validation API（规范）和 Hibernate Validator（实现）两部分。

Bean Validation 是 Java 定义的一套基于注解/XML 的数据校验规范，目前已经从 JSR 303 的 Bean Validation 1.0 版本升级到 JSR 349 的 Bean Validation 1.1 版本，再到 JSR 380 的 Bean Validation 2.0 版本。

2018 年，Oracle（甲骨文）公司决定把 Java EE 移交给开源组织 Eclipse 基金会，正式改名为 Jakarta EE。Bean Validation 也就自然命名为 Jakarta Bean Validation。在编写本书时，其最新版本是 Jakarta Bean Validation 3.0。

对于 Jakarta Bean Validation 验证，可以使用它实现 Hibernate Validator，注意 Hibernate Validator 和 Hibernate 无关，只是使用 Hibernate Validator 进行数据验证。

本节使用 Hibernate Validator 对表单进行验证，因为 spring-boot-starter-web 不再依赖 hibernate-validator 的 JAR 包，所以在 Spring Boot 的 Web 应用中使用 Hibernate Validator 对表单进行验证时，需要加载 Hibernate Validator 所依赖的 JAR 包，示例代码如下：

```xml
<dependency>
    <groupId>org.hibernate.validator</groupId>
    <artifactId>hibernate-validator</artifactId>
</dependency>
```

在使用 Hibernate Validator 验证表单时，需要使用它的标注类型在实体模型的属性上嵌入约束，标注类型具体如下。

❶ 空检查

@Null：验证对象是否为 null。

@NotNull：验证对象是否不为 null，无法检查长度为 0 的字符串。

@NotBlank：检查约束字符串是不是 null，以及被 trim 后的长度是否大于 0，只针对字符串，且应去掉前后空格。

@NotEmpty：检查约束元素是否为 null 或者是 empty。

示例如下：

```
@NotBlank(message="{goods.gname.required}")    //goods.gname.required 为属性文件的错误代码
private String gname;
```

❷ **boolean 检查**

@AssertTrue：验证 boolean 属性是否为 true。

@AssertFalse：验证 boolean 属性是否为 false。

示例如下：

```
@AssertTrue
private boolean isLogin;
```

❸ **长度检查**

@Size(min＝，max＝)：验证对象（Array、Collection、Map、String）的长度是否在给定的范围之内。

@Length(min＝，max＝)：验证字符串的长度是否在给定的范围之内。

示例如下：

```
@Length(min=1,max=100)
private String gdescription;
```

❹ **日期检查**

@Past：验证 Date 和 Calendar 对象是否在当前时间之前。

@Future：验证 Date 和 Calendar 对象是否在当前时间之后。

@Pattern：验证 String 对象是否符合正则表达式的规则。

示例如下：

```
@Past(message="{gdate.invalid}")
private Date gdate;
```

❺ **数值检查**

@Min：验证 Number 和 String 对象是否大于或等于指定的值。

@Max：验证 Number 和 String 对象是否小于或等于指定的值。

@DecimalMax：被标注的值必须不大于约束中指定的最大值，这个约束的参数是一个通过 BigDecimal 定义的最大值的字符串表示，小数存在精度。

@DecimalMin：被标注的值必须不小于约束中指定的最小值，这个约束的参数是一个通过 BigDecimal 定义的最小值的字符串表示，小数存在精度。

@Digits：验证 Number 和 String 的构成是否合法。

@Digits(integer＝,fraction＝)：验证字符串是否符合指定格式的数字，integer 指定整数的精度，fraction 指定小数的精度。

@Range(min＝，max＝)：检查数字是否介于 min 和 max 之间。

@Valid：对关联对象进行校验，如果关联对象是一个集合或者数组，那么对其中的元素进行校验；如果是一个 map，则对其中的值部分进行校验。

@CreditCardNumber：信用卡验证。

@Email：验证是否为邮件地址，如果为 null，不进行验证，通过验证。

示例如下:

```
@Range(min=0,max=100,message="{gprice.invalid}")
private double gprice;
```

下面通过实例讲解使用 Hibernate Validator 验证表单的过程。

【例 7-4】 使用 Hibernate Validator 验证表单的过程。

其具体实现步骤如下。

❶ 创建表单实体模型

在 ch7_1 应用的 com.ch.ch7_1.model 包中创建表单实体模型类 Goods。在该类中使用 Jakarta Bean Validation 的标注类型对属性进行分组验证,核心代码如下:

```
@Data
public class Goods {
    //add 组
    public interface Add{}
    //update 组
    public interface Update{}
    @NotBlank(groups = {Add.class, Update.class}, message="商品名必须输入")
    @Length(groups = {Add.class, Update.class}, min=1, max=5, message="商品名的长度为 1~5")
    private String gname;
    @Range(groups = {Add.class}, min=0, max=100, message="商品价格为 0~100")
    private double gprice;
}
```

❷ 创建控制器类

在 ch7_1 应用的 com.ch.ch7_1.controller 包中创建控制器类 TestValidatorController。在该类中有两个处理方法,一个是界面初始化处理方法 testValidator,另一个是添加请求处理方法 add,在 add 方法中使用@Validated 注解使验证生效。核心代码如下:

```
@Controller
public class TestValidatorController {
    @GetMapping("/testValidator")
    public String testValidator(@ModelAttribute("goodsInfo") Goods goods){
        goods.setGname("商品名初始化");
        goods.setGprice(0.0);
        return "testValidator";
    }
    @PostMapping(value="/add")
    //@Validated({Goods.Add.class})验证 add 组,可以同时验证多组@Validated({Goods.Add.class,
Goods.Update.class})
    public String add(@Validated({Goods.Add.class}) @ModelAttribute("goodsInfo") Goods
goods,BindingResult rs){
        //@ModelAttribute("goodsInfo")与 th:object="${goodsInfo}"相对应
        if(rs.hasErrors()){                                                    //验证失败
            return "testValidator";
        }
        //验证成功,可以到任意地方,在这里直接到 testValidator 界面
        return "testValidator";
    }
}
```

❸ 创建视图页面

在 ch7_1 应用的 src/main/resources/templates 目录下创建视图页面 testValidator.html。在该视图页面中直接读取 ModelAttribute 里面注入的数据,然后通过 th:errors="*{xxx}"获得验证错误信息。核心代码如下:

```html
<body>
    <h2>通过 th:object 访问对象的方式</h2>
    <div th:object="${goodsInfo}">
        <p th:text="* {gname}"></p>
        <p th:text="* {gprice}"></p>
    </div>
    <h1>表单提交</h1>
    <!-- 表单提交用户信息,注意表单参数的设置,直接是 * {} -->
    <form   th:action="@{/add}" th:object="${goodsInfo}" method="post">
        <div><span>商品名</span><input type="text" th:field="* {gname}"/><span th:errors="* {gname}"></span></div>
        <div><span>商品价格</span><input type="text" th:field="* {gprice}"/><span th:errors="* {gprice}"></span></div>
        <input type="submit" />
    </form>
</body>
```

❹ 运行

首先运行 Ch71Application 主类,然后访问 http://localhost:8080/ch7_1/testValidator,测试效果如图 7.7 所示。

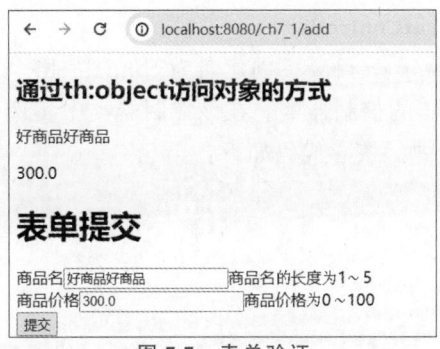

图 7.7　表单验证

▶7.2.6　基于 Thymeleaf 与 BootStrap 的 Web 开发实例

在本书后续的 Web 应用开发中尽量使用前端开发工具包 BootStrap、JavaScript 框架 jQuery 和 Spring MVC 框架。对于 BootStrap 和 jQuery 的相关知识,请读者自行学习。下面通过一个实例讲解如何创建基于 Thymeleaf 模板引擎的 Spring Boot Web 应用 ch7_2。

【例 7-5】 创建基于 Thymeleaf 模板引擎的 Spring Boot Web 应用 ch7_2。

其具体实现步骤如下。

❶ 创建基于 **Thymeleaf** 模板引擎的 **Spring Boot Web** 应用 **ch7_2**

在 IDEA 中创建基于 Thymeleaf、Lombok 和 Spring Web 依赖的应用 ch7_2。

❷ 设置 **Web** 应用 **ch7_2** 的上下文路径

在 ch7_2 的 application.properties 文件中配置如下内容:

```
server.servlet.context-path=/ch7_2
```

❸ 创建实体类 **Book**

在应用 ch7_2 的 src/main/java 目录下创建名为 com.ch.ch7_2.model 的包,并在该包中创建名为 Book 的实体类。该实体类用在模板页面中展示数据,具体代码如下:

```
package com.ch.ch7_2.model;
import lombok.Data;
@Data
public class Book {
    String isbn;
    Double price;
    String bname;
    String publishing;
    String author;
    String picture;
    public Book(String isbn, Double price, String bname, String publishing, String author, String picture) {
        super();
        this.isbn = isbn;
        this.price = price;
        this.bname = bname;
        this.publishing = publishing;
        this.author = author;
        this.picture = picture;
    }
}
```

❹ 创建控制器类 ThymeleafController

在应用 ch7_2 的 src/main/java 目录下创建名为 com.ch.ch7_2.controller 的包，并在该包中创建名为 ThymeleafController 的控制器类。在该控制器类中实例化 Book 类的多个对象，并保存到集合 ArrayList＜Book＞中。核心代码如下：

```
@Controller
public class ThymeleafController {
    @GetMapping("/")
    public String index(Model model) {
        Book teacherGeng = new Book(
                "9787302629443",59.8,
                "Vue.js 3.x 从入门到实战(微课视频版)",
                "清华大学出版社","陈恒","093127-01.jpg"
        );
        List<Book> chenHeng = new ArrayList<Book>();
        Book b1 = new Book(
                "9787302529118", 69.8,
                "Java Web 开发从入门到实战(微课版)",
                "清华大学出版社","陈恒","082526-01.jpg"
        );
        chenHeng.add(b1);
        Book b2 = new Book(
                "9787302651192", 59.8,
                "Java EE 框架整合开发入门到实战——Spring+Spring MVC+MyBatis(第 2 版·微课视频)",
                "清华大学出版社","陈恒","101137-01.jpg");
        chenHeng.add(b2);
        model.addAttribute("aBook", teacherGeng);
        model.addAttribute("books", chenHeng);
        return "index";
    }
}
```

❺ 整理脚本、样式等静态文件

JS 脚本、CSS 样式、图片等静态文件默认放置在应用 ch7_2 的 src/main/resources/static 目录下。

❻ 创建视图页面

Tymeleaf 模板默认将视图页面放在 src/main/resources/templates 目录下,因此在 src/main/resources/templates 目录下新建 HTML 页面文件 index.html。在该页面中使用 Tymeleaf 模板显示控制器类 TestThymeleafController 中的 model 对象数据,核心代码如下:

```html
<body>
    <!-- 面板 -->
    <div class="panel panel-primary">
        <!-- 面板头信息 -->
        <div class="panel-heading">
            <!-- 面板标题 -->
            <h3 class="panel-title">第一个基于 Thymeleaf 与 BootStrap 的 Spring Boot Web 应用</h3>
        </div>
    </div>
    <!-- 容器 -->
    <div class="container">
        <div>
            <h4>图书列表</h4>
        </div>
        <div class="row">
            <!-- col-md 针对桌面显示器, col-sm 针对平板 -->
            <div class="col-md-4 col-sm-6">
                <a href="">
                    <img th:src="'images/' + ${aBook.picture}" alt="图书封面"
                         style="height: 180px; width: 40%;"/>
                </a>
                <!-- caption 容器中放置其他基本信息,例如标题、文本描述等 -->
                <div class="caption">
                    <h4 th:text="${aBook.bname}"></h4>
                    <p th:text="${aBook.author}"></p>
                    <p th:text="${aBook.isbn}"></p>
                    <p th:text="${aBook.price}"></p>
                    <p th:text="${aBook.publishing}"></p>
                </div>
            </div>
            <!-- 循环取出集合数据 -->
            <div class="col-md-4 col-sm-6" th:each="book:${books}">
                <a href="">
                    <img th:src="'images/' + ${book.picture}" alt="图书封面"
                         style="height: 180px; width: 40%;"/>
                </a>
                <div class="caption">
                    <h4 th:text="${book.bname}"></h4>
                    <p th:text="${book.author}"></p>
                    <p th:text="${book.isbn}"></p>
                    <p th:text="${book.price}"></p>
                    <p th:text="${book.publishing}"></p>
                </div>
            </div>
        </div>
    </div>
</body>
```

❼ 运行

首先运行 Ch72Application 主类,然后访问 http://localhost:8080/ch7_2/,运行效果如图 7.8 所示。

图 7.8　例 7-5 的运行效果

7.3　用 Spring Boot 处理 JSON 数据

7.3.1　JSON 数据结构

JSON(JavaScript Object Notation,JS 对象标记)是一种轻量级的数据交换格式。与 XML 一样,JSON 也是基于纯文本的数据格式,有对象结构和数组结构两种。

❶ 对象结构

对象结构以"{"开始,以"}"结束,中间部分由 0 个或多个以英文","分隔的 key-value 对构成,key 和 value 之间以英文":"分隔。对象结构的语法如下:

```
{
    key1:value1,
    key2:value2,
    …
}
```

其中,key 必须为 String 类型,value 可以是 String、Number、Object、Array 等数据类型。例如,一个 Person 对象包含姓名、密码、年龄等信息,使用 JSON 的表示形式如下:

```
{
    "pname":"陈恒",
    "password":"123456",
    "page":40
}
```

❷ 数组结构

数组结构以"["开始,以"]"结束,中间部分由 0 个或多个以英文","分隔的值的列表组成。数组结构的语法结构如下:

```
[
    value1,
    value2,
    …
]
```

上述两种（对象、数组）数据结构也可以分别组合，构成更为复杂的数据结构。例如，一个 student 对象包含 sno、sname、hobby 和 college 对象，其 JSON 的表示形式如下：

```json
{
    "sno":"201802228888",
    "sname":"张三",
    "hobby":["篮球","足球"],
    "college":{
        "cname":"清华大学",
        "city":"北京"
    }
}
```

7.3.2 JSON 数据转换

在 Spring Boot 的 Web 应用中内置了 JSON 数据的解析功能，默认使用 Jackson 自动完成解析（不需要加载 Jackson 依赖包），当控制器返回一个 Java 对象或集合数据时，Spring Boot 自动将其转换成 JSON 数据，使用起来方便、简洁。

在 Spring Boot 处理 JSON 数据时，需要用到两个重要的 JSON 格式转换注解，分别是 @RequestBody 和 @ResponseBody。

- @RequestBody：用于将请求体中的数据绑定到方法的形参中，该注解应用在方法的形参上。
- @ResponseBody：用于直接返回 JSON 对象，该注解应用在方法上。

下面通过一个实例讲解 Spring Boot 处理 JSON 数据的过程，该实例针对返回实体对象、ArrayList 集合、Map＜String，Object＞集合以及 List＜Map＜String，Object＞＞集合分别处理。

【例 7-6】 Spring Boot 处理 JSON 数据的过程。

其具体实现步骤如下。

❶ 创建实体类

在 ch7_2 应用的 com.ch.ch7_2.model 包中创建实体类 Person，具体代码如下：

```java
package com.ch.ch7_2.model;
import lombok.Data;
@Data
public class Person {
    private String pname;
    private String password;
    private Integer page;
}
```

❷ 创建视图页面

在 ch7_2 应用的 src/main/resources/templates 目录下创建视图页面 input.html。在 input.html 页面中引入 jQuery 框架，并使用它的 ajax 方法进行异步请求。核心代码如下：

```html
<link rel="stylesheet" th:href="@{css/bootstrap.min.css}" />
<script type="text/javascript" th:src="@{js/jquery-3.6.0.min.js}"></script>
<script type="text/javascript">
    function testJson() {
        //获取输入的值 pname 为 id
        var pname = $("#pname").val();
        var password = $("#password").val();
        var page = $("#page").val();
```

```javascript
            alert(password);
            $.ajax({
                //发送请求的 URL 字符串
                url: "testJson",
                //定义回调响应的数据格式为 JSON 字符串,该属性可以省略
                dataType: "json",
                //请求类型
                type: "post",
                //定义发送请求的数据格式为 JSON 字符串
                contentType: "application/json",
                //data 表示发送的数据
                data: JSON.stringify({pname:pname,password:password,page:page}),
                //成功响应的结果
                success: function(data){
                    if(data != null){
                        //返回一个 Person 对象
                        //alert("输入的用户名:" + data.pname + ",密码:" + data.password +
                        //",年龄:" +data.page);
                        //ArrayList<Person>对象
                        /**for(var i = 0; i < data.length; i++){
                            alert(data[i].pname);
                        }**/
                        //返回一个 Map<String, Object>对象
                        //alert(data.pname);                                    //pname 为 key
                        //返回一个 List<Map<String, Object>>对象
                        for(var i = 0; i < data.length; i++){
                            alert(data[i].pname);
                        }
                    }
                },
                //请求出错
                error:function(){
                    alert("数据发送失败");
                }
            });
        }
    </script>
</head>
<body>
    <div class="panel panel-primary">
        <div class="panel-heading">
            <h3 class="panel-title">处理 JSON 数据</h3>
        </div>
    </div>
    <div class="container">
        <div>
        <h4>添加用户</h4>
        </div>
        <div class="row">
            <div class="col-md-6 col-sm-6">
                <form class="form-horizontal" action="">
                    <div class="form-group">
                        <div class="input-group col-md-6">
                            <span class="input-group-addon">
                                <i class="glyphicon glyphicon-pencil"></i>
                            </span>
                            <input class="form-control" type="text"
                              id="pname" th:placeholder="请输入用户名"/>
                        </div>
                    </div>
                    <div class="form-group">
```

```html
                    <div class="input-group col-md-6">
                        <span class="input-group-addon">
                            <i class="glyphicon glyphicon-pencil"></i>
                        </span>
                        <input class="form-control" type="password"
                         id="password" th:placeholder="请输入密码"/>
                    </div>
                </div>
                <div class="form-group">
                    <div class="input-group col-md-6">
                        <span class="input-group-addon">
                            <i class="glyphicon glyphicon-pencil"></i>
                        </span>
                        <input class="form-control" type="text"
                         id="page" th:placeholder="请输入年龄"/>
                    </div>
                </div>
                <div class="form-group">
                    <div class="col-md-6">
                        <div class="btn-group btn-group-justified">
                            <div class="btn-group">
                                <button type="button" onclick="testJson()"
                                    class="btn btn-success">
                                    <span class="glyphicon glyphicon-share"></span>
                                     测试
                                </button>
                            </div>
                        </div>
                    </div>
                </div>
            </form>
        </div>
    </div>
</div>
</body>
```

❸ 创建控制器类

在 ch7_2 应用的 com.ch.ch7_2.controller 包中创建控制器类 TestJsonController。在该类中有两个处理方法，一个是界面导航方法 input，另一个是接收页面请求的方法。核心代码如下：

```
@Controller
public class TestJsonController {
    //进入视图页面
    @GetMapping("/input")
    public String input() {
        return "input";
    }
    //接收页面请求的 JSON 数据
    @RequestMapping("/testJson")
    @ResponseBody
    /* @RestController 注解相当于@ResponseBody+ @Controller 的作用。
    ①如果只是使用@RestController 注解 Controller，则 Controller 中的方法无法返回 JSP 页面或者
HTML，返回的内容就是 return 的内容。
    ②如果需要返回到指定页面，则需要用@Controller 注解。如果需要返回 JSON、XML 或自定义 mediaType
内容到页面，则需要在对应的方法上加@ResponseBody 注解。
    */
    public List<Map<String, Object>> testJson(@RequestBody Person user) {
        //打印接收的 JSON 格式数据
        System.out.println("pname=" + user.getPname() +
            ", password=" + user.getPassword() + ",page=" + user.getPage());
```

```
        //返回 Person 对象
        //return user;
        /**ArrayList<Person> allp = new ArrayList<Person>();
        Person p1 = new Person();
        p1.setPname("陈恒 1");
        p1.setPassword("123456");
        p1.setPage(80);
        allp.add(p1);
        Person p2 = new Person();
        p2.setPname("陈恒 2");
        p2.setPassword("78910");
        p2.setPage(90);
        allp.add(p2);
        //返回 ArrayList<Person>对象
        return allp;
        **/
        Map<String, Object> map = new HashMap<String, Object>();
        map.put("pname", "陈恒 2");
        map.put("password", "123456");
        map.put("page", 25);
        //返回一个 Map<String, Object>对象
        //return map;
        //返回一个 List<Map<String, Object>>对象
        List<Map<String, Object>> allp = new ArrayList<Map<String, Object>>();
        allp.add(map);
        Map<String, Object> map1 = new HashMap<String, Object>();
        map1.put("pname", "陈恒 3");
        map1.put("password", "54321");
        map1.put("page", 55);
        allp.add(map1);
        return allp;
    }
}
```

❹ 运行

首先运行 Ch72Application 主类，然后访问 http://localhost:8080/ch7_2/input，运行效果如图 7.9 所示。

图 7.9　input.html 的运行效果

7.4　Spring Boot 文件的上传与下载

上传与下载文件是 Web 应用开发中常用的功能之一。本节将讲解如何在 Spring Boot 的 Web 应用开发中实现文件的上传与下载。

Spring Boot 的 spring-boot-starter-web 已经集成了 Spring MVC（基于 Spring MVC 的文件

第 7 章 Spring Boot 的 Web 开发

的上传见本书中的 2.6 节），所以使用 Spring Boot 实现文件的上传更加便捷。

下面通过一个实例讲解 Spring Boot 文件上传与下载的实现过程。

【例 7-7】 Spring Boot 文件的上传与下载。

其具体实现步骤如下。

❶ 设置上传文件大小的限制

在 Web 应用 ch7_2 的配置文件 application.properties 中添加如下配置限制上传文件的大小。

```
#在上传文件时,默认单个上传文件的大小是 1MB,max-file-size 设置单个上传文件的大小
spring.servlet.multipart.max-file-size=50MB
#默认总文件大小是 10MB,max-request-size 设置总上传文件的大小
spring.servlet.multipart.max-request-size=500MB
```

❷ 创建选择文件视图页面

在 ch7_2 应用的 src/main/resources/templates 目录下创建选择文件视图页面 uploadFile.html。在该页面中有一个 enctype 属性值为 multipart/form-data 的 form 表单，核心代码如下：

```html
<body>
<div class="panel panel-primary">
    <div class="panel-heading">
        <h3 class="panel-title">文件上传示例</h3>
    </div>
</div>
<div class="container">
    <div class="row">
        <div class="col-md-6 col-sm-6">
            <form class="form-horizontal" action="upload"
            method="post" enctype="multipart/form-data">
                <div class="form-group">
                    <div class="input-group col-md-6">
                        <span class="input-group-addon">
                            <i class="glyphicon glyphicon-pencil"></i>
                        </span>
                        <input class="form-control" type="text"
                         name="description" th:placeholder="文件描述"/>
                    </div>
                </div>
                <div class="form-group">
                    <div class="input-group col-md-6">
                        <span class="input-group-addon">
                            <i class="glyphicon glyphicon-search"></i>
                        </span>
                        <input class="form-control" type="file"
                         name="myfile" th:placeholder="请选择文件"/>
                    </div>
                </div>
                <div class="form-group">
                    <div class="col-md-6">
                        <div class="btn-group btn-group-justified">
                            <div class="btn-group">
                                <button type="submit" class="btn btn-success">
                                    <span class="glyphicon glyphicon-share"></span>
                                     上传文件
                                </button>
                            </div>
                        </div>
                    </div>
```

```
                </div>
            </form>
        </div>
    </div>
</body>
```

❸ 创建实体类

在 ch7_2 应用的 com.ch.ch7_2.model 包中创建实体类 MyFile 封装文件对象,具体代码如下:

```java
package com.ch.ch7_2.model;
import lombok.Data;
@Data
public class MyFile {
    int fno;
    String fname;
}
```

❹ 创建控制器类

在 ch7_2 应用的 com.ch.ch7_2.controller 包中创建控制器类 TestFileUpload。在该类中有 4 个处理方法,分别是界面导航方法 uploadFile、实现文件上传的 upload 方法、显示将要被下载文件的 showDownLoad 方法、实现下载功能的 download 方法。核心代码如下:

```java
@Controller
public class TestFileUpload {
    //进入文件选择页面
    @RequestMapping("/uploadFile")
    public String uploadFile() {
        return "uploadFile";
    }
    //上传文件自动绑定到 MultipartFile 对象中,在这里使用处理方法的形参接收请求参数
    @RequestMapping("/upload")
    public String upload(HttpServletRequest request, @RequestParam("description") String description,
            @RequestParam("myfile") MultipartFile myfile) throws IllegalStateException, IOException {
        //如果选择了上传文件,则将文件上传到指定的目录 uploadFiles
        if(!myfile.isEmpty()) {
            //上传文件的路径
            String path = request.getServletContext().getRealPath("/uploadFiles/");
            //获得上传文件的原名
            String fileName = myfile.getOriginalFilename();
            File filePath = new File(path + File.separator + fileName);
            //如果文件目录不存在,创建目录
            if(!filePath.getParentFile().exists()) {
                filePath.getParentFile().mkdirs();
            }
            //将上传文件保存到一个目标文件中
            myfile.transferTo(filePath);
        }
        //转发到一个请求处理方法,查询将要下载的文件
        return "forward:/showDownLoad";
    }
    //显示要下载的文件
    @RequestMapping("/showDownLoad")
    public String showDownLoad(HttpServletRequest request, Model model) {
        String path = request.getServletContext().getRealPath("/uploadFiles/");
```

```java
        File fileDir = new File(path);
        //从指定目录获得文件列表
        File filesList[] = fileDir.listFiles();
        ArrayList<MyFile> myList = new ArrayList<MyFile>();
        for (int i = 0; i < filesList.length; i++) {
            MyFile mf = new MyFile();
            mf.setFno(i + 1);
            mf.setFname(filesList[i].getName());
            myList.add(mf);
        }
        model.addAttribute("filesList", myList);
        return "showFile";
    }
    //实现下载功能
    @RequestMapping("/download")
    public void download(
            HttpServletRequest request, HttpServletResponse response,
            @RequestParam("filename") String filename) throws IOException {
        //下载文件的路径
        String path = request.getServletContext().getRealPath("/uploadFiles/");
        //创建将要下载的文件对象
        File downFile = new File(path + File.separator + filename);
        //使用 URLEncoder.encode 对文件名进行编码
        filename = URLEncoder.encode(filename,"UTF-8");
        response.setHeader("Content-Type", "application/x-msdownload");
        response.setHeader("Content-Disposition", "attachment; filename=" + filename);
        FileInputStream in = null;           //输入流
        ServletOutputStream out = null;      //输出流
        //读入文件
        in = new FileInputStream(downFile);
        //得到响应对象的输出流,用于向客户端输出二进制数据
        out = response.getOutputStream();
        out.flush();
        int aRead = 0;
        byte b[] = new byte[1024];
        while ((aRead = in.read(b)) != -1 & in != null) {
            out.write(b, 0, aRead);
        }
        out.flush();
        in.close();
        out.close();
    }
}
```

❺ **创建文件下载视图页面**

在 ch7_2 应用的 src/main/resources/templates 目录下创建文件下载视图页面 showFile.html,核心代码如下：

```html
<body>
<div class="panel panel-primary">
  <div class="panel-heading">
    <h3 class="panel-title">文件下载示例</h3>
  </div>
</div>
<div class="container">
  <div class="panel panel-primary">
    <div class="panel-heading">
      <h3 class="panel-title">文件列表</h3>
    </div>
    <div class="panel-body">
```

```html
          <div class="table table-responsive">
            <table class="table table-bordered table-hover">
              <tbody class="text-center">
                <tr th:each="myFile:${filesList}">
                  <td>
                    <span th:text="${myFile.fno}"></span>
                  </td>
                  <td>
                    <a th:href="@{download(filename=${myFile.fname})}">
                      <span th:text="${myFile.fname}"></span>
                    </a>
                  </td>
                </tr>
              </tbody>
            </table>
          </div>
        </div>
      </div>
    </div>
  </body>
```

❻ 运行

首先运行 Ch72Application 主类，然后访问 http://localhost:8080/ch7_2/uploadFile，运行效果如图 7.10 所示。

图 7.10　文件选择界面

在图 7.10 中输入文件描述，并选择上传文件，然后单击"上传文件"按钮，实现文件的上传。在文件上传成功后，打开如图 7.11 所示的下载文件列表页面。

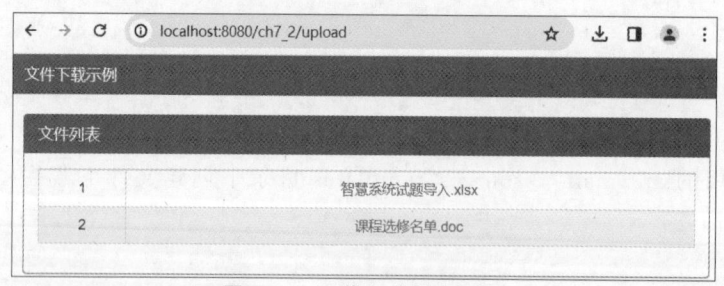

图 7.11　下载文件列表页面

单击图 7.11 中的文件名即可下载文件，至此文件的上传与下载示例演示完毕。

7.5　Spring Boot 的异常统一处理

在 Spring Boot 应用的开发中，不管是对底层数据库操作，还是对业务层操作，或者是对控制层操作，都不可避免地会遇到各种可预知的、不可预知的异常需要处理。如果每个过程都单独处理异常，那么系统的代码耦合度高，工作量大且不好统一，以后维护的工作量也很大。

如果能将所有类型的异常处理从各层中解耦出来，既保证了相关处理过程的功能较单一，也实现了异常信息的统一处理和维护。幸运的是，Spring 框架支持这样的实现。本节将从 @ExceptionHandler 与 @ControllerAdvice 注解两种方式讲解 Spring Boot 应用的异常统一处理。

▶7.5.1 自定义 error 页面

在 Spring Boot Web 应用的 src/main/resources/templates 目录下添加 error.html 页面，当访问发生错误或异常时，Spring Boot 将自动找到该页面作为错误页面。Spring Boot 为错误页面提供了以下属性。

- timestamp：错误发生时间。
- status：HTTP 状态码。
- error：错误原因。
- exception：异常的类名。
- message：异常消息（如果这个错误是由异常引起的）。
- errors：BindingResult 异常中的各种错误（如果这个错误是由异常引起的）。
- trace：异常跟踪信息（如果这个错误是由异常引起的）。
- path：错误发生时请求的 URL 路径。

下面通过一个实例讲解在 Spring Boot 应用的开发中如何使用自定义 error 页面。

【例 7-8】 使用自定义 error 页面。

其具体实现步骤如下。

❶ 创建基于 Thymeleaf 模板引擎的 Spring Boot Web 应用 ch7_3

创建基于 Thymeleaf 模板引擎的 Spring Boot Web 应用 ch7_3。

❷ 设置 Web 应用 ch7_3 的上下文路径

在 ch7_3 的 application.properties 文件中配置如下内容：

```
server.servlet.context-path=/ch7_3
```

❸ 创建自定义异常类 MyException

在应用 ch7_3 的 src/main/java 目录下创建名为 com.ch.ch7_3.exception 的包，并在该包中创建名为 MyException 的异常类，具体代码如下：

```
package com.ch.ch7_3.exception;
public class MyException extends Exception {
    public MyException() {
        super();
    }
    public MyException(String message) {
        super(message);
    }
}
```

❹ 创建控制器类 TestHandleExceptionController

在应用 ch7_3 的 src/main/java 目录下创建名为 com.ch.ch7_3.controller 的包，并在该包中创建名为 TestHandleExceptionController 的控制器类。在该控制器类中有 4 个请求处理方法，一个导航到 index.html，另外三个分别抛出不同的异常（并没有处理异常）。核心代码如下：

```
@Controller
public class TestHandleExceptionController {
    @GetMapping("/")
```

```
    public String index() {
        return "index";
    }
    @GetMapping("/db")
    public void db() throws SQLException {
        throw new SQLException("数据库异常");
    }
    @GetMapping("/my")
    public void my() throws MyException {
        throw new MyException("自定义异常");
    }
    @GetMapping("/no")
    public void no() throws Exception {
        throw new Exception("未知异常");
    }
}
```

❺ 整理脚本、样式等静态文件

JS脚本、CSS样式、图片等静态文件默认放置在src/main/resources/static目录下,ch7_3应用引入了与ch7_2一样的BootStrap和jQuery。

❻ 创建视图页面

Tymeleaf模板默认将视图页面放在src/main/resources/templates目录下,因此在src/main/resources/templates目录下新建HTML页面文件index.html和error.html。

在index.html页面中有4个超链接请求,三个请求在控制器中有对应处理,另一个请求是404错误。核心代码如下:

```
<body>
    <div class="panel panel-primary">
        <div class="panel-heading">
            <h3 class="panel-title">异常处理示例</h3>
        </div>
    </div>
    <div class="container">
        <div class="row">
            <div class="col-md-4 col-sm-6">
                <a th:href="@{db}">处理数据库异常</a><br>
                <a th:href="@{my}">处理自定义异常</a><br>
                <a th:href="@{no}">处理未知错误</a>
                <hr>
                <a th:href="@{nofound}">404错误</a>
            </div>
        </div>
    </div>
</body>
```

在error.html页面中,使用Spring Boot为错误页面提供的属性显示错误消息。核心代码如下:

```
<body>
    <div class="panel-1 container clearfix">
        <div class="error">
            <p class="title"><span class="code" th:text="${status}"></span>非常抱歉,没有找到您要查看的页面</p>
            <div class="common-hint-word">
                <div th:text="${#dates.format(timestamp,'yyyy-MM-dd HH:mm:ss')}"></div>
                <div th:text="${error}"></div>
            </div>
```

```
        </div>
    </div>
</body>
```

❼ 运行

首先运行 Ch73Application 主类，然后访问 http://localhost:8080/ch7_3/打开 index.html 页面，运行效果如图 7.12 所示。

单击图 7.12 中的超链接，Spring Boot 应用将根据链接请求到控制器中找对应的处理。例如，单击图 7.12 中的"处理数据库异常"链接，将执行控制器中的 public void db() throws SQLException 方法，而该方法仅抛出了 SQLException 异常，并没有处理异常。当 Spring Boot 发现有异常抛出并且没有处理时，将自动在 src/main/resources/templates 目录下找到 error.html 页面显示异常信息，效果如图 7.13 所示。

图 7.12　index.html 页面　　　图 7.13　error.html 页面

从例 7-8 的运行结果可以看出，使用自定义 error 页面并没有真正地处理异常，只是将异常或错误信息显示给客户端，因为在服务器控制台上同样抛出了异常，如图 7.14 所示。

图 7.14　异常信息

7.5.2　@ExceptionHandler 注解

在 7.5.1 节中使用自定义 error 页面并没有真正地处理异常，在本节使用@ExceptionHandler 注解处理异常。如果在 Controller 中有一个使用@ExceptionHandler 注解修饰的方法，那么当 Controller 的其他任何方法抛出异常时都由该方法处理异常。

下面通过实例讲解如何使用@ExceptionHandler 注解处理异常。

【例 7-9】　使用@ExceptionHandler 注解处理异常。

其具体实现步骤如下。

❶ 在控制器类中添加使用@ExceptionHandler 注解修饰的方法

在例 7-8 的控制器类 TestHandleExceptionController 中添加一个使用@ExceptionHandler 注解修饰的方法，具体代码如下：

```
@ExceptionHandler(value=Exception.class)
public String handlerException(Exception e) {
    //数据库异常
    if (e instanceof SQLException) {
        return "sqlError";
    } else if (e instanceof MyException) {          //自定义异常
```

```
            return "myError";
    } else {                                         //未知异常
            return "noError";
    }
}
```

❷ 创建 sqlError、myError 和 noError 页面

在 ch7_3 的 src/main/resources/templates 目录下创建 sqlError、myError 和 noError 页面。当发生 SQLException 异常时，Spring Boot 处理后显示 sqlError 页面；当发生 MyException 异常时，Spring Boot 处理后显示 myError 页面；当发生未知异常时，Spring Boot 处理后显示 noError 页面。具体代码略。

❸ 运行

再次运行 Ch73Application 主类，然后访问 http://localhost:8080/ch7_3/打开 index.html 页面，当单击"处理数据库异常"链接时，执行控制器中的 public void db() throws SQLException 方法，该方法抛出了 SQLException，这时 Spring Boot 会自动执行使用@ExceptionHandler 注解修饰的方法 public String handlerException(Exception e)进行异常处理并打开 sqlError.html 页面，同时观察到控制台没有抛出异常信息。注意，当单击"404 错误"链接时，还是由自定义 error 页面显示错误信息，这是因为没有执行控制器中抛出异常的方法，进而不会执行使用@ExceptionHandler 注解修饰的方法。

从例 7-9 可以看出，在控制器中添加使用@ExceptionHandler 注解修饰的方法才能处理异常。在一个 Spring Boot 应用中往往存在多个控制器，不太合适在每个控制器中添加使用@ExceptionHandler 注解修饰的方法进行异常处理。可以将使用@ExceptionHandler 注解修饰的方法放到一个父类中，然后让所有需要处理异常的控制器继承该类。例如，将例 7-9 中使用@ExceptionHandler 注解修饰的方法移到一个父类 BaseController 中，然后让控制器类 TestHandleExceptionController 继承该父类即可处理异常。

▶7.5.3 @ControllerAdvice 注解

使用 7.5.2 节中的父类 Controller 进行异常处理，有其自身的缺点，那就是代码耦合性太高，可以使用@ControllerAdvice 注解降低这种父子耦合关系。

@ControllerAdvice 注解，顾名思义，是一个增强的 Controller。使用该 Controller，可以实现三个方面的功能，即全局异常处理、全局数据绑定以及全局数据预处理。本节学习如何使用@ControllerAdvice 注解进行全局异常处理。

使用@ControllerAdvice 注解的类是当前 Spring Boot 应用中所有类的统一异常处理类，在该类中使用@ExceptionHandler 注解的方法来统一处理异常，不需要在每个 Controller 中逐一定义异常处理方法，这是因为它对所有注解了@RequestMapping 的控制器方法有效。

下面通过实例讲解如何使用@ControllerAdvice 注解进行全局异常处理。

【例 7-10】 使用@ControllerAdvice 注解进行全局异常处理。

其具体实现步骤如下。

❶ 创建使用@ControllerAdvice 注解的类

在 ch7_3 的 com.ch.ch7_3.controller 包中创建名为 GlobalExceptionHandlerController 的类。使用@ControllerAdvice 注解修饰该类，并将例 7-9 中使用@ExceptionHandler 注解修饰的方法移到该类中，核心代码如下：

```
@ControllerAdvice
public class GlobalExceptionHandlerController {
    @ExceptionHandler(value=Exception.class)
    public String handlerException(Exception e) {
        //数据库异常
        if (e instanceof SQLException) {
            return "sqlError";
        } else if (e instanceof MyException) {      //自定义异常
            return "myError";
        } else {                                     //未知异常
            return "noError";
        }
    }
}
```

❷ 运行

再次运行 Ch73Application 主类，然后访问 http://localhost：8080/ch7_3/打开 index.html 页面测试即可。

7.6　Spring Boot 对 JSP 的支持

尽管 Spring Boot 建议使用 HTML 完成动态页面，但也有一部分 Java Web 应用使用 JSP 完成动态页面。遗憾的是 Spring Boot 官方不推荐使用 JSP 技术，但考虑到这是常用的技术，本节将介绍 Spring Boot 如何集成 JSP 技术。

下面通过实例讲解 Spring Boot 如何集成 JSP 技术。

【例 7-11】　Spring Boot 集成 JSP 技术。

其具体实现步骤如下：

❶ 创建 Spring Boot Web 应用 ch7_4

在 IDEA 中创建基于 Lombok 和 Spring Web 依赖的 Spring Boot Web 应用 ch7_4。

❷ 修改 pom.xml 文件，添加 Servlet、Tomcat 和 JSTL 依赖

因为在 JSP 页面中使用 EL 和 JSTL 标签显示数据，所以在 pom.xml 文件中除了添加 Servlet 和 Tomcat 依赖外，还需要添加 JSTL 依赖，具体代码如下：

```xml
<!-- 添加 Servlet 依赖 -->
<dependency>
    <groupId>jakarta.servlet</groupId>
    <artifactId>jakarta.servlet-api</artifactId>
    <version>6.0.0</version>
    <scope>provided</scope>
</dependency>
<!-- 添加 Tomcat 依赖 -->
<dependency>
    <groupId>org.springframework.boot</groupId>
    <artifactId>spring-boot-starter-tomcat</artifactId>
    <scope>provided</scope>
</dependency>
<!-- Jasper 是 Tomcat 使用的引擎,使用 tomcat-embed-jasper 可以将 Web 应用在内嵌的 Tomcat 下运
行 -->
<dependency>
    <groupId>org.apache.tomcat.embed</groupId>
    <artifactId>tomcat-embed-jasper</artifactId>
    <scope>provided</scope>
</dependency>
```

```xml
<!-- 添加 JSTL 依赖 -->
<dependency>
    <groupId>org.glassfish.web</groupId>
    <artifactId>jakarta.servlet.jsp.jstl</artifactId>
    <version>3.0.0</version>
</dependency>
```

❸ 设置 Web 应用 ch7_4 的上下文路径及页面配置信息

在 ch7_4 的 application.properties 文件中配置如下内容：

```
server.servlet.context-path=/ch7_4
#设置页面前缀目录
spring.mvc.view.prefix=/WEB-INF/jsp/
#设置页面后缀
spring.mvc.view.suffix=.jsp
```

❹ 创建实体类 Book

创建名为 com.ch.ch7_4.model 的包，并在该包中创建名为 Book 的实体类。该实体类用于模板页面展示数据，代码与例 7-5 中的 Book 一样，这里不再赘述。

❺ 创建控制器类 ThymeleafController

创建名为 com.ch.ch7_4.controller 的包，并在该包中创建名为 ThymeleafController 的控制器类。在该控制器类中实例化 Book 类的多个对象，并保存到集合 ArrayList<Book> 中。其代码与例 7-5 中的 ThymeleafController 一样，这里不再赘述。

❻ 整理脚本、样式等静态文件

JS 脚本、CSS 样式、图片等静态文件默认放置在 src/main/resources/static 目录下，ch7_4 应用引入的 BootStrap 和 jQuery 与例 7-5 中的一样，这里不再赘述。

❼ 创建视图页面

从 application.properties 配置文件中可知，将 JSP 文件路径指定到/WEB-INF/jsp/目录。因此需要在 src/main 目录下创建目录 webapp/WEB-INF/jsp/，并在该目录下创建 JSP 文件 index.jsp，具体代码如下：

```jsp
<%@ page language="java" contentType="text/html; charset=UTF-8" pageEncoding="UTF-8"%>
<!-- 引入 JSTL 标签 -->
<%@ taglib prefix="c" uri="http://java.sun.com/jsp/jstl/core" %>
<%
    String path = request.getContextPath();
    String basePath = request.getScheme() + "://" + request.getServerName() + ":" + request.getServerPort() + path + "/";
%>
<!DOCTYPE html>
<html>
<head>
<base href="<%=basePath%>">
<meta charset="UTF-8">
<title>JSP测试</title>
<link href="css/bootstrap.min.css" rel="stylesheet">
</head>
<body>
    <div class="panel panel-primary">
        <div class="panel-heading">
            <h3 class="panel-title">第一个基于JSP技术的Spring Boot Web应用</h3>
        </div>
    </div>
```

```html
        <div class="container">
            <div>
                <h4>图书列表</h4>
            </div>
            <div class="row">
                <div class="col-md-4 col-sm-6">
                    <!-- 使用 EL 表达式 -->
                    <a href="">
                        <img src="images/${aBook.picture}" alt="图书封面" style="height: 180px; width: 40%;"/>
                    </a>
                    <div class="caption">
                        <h4>${aBook.bname}</h4>
                        <p>${aBook.author}</p>
                        <p>${aBook.isbn}</p>
                        <p>${aBook.price}</p>
                        <p>${aBook.publishing}</p>
                    </div>
                </div>
                <!-- 使用 JSTL 标签 forEach 循环取出集合数据 -->
                <c:forEach var="book" items="${books}">
                    <div class="col-md-4 col-sm-6">
                        <a href="">
                            <img src="images/${book.picture}" alt="图书封面" style="height: 180px; width: 40%;"/>
                        </a>
                        <div class="caption">
                            <h4>${book.bname}</h4>
                            <p>${book.author}</p>
                            <p>${book.isbn}</p>
                            <p>${book.price}</p>
                            <p>${book.publishing}</p>
                        </div>
                    </div>
                </c:forEach>
            </div>
        </div>
</body>
</html>
```

❽ 运行

首先运行 Ch74Application 主类，然后访问 http://localhost:8080/ch7_4/，运行效果如图 7.15 所示。

图 7.15　例 7-11 的运行效果

本章小结

本章首先介绍了 Spring Boot 的 Web 开发支持，然后详细讲述了 Spring Boot 推荐使用的 Thymeleaf 模板引擎，包括 Thymeleaf 的基础语法、常用属性以及国际化。同时，本章还介绍了 Spring Boot 对 JSON 数据的处理、文件的上传与下载、异常统一处理和 Spring Boot 对 JSP 的支持等 Web 应用开发的常用功能。

习题 7

使用 Hibernate Validator 验证如图 7.16 所示的表单信息，具体要求如下：
(1) 用户名必须输入，并且长度为 5～20。
(2) 年龄为 18～60。
(3) 工作日期在系统时间之前。

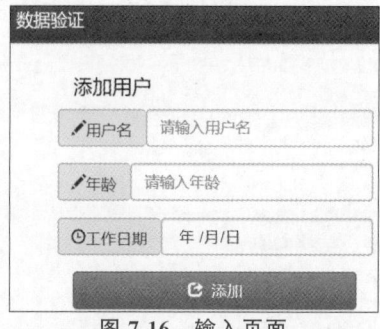

图 7.16 输入页面

第 8 章　Spring Boot 的数据访问

学习目的与要求

本章将详细介绍 Spring Boot 访问数据库的解决方案。通过本章的学习，掌握 Spring Boot 访问关系数据库及非关系数据库的解决方案。

本章主要内容

- Spring Data JPA
- Spring Boot 整合 REST
- Spring Boot 整合 MongoDB
- Spring Boot 整合 Redis
- 数据缓存 Cache
- Spring Boot 整合 MyBatis

Spring Data 是 Spring 访问数据库的解决方案，是一个伞形项目，包含大量关系数据库及非关系数据库的数据访问解决方案。本章将详细介绍 Spring Data JPA、Spring Data REST、Spring Data MongoDB、Spring Data Redis 等 Spring Data 的子项目。

8.1 Spring Data JPA

Spring Data JPA 是 Spring Data 的子项目，在讲解 Spring Data JPA 之前先了解一下 Hibernate，这是因为 Spring Data JPA 由 Hibernate 默认实现。

Hibernate 是一个开源的对象关系映射框架，它对 JDBC 进行了非常轻量级的对象封装，将 POJO(Plain Ordinary Java Object)简单的 Java 对象与数据库表建立映射关系，是一个全自动的 ORM(Object Relational Mapping)框架。Hibernate 可以自动生成 SQL 语句、自动执行，使得 Java 开发人员可以随心所欲地使用对象编程思维来操纵数据库。

JPA(Java Persistence API)是官方提出的 Java 持久化规范。JPA 通过注解或 XML 描述对象-关系(表)的映射关系，并将内存中的实体对象持久化到数据库。

Spring Data JPA 通过提供基于 JPA 的 Repository 极大地简化了 JPA 的写法，在几乎不写实现的情况下实现数据库的访问和操作。使用 Spring Data JPA 建立数据访问层十分方便，只需要定义一个继承 JpaRepository 接口的接口即可。

继承了 JpaRepository 接口的自定义数据访问接口具有 JpaRepository 接口的所有数据访问操作方法。

JpaRepository 接口提供的常用方法如下。

void flush()：将缓存的对象数据操作更新到数据库。

<S extends T> S saveAndFlush(S entity)：在保存对象的同时立即更新到数据库。

<S extends T> List<S> saveAllAndFlush(Iterable<S> entities)：在保存多个对象的同时立即更新到数据库。

void deleteAllInBatch(Iterable<T> entities)：批量删除提供的实体对象。

void deleteAllByIdInBatch(Iterable<ID> ids)：根据 id 批量删除提供的实体对象。

void deleteAllInBatch()：批量删除所有的实体对象。

T getReferenceById(ID id)：根据 id 获得对应的实体对象。

<S extends T> List<S> findAll(Example<S> example)：根据提供的 example 实例查询实体对象数据。

<S extends T> List<S> findAll(Example<S> example, Sort sort)：根据提供的 example 实例按照指定规则查询实体对象数据。

8.1.1 Spring Boot 的支持

在 Spring Boot 应用中，如果需要使用 Spring Data JPA 访问数据库，那么可以通过 IDEA 在创建 Spring Boot 应用时选择 Spring Data JPA 模块依赖。

❶ JDBC 的自动配置

Spring Data JPA 模块的依赖关系如图 8.1 所示。从图 8.1 所示的依赖关系可知，spring-boot-starter-data-jpa 依赖于 spring-boot-starter-jdbc，而 Spring Boot 对 spring-boot-starter-jdbc 做了自动配置。JDBC 自动配置的源代码位于 org.springframework.boot.autoconfigure.jdbc 包下。从 DataSourceProperties 类可以看出，可以使用"spring.datasource"为前缀的属性在 application.properties 配置文件中配置 datasource。

图 8.1 Spring Data JPA 模块的依赖关系

❷ JPA 的自动配置

Spring Boot 对 JPA 的自动配置位于 org.springframework.boot.autoconfigure.orm.jpa 包下。从 HibernateJpaAutoConfiguration 类可以看出，Spring Boot 对 JPA 的默认实现是 Hibernate。从 JpaProperties 类可以看出，可以使用以"spring.jpa"为前缀的属性在 application.properties 配置文件中配置 JPA。

❸ Spring Data JPA 的自动配置

Spring Boot 对 Spring Data JPA 的自动配置位于 org.springframework.boot.autoconfigure.data.jpa 包下。从 JpaRepositoriesAutoConfiguration 类可以看出，JpaRepositoriesAutoConfiguration 依赖于 HibernateJpaAutoConfiguration 配置；从 JpaRepositoriesRegistrar 类可以看出，Spring Boot 自动开启了对 Spring Data JPA 的支持，即开发人员无须在配置类中显式声明 @EnableJpaRepositories。

❹ Spring Boot 应用的 Spring Data JPA

从上述分析可知，在 Spring Boot 应用中使用 Spring Data JPA 访问数据库时，除了添加 spring-boot-starter-data-jpa 依赖外，只需定义 DataSource、持久化实体类和数据访问层，并在需要使用数据访问的地方（如 Service 层）依赖注入数据访问层即可。

8.1.2 简单条件查询

从前面的学习可知，只需定义一个继承 JpaRepository 接口的接口即可使用 Spring Data JPA 建立数据访问层。因此，自定义的数据访问接口完全继承了 JpaRepository 的接口方法。更重要的是，在自定义的数据访问接口中可以根据查询关键字定义查询方法，这些查询方法需要符合它的命名规则，一般是根据持久化实体类的属性名来确定的。

❶ 查询关键字

目前,Spring Data JPA 支持的查询关键字如表 8.1 所示。

表 8.1 查询关键字

关 键 字	示 例	JPQL 代码段
And	findByLastnameAndFirstname	… where x.lastname = ?1 and x.firstname = ?2
Or	findByLastnameOrFirstname	… where x.lastname = ?1 or x.firstname = ?2
Is、Equals	findByFirstname,findByFirstnameIs,findByFirstnameEquals	… where x.firstname = ?1
Between	findByStartDateBetween	… where x.startDate between ?1 and ?2
LessThan	findByAgeLessThan	… where x.age < ?1
LessThanEqual	findByAgeLessThanEqual	… where x.age <= ?1
GreaterThan	findByAgeGreaterThan	… where x.age > ?1
GreaterThanEqual	findByAgeGreaterThanEqual	… where x.age >= ?1
After	findByStartDateAfter	… where x.startDate > ?1
Before	findByStartDateBefore	… where x.startDate < ?1
IsNull	findByAgeIsNull	… where x.age is null
IsNotNull、NotNull	findByAge(Is)NotNull	… where x.age not null
Like	findByFirstnameLike	… where x.firstname like ?1
NotLike	findByFirstnameNotLike	… where x.firstname not like ?1
StartingWith	findByFirstnameStartingWith	… where x.firstname like ?1 参数后加%,即以参数开头的模糊查询
EndingWith	findByFirstnameEndingWith	… where x.firstname like ?1 参数前加%,即以参数结尾的模糊查询
Containing	findByFirstnameContaining	… where x.firstname like ?1 参数两边加%,即包含参数的模糊查询
OrderBy	findByAgeOrderByLastnameDesc	… where x.age = ?1 order by x.lastname desc
Not	findByLastnameNot	… where x.lastname <> ?1
In	findByAgeIn(Collection<Age> ages)	… where x.age in ?1
NotIn	findByAgeNotIn(Collection<Age> ages)	… where x.age not in ?1
True	findByActiveTrue()	… where x.active = true
False	findByActiveFalse()	… where x.active = false
IgnoreCase	findByFirstnameIgnoreCase	… where UPPER(x.firstame) = UPPER(?1)

❷ 限制查询结果的数量

在 Spring Data JPA 中,使用 Top 和 First 关键字限制查询结果的数量。示例如下:

```
public interface UserRepository extends JpaRepository<MyUser, Integer>{
    /**
     * 获得符合查询条件的前 10 条
     */
    public List<MyUser> findTop10ByUnameLike(String uname);
    /**
     * 获得符合查询条件的前 15 条
     */
```

```
    public List<MyUser> findFirst15ByUnameLike(String uname);
}
```

❸ 简单条件查询示例

下面通过实例讲解在 Spring Boot Web 应用中如何使用 Spring Data JPA 进行简单条件查询。

【例 8-1】 使用 Spring Data JPA 进行简单条件查询。

其具体实现步骤如下。

1）创建数据库

本书使用的关系数据库是 MySQL 8,为了演示本例,首先通过命令"CREATE DATABASE springbootjpa;"创建名为 springbootjpa 的数据库。

2）创建 Spring Boot Web 应用 ch8_1

在 IDEA 中创建基于 Lombok、Thymeleaf、MySQL Driver、Spring Web 和 Spring Data JPA 的 Spring Boot Web 应用 ch8_1。

3）设置 Web 应用 ch8_1 的上下文路径及数据源配置信息

在 ch8_1 的 application.properties 文件中配置如下内容:

```
server.servlet.context-path=/ch8_1
#数据库地址
spring.datasource.url=jdbc:mysql://localhost:3306/springbootjpa?useUnicode=
true&characterEncoding=UTF-8&allowMultiQueries=true&serverTimezone=GMT%2B8
#数据库用户名
spring.datasource.username=root
#数据库密码
spring.datasource.password=root
#数据库驱动
spring.datasource.driver-class-name=com.mysql.cj.jdbc.Driver
####
#JPA持久化配置
####
#指定数据库类型
spring.jpa.database=MYSQL
#指定是否在日志中显示SQL语句
spring.jpa.show-sql=true
#指定自动创建、更新数据库表等配置,update表示如果数据库中存在持久化类对应的表就不创建,
#不存在就创建
spring.jpa.hibernate.ddl-auto=update
#让控制器输出的JSON字符串格式更美观
spring.jackson.serialization.indent-output=true
```

4）创建持久化实体类 MyUser

创建名为 com.ch.ch8_1.entity 的包,并在该包中创建名为 MyUser 的持久化实体类。核心代码如下:

```
@Entity
@Table(name = "user_table")
@Data
public class MyUser implements Serializable{
    private static final long serialVersionUID = 1L;
    @Id
    @GeneratedValue(strategy = GenerationType.IDENTITY)
    private int id;                                                    //主键
    /**使用@Column注解,可以配置列相关属性(列名、长度等),
```

```
     * 可以省略,默认为属性名小写,如果属性名是词组,将在中间加上"_"。
     */
    private String uname;
    private String usex;
    private int age;
}
```

在持久化类中,@Entity 注解表明该实体类是一个与数据库表映射的实体类。@Table 表示实体类与哪个数据库表映射,如果没有通过 name 属性指定表名,默认为小写的类名。如果类名为词组,将在中间加上"_"(如 MyUser 类对应的表名为 my_user)。@Id 注解的属性表示该属性映射为数据库表的主键。@GeneratedValue 注解默认使用的主键生成方式为自增,如果是 MySQL、SQL Server 等关系数据库,可映射成一个递增的主键;如果是 Oracle 等关系数据库,Hibernate 将自动生成一个名为 HIBERNATE_SEQUENCE 的序列。

5)创建数据访问层

创建名为 com.ch.ch8_1.repository 的包,并在该包中创建名为 UserRepository 的接口。该接口继承 JpaRepository 接口,核心代码如下:

```
/**
 * 这里不需要使用@Repository 注解数据访问层,
 * 因为 Spring Boot 自动配置了 JpaRepository
 */
public interface UserRepository extends JpaRepository<MyUser, Integer>{
    public MyUser findByUname(String uname);
    public List<MyUser> findByUnameLike(String uname);
}
```

由于 UserRepository 接口继承了 JpaRepository 接口,所以在 UserRepository 接口中除了上述自定义的两个接口方法外(方法名的命名规范参照表 8.1),还拥有 JpaRepository 的接口方法。

6)创建业务层

创建名为 com.ch.ch8_1.service 的包,并在该包中创建 UserService 接口和接口的实现类 UserServiceImpl。UserService 接口的代码略。

UserServiceImpl 实现类的核心代码如下:

```
@Service
public class UserServiceImpl implements UserService{
    @Autowired                                            //依赖注入数据访问层
    private UserRepository userRepository;
    @Override
    public void saveAll() {
        MyUser mu1 = new MyUser();
        mu1.setUname("陈恒 1");
        mu1.setUsex("男");
        mu1.setAge(88);
        MyUser mu2 = new MyUser();
        mu2.setUname("陈恒 2");
        mu2.setUsex("女");
        mu2.setAge(18);
        MyUser mu3 = new MyUser();
        mu3.setUname("陈恒 3");
        mu3.setUsex("男");
        mu3.setAge(99);
        List<MyUser> users = new ArrayList<MyUser>();
```

```java
        users.add(mu1);
        users.add(mu2);
        users.add(mu3);
        //调用父接口中的 saveAllAndFlush 方法
        userRepository.saveAllAndFlush(users);
    }
    @Override
    public List<MyUser> findAll() {
        //调用父接口中的 findAll 方法
        return userRepository.findAll();
    }
    @Override
    public MyUser findByUname(String uname) {
        return userRepository.findByUname(uname);
    }
    @Override
    public List<MyUser> findByUnameLike(String uname) {
        return userRepository.findByUnameLike("%" + uname + "%");
    }
    @Override
    public MyUser getOne(int id) {
        //调用父接口中的 getReferenceById 方法
        return userRepository.getReferenceById(id);
    }
}
```

7）创建控制器类 UserTestController

创建名为 com.ch.ch8_1.controller 的包，并在该包中创建名为 UserTestController 的控制器类。UserTestController 的核心代码如下：

```java
@Controller
public class UserTestController {
    @Autowired
    private UserService userService;
    @GetMapping("/save")
    @ResponseBody
    public String save() {
        userService.saveAll();
        return "保存用户成功!";
    }
    @GetMapping("/findByUname")
    public String findByUname(@RequestParam("uname") String uname, Model model) {
        model.addAttribute("title", "根据用户名查询一个用户");
        model.addAttribute("auser", userService.findByUname(uname));
        return "showAuser";
    }
    @GetMapping("/getOne")
    public String getOne(@RequestParam("id") int id, Model model) {
        model.addAttribute("title", "根据用户 id 查询一个用户");
        model.addAttribute("auser", userService.getOne(id));
        return "showAuser";
    }
    @GetMapping("/findAll")
    public String findAll(Model model){
        model.addAttribute("title", "查询所有用户");
        model.addAttribute("allUsers", userService.findAll());
        return "showAll";
    }
    @GetMapping("/findByUnameLike")
    public String findByUnameLike(@RequestParam("uname") String uname, Model model){
```

```
            model.addAttribute("title","根据用户名模糊查询所有用户");
            model.addAttribute("allUsers",userService.findByUnameLike(uname));
            return "showAll";
    }
}
```

8）整理脚本、样式等静态文件

JS 脚本、CSS 样式、图片等静态文件默认放置在 src/main/resources/static 目录下，ch8_1 应用引入的 BootStrap 和 jQuery 与例 7-5 中的一样，这里不再赘述。

9）创建视图页面

在 src/main/resources/templates 目录下创建视图页面 showAll.html 和 showAuser.html。showAll.html 的核心代码如下：

```
<body>
    <div class="panel panel-primary">
        <div class="panel-heading">
            <h3 class="panel-title">Spring Data JPA 简单查询</h3>
        </div>
    </div>
    <div class="container">
        <div class="panel panel-primary">
            <div class="panel-heading">
                <h3 class="panel-title"><span th:text="${title}"></span></h3>
            </div>
            <div class="panel-body">
                <div class="table table-responsive">
                    <table class="table table-bordered table-hover">
                        <tbody class="text-center">
                            <tr th:each="user:${allUsers}">
                                <td><span th:text="${user.id}"></span></td>
                                <td><span th:text="${user.uname}"></span></td>
                                <td><span th:text="${user.usex}"></span></td>
                                <td><span th:text="${user.age}"></span></td>
                            </tr>
                        </tbody>
                    </table>
                </div>
            </div>
        </div>
    </div>
</body>
```

showAuser.html 的核心代码如下：

```
<body>
    <div class="panel panel-primary">
        <div class="panel-heading">
            <h3 class="panel-title">Spring Data JPA 简单查询</h3>
        </div>
    </div>
    <div class="container">
        <div class="panel panel-primary">
            <div class="panel-heading">
                <h3 class="panel-title"><span th:text="${title}"></span></h3>
            </div>
            <div class="panel-body">
```

```html
            <div class="table table-responsive">
                <table class="table table-bordered table-hover">
                    <tbody class="text-center">
                        <tr>
                            <td><span th:text="${auser.id}"></span></td>
                            <td><span th:text="${auser.uname}"></span></td>
                            <td><span th:text="${auser.usex}"></span></td>
                            <td><span th:text="${auser.age}"></span></td>
                        </tr>
                    </tbody>
                </table>
            </div>
        </div>
    </div>
</body>
```

10）运行

首先运行 Ch81Application 主类，然后访问"http://localhost:8080/ch8_1/save"，运行效果如图 8.2 所示。

图 8.2 保存用户

"http://localhost:8080/ch8_1/save"成功运行后，将在 MySQL 的 springbootjpa 数据库中创建一张名为 user_table 的数据库表，并插入三条记录。

通过访问"http://localhost:8080/ch8_1/findAll"查询所有用户，运行效果如图 8.3 所示。

图 8.3 查询所有用户

通过访问"http://localhost:8080/ch8_1/findByUnameLike? uname=陈"模糊查询所有陈姓用户，运行效果与图 8.3 一样。

通过访问"http://localhost:8080/ch8_1/findByUname? uname=陈恒2"查询一个名为陈恒 2 的用户的信息，运行效果如图 8.4 所示。

图 8.4 查询一个名为陈恒 2 的用户的信息

通过访问"http://localhost:8080/ch8_1/getOne?id=1"查询一个 id 为 1 的用户的信息，运

行效果如图 8.5 所示。

图 8.5 查询一个 id 为 1 的用户的信息

▶8.1.3 关联查询

在 Spring Data JPA 中有一对一、一对多、多对多等关系映射。本节将针对这些关系映射进行讲解。

❶ @OneToOne

一对一关系在现实生活中是十分常见的，例如一个大学生只有一张一卡通，一张一卡通只属于一个大学生。再如人与身份证的关系也是一对一关系。

在 Spring Data JPA 中可用两种方式描述一对一关系映射，一种是通过外键的方式（一个实体通过外键关联到另一个实体的主键）；另一种是通过一张关联表来保存两个实体一对一的关系。下面通过外键的方式讲解一对一关系映射。

【例 8-2】 使用 Spring Data JPA 实现人与身份证的一对一关系映射。

首先为例 8-2 创建基于 Lombok、MySQL Driver 以及 Spring Data JPA 依赖的 Spring Boot Web 应用 ch8_2。ch8_2 应用的数据库、pom.xml 以及 application.properties 与 ch8_1 应用基本一样，这里不再赘述。

其他内容的具体实现步骤如下。

1）创建持久化实体类

创建名为 com.ch.ch8_2.entity 的包，并在该包中创建名为 Person 和 IdCard 的持久化实体类。

Person 的核心代码如下：

```java
@Entity
@Table(name="person_table")
/**解决 No serializer found for class org.hibernate.proxy.pojo.bytebuddy.ByteBuddyInterceptor 异常 */
@JsonIgnoreProperties(value = {"hibernateLazyInitializer"})
@Data
public class Person implements Serializable{
    @Id
    @GeneratedValue(strategy = GenerationType.IDENTITY)
    private int id;                                    //自动递增的主键
    private String pname;
    private String psex;
    private int page;
    @OneToOne(
            optional = true,
            fetch = FetchType.LAZY,
            targetEntity = IdCard.class,
            cascade = CascadeType.ALL
    )
    /*指明 Person 对应表的 id_Card_id 列作为外键与 IdCard 对应表的 id 列进行关联，unique = true 指明 id_Card_id 列的值不可重复 */
    @JoinColumn(
```

```
            name = "id_Card_id",
            referencedColumnName = "id",
            unique = true
    )
    @JsonIgnore
    /* 如果 A 对象持有 B 的引用，B 对象持有 A 的引用，这样就形成了循环引用，如果直接使用 JSON 转换会报
错，使用@JsonIgnore 解决该错误。 */
    private IdCard idCard;
}
```

在上述实体类 Person 中，@OneToOne 注解有 5 个属性，分别为 targetEntity、cascade、fetch、optional 和 mappedBy。

targetEntity 属性：class 类型属性，定义关系类的类型，默认是该成员属性对应的类类型，所以通常不需要提供定义。

cascade 属性：CascadeType[]类型。该属性定义类和类之间的级联关系。定义的级联关系将被容器视为对当前类对象及其关联类对象采取相同的操作，而且这种关系是递归调用的。cascade 的值从 CascadeType.PERSIST（级联新建）、CascadeType.REMOVE（级联删除）、CascadeType.REFRESH（级联刷新）、CascadeType.MERGE（级联更新）中选择一个或多个，还有一个选择是使用 CascadeType.ALL，表示选择全部 4 项。

FetchType.LAZY 为懒加载，在加载一个实体时，定义懒加载的属性不会马上从数据库中加载。FetchType.EAGER 为急加载，在加载一个实体时，定义急加载的属性会立即从数据库中加载。

optional = true，表示 idCard 属性可以为 null，也就是允许没有身份证，如未成年人没有身份证。

mappedBy 标签一定是定义在关系的被维护端，它指向关系的维护端；只有@OneToOne、@OneToMany、@ManyToMany 上才有 mappedBy 属性，ManyToOne 不存在该属性。拥有 mappedBy 注解的实体类为关系的被维护端。

IdCard 的核心代码如下：

```
@Entity
@Table(name = "idcard_table")
@JsonIgnoreProperties(value = { "hibernateLazyInitializer"})
@Data
public class IdCard implements Serializable{
    @Id
    @GeneratedValue(strategy = GenerationType.IDENTITY)
    private int id;                                      //自动递增的主键
    private String code;
    /**
     * @Temporal 主要用来指明 java.util.Date 或 java.util.Calendar 类型的属性具体
     * 与数据库(date、time、timestamp)三个类型中的哪一个进行映射
     */
    @Temporal(value = TemporalType.DATE)
    private Calendar birthday;
    private String address;
    /**
     * optional = false 设置 person 属性值不能为 null，也就是身份证必须有对应的主人。
     * mappedBy = "idCard"与 Person 类中的 idCard 属性一致
     *拥有 mappedBy 注解的实体类为关系的被维护端。
     */
    @OneToOne(
```

```
            optional = false,
            fetch = FetchType.LAZY,
            targetEntity = Person.class,
            mappedBy = "idCard",
            cascade = CascadeType.ALL
    )
    private Person person;                              //对应的人
}
```

2）创建数据访问层

创建名为 com.ch.ch8_2.repository 的包，并在该包中创建名为 IdCardRepository 和 PersonRepository 的接口。

IdCardRepository 的核心代码如下：

```
public interface IdCardRepository extends JpaRepository<IdCard, Integer>{
    /**
     * 根据人员ID查询身份信息(关联查询,根据person属性的id)
     * 相当于JPQL语句:select ic from IdCard ic where ic.person.id = ?1
     */
    public IdCard findByPerson_id(Integer id);
    /**
     * 根据地址和身份证号查询身份信息
     * 相当于JPQL语句:select ic from IdCard ic where ic.address = ?1 and ic.code =?2
     */
    public List<IdCard> findByAddressAndCode(String address, String code);
}
```

按照 Spring Data JPA 的规则，查询两个有关联关系的对象，可以通过方法名中的"_"（下画线）来标识。如根据人员 ID 查询身份信息 findByPerson_id。JPQL（Java Persistence Query Language）是一种和 SQL 非常类似的中间性和对象化查询语言，它最终被编译成针对不同底层数据库的 SQL 查询，从而屏蔽不同数据库的差异。JPQL 语句可以是 select 语句、update 语句或 delete 语句，它们都通过 Query 接口封装执行。JPQL 的具体内容不是本书的重点，需要学习的读者请参考相关内容。

PersonRepository 的核心代码如下：

```
public interface PersonRepository extends JpaRepository<Person, Integer>{
    /**
     * 根据身份ID查询人员信息(关联查询,根据idCard属性的id)
     * 相当于JPQL语句:select p from Person p where p.idCard.id = ?1
     */
    public Person findByIdCard_id(Integer id);
    /**
     * 根据人名和性别查询人员信息
     * 相当于JPQL语句:select p from Person p where p.pname = ?1 and p.psex = ?2
     */
    public List<Person> findByPnameAndPsex(String pname, String psex);
}
```

3）创建业务层

创建名为 com.ch.ch8_2.service 的包，并在该包中创建名为 PersonAndIdCardService 的接口和接口实现类 PersonAndIdCardServiceImpl。PersonAndIdCardService 接口的代码略。

PersonAndIdCardServiceImpl 的核心代码如下：

```
@Service
public class PersonAndIdCardServiceImpl implements PersonAndIdCardService{
```

```java
    @Autowired
    private IdCardRepository idCardRepository;
    @Autowired
    private PersonRepository personRepository;
    @Override
    public void saveAll() {
        //保存身份证
        IdCard ic1 = new IdCard();
        ic1.setCode("123456789");
        ic1.setAddress("北京");
        Calendar c1 = Calendar.getInstance();
        c1.set(2023, 8, 13);
        ic1.setBirthday(c1);
        IdCard ic2 = new IdCard();
        ic2.setCode("000123456789");
        ic2.setAddress("上海");
        Calendar c2 = Calendar.getInstance();
        c2.set(2023, 8, 14);
        ic2.setBirthday(c2);
        IdCard ic3 = new IdCard();
        ic3.setCode("1111123456789");
        ic3.setAddress("广州");
        Calendar c3 = Calendar.getInstance();
        c3.set(2023, 8, 15);
        ic3.setBirthday(c3);
        List<IdCard> idCards = new ArrayList<IdCard>();
        idCards.add(ic1);
        idCards.add(ic2);
        idCards.add(ic3);
        idCardRepository.saveAllAndFlush(idCards);
        //保存人员
        Person p1 = new Person();
        p1.setPname("陈恒 1");
        p1.setPsex("男");
        p1.setPage(88);
        p1.setIdCard(ic1);
        Person p2 = new Person();
        p2.setPname("陈恒 2");
        p2.setPsex("女");
        p2.setPage(99);
        p2.setIdCard(ic2);
        Person p3 = new Person();
        p3.setPname("陈恒 3");
        p3.setPsex("女");
        p3.setPage(18);
        p3.setIdCard(ic3);
        List<Person> persons = new ArrayList<Person>();
        persons.add(p1);
        persons.add(p2);
        persons.add(p3);
        personRepository.saveAllAndFlush(persons);
    }
    @Override
    public List<Person> findAllPerson() {
        return personRepository.findAll();
    }
    @Override
    public List<IdCard> findAllIdCard() {
        return idCardRepository.findAll();
    }
    /**
```

```java
     * 根据人员 ID 查询身份信息(关联查询)
     */
    @Override
    public IdCard findByPerson_id(Integer id) {
        return idCardRepository.findByPerson_id(id);
    }
    @Override
    public List<IdCard> findByAddressAndCode(String address, String code) {
        return idCardRepository.findByAddressAndCode(address, code);
    }
    /**
     * 根据身份 ID 查询人员信息(关联查询)
     */
    @Override
    public Person findByIdCard_id(Integer id) {
        return personRepository.findByIdCard_id(id);
    }
    @Override
    public List<Person> findByPnameAndPsex(String pname, String psex) {
        return personRepository.findByPnameAndPsex(pname, psex);
    }
    @Override
    public IdCard getOneIdCard(Integer id) {
        return idCardRepository.getReferenceById(id);
    }
    @Override
    public Person getOnePerson(Integer id) {
        return personRepository.getReferenceById(id);
    }
}
```

4）创建控制器类

创建名为 com.ch.ch8_2.controller 的包,并在该包中创建名为 TestOneToOneController 的控制器类。

TestOneToOneController 的核心代码如下:

```java
@RestController
public class TestOneToOneController {
    @Autowired
    private PersonAndIdCardService personAndIdCardService;
    @GetMapping("/save")
    public String save() {
        personAndIdCardService.saveAll();
        return "人员和身份保存成功!";
    }
    @GetMapping("/findAllPerson")
    public List<Person> findAllPerson() {
        return personAndIdCardService.findAllPerson();
    }
    @GetMapping("/findAllIdCard")
    public List<IdCard> findAllIdCard() {
        return personAndIdCardService.findAllIdCard();
    }
    /**
     * 根据人员 ID 查询身份信息(关联查询)
     */
    @GetMapping("/findByPerson_id")
    public IdCard findByPerson_id(@RequestParam("id") Integer id) {
        return personAndIdCardService.findByPerson_id(id);
```

```java
    }
    @GetMapping("/findByAddressAndCode")
    public List<IdCard> findByAddressAndCode(@RequestParam("address") String address,
                                             @RequestParam("code") String code){
        return personAndIdCardService.findByAddressAndCode(address, code);
    }
    /**
     * 根据身份ID查询人员信息(关联查询)
     */
    @GetMapping("/findByIdCard_id")
    public Person findByIdCard_id(@RequestParam("id") Integer id) {
        return personAndIdCardService.findByIdCard_id(id);
    }
    @GetMapping("/findByPnameAndPsex")
    public List<Person> findByPnameAndPsex(@RequestParam("pname") String pname,
                                           @RequestParam("psex") String psex) {
        return personAndIdCardService.findByPnameAndPsex(pname, psex);
    }
    @GetMapping("/getOneIdCard")
    public IdCard getOneIdCard(@RequestParam("id") Integer id) {
        return personAndIdCardService.getOneIdCard(id);
    }
    @GetMapping("/getOnePerson")
    public Person getOnePerson(@RequestParam("id") Integer id) {
        return personAndIdCardService.getOnePerson(id);
    }
}
```

5）运行

首先运行 Ch82Application 主类，然后访问"http://localhost:8080/ch8_2/save"，运行效果如图 8.6 所示。

"http://localhost:8080/ch8_2/save"成功运行后，将在 MySQL 的 springbootjpa 数据库中创建名为 idcard_table 和 person_table 的数据库表(实体类成功加载后就已经创建好数据表)，并分别插入三条记录。

图 8.6 保存数据

通过"http://localhost:8080/ch8_2/findByIdCard_id?id=1"查询身份证 id 为 1 的人员信息(关联查询)，运行效果如图 8.7 所示。

通过"http://localhost:8080/ch8_2/findByPerson_id?id=1"查询人员 id 为 1 的身份证信息(关联查询)，运行效果如图 8.8 所示。

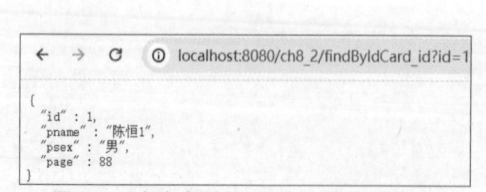

图 8.7　查询身份证 id 为 1 的人员信息

图 8.8　查询人员 id 为 1 的身份证信息

❷ **@OneToMany 和@ManyToOne**

在实际生活中，作者和文章是一对多的双向关系。那么在 Spring Data JPA 中如何描述一对多的双向关系？

在 Spring Data JPA 中使用@OneToMany 和@ManyToOne 来表示一对多的双向关系。例如，一端(Author)使用@OneToMany，多端(Article)使用@ManyToOne。

在 JPA 规范中，一对多的双向关系由多端(如 Article)来维护。也就是说多端为关系的维护端，负责关系的增、删、改、查；一端为关系的被维护端，不能维护关系。

一端(Author)使用@OneToMany 注解的 mappedBy＝"author"属性表明是关系的被维护端。多端(Article)使用@ManyToOne 和@JoinColumn 来注解 author 属性，@ManyToOne 表明 Article 是多端，@JoinColumn 设置在 article 表的关联字段(外键)上。

【例 8-3】 使用 Spring Data JPA 实现 Author 与 Article 的一对多关系映射。

在 ch8_2 应用中实现例 8-3，具体实现步骤如下。

1）创建持久化实体类

在 com.ch.ch8_2.entity 包中创建名为 Author 和 Article 的持久化实体类。

Author 的核心代码如下：

```
@Entity
@Table(name = "author_table")
@JsonIgnoreProperties(value = {"hibernateLazyInitializer"})
@Data
public class Author implements Serializable{
    @Id
    @GeneratedValue(strategy = GenerationType.IDENTITY)
    private int id;
    //作者名
    private String aname;
    //文章列表,作者与文章是一对多的关系
    @OneToMany(
            mappedBy = "author",
            cascade=CascadeType.ALL,
            targetEntity = Article.class,
            fetch=FetchType.LAZY
    )
    private List<Article> articleList;
}
```

Article 的核心代码如下：

```
@Entity
@Table(name = "article_table")
@JsonIgnoreProperties(value = {"hibernateLazyInitializer"})
@Data
public class Article implements Serializable{
    @Id
    @GeneratedValue(strategy = GenerationType.IDENTITY)
    private int id;
    //标题
    @Column(nullable = false, length = 50)
    private String title;
    //文章内容
    @Lob                                          //大对象,映射为 MySQL 的 Long 文本类型
    @Basic(fetch = FetchType.LAZY)
    @Column(nullable = false)
    private String content;
    //所属作者,文章与作者是多对一的关系
    @ManyToOne(cascade={CascadeType.MERGE,CascadeType.REFRESH},optional=false)
    //可选属性 optional=false,表示 author 不能为空。删除文章,不影响用户
    @JoinColumn(name="id_author_id")              //设置在 article 表中的关联字段(外键)
```

```
    @JsonIgnore
    private Author author;
}
```

2）创建数据访问层

在 com.ch.ch8_2.repository 包中创建名为 AuthorRepository 和 ArticleRepository 的接口。AuthorRepository 的核心代码如下：

```
public interface AuthorRepository extends JpaRepository<Author, Integer>{
    /**
     * 根据文章标题包含的内容查询作者(关联查询)
     * 相当于 JPQL 语句:select a from Author a inner join a.articleList t where t.title like %?1%
     */
    public Author findByArticleList_titleContaining(String title);
}
```

ArticleRepository 的核心代码如下：

```
public interface ArticleRepository extends JpaRepository<Article, Integer>{
    /**
     * 根据作者 id 查询文章信息(关联查询,根据 author 属性的 id)
     * 相当于 JPQL 语句:select a from Article a where a.author.id = ?1
     */
    public List<Article> findByAuthor_id(Integer id);
    /**
     * 根据作者名查询文章信息(关联查询,根据 author 属性的 aname)
     * 相当于 JPQL 语句:select a from Article a where a.author.aname = ?1
     */
    public List<Article> findByAuthor_aname(String aname);
}
```

3）创建业务层

在 com.ch.ch8_2.service 包中创建名为 AuthorAndArticleService 的接口和接口实现类 AuthorAndArticleServiceImpl。AuthorAndArticleService 接口的代码略。

AuthorAndArticleServiceImpl 的核心代码如下：

```
@Service
public class AuthorAndArticleServiceImpl implements AuthorAndArticleService{
    @Autowired
    private AuthorRepository authorRepository;
    @Autowired
    private ArticleRepository articleRepository;
    @Override
    public void saveAll() {
        //保存作者(先保存"一"的一端)
        Author a1 = new Author();
        a1.setAname("陈恒 1");
        Author a2 = new Author();
        a2.setAname("陈恒 2");
        ArrayList<Author> allAuthor = new ArrayList<Author>();
        allAuthor.add(a1);
        allAuthor.add(a2);
        authorRepository.saveAll(allAuthor);
        //保存文章
        Article at1 = new Article();
        at1.setTitle("JPA 的一对多 111");
        at1.setContent("其实一对多映射关系很常见 111。");
        //设置关系
```

```
        at1.setAuthor(a1);
        Article at2 = new Article();
        at2.setTitle("JPA 的一对多 222");
        at2.setContent("其实一对多映射关系很常见 222。");
        //设置关系
        at2.setAuthor(a1);                          //文章 2 与文章 1 的作者相同
        Article at3 = new Article();
        at3.setTitle("JPA 的一对多 333");
        at3.setContent("其实一对多映射关系很常见 333。");
        //设置关系
        at3.setAuthor(a2);
        Article at4 = new Article();
        at4.setTitle("JPA 的一对多 444");
        at4.setContent("其实一对多映射关系很常见 444。");
        //设置关系
        at4.setAuthor(a2);                          //文章 3 与文章 4 的作者相同
        ArrayList<Article> allAt = new ArrayList<Article>();
        allAt.add(at1);
        allAt.add(at2);
        allAt.add(at3);
        allAt.add(at4);
        articleRepository.saveAll(allAt);
    }
    @Override
    public List<Article> findByAuthor_id(Integer id) {
        return articleRepository.findByAuthor_id(id);
    }
    @Override
    public List<Article> findByAuthor_aname(String aname) {
        return articleRepository.findByAuthor_aname(aname);
    }
    @Override
    public Author findByArticleList_titleContaining(String title) {
        return authorRepository.findByArticleList_titleContaining(title);
    }
}
```

4）创建控制器类

在 com.ch.ch8_2.controller 包中创建名为 TestOneToManyController 的控制器类。TestOneToManyController 的核心代码如下：

```
@RestController
public class TestOneToManyController {
    @Autowired
    private AuthorAndArticleService authorAndArticleService;
    @GetMapping("/saveOneToMany")
    public String save() {
        authorAndArticleService.saveAll();
        return "作者和文章保存成功!";
    }
    @GetMapping("/findArticleByAuthor_id")
    public List<Article> findByAuthor_id(@RequestParam("id") Integer id) {
        return authorAndArticleService.findByAuthor_id(id);
    }
    @GetMapping("/findArticleByAuthor_aname")
    public List<Article> findByAuthor_aname(@RequestParam("aname") String aname) {
        return authorAndArticleService.findByAuthor_aname(aname);
    }
    @GetMapping("/findByArticleList_titleContaining")
```

```
    public Author findByArticleList_titleContaining(@RequestParam("title") String title) {
        return authorAndArticleService.findByArticleList_titleContaining(title);
    }
}
```

5）运行

首先运行 Ch82Application 主类，然后访问"http://localhost:8080/ch8_2/saveOneToMany"，运行效果如图 8.9 所示。

图 8.9　保存数据

"http://localhost:8080/ch8_2/saveOneToMany"成功运行后，将在 MySQL 的 springbootjpa 数据库中创建名为 author_table 和 article_table 的数据库表，并在 author_table 表中插入两条记录，同时在 article_table 表中插入 4 条记录。

通过"http://localhost:8080/ch8_2/findArticleByAuthor_id?id=2"查询作者 id 为 2 的文章列表（关联查询），运行效果如图 8.10 所示。

图 8.10　查询作者 id 为 2 的文章列表

通过"http://localhost:8080/ch8_2/findArticleByAuthor_aname?aname=陈恒1"查询作者名为陈恒 1 的文章列表（关联查询），运行效果如图 8.11 所示。

图 8.11　查询作者名为陈恒 1 的文章列表

通过"http://localhost:8080/ch8_2/findByArticleList_titleContaining?title=对多一"查询文章标题中包含"对多一"的作者（关联查询），运行效果如图 8.12 所示。

图 8.12　查询文章标题中包含"对多一"的作者

❸ @ManyToMany

在实际生活中,用户和权限是多对多的关系。一个用户可以有多个权限,一个权限也可以被很多用户拥有。

在 Spring Data JPA 中使用@ManyToMany 来注解多对多的映射关系,由一个关联表来维护。关联表的名称默认是主表名+下画线+从表名(主表是指关系维护端对应的表,从表是指关系被维护端对应的表)。关联表只有两个外键字段,分别指向主表 ID 和从表 ID。字段的名称默认为主表名+下画线+主表中的主键列名,从表名+下画线+从表中的主键列名。需要注意的是,在多对多关系中一般不设置级联保存、级联删除、级联更新等操作。

【例 8-4】 使用 Spring Data JPA 实现用户(User)与权限(Authority)的多对多关系映射。

在 ch8_2 应用中实现例 8-4,具体实现步骤如下。

1) 创建持久化实体类

在 com.ch.ch8_2.entity 包中创建名为 User 和 Authority 的持久化实体类。

User 的核心代码如下:

```java
@Entity
@Table(name = "user")
@JsonIgnoreProperties(value = {"hibernateLazyInitializer"})
@Data
public class User implements Serializable{
    @Id
    @GeneratedValue(strategy = GenerationType.IDENTITY)
    private int id;
    private String username;
    private String password;
    @ManyToMany
    @JoinTable(name = "user_authority",joinColumns = @JoinColumn(name = "user_id"),
        inverseJoinColumns = @JoinColumn(name = "authority_id"))
    /**1.关系维护端,负责多对多关系的绑定和解除。
     2.@JoinTable注解的name属性指定关联表的名字,joinColumns指定外键的名字,关联到关系维护端(User)。
     3.inverseJoinColumns指定外键的名字,需要关联的关系,称为被维护端(Authority)。
     4.其实可以不使用@JoinTable注解,默认生成的关联表的名称为主表表名+下画线+从表表名,即表名为user_authority。
     关联到主表的外键名:主表名+下画线+主表中的主键列名,即 user_id。
     关联到从表的外键名:主表中用于关联的属性名+下画线+从表的主键列名,即 authority_id。主表就是关系维护端对应的表,从表就是关系被维护端对应的表
     */
    private List<Authority> authorityList;
}
```

Authority 的核心代码如下:

```java
@Entity
@Table(name = "authority")
@JsonIgnoreProperties(value = {"hibernateLazyInitializer"})
@Data
public class Authority implements Serializable{
    @Id
    @GeneratedValue(strategy = GenerationType.IDENTITY)
    private int id;
    @Column(nullable = false)
    private String name;
    @ManyToMany(mappedBy = "authorityList")
    @JsonIgnore
```

```
    private List<User> userList;
}
```

2）创建数据访问层

在 com.ch.ch8_2.repository 包中创建名为 UserRepository 和 AuthorityRepository 的接口。UserRepository 的核心代码如下：

```java
public interface UserRepository extends JpaRepository<User, Integer>{
    /**
     * 根据权限id查询拥有该权限的用户(关联查询)
     * 相当于 JPQL 语句:select u from User u inner join u.authorityList a where a.id = ?1
     */
    public List<User> findByAuthorityList_id(int id);
    /**
     * 根据权限名查询拥有该权限的用户(关联查询)
     * 相当于 JPQL 语句:select u from User u inner join u.authorityList a where a.name = ?1
     */
    public List<User> findByAuthorityList_name(String name);
}
```

AuthorityRepository 的核心代码如下：

```java
public interface AuthorityRepository extends JpaRepository<Authority, Integer>{
    /**
     * 根据用户id查询用户所拥有的权限(关联查询)
     * 相当于 JPQL 语句:select a from Authority a inner join a.userList u where u.id = ?1
     */
    public List<Authority> findByUserList_id(int id);
    /**
     * 根据用户名查询用户所拥有的权限(关联查询)
     * 相当于 JPQL 语句:select a from Authority a inner join a.userList u where u.username = ?1
     */
    public List<Authority> findByUserList_Username(String username);
}
```

3）创建业务层

在 com.ch.ch8_2.service 包中创建名为 UserAndAuthorityService 的接口和接口实现类 UserAndAuthorityServiceImpl。UserAndAuthorityService 接口的代码略。

UserAndAuthorityServiceImpl 的核心代码如下：

```java
@Service
public class UserAndAuthorityServiceImpl implements UserAndAuthorityService{
    @Autowired
    private AuthorityRepository authorityRepository;
    @Autowired
    private UserRepository userRepository;
    @Override
    public void saveAll() {
        //添加权限 1
        Authority at1 = new Authority();
        at1.setName("增加");
        authorityRepository.save(at1);
        //添加权限 2
        Authority at2 = new Authority();
        at2.setName("修改");
        authorityRepository.save(at2);
        //添加权限 3
        Authority at3 = new Authority();
```

```
            at3.setName("删除");
            authorityRepository.save(at3);
            //添加权限 4
            Authority at4 = new Authority();
            at4.setName("查询");
            authorityRepository.save(at4);
            //添加用户 1
            User u1 = new User();
            u1.setUsername("陈恒 1");
            u1.setPassword("123");
            ArrayList<Authority> authorityList1 = new ArrayList<Authority>();
            authorityList1.add(at1);
            authorityList1.add(at2);
            authorityList1.add(at3);
            u1.setAuthorityList(authorityList1);
            userRepository.save(u1);
            //添加用户 2
            User u2 = new User();
            u2.setUsername("陈恒 2");
            u2.setPassword("234");
            ArrayList<Authority> authorityList2 = new ArrayList<Authority>();
            authorityList2.add(at2);
            authorityList2.add(at3);
            authorityList2.add(at4);
            u2.setAuthorityList(authorityList2);
            userRepository.save(u2);
    }
    @Override
    public List<User> findByAuthorityList_id(int id) {
            return userRepository.findByAuthorityList_id(id);
    }
    @Override
    public List<User> findByAuthorityList_name(String name) {
            return userRepository.findByAuthorityList_name(name);
    }
    @Override
    public List<Authority> findByUserList_id(int id) {
            return authorityRepository.findByUserList_id(id);
    }
    @Override
    public List<Authority> findByUserList_Username(String username) {
            return authorityRepository.findByUserList_Username(username);
    }
}
```

4）创建控制器类

在 com.ch.ch8_2.controller 包中创建名为 TestManyToManyController 的控制器类。TestManyToManyController 的核心代码如下：

```
@RestController
public class TestManyToManyController {
    @Autowired
    private UserAndAuthorityService userAndAuthorityService;
    @GetMapping("/saveManyToMany")
    public String save() {
        userAndAuthorityService.saveAll();
        return "权限和用户保存成功!";
    }
    @GetMapping("/findByAuthorityList_id")
    public List<User> findByAuthorityList_id(@RequestParam("id") int id) {
```

```java
        return userAndAuthorityService.findByAuthorityList_id(id);
    }
    @GetMapping("/findByAuthorityList_name")
    public List<User> findByAuthorityList_name(@RequestParam("name") String name) {
        return userAndAuthorityService.findByAuthorityList_name(name);
    }
    @GetMapping("/findByUserList_id")
    public List<Authority> findByUserList_id(@RequestParam("id") (int id) {
        return userAndAuthorityService.findByUserList_id(id);
    }
    @GetMapping("/findByUserList_Username")
    public List< Authority > findByUserList_Username (@RequestParam ("username") String username) {
        return userAndAuthorityService.findByUserList_Username(username);
    }
}
```

5）运行

首先运行 Ch82Application 主类，然后访问"http://localhost:8080/ch8_2/saveManyToMany"，运行效果如图 8.13 所示。

通过"http://localhost:8080/ch8_2/findByAuthorityList_id?id=1"查询拥有 id 为 1 的权限的用户列表（关联查询），运行效果如图 8.14 所示。

图 8.13　保存数据

图 8.14　查询拥有 id 为 1 的权限的用户列表

通过"http://localhost:8080/ch8_2/findByAuthorityList_name？name＝修改"查询拥有"修改"权限的用户列表（关联查询），运行效果如图 8.15 所示。

图 8.15　查询拥有"修改"权限的用户列表

通过"http://localhost:8080/ch8_2/findByUserList_id?id=2"查询 id 为 2 的用户的权限列表(关联查询)，运行效果如图 8.16 所示。

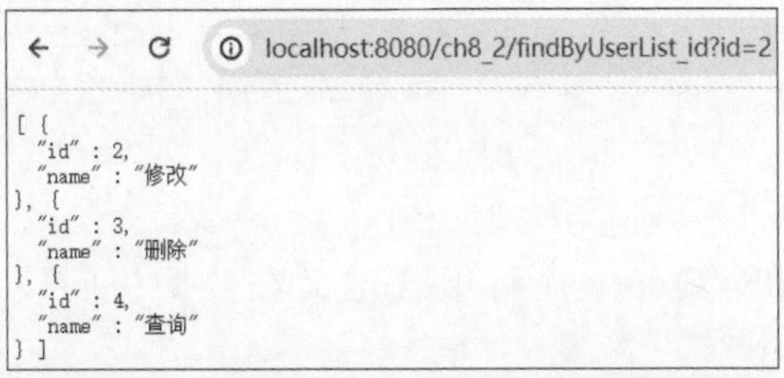

图 8.16　查询 id 为 2 的用户的权限列表

通过"http://localhost:8080/ch8_2/findByUserList_Username? username＝陈恒 2"查询用户名为"陈恒 2"的用户的权限列表(关联查询)，运行效果如图 8.17 所示。

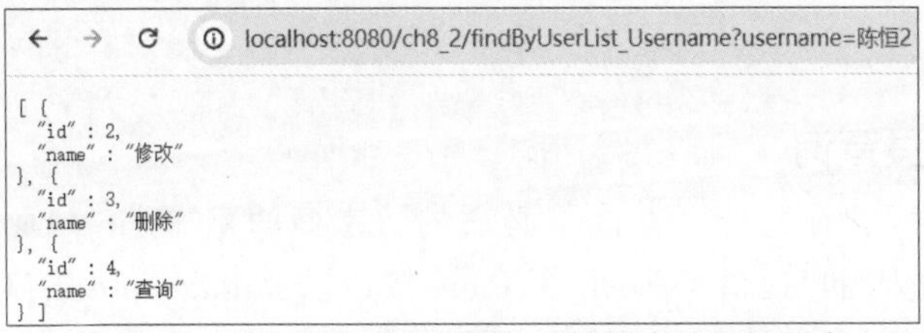

图 8.17　查询用户名为"陈恒 2"的用户的权限列表

8.1.2 节和 8.1.3 节中的查询方法必须严格按照 Spring Data JPA 的查询关键字的命名规范进行命名。如何才能摆脱查询关键字和关联查询命名规范的约束呢？可以通过@Query、@NamedQuery 直接定义 JPQL 语句进行数据的访问操作。

▶8.1.4　@Query 和@Modifying 注解

❶ @Query 注解

使用@Query 注解可以将 JPQL 语句直接定义在数据访问接口方法上，并且接口方法名不受查询关键字和关联查询命名规范的约束。示例代码如下：

```java
public interface AuthorityRepository extends JpaRepository<Authority, Integer>{
    /**
     * 根据用户名查询用户所拥有的权限(关联查询)
     */
    @Query("select a from Authority a inner join a.userList u where u.username = ?1")
    public List<Authority> findByUserListUsername(String username);
}
```

使用@Query 注解定义 JPQL 语句，可以直接返回 List<Map<String，Object>>对象。示例代码如下：

```java
/**
 * 根据作者 id 查询文章信息(标题和内容)
 */
@Query("select new Map(a.title as title, a.content as content) from Article a where a.author.id = ?1")
public List<Map<String, Object>> findTitleAndContentByAuthorId(Integer id);
```

使用@Query 注解定义 JPQL 语句，之前的方法是使用参数位置("？1"指代的是获取方法形参列表中的第 1 个参数值，1 代表的是参数位置，以此类推)来获取参数值。除此之外，Spring Data JPA 还支持使用名称来获取参数值，使用格式为"：参数名称"。示例代码如下：

```
/**
 * 根据作者名和作者 id 查询文章信息
 */
@Query("select a from Article a where a.author.aname = :aname1 and a.author.id = :id1")
public List<Article> findArticleByAuthorAnameAndId(@Param("aname1") String aname, @Param
("id1") Integer id);
```

❷ @Modifying 注解

可以使用@Modifying 和@Query 注解组合定义在数据访问接口方法上，进行更新查询操作，示例代码如下：

```
/**
 * 根据作者 id 删除作者
 */
@Modifying
@Query("delete from Author a where a.id = ?1")
public int deleteAuthorByAuthorId(int id);
```

8.1.5 排序与分页查询

在实际应用开发中，排序与分页查询是必需的。幸运的是 Spring Data JPA 充分考虑了排序与分页查询的场景，为用户提供了 Sort 类、Page 接口以及 Pageable 接口。

例如，以下数据访问接口：

```
public interface AuthorRepository extends JpaRepository<Author, Integer>{
    List<Author> findByAnameContaining(String aname, Sort sort);
}
```

那么在 Service 层可以这样使用排序：

```
public List<Author> findByAnameContaining(String aname, String sortColumn) {
    //按 sortColumn 降序排序
    return authorRepository.findByAnameContaining(aname, new Sort(Direction.DESC,
sortColumn));
}
```

可以使用 Pageable 接口的实现类 PageRequest 的 of 方法构造分页查询对象，示例代码如下：

```
Page<Author> pageData = authorRepository.findAll(PageRequest.of(page-1, size, new Sort
(Direction.DESC, "id")));
```

其中 Page 接口可以获得当前页面的记录、总页数、总记录数等信息，示例代码如下：

```
//获得当前页面的记录
List<Author> allAuthor = pageData.getContent();
model.addAttribute("allAuthor",allAuthor);
//获得总记录数
model.addAttribute("totalCount", pageData.getTotalElements());
//获得总页数
model.addAttribute("totalPage", pageData.getTotalPages());
```

下面通过实例讲解 Spring Data JPA 的排序与分页查询的使用方法。

【例 8-5】 排序与分页查询的使用方法。

首先为例 8-5 创建基于 Thymeleaf、MySQL Driver、Lombok 和 Spring Data JPA 的 Spring Boot Web 应用 ch8_3。ch8_3 应用的数据库、pom.xml、application.properties 以及静态资源等内容与 ch8_1 应用基本一样，这里不再赘述。

其他内容的具体实现步骤如下。

❶ 创建持久化实体类

在应用 ch8_3 的 src/main/java 目录下创建名为 com.ch.ch8_3.entity 的包，并在该包中创建名为 Article 和 Author 的持久化实体类，具体代码分别与 ch8_2 应用的 Article 和 Author 的代码一样，这里不再赘述。

❷ 创建数据访问层

在应用 ch8_3 的 src/main/java 目录下创建名为 com.ch.ch8_3.repository 的包，并在该包中创建名为 AuthorRepository 的接口。

AuthorRepository 的核心代码如下：

```java
public interface AuthorRepository extends JpaRepository<Author, Integer>{
    /**
     * 查询作者名中含有 name 的作者列表,并排序
     */
    List<Author> findByAnameContaining(String aname, Sort sort);
}
```

❸ 创建业务层

在应用 ch8_3 的 src/main/java 目录下创建名为 com.ch.ch8_3.service 的包，并在该包中创建名为 ArticleAndAuthorService 的接口和接口实现类 ArticleAndAuthorServiceImpl。

ArticleAndAuthorService 的核心代码如下：

```java
public interface ArticleAndAuthorService {
    /**
     * name 代表作者名的一部分(模糊查询),sortColumn 代表排序列
     */
    List<Author> findByAnameContaining(String aname, String sortColumn);
    /**
     * 分页查询作者,page 代表第几页
     */
    public String findAllAuthorByPage(Integer page, Model model);
}
```

ArticleAndAuthorServiceImpl 的核心代码如下：

```java
@Service
public class ArticleAndAuthorServiceImpl implements ArticleAndAuthorService{
    @Autowired
    private AuthorRepository authorRepository;
    @Override
    public List<Author> findByAnameContaining(String aname, String sortColumn) {
        //按 sortColumn 降序排序
         return authorRepository.findByAnameContaining(aname, Sort.by(Direction.DESC, sortColumn));
    }
    @Override
    public String findAllAuthorByPage(Integer page, Model model) {
        if(page == null) {                    //第一次访问 findAllAuthorByPage 方法时
            page = 1;
        }
        int size = 2;                         //每页显示两条
        //分页查询,of 方法的第一个参数代表第几页(比实际小 1),
        //第二个参数代表页面大小,第三个参数代表排序规则
        Page<Author> pageData =
        authorRepository.findAll(PageRequest.of(page-1, size, Sort.by(Direction.DESC, "id")));
```

```
        //获得当前页面数据并转换成 List<Author>,转发到视图页面显示
        List<Author> allAuthor = pageData.getContent();
        model.addAttribute("allAuthor",allAuthor);
        //共多少条记录
        model.addAttribute("totalCount", pageData.getTotalElements());
        //共多少页
        model.addAttribute("totalPage", pageData.getTotalPages());
        //当前页
        model.addAttribute("page", page);
        return "index";
    }
}
```

❹ 创建控制器类

在应用 ch8_3 的 src/main/java 目录下创建名为 com.ch.ch8_3.controller 的包,并在该包中创建名为 TestSortAndPage 的控制器类。

TestSortAndPage 的核心代码如下:

```
@Controller
public class TestSortAndPage {
    @Autowired
    private ArticleAndAuthorService articleAndAuthorService;
    @GetMapping("/findByAnameContaining")
    @ResponseBody
    public List<Author> findByAnameContaining(@RequestParam("aname") String aname,
                                              @RequestParam("sortColumn") String sortColumn){
        return articleAndAuthorService.findByAnameContaining(aname, sortColumn);
    }
    @GetMapping("/findAllAuthorByPage")
    public String findAllAuthorByPage(@RequestParam("page") Integer page, Model model){
        return articleAndAuthorService.findAllAuthorByPage(page, model);
    }
}
```

❺ 创建视图页面

在 src/main/resources/templates 目录下创建视图页面 index.html。

index.html 的核心代码如下:

```
<body>
    <div class="panel panel-primary">
        <div class="panel-heading">
            <h3 class="panel-title">Spring Data JPA 分页查询</h3>
        </div>
    </div>
    <div class="container">
        <div class="panel panel-primary">
            <div class="panel-body">
                <div class="table table-responsive">
                    <table class="table table-bordered table-hover">
                        <tbody class="text-center">
                            <tr th:each="author:${allAuthor}">
                                <td><span th:text="${author.id}"></span></td>
                                <td><span th:text="${author.aname}"></span></td>
                            </tr>
                            <tr>
                                <td colspan="2" align="right">
                                    <ul class="pagination">
                                        <li><a>第<span th:text="${page}"></span>页</a></li>
```

```html
                        <li><a>共<span th:text="${totalPage}"></span>页</a></li>
                        <li><a>共<span th:text="${totalCount}"></span>条</a></li>
                        <li>
        <a th:href="@{findAllAuthorByPage(page=${page-1})}" th:if="${page != 1}">上一页</a>
                        </li>
                        <li>
        <a th:href="@{findAllAuthorByPage(page=${page+1})}" th:if="${page != totalPage}">下一页
        </a>
                        </li>
                    </ul>
                </td>
            </tr>
        </tbody>
    </table>
            </div>
        </div>
    </div>
</body>
```

❻ 运行

首先运行 Ch83Application 主类，然后通过"http://localhost：8080/ch8_3/findByAnameContaining?aname=陈&sortColumn=id"查询作者名中含有"陈"的作者列表，并按照 id 降序。

通过"http://localhost：8080/ch8_3/findAllAuthorByPage? page=1"分页查询作者，并按照 id 降序，运行效果如图 8.18 所示。

图 8.18 分页查询作者

8.2 REST

本节将介绍 RESTful 风格接口，并通过 Spring Boot 来实现 RESTful。

▶ 8.2.1 REST 简介

REST 即表现层状态转化（英文为 Representational State Transfer），它是 Roy Thomas Fielding 博士在 2000 年的博士论文中提出来的一种软件架构风格。REST 是一种针对网络应用的设计和开发方式，可以降低开发的复杂性，提高系统的可伸缩性。目前在三种主流的 Web 服务实现方案中，因为 REST 模式的 Web 服务与复杂的 SOAP 和 XML-RPC 相比更加简洁，越来越多的 Web 服务开始采用 REST 风格设计和实现。

REST 是一组架构约束条件和原则，这些约束条件和原则如下。

（1）使用客户/服务器模型：客户和服务器之间通过一个统一的接口来互相通信。

(2) 层次化的系统：在一个 REST 系统中，客户端并不会固定地与一个服务器打交道。

(3) 无状态：在一个 REST 系统中，服务端并不会保存有关客户的任何状态。也就是说，客户端自身负责用户状态的维持，并在每次发送请求时都需要提供足够的信息。

(4) 可缓存：REST 系统需要能够恰当地缓存请求，以尽量减少服务器端和客户端之间的信息传输，从而提高性能。

(5) 统一的接口：一个 REST 系统需要使用一个统一的接口来完成子系统之间以及服务与用户之间的交互，这使得 REST 系统中的各个子系统可以独自完成演化。

满足这些约束条件和原则的应用程序或设计就是 RESTful。需要注意的是，REST 是设计风格而不是标准。REST 通常基于 HTTP、URI、XML 以及 HTML 这些现有的广泛流行的协议和标准。

理解 RESTful 架构，应该先理解 Representational State Transfer 这个词组到底是什么意思，它的每一个词表达了什么含义。

- 资源（Resources）："表现层状态转化"中的"表现层"其实指的是"资源"的"表现层"。

"资源"就是网络上的一个实体，或者说是网络上的一个具体信息。"资源"可以是一段文本、一张图片、一段视频，总之就是一个具体的实体。可以使用一个 URI（统一资源定位符）指向资源，每种资源对应一个特定的 URI。当需要获取资源时，访问它的 URI 即可，因此 URI 是每个资源的地址或独一无二的标识符。REST 风格的 Web 服务是通过一个简洁、清晰的 URI 来提供资源链接，客户端通过对 URI 发送 HTTP 请求获得这些资源，而获取和处理资源的过程让客户端应用的状态发生改变。

- 表现层（Representation）："资源"是一种信息实体，可以有多种外在的表现形式。通常将"资源"呈现出来的形式称为它的"表现层"。例如，文本可以使用 TXT 格式表现，也可以使用 XML 格式、JSON 格式表现。

- 状态转化（State Transfer）：客户端访问一个网站，就代表了它和服务器的一个互动过程，在这个互动过程中将涉及数据和状态的变化。HTTP 协议是一个无状态的通信协议，这意味着所有状态都保存在服务器端。因此，如果客户端操作服务器，需要通过某种手段（如 HTTP 协议）让服务器端发生"状态变化"。这种转化是建立在表现层之上的，所以就是"表现层状态转化"。

在流行的各种 Web 框架中（包括 Spring Boot）都支持 REST 开发。REST 并不是一种技术或者规范，而是一种架构风格，包括了如何标识资源、如何标识操作接口及操作的版本、如何标识操作的结果等，主要内容如下。

❶ 使用"api"作为上下文

在 REST 架构中，建议使用"api"作为上下文，示例如下。

```
http://localhost:8080/api
```

❷ 增加一个版本标识

在 REST 架构中，可以通过 URL 标识版本信息，示例如下：

```
http://localhost:8080/api/v1.0
```

❸ 标识资源

在 REST 架构中，可以将资源名称放到 URL 中，示例如下：

```
http://localhost:8080/api/v1.0/user
```

❹ 确定 HTTP Method

HTTP 协议有 5 个常用的表示操作方式的动词，即 GET、POST、PUT、DELETE、PATCH。它们分别对应 5 种基本操作，GET 用来获取资源，POST 用来增加资源（也可以用于更新资源），PUT 用来更新资源，DELETE 用来删除资源，PATCH 用来更新资源的部分属性。示例如下：

（1）新增用户：

```
POST http://localhost:8080/api/v1.0/user
```

（2）查询 id 为 123 的用户：

```
GET http://localhost:8080/api/v1.0/user/123
```

（3）更新 id 为 123 的用户：

```
PUT http://localhost:8080/api/v1.0/user/123
```

（4）删除 id 为 123 的用户：

```
DELETE http://localhost:8080/api/v1.0/user/123
```

（5）更新 id 为 123 的用户的 email 属性值：

```
PATCH http://localhost:8080/api/v1.0/user/123?email=gogo@126.com
```

❺ 确定 HTTP Status

常用的服务器向用户返回的状态码和提示信息如下。

（1）200 OK -［GET］：服务器成功返回用户请求的数据。

（2）201 CREATED -［POST/PUT/PATCH］：用户新建或修改数据成功。

（3）202 Accepted -［*］：表示一个请求已经进入后台排队（异步任务）。

（4）204 NO CONTENT -［DELETE］：用户删除数据成功。

（5）400 INVALID REQUEST -［POST/PUT/PATCH］：用户发出的请求有错误，服务器没有进行新建或修改数据的操作。

（6）401 Unauthorized -［*］：表示用户没有权限（令牌、用户名、密码错误）。

（7）403 Forbidden -［*］：表示用户得到授权（与 401 错误相对），但是访问是被禁止的。

（8）404 NOT FOUND -［*］：用户发出的请求针对的是不存在的记录，服务器没有进行操作。

（9）406 Not Acceptable -［GET］：用户请求的格式不可得（例如用户请求 JSON 格式，但是只有 XML 格式）。

（10）410 Gone -［GET］：用户请求的资源被永久删除，且不会再得到。

（11）422 Unprocesable entity -［POST/PUT/PATCH］：当创建一个对象时发生一个验证错误。

（12）500 INTERNAL SERVER ERROR -［*］：服务器发生错误，用户将无法判断发出的请求是否成功。

▶8.2.2　Spring Boot 整合 REST

在 Spring Boot 的 Web 应用中自动支持 REST。也就是说，只要 spring-boot-starter-web 依

赖在 POM 中，就支持 REST。

【例 8-6】 一个 RESTful 应用示例。

假如在 ch8_2 应用的控制器类中有如下处理方法：

```
@GetMapping("/findArticleByAuthor_id/{id}")
    public List<Article> findByAuthor_id1(@PathVariable("id") Integer id) {
    return authorAndArticleService.findByAuthor_id(id);
}
```

那么可以使用如下 REST 风格的 URL 访问上述处理方法：

```
http://localhost:8080/ch8_2/findArticleByAuthor_id/2
```

在例 8-6 中使用了 URL 模板模式映射@RequestMapping("/findArticleByAuthor_id/{id}")，其中{XXX}为占位符，请求的 URL 可以是"/findArticleByAuthor_id/1"或"/findArticleByAuthor_id/2"。通过在处理方法中使用@PathVariable 获取{XXX}中的 XXX 变量值。@PathVariable 用于将请求 URL 中的模板变量映射到功能处理方法的参数上。如果{XXX}中的变量名 XXX 和形参的名称一致，则@PathVariable 不用指定名称。

▶ 8.2.3 Spring Data REST

Spring Data JPA 基于 Spring Data 的 repository 之上，可以将 repository 自动输出为 REST 资源。目前，Spring Data REST 支持将 Spring Data JPA、Spring Data MongoDB、Spring Data Neo4j、Spring Data GemFire 以及 Spring Data Cassandra 的 repository 自动转换成 REST 服务。

Spring Boot 对 Spring Data REST 的自动配置存放在 org.springframework.boot.autoconfigure.data.rest 包中。通过 SpringBootRepositoryRestConfigurer 类的源代码可以得出，Spring Boot 已经自动配置了 RepositoryRestConfiguration，所以在 Spring Boot 应用中使用 Spring Data REST 只需引入 spring-boot-starter-data-rest 的依赖即可。

下面通过实例讲解 Spring Data REST 的构建过程。

【例 8-7】 Spring Data REST 的构建过程。

其具体实现步骤如下。

❶ **创建 Spring Boot 应用 ch8_4**

创建 Spring Boot 应用 ch8_4，依赖为 Lombok、MySQL Driver、Spring Data JPA 和 Rest Repositories。

❷ **设置应用 ch8_4 的上下文路径及数据源配置信息**

在 ch8_4 的 application.properties 文件中配置如下内容：

```
server.servlet.context-path=/api
#数据库地址
spring.datasource.url=jdbc:mysql://localhost:3306/springbootjpa?useUnicode=true&characterEncoding=UTF-8&allowMultiQueries=true&serverTimezone=GMT%2B8
#数据库用户名
spring.datasource.username=root
#数据库密码
spring.datasource.password=root
#数据库驱动
spring.datasource.driver-class-name=com.mysql.cj.jdbc.Driver
#指定数据库类型
spring.jpa.database=MYSQL
#指定是否在日志中显示 SQL 语句
spring.jpa.show-sql=true
```

```properties
#指定自动创建、更新数据库表等配置,update 表示如果数据库中存在持久化类对应的表就不创建,不存在就
#创建
spring.jpa.hibernate.ddl-auto=update
#让控制器输出的 JSON 字符串格式更美观
spring.jackson.serialization.indent-output=true
```

❸ 创建持久化实体类 Student

在应用 ch8_4 的 src/main/java 目录下创建名为 com.ch.ch8_4.entity 的包,并在该包中创建名为 Student 的持久化实体类。核心代码如下:

```java
@Entity
@Table(name = "student_table")
@Data
public class Student implements Serializable{
    @Id
    @GeneratedValue(strategy = GenerationType.IDENTITY)
    private int id;                                                          //主键
    private String sno;
    private String sname;
    private String ssex;
    public Student() {
        super();
    }
    public Student(int id, String sno, String sname, String ssex) {
        super();
        this.id = id;
        this.sno = sno;
        this.sname = sname;
        this.ssex = ssex;
    }
}
```

❹ 创建数据访问层

在应用 ch8_4 的 src/main/java 目录下创建名为 com.ch.ch8_4.repository 的包,并在该包中创建名为 StudentRepository 的接口,该接口继承 JpaRepository 接口,核心代码如下:

```java
public interface StudentRepository extends JpaRepository<Student, Integer>{
    /**
     * 自定义接口查询方法,暴露为 REST 资源
     */
    @RestResource(path = "snameStartsWith", rel = "snameStartsWith")
    List<Student> findBySnameStartsWith(@Param("sname") String sname);
}
```

在上述数据访问接口中定义了 findBySnameStartsWith,并使用@RestResource 注解将该方法暴露为 REST 资源。

至此,基于 Spring Data 的 REST 资源服务已经构建完毕,接下来就是使用 REST 客户端测试此服务。

▶8.2.4 REST 服务测试

在 Web 和移动端开发时,经常会调用服务器端的 RESTful 接口进行数据请求,为了调试,一般会先用工具进行测试,通过测试后才开始在开发中使用。本节将介绍如何使用 Postman 进行 8.2.3 节的 RESTful 接口请求测试。

Postman 是一个接口测试工具,在做接口测试时,Postman 相当于一个客户端,它可以模拟用户发起的各类 HTTP 请求,将请求数据发送至服务器端,获取对应的响应结果,从而验证响应

中的结果数据是否和预期值相匹配。

Postman主要用来模拟各种HTTP请求（如get、post、delete、put等），Postman与浏览器的区别在于有的浏览器不能输出JSON格式，而Postman可以更直观地看到接口返回的结果。

用户可以从官网https://www.postman.com/下载对应的Postman安装程序。在安装成功后，不需要创建账号即可使用。

❶ 获得列表数据

在RESTful架构中，每个网址代表一种资源（resource），所以在网址中不能有动词，只能有名词，而且所用的名词往往与实体名对应。一般来说，数据库中的表都是同种记录的"集合"（collection），所以API中的名词也应该使用复数，如students。

运行ch8_4的主类Ch84Application，然后在student_table中添加几条学生信息，在Postman中使用GET方式访问"http://localhost:8080/api/students"请求路径获得所有学生信息，如图8.19所示。

图8.19 获得所有学生信息界面

❷ 获得单一对象

在Postman中使用GET方式访问"http://localhost:8080/api/students/1"请求路径获得id为1的学生信息，如图8.20所示。

❸ 查询

在Postman中search调用自定义的接口查询方法，因此可以使用GET访问"http://localhost:8080/api/students/search/snameStartsWith?sname=陈"请求路径调用List＜Student＞findBySnameStartsWith(@Param("sname") String sname)接口方法，获得姓名前缀为"陈"的学生信息，如图8.21所示。

❹ 分页查询

在Postman中使用GET方式访问"http://localhost:8080/api/students?page=0&size=2"请求路径获得第一页的学生信息（page=0即第一页，size=2即每页的记录数量为2），如图8.22所示。

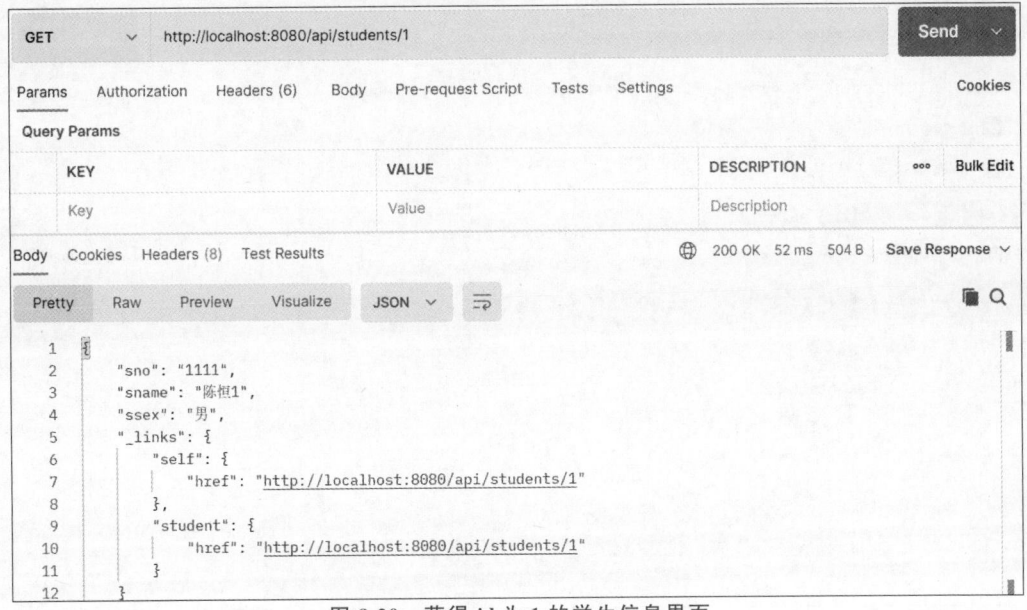

图 8.20　获得 id 为 1 的学生信息界面

图 8.21　获得姓名前缀为"陈"的学生信息界面

从图 8.22 所示的结果可以看出，不仅获得了当前分页的数据，而且给出了上一页、下一页、第一页、最后一页的 REST 资源路径。

❺ 排序

在 Postman 中使用 GET 方式访问"http://localhost:8080/api/students? sort＝sno,desc"请求路径获得按照 sno 属性倒序的列表，如图 8.23 所示。

❻ 保存

在 Postman 中发起 POST 方式请求"http://localhost:8080/api/students"实现新增功能，将要保存的数据放置在请求体中，数据类型为 JSON，如图 8.24 所示。

从图 8.24 可以看出，保存成功后，新数据的 id 为 4。

图 8.22　分页查询界面

图 8.23　排序查询界面

❼ 更新

假如需要更新新增的 id 为 4 的数据，可以在 Postman 中使用 PUT 方式访问"http://localhost:8080/api/students/4"，修改提交的数据，如图 8.25 所示。

❽ 删除

假如需要删除新增的 id 为 4 的数据，可以在 Postman 中使用 DELETE 方式访问"http://localhost:8080/api/students/4"，删除数据，如图 8.26 所示。

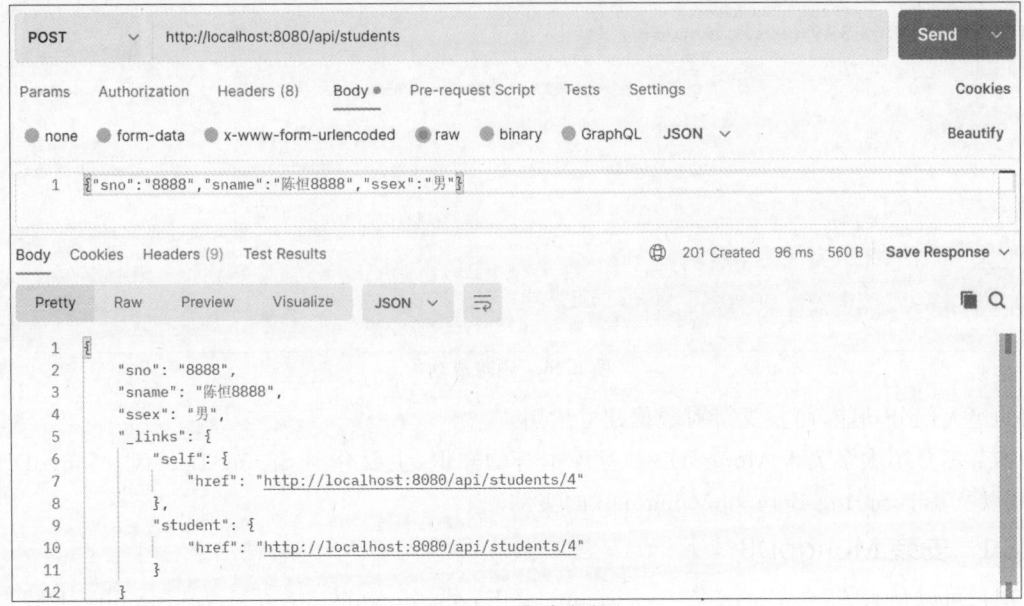

图 8.24　保存数据

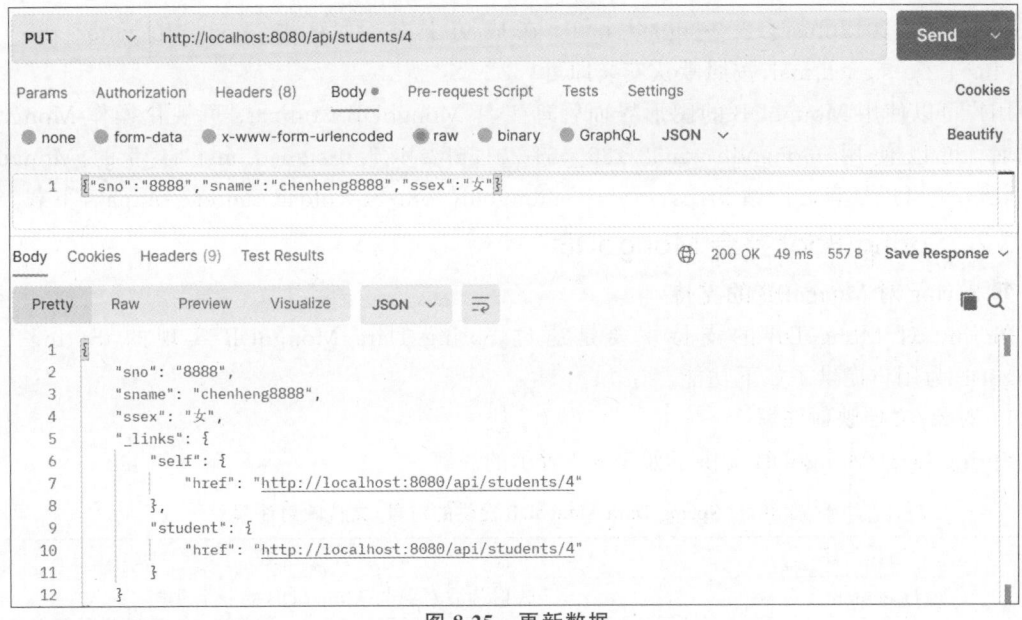

图 8.25　更新数据

8.3　MongoDB

　　MongoDB 是一个基于分布式文件存储的 NoSQL 数据库，用 C++ 语言编写，旨在为 Web 应用提供可扩展的高性能数据存储解决方案。

　　MongoDB 是一个介于关系数据库和非关系数据库之间的产品，是非关系数据库中功能最丰富、最像关系数据库的。它支持的数据结构非常松散，是类似 JSON 的 BSON（Binary JSON，二进制 JSON）格式，因此可以存储比较复杂的数据类型。MongoDB 最大的特点是它支持的查询语言非常强大，其语法有点类似于面向对象的查询语言，几乎可以实现类似关系型数据库单表

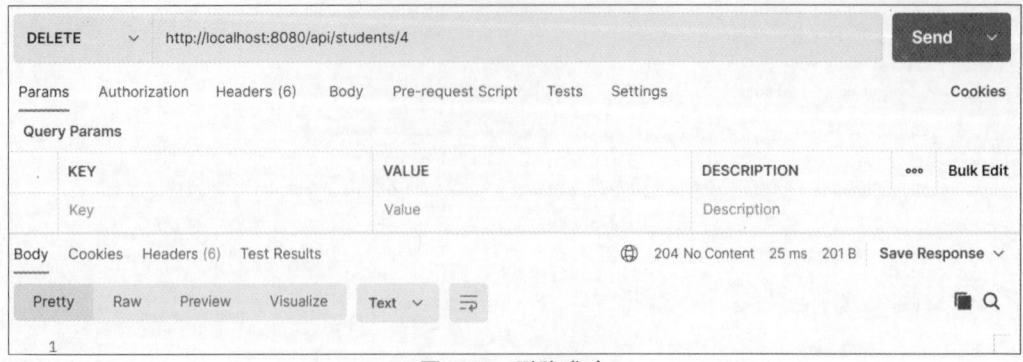

图 8.26　删除成功

查询的绝大部分功能,而且支持对数据建立索引。

本节不介绍太多关于 MongoDB 数据库本身的知识,主要介绍 Spring Boot 对 MongoDB 的支持,以及基于 Spring Boot 和 MongoDB 的实例。

▶8.3.1　安装 MongoDB

用户可以从官方网站 https://www.mongodb.com/download-center/community 下载自己计算机的操作系统所对应版本的 MongoDB,作者在编写本书时使用的 MongoDB 是 mongodb-win32-x86_64-2012plus-4.2.0-signed.msi。在成功下载后,双击 mongodb-win32-x86_64-2012plus-4.2.0-signed.msi,按照默认安装即可。

用户可以使用 MongoDB 的图形界面管理工具 MongoDB Compass 可视化操作 MongoDB 数据库,可以使用 mongodb-win32-x86_64-2012plus-4.2.0-signed.msi 自带的 MongoDB Compass,也可以从官方网站 https://www.mongodb.com/download-center/compass 下载。

▶8.3.2　Spring Boot 整合 MongoDB

❶ Spring 对 MongoDB 的支持

Spring 对 MongoDB 的支持主要是通过 Spring Data MongoDB 实现的,Spring Data MongoDB 为用户提供了如下功能。

1)对象/文档映射注解

Spring Data MongoDB 提供了如表 8.2 所示的注解。

表 8.2　Spring Data MongoDB 提供的对象/文档映射注解

注　　解	含　　义
@Document	映射领域对象与 MongoDB 的一个文档
@Id	映射当前属性是文档对象 id
@DBRef	当前属性将参考其他文档
@Field	为文档的属性定义名称
@Version	将当前属性作为版本

2)MongoTemplate

与 JdbcTemplate 一样,Spring Data MongoDB 为用户提供了一个 MongoTemplate,而 MongoTemplate 提供了数据访问的方法。

3)Repository

类似于 Spring Data JPA,Spring Data MongoDB 为用户提供了 Repository 的支持,其使用

方式和 Spring Data JPA 一样,示例如下:

```
public interface PersonRepository extends MongoRepository<Person, String>{
}
```

❷ Spring Boot 对 MongoDB 的支持

Spring Boot 对 MongoDB 的自动配置位于 org.springframework.boot.autoconfigure.mongo 包中,主要配置了数据库连接、MongoTemplate,用户可以在配置文件中使用以"spring.data.mongodb"为前缀的属性来配置 MongoDB 的相关信息。Spring Boot 对 MongoDB 提供了一些默认属性,如默认端口号为 27017、默认服务器为 localhost、默认数据库为 test、默认无用户名和无密码访问方式,并默认开启了对 Repository 的支持。因此,用户在 Spring Boot 应用中只需引入 spring-boot-starter-data-mongodb 依赖即可按照默认配置操作 MongoDB 数据库。

▶8.3.3 增、删、改、查

本节通过实例讲解如何在 Spring Boot 应用中对 MongoDB 数据库进行增、删、改、查。

【例 8-8】 在 Spring Boot 应用中对 MongoDB 数据库进行增、删、改、查。

其具体实现步骤如下。

❶ 创建基于 spring-boot-starter-data-mongodb 依赖的 Spring Boot Web 应用 ch8_5

在 IDEA 中创建基于 Lombok、Spring Web 与 Spring Data Mongo 依赖的 Spring Boot Web 应用 ch8_5。

❷ 配置 application.properties 文件

在 Spring Boot Web 应用 ch8_5 中使用 MongoDB 的默认数据库连接,因此不需要在 application.properties 文件中配置数据库连接信息。application.properties 文件的具体内容如下:

```
server.servlet.context-path=/ch8_5
#让控制器输出的JSON字符串格式更美观
spring.jackson.serialization.indent-output=true
```

❸ 创建领域模型

在应用 ch8_5 的 src/main/java 目录中创建名为 com.ch.ch8_5.domain 的包,并在该包中创建领域模型 Person(人)以及 Person 去过的 Location(地点)。在 Person 类中使用@Document 注解将 Person 领域模型和 MongoDB 的文档进行映射。

Person 的核心代码如下:

```
@Document
@Data
public class Person {
    @Id
    private String pid;
    private String pname;
    private Integer page;
    private String psex;
    @Field("plocs")
    private List<Location> locations = new ArrayList<Location>();
    public Person() {
        super();
    }
    public Person(String pname, Integer page, String psex) {
        super();
```

```
        this.pname = pname;
        this.page = page;
        this.psex = psex;
    }
}
```

Location 的核心代码如下:

```
@Data
public class Location {
    private String locName;
    private String year;
    public Location() {
        super();
    }
    public Location(String locName, String year) {
        super();
        this.locName = locName;
        this.year = year;
    }
}
```

❹ 创建数据访问接口

在应用 ch8_5 的 src/main/java 目录中创建名为 com.ch.ch8_5.repository 的包,并在该包中创建数据访问接口 PersonRepository,该接口继承 MongoRepository 接口。PersonRepository 接口的核心代码如下:

```
public interface PersonRepository extends MongoRepository<Person, String>{
    Person findByPname(String pname);              //支持方法名查询,方法名的命名规则参照表 8.1
    @Query("{'psex':?0}")                          //JSON 字符串
    List<Person> selectPersonsByPsex(String psex);
}
```

❺ 创建控制器层

由于本实例业务简单,直接在控制器层调用数据访问层。创建名为 com.ch.ch8_5.controller 的包,并在该包中创建控制器类 TestMongoDBController。

TestMongoDBController 的核心代码如下:

```
@RestController
public class TestMongoDBController {
    @Autowired
    private PersonRepository personRepository;
    @RequestMapping("/save")
    public List<Person> save() {
        List<Location> locations1 = new ArrayList<Location>();
        Location loc1 = new Location("北京","2023");
        Location loc2 = new Location("上海","2024");
        locations1.add(loc1);
        locations1.add(loc2);
        List<Location> locations2 = new ArrayList<Location>();
        Location loc3 = new Location("广州","2025");
        Location loc4 = new Location("深圳","2026");
        locations2.add(loc3);
        locations2.add(loc4);
        List<Person> persons = new ArrayList<Person>();
        Person p1 = new Person("陈恒 1", 88, "男");
        p1.setLocations(locations1);
        Person p2 = new Person("陈恒 2", 99, "女");
```

```java
        p2.setLocations(locations2);
        persons.add(p1);
        persons.add(p2);
        return personRepository.saveAll(persons);
    }
    @RequestMapping("/findByPname")
    public Person findByPname(String pname) {
        return personRepository.findByPname(pname);
    }
    @RequestMapping("/selectPersonsByPsex")
    public List<Person> selectPersonsByPsex(String psex) {
        return personRepository.selectPersonsByPsex(psex);
    }
    @RequestMapping("/updatePerson")
    public Person updatePerson(String oldPname, String newPname) {
        Person p1 = personRepository.findByPname(oldPname);
        if(p1 != null)
            p1.setPname(newPname);
        return personRepository.save(p1);
    }
    @RequestMapping("/deletePerson")
    public void updatePerson(String pname) {
        Person p1 = personRepository.findByPname(pname);
        personRepository.delete(p1);
    }
}
```

❻ 运行

首先运行 Ch85Application 主类，然后访问"http://localhost:8080/ch8_5/save"测试保存数据，运行效果如图 8.27 所示。

图 8.27　测试保存

在保存成功后，使用 MongoDB 的图形界面管理工具 MongoDB Compass 打开查看已保存的数据，如图 8.28 所示。

通过"http://localhost:8080/ch8_5/findByPname? pname＝陈恒 1"查询人名为"陈恒 1"的文档数据。

通过"http://localhost:8080/ch8_5/selectPersonsByPsex? psex＝女"查询性别为"女"的文档数据。

通过"http://localhost:8080/ch8_5/updatePerson? oldPname＝陈恒 1&newPname＝陈恒

图8.28 查看已保存的数据

111"将人名为"陈恒 1"的数据修改成人名为"陈恒 111"。

通过"http://localhost:8080/ch8_5/deletePerson？pname＝陈恒 111"将人名为"陈恒 111"的文档数据删除。

至此，通过 Spring Boot Web 应用对 MongoDB 数据库的操作演示完毕。

8.4 Redis

Redis 是一个开源的使用 ANSI C 语言编写、支持网络、可基于内存也可持久化的日志型、Key-Value 数据库，并提供多种语言的 API。它支持字符串、哈希表、列表、集合、有序集合、位图、地理空间信息等数据类型，同时也可以作为高速缓存和消息队列代理。Redis 在内存中存储数据，因此存放在 Redis 中的数据不应该大于内存容量，否则会导致操作系统的性能降低。

本节不介绍太多关于 Redis 数据库本身的知识，主要介绍 Spring Boot 对 Redis 的支持，以及基于 Spring Boot 和 Redis 的实例。

▶ 8.4.1 安装 Redis

❶ 下载 Redis

在编写本书时，Redis 官方网站只提供 Linux 版本的下载，因此作者只能通过 https://github.com/MSOpenTech/redis/tags 从 github 上下载 Redis，本书下载的版本是 Redis-x64-3.2.100.zip。在"运行"对话框中输入 cmd，然后把目录指向解压的 Redis 目录，如图 8.29 所示。

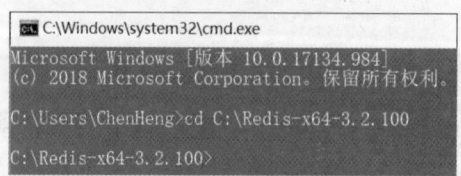

图 8.29 将目录指向解压的 Redis 目录

❷ 启动 Redis 服务

使用 redis-server redis.windows.conf 命令行启动 Redis 服务，出现如图 8.30 所示的显示，表示成功启动了 Redis 服务。

虽然启动了 Redis 服务，但关闭 cmd 窗口 Redis 服务就会消失，所以需要把 Redis 设置成

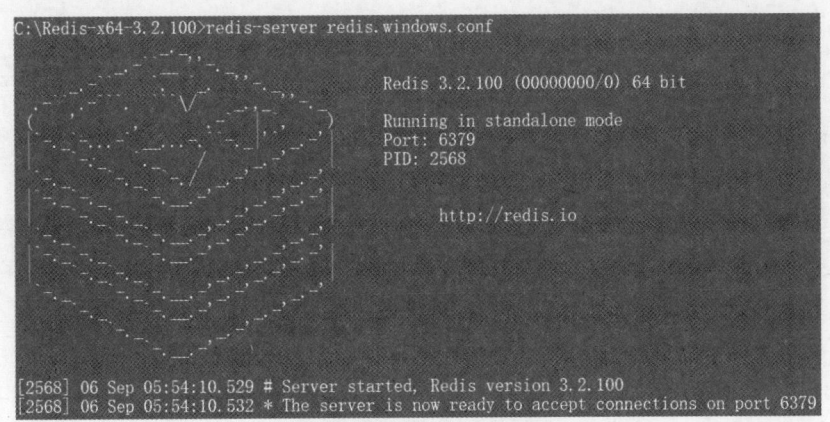

图 8.30　启动 Redis 服务

Windows 下的服务。

关闭 cmd 窗口重新打开，进入 Redis 解压目录，执行设置服务命令 redis-server --service-install redis.windows-service.conf --loglevel verbose，如图 8.31 所示。

图 8.31　将 Redis 服务设置成 Windows 下的服务

图 8.31 中没有报错，表示成功设置成 Windows 下的服务，刷新服务，可以看到 Redis 服务，如图 8.32 所示。

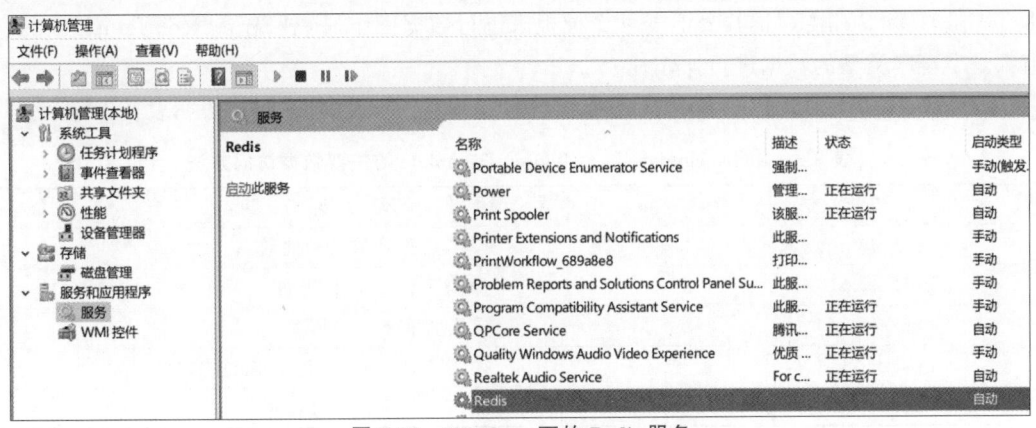

图 8.32　Windows 下的 Redis 服务

❸ 常用的 Redis 服务命令

卸载服务：redis-server --service-uninstall

开启服务：redis-server --service-start

停止服务：redis-server --service-stop

❹ 启动 Redis 服务

用户可以在 cmd 窗口中使用 redis-server --service-start 命令启动 Redis 服务，如图 8.33 所示，也可以在图 8.32 中启动 Redis 服务。

图 8.33 启动 Redis 服务

❺ 操作测试 Redis

如图 8.34 所示,在启动 Redis 服务后,首先使用 redis-cli.exe -h 127.0.0.1 -p 6379 命令创建一个地址为 127.0.0.1、端口号为 6379 的 Redis 数据库服务,然后使用 set key value 和 get key 命令保存和获得数据。

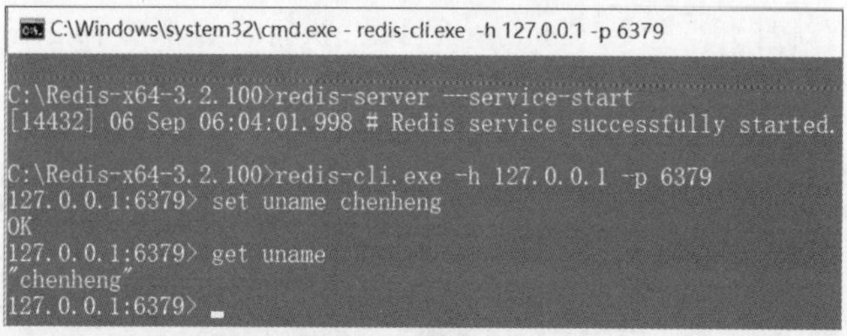

图 8.34 测试 Redis

8.4.2 Spring Boot 整合 Redis

❶ Spring Data Redis

Spring 对 Redis 的支持是通过 Spring Data Redis 来实现的。Spring Data Redis 为用户提供了 RedisTemplate 和 StringRedisTemplate 两个模板来进行数据操作,其中 StringRedisTemplate 只针对键值都是字符串类型的数据进行操作。

RedisTemplate 和 StringRedisTemplate 模板提供的主要数据访问方法如表 8.3 所示。

表 8.3 RedisTemplate 和 StringRedisTemplate 的主要数据访问方法

方　法	说　明
opsForValue()	操作只有简单属性的数据
opsForList()	操作含有 List 的数据
opsForSet()	操作含有 Set 的数据
opsForZSet()	操作含有 Zset(有序的 Set)的数据
opsForHash()	操作含有 Hash 的数据

❷ Serializer

当数据存储到 Redis 时,键和值都是通过 Spring 提供的 Serializer 序列化到数据的。RedisTemplate 默认使用 JdkSerializationRedisSerializer 序列化,StringRedisTemplate 默认使用 StringRedisSerializer 序列化。

❸ Spring Boot 的支持

Spring Boot 对 Redis 的支持位于 org.springframework.boot.autoconfigure.data.redis 包下,在 RedisAutoConfiguration 配置类中为用户默认配置了 RedisTemplate 和 StringRedisTemplate,让用户

可以直接使用 Redis 存储数据。

在 RedisProperties 类中看到可以使用以 "spring.redis" 为前缀的属性在 application.properties 中配置 Redis，主要属性默认配置如下：

```
spring.redis.database = 0           #数据库名 db0
spring.redis.host = localhost       #服务器地址
spring.redis.port = 6379            #连接端口号
spring.redis.max-idle = 8           #连接池的最大连接数
spring.redis.min-idle = 0           #连接池的最小连接数
spring.redis.max-active = 8         #在给定时间连接池可以分配的最大连接数
spring.redis.max-wait = -1          #当池被耗尽时，抛出异常之前连接分配应该阻塞的最大时间量（以毫秒为
                                    #单位）。使用负值表示无限期地阻止
```

从上述默认属性值可以看出配置了数据库名为 db0、服务器地址为 localhost、端口号为 6379 的 Redis。

因此，在 Spring Boot 应用中只要引入 spring-boot-starter-data-redis 依赖就可以使用默认配置的 Redis 进行数据操作。

▶ 8.4.3 使用 StringRedisTemplate 和 RedisTemplate

本节通过实例讲解如何在 Spring Boot 应用中使用 StringRedisTemplate 和 RedisTemplate 模板访问操作 Redis 数据库。

【例 8-9】 在 Spring Boot 应用中使用 StringRedisTemplate 和 RedisTemplate 模板访问操作 Redis 数据库。

其具体实现步骤如下。

❶ 创建基于 spring-boot-starter-data-redis 依赖的 Spring Boot Web 应用 ch8_6

在 IDEA 中创建基于 Lombok、Spring Web 与 Spring Data Redis 依赖的 Spring Boot Web 应用 ch8_6。

❷ 配置 application.properties 文件

在 Spring Boot 应用 ch8_6 中使用 Redis 的默认数据库连接，所以不需要在 application.properties 文件中配置数据库连接信息。

❸ 创建实体类

在应用 ch8_6 的 src/main/java 目录中创建名为 com.ch.ch8_6.entity 的包，并在该包中创建名为 Student 的实体类。该类必须实现序列化接口，这是因为使用 Jackson 做序列化需要一个空构造。

Student 的核心代码如下：

```
@Data
public class Student implements Serializable{
    private String sno;
    private String sname;
    private Integer sage;
    public Student() {
        super();
    }
    public Student(String sno, String sname, Integer sage) {
        super();
        this.sno = sno;
        this.sname = sname;
        this.sage = sage;
```

        }
    }

❹ 创建数据访问层

在应用 ch8_6 的 src/main/java 目录中创建名为 com.ch.ch8_6.repository 的包,并在该包中创建名为 StudentRepository 的类,该类使用@Repository 注解标注为数据访问层。

StudentRepository 的核心代码如下:

```
@Repository
public class StudentRepository{
    @SuppressWarnings("unused")
    @Autowired
    private StringRedisTemplate stringRedisTemplate;
    @SuppressWarnings("unused")
    @Autowired
    private RedisTemplate<Object, Object> redisTemplate;
    /**
     * 使用@Resource 注解指定 stringRedisTemplate,可注入基于字符串的简单属性操作方法
     * ValueOperations<String, String> valueOpsStr = stringRedisTemplate.opsForValue();
     */
    @Resource(name="stringRedisTemplate")
    ValueOperations<String, String> valueOpsStr;
    /**
     * 使用@Resource 注解指定 redisTemplate,可注入基于对象的简单属性操作方法
     * ValueOperations<Object, Object> valueOpsObject = redisTemplate.opsForValue();
     */
    @Resource(name="redisTemplate")
    ValueOperations<Object, Object> valueOpsObject;
    /**
     * 保存字符串到 Redis
     */
    public void saveString(String key, String value) {
        valueOpsStr.set(key, value);
    }
    /**
     * 保存对象到 Redis
     */
    public void saveStudent(Student stu) {
        valueOpsObject.set(stu.getSno(), stu);
    }
    /**
     * 保存 List 数据到 Redis
     */
    public void saveMultiStudents(Object key, List<Student> stus) {
        valueOpsObject.set(key, stus);
    }
    /**
     * 从 Redis 中获得字符串数据
     */
    public String getString(String key) {
        return valueOpsStr.get(key);
    }
    /**
     * 从 Redis 中获得对象数据
     */
    public Object getObject(Object key) {
        return valueOpsObject.get(key);
    }
}
```

第 8 章 Spring Boot 的数据访问

❺ 创建控制器层

由于本实例业务简单，直接在控制器层调用数据访问层。在应用 ch8_6 的 src/main/java 目录中创建名为 com.ch.ch8_6.controller 的包，并在该包中创建控制器类 TestRedisController。TestRedisController 的核心代码如下：

```java
@RestController
public class TestRedisController {
    @Autowired
    private StudentRepository studentRepository;
    @GetMapping("/save")
    public void save() {
        studentRepository.saveString("uname", "陈恒");
        Student s1 = new Student("111", "陈恒 1", 77);
        studentRepository.saveStudent(s1);
        Student s2 = new Student("222", "陈恒 2", 88);
        Student s3 = new Student("333", "陈恒 3", 99);
        List<Student> stus = new ArrayList<Student>();
        stus.add(s2);
        stus.add(s3);
        studentRepository.saveMultiStudents("multiStus", stus);
    }
    @GetMapping("/getUname")
    public String getUname(String key) {
        return studentRepository.getString(key);
    }
    @GetMapping("/getStudent")
    public Student getStudent(String key) {
        return (Student) studentRepository.getObject(key);
    }
    @GetMapping("/getMultiStus")
    public List<Student> getMultiStus(String key) {
        return (List<Student>) studentRepository.getObject(key);
    }
}
```

❻ 修改配置类 Ch89Application

RedisTemplate 默认使用 JdkSerializationRedisSerializer 序列化数据，这对使用 Redis Client 查看数据很不直观，因为 JdkSerializationRedisSerializer 使用二进制形式存储数据。所以，在此自己配置 RedisTemplate，并定义 Serializer。

修改后的配置类 Ch86Application 的核心代码如下：

```java
@SpringBootApplication
public class Ch86Application {
    public static void main(String[] args) {
        SpringApplication.run(Ch86Application.class, args);
    }
    @Bean
    public RedisTemplate < Object, Object > redisTemplate ( RedisConnectionFactory redisConnectionFactory ) {
        RedisTemplate<Object, Object> rTemplate = new RedisTemplate<Object, Object>();
        rTemplate.setConnectionFactory(redisConnectionFactory);
        @SuppressWarnings({ "unchecked", "rawtypes" })
        Jackson2JsonRedisSerializer<Object> jackson2JsonRedisSerializer =
            new Jackson2JsonRedisSerializer(Object.class);
        ObjectMapper om = new ObjectMapper();
        om.setVisibility(PropertyAccessor.ALL, JsonAutoDetect.Visibility.ANY);
        om.enableDefaultTyping(ObjectMapper.DefaultTyping.NON_FINAL);
        jackson2JsonRedisSerializer.setObjectMapper(om);
```

```
        //设置值的序列化采用 Jackson2JsonRedisSerializer
        rTemplate.setValueSerializer(jackson2JsonRedisSerializer);
        //设置键的序列化采用 StringRedisSerializer
        rTemplate.setKeySerializer(new StringRedisSerializer());
        return rTemplate;
    }
}
```

❼ 运行测试

首先运行 Ch86Application 主类,然后通过"http://localhost:8080/save"测试存储数据。

通过"http://localhost:8080/getUname?key=uname"查询 key 为 uname 的字符串值,如图 8.35 所示。

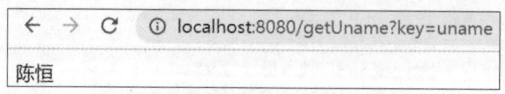

图 8.35 根据 key 查询字符串值

通过"http://localhost:8080/getStudent?key=111"查询 key 为 111 的 Student 对象值,如图 8.36 所示。

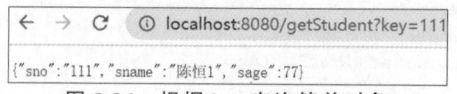

图 8.36 根据 key 查询简单对象

通过"http://localhost:8080/getMultiStus?key=multiStus"查询 key 为 multiStus 的 List 集合,如图 8.37 所示。

图 8.37 根据 key 查询 List 集合

8.5 数据缓存 Cache

内存的读取速度远大于硬盘的读取速度,当需要重复地获取相同数据时,一次一次地请求数据库或者远程服务,会导致大量的时间消耗在数据库查询或者远程方法的调用上,最终导致程序的性能降低,这就是数据缓存要解决的问题。

本节将介绍 Spring Boot 应用中 Cache 的一般概念,Spring Cache 对 Cache 进行了抽象,提供了 CacheManager 和 Cache 接口,并提供了 @Cacheable、@CachePut、@CacheEvict、@Caching、@CacheConfig 等注解。Spring Boot 应用基于 Spring Cache,既提供了基于内存实现的缓存管理器用于单体应用系统,也集成了 Redis、EhCache 等缓存服务器用于大型系统或分布式系统。

▶8.5.1 Spring 缓存支持

Spring 框架定义了 org.springframework.cache.CacheManager 和 org.springframework.cache.Cache 接口来统一不同的缓存技术。针对不同的缓存技术,需要实现不同的 CacheManager。例如,在使用 EhCache 作为缓存技术时,需要注册实现 CacheManager 的 Bean,示例代码如下:

```
@Bean
public EhCacheCacheManager cacheManager(CacheManager ehCacheCacheManager) {
    return new EhCacheCacheManager(ehCacheCacheManager);
}
```

CacheManager 的常用实现如表 8.4 所示。

表 8.4 CacheManager 的常用实现

CacheManager	描 述
SimpleCacheManager	使用简单的 Collection 来存储缓存，主要用于测试
NoOpCacheManager	仅用于测试，不会实际存储缓存
ConcurrentMapCacheManager	使用 ConcurrentMap 存储缓存，Spring 默认采用此技术存储缓存
EhCacheCacheManager	使用 EhCache 作为缓存技术
JCacheCacheManager	支持 JCache(JSR-107)标准实现作为缓存技术，如 Apache Commons JCS
RedisCacheManager	使用 Redis 作为缓存技术
HazelcastCacheManager	使用 Hazelcast 作为缓存技术

一旦配置好 Spring 缓存支持，就可以在 Spring 容器管理的 Bean 中使用缓存注解（基于 AOP 原理），在一般情况下，都是在业务层（Service 类）使用这些注解。

❶ @Cacheable

@Cacheable 可以标记在一个方法上，也可以标记在一个类上。当标记在一个方法上时表示该方法是支持缓存的，当标记在一个类上时表示该类的所有方法都是支持缓存的。对于一个支持缓存的方法，在方法执行前，Spring 先检查缓存中是否存在方法返回的数据，如果存在，则直接返回缓存数据；如果不存在，则调用方法并将方法的返回值存入缓存。

@Cacheable 注解经常使用 value、key、condition 等属性。

value：缓存的名称，指定一个或多个缓存名称。例如@Cacheable(value="mycache")或者@Cacheable(value={"cache1","cache2"})。该属性与 cacheNames 属性的意义相同。

key：缓存的 key，可以为空，如果指定需要按照 SpEL 表达式编写；如果不指定，则默认按照方法的所有参数进行组合。例如@Cacheable(value="testcache",key="#student.id")。

condition：缓存的条件，可以为空，如果指定需要按照 SpEL 编写，返回 true 或者 false，只有为 true 才进行缓存。例如@Cacheable(value="testcache",condition="#student.id>2")。该属性与 unless 相反，当条件成立时不进行缓存。

❷ @CacheEvict

@CacheEvict 是用来标记在需要清除缓存元素的方法或类上的。当标记在一个类上时表示其中所有方法的执行都会触发缓存的清除操作。@CacheEvict 可以指定的属性有 value、key、condition、allEntries 和 beforeInvocation。其中 value、key 和 condition 的语义与@Cacheable 对应的属性类似。

allEntries：是否清空所有缓存内容，默认为 false，如果指定为 true，则方法调用后将立即清空所有缓存。例如@CacheEvict(value="testcache"，allEntries=true)。

beforeInvocation：是否在方法执行前就清空，默认为 false，如果指定为 true，则在方法还没有执行时就清空缓存。在默认情况下，如果方法执行抛出异常，则不会清空缓存。

❸ @CachePut

@CachePut 也可以声明一个方法支持缓存功能，与@Cacheable 不同的是，使用@CachePut

标记的方法在执行前不会去检查缓存中是否存在之前执行过的结果,而是每次都会执行该方法,并将执行结果以键-值对的形式存入指定的缓存中。

@CachePut 也可以标记在类上和方法上。@CachePut 的属性与@Cacheable 的属性一样。

❹ @Caching

@Caching 注解可以让用户在一个方法或者类上同时指定多个 Spring Cache 相关的注解。其拥有 cacheable、put 和 evict 三个属性,分别用于指定@Cacheable、@CachePut 和@CacheEvict。示例如下:

```
@Caching(
cacheable = @Cacheable("cache1"),
evict = { @CacheEvict("cache2"),@CacheEvict(value = "cache3", allEntries = true) }
)
```

❺ @CacheConfig

所有的 Cache 注解都需要提供 Cache 名称,如果每个 Service 方法上都包含相同的 Cache 名称,写起来可能重复。此时可以使用@CacheConfig 注解作用在类上,设置当前缓存的一些公共设置。

▶8.5.2 Spring Boot 缓存支持

在 Spring 中使用缓存技术的关键是配置缓存管理器 CacheManager,而 Spring Boot 为用户自动配置了多个 CacheManager 的实现。Spring Boot 的 CacheManager 的自动配置位于 org.springframework.boot.autoconfigure.cache 包中。Spring Boot 自动配置了 EhCacheCacheConfiguration、GenericCacheConfiguration、HazelcastCacheConfiguration、HazelcastJCacheCustomizationConfiguration、InfinispanCacheConfiguration、JcacheCacheConfiguration、NoOpCacheConfiguration、RedisCacheConfiguration 和 SimpleCacheConfiguration。在默认情况下,Spring Boot 使用的是 SimpleCacheConfiguration,即使用 ConcurrentMapCacheManager。Spring Boot 支持以"spring.cache"为前缀的属性来进行缓存的相关配置。

在 Spring Boot 应用中,使用缓存技术只需在应用中引入相关缓存技术的依赖,并在配置类中使用@EnableCaching 注解开启缓存支持即可。

下面通过实例讲解如何在 Spring Boot 应用中使用默认的缓存技术 ConcurrentMapCacheManager。

【例 8-10】 在 Spring Boot 应用中使用默认的缓存技术 ConcurrentMapCacheManager。

其具体实现步骤如下。

❶ 创建基于 Lombok、spring-boot-starter-cache 和 spring-boot-starter-data-jpa 依赖的 Spring Boot Web 应用 ch8_7

在 IDEA 中创建基于 Lombok、MySQL Driver、Spring Web、Spring cache abstraction 与 Spring Data JPA 依赖的 Spring Boot Web 应用 ch8_7。

❷ 配置 application.properties 文件

在应用 ch8_7 中使用 Spring Data JPA 访问 MySQL 数据库,所以在 application.properties 文件中配置数据库连接信息,但因为使用默认的缓存技术 ConcurrentMapCacheManager,所以不需要缓存的相关配置。

应用 ch8_7 的 application.properties 文件内容与例 8-1 中的基本一样,这里不再赘述。

❸ 创建持久化实体类

在应用 ch8_7 的 src/main/java 目录中创建名为 com.ch.ch8_7.entity 的包,并在该包中创建

持久化实体类 Student。

Student 的核心代码如下：

```java
@Entity
@Table(name = "student_table")
@JsonIgnoreProperties(value = {"hibernateLazyInitializer"})
@Data
public class Student implements Serializable{
    private static final long serialVersionUID = 1L;
    @Id
    @GeneratedValue(strategy = GenerationType.IDENTITY)
    private int id;                                       //主键
    private String sno;
    private String sname;
    private String ssex;
    public Student() {
        super();
    }
    public Student(int id, String sno, String sname, String ssex) {
        super();
        this.id = id;
        this.sno = sno;
        this.sname = sname;
        this.ssex = ssex;
    }
}
```

❹ 创建数据访问接口

在应用 ch8_7 的 src/main/java 目录中创建名为 com.ch.ch8_7.repository 的包，并在该包中创建名为 StudentRepository 的数据访问接口。

StudentRepository 的核心代码如下：

```java
public interface StudentRepository extends JpaRepository<Student, Integer>{}
```

❺ 创建业务层

在应用 ch8_7 的 src/main/java 目录中创建名为 com.ch.ch8_7.service 的包，并在该包中创建 StudentService 接口和该接口的实现类 StudentServiceImpl。StudentService 接口的代码略。

StudentServiceImpl 的核心代码如下：

```java
@Service
public class StudentServiceImpl implements StudentService{
    @Autowired
    private StudentRepository studentRepository;
    @Override
    @CachePut(value = "student", key="#student.id")
    public Student saveStudent(Student student) {
        Student s = studentRepository.save(student);
        System.out.println("为 key=" + student.getId() + "数据做了缓存");
        return s;
    }
    @Override
    @CacheEvict(value = "student", key="#student.id")
    public void deleteCache(Student student) {
        System.out.println("删除了 key=" + student.getId() + "的数据缓存");
    }
    @Override
    @Cacheable(value = "student")
```

```
    public Student selectOneStudent(Integer id) {
        Student s = studentRepository.getOne(id);
        System.out.println("为 key=" + id + "数据做了缓存");
        return s;
    }
}
```

在上述 Service 的实现类中,使用@CachePut 注解将新增的或更新的数据保存到缓存,其中缓存名为 student,数据的 key 是 student 的 id;使用@CacheEvict 注解从缓存 student 中删除 key 为 student 的 id 的数据;使用@Cacheable 注解将 key 为 student 的 id 的数据缓存到名为 student 的缓存中。

❻ 创建控制器层

在应用 ch8_7 的 src/main/java 目录中创建名为 com.ch.ch8_7.controller 的包,并在该包中创建名为 TestCacheController 的控制器类。

TestCacheController 的核心代码如下:

```
@RestController
public class TestCacheController {
    @Autowired
    private StudentService studentService;
    @GetMapping("/savePut")
    public Student save(Student student) {
        return studentService.saveStudent(student);
    }
    @GetMapping("/selectAble")
    public Student select(Integer id) {
        return studentService.selectOneStudent(id);
    }
    @GetMapping("/deleteEvict")
    public String deleteCache(Student student) {
        studentService.deleteCache(student);
        return "ok";
    }
}
```

❼ 开启缓存支持

在应用的主类 Ch87Application 中使用@EnableCaching 注解开启缓存支持,核心代码如下:

```
@EnableCaching
@SpringBootApplication
public class Ch87Application {}
```

❽ 运行测试

1)测试@Cacheable

启动应用程序的主类,第一次访问 http://localhost:8080/ch8_7/selectAble?id=3,在第一次访问时将调用方法查询数据库,并将查询到的数据存储到缓存 student 中。此时控制台输出结果如图 8.38 所示。

页面显示数据如图 8.39 所示。

再次访问 http://localhost:8080/ch8_7/selectAble?id=3,此时控制台没有输出"为 key=3 数据做了缓存"以及 Hibernate 的查询语句,这表明没有调用查询方法,页面数据直接从数据缓存中获得。

图 8.38　第一次访问查询时控制台的输出结果

图 8.39　第一次访问查询时的页面数据

2）测试@CachePut

重启应用程序的主类，访问 http://localhost:8080/ch8_7/savePut?sname＝陈恒 5&sno＝555&ssex＝男，此时控制台输出结果如图 8.40 所示，页面数据如图 8.41 所示。

图 8.40　测试@CachePut 时控制台的输出结果

图 8.41　测试@CachePut 时的页面数据

这时访问 http://localhost:8080/ch8_7/selectAble?id＝8，控制台无输出，从缓存直接获得数据，页面数据如图 8.41 所示。

3）测试@CacheEvict

重启应用程序的主类，首先访问 http://localhost:8080/ch8_7/selectAble?id＝1，为 key 为 1 的数据做缓存，再次访问 http://localhost:8080/ch8_7/selectAble?id＝1，确认数据已从缓存中获取。然后访问 http://localhost:8080/ch8_7/deleteEvict?id＝1，从缓存 student 中删除 key 为 1 的数据，此时控制台输出结果如图 8.42 所示。

图 8.42　测试@CacheEvict 删除缓存数据

最后，再次访问 http://localhost:8080/ch8_7/selectAble?id＝1，此时重新为 key＝1 的数据做了缓存。

8.5.3　使用 Redis Cache

在 Spring Boot 中使用 Redis Cache，只需要添加 spring-boot-starter-data-redis 依赖即可。

下面通过实例测试 Redis Cache。

【例 8-11】 在 8.4.3 节中例 8-9 的基础上测试 Redis Cache。

其具体实现步骤如下。

❶ 使用@Cacheable 注解修改控制器方法

将控制器中的 getUname 方法修改如下：

```
@GetMapping("/getUname")
@Cacheable(value = "myuname")
public String getUname(String key) {
    System.out.println("测试缓存");
    return studentRepository.getString(key);
}
```

❷ 使用@EnableCaching 注解开启缓存支持

在应用程序的主类 Ch86Application 上使用@EnableCaching 注解开启缓存支持。

❸ 测试 Redis Cache

在启动应用程序的主类后，多次访问"http://localhost:8080/getUname?key=uname"，但"测试缓存"字样在控制台中仅打印一次，页面查询结果不变。这说明只有在第一次访问时调用了查询方法，后面多次访问都是从缓存中直接获得数据。

8.6 Spring Boot 整合 MyBatis

MyBatis 本是 Apache Software Foundation 的一个开源项目 iBatis，2010 年这个项目由 Apache Software Foundation 迁移到 Google Code，并改名为 MyBatis。

MyBatis 是一个基于 Java 的持久层框架。MyBatis 提供的持久层框架包括 SQL Maps 和 Data Access Objects（DAO），它消除了几乎所有的 JDBC 代码和参数的手工设置以及结果集的检索。MyBatis 将简单的 XML 或注解用于配置和原始映射，将接口和 Java 的 POJOs（Plain Old Java Objects，普通的 Java 对象）映射成数据库中的记录。

目前，Java 的持久层框架产品有许多，常见的有 Hibernate 和 MyBatis。MyBatis 是一个半自动映射的框架，因为 MyBatis 需要手动匹配 POJO、SQL 和映射关系；Hibernate 是一个全表映射的框架，只需提供 POJO 和映射关系即可。MyBatis 是一个小巧、方便、高效、简单、直接、半自动化的持久层框架；Hibernate 是一个强大、方便、高效、复杂、间接、全自动化的持久化框架。两个持久层框架各有优缺点，开发者可根据实际应用选择它们。

下面通过实例讲解如何在 Spring Boot 应用中使用 MyBatis 框架操作数据库（基于 XML 的映射配置）。

【例 8-12】 在 Spring Boot 应用中使用 MyBatis 框架操作数据库（基于 XML 的映射配置）。

其具体实现步骤如下。

❶ 创建 Spring Boot Web 应用

在创建 Spring Boot Web 应用 ch8_8 时选择 MyBatis Framework（mybatis-spring-boot-starter）、MySQL Driver、Lombok、Spring Web 等依赖。在该应用中操作的数据库是 springtest，操作的数据表是 user 表。

❷ 设置 Web 应用 ch8_8 的上下文路径及数据源配置信息

在应用 ch8_8 的 application.properties 文件中配置如下内容：

```
server.servlet.context-path=/ch8_8
#数据库地址
spring.datasource.url = jdbc:mysql://localhost:3306/springtest?useUnicode=true&characterEncoding=UTF-8&allowMultiQueries=true&serverTimezone=GMT%2B8
#数据库用户名
spring.datasource.username=root
#数据库密码
spring.datasource.password=root
#数据库驱动
spring.datasource.driver-class-name=com.mysql.cj.jdbc.Driver
#设置包的别名(在Mapper映射文件中直接使用实体类名)
mybatis.type-aliases-package=com.ch.ch8_8.entity
#告诉系统到哪里去找mapper.xml文件(映射文件)
mybatis.mapperLocations=classpath:mappers/*.xml
#在控制台中输出SQL语句日志
logging.level.com.ch.ch8_8.repository=debug
#让控制器输出的JSON字符串格式更美观
spring.jackson.serialization.indent-output=true
```

❸ 创建实体类

创建名为 com.ch.ch8_8.entity 的包，并在该包中创建 MyUser 实体类，核心代码如下：

```
@Data
public class MyUser {
    private Integer uid;                    //与数据表中的字段名相同
    private String uname;
    private String usex;
}
```

❹ 创建数据访问接口

创建名为 com.ch.ch8_8.repository 的包，并在该包中创建 MyUserRepository 接口。MyUserRepository 的核心代码如下：

```
/*
 * @Repository可有可无,但有时提示依赖注入找不到(不影响运行),
 * 加上后可以消去依赖注入的报错信息。
 * 这里不再需要@Mapper,是因为在启动类中使用@MapperScan注解,
 * 将数据访问层的接口都注解为Mapper接口的实现类,
 * @Mapper与@MapperScan两者用一个即可
 */
@Repository
public interface MyUserRepository {
    List<MyUser> findAll();
}
```

❺ 创建 Mapper 映射文件

在 src/main/resources 目录下创建名为 mappers 的包，并在该包中创建 SQL 映射文件 MyUserMapper.xml，具体代码如下：

```
<?xml version="1.0" encoding="UTF-8"?>
<!DOCTYPE mapper
PUBLIC "-//mybatis.org//DTD Mapper 3.0//EN"
"http://mybatis.org/dtd/mybatis-3-mapper.dtd">
<mapper namespace="com.ch.ch8_8.repository.MyUserRepository">
    <select id="findAll" resultType="MyUser">
        select * from user
    </select>
</mapper>
```

❻ 创建业务层

创建名为 com.ch.ch8_8.service 的包,并在该包中创建 MyUserService 接口和 MyUserServiceImpl 实现类。MyUserService 的代码略。

MyUserServiceImpl 的核心代码如下:

```java
@Service
public class MyUserServiceImpl implements MyUserService{
    @Autowired
    private MyUserRepository myUserRepository;
    @Override
    public List<MyUser> findAll() {
        return myUserRepository.findAll();
    }
}
```

❼ 创建控制器类 MyUserController

创建名为 com.ch.ch8_8.controller 的包,并在该包中创建控制器类 MyUserController。MyUserController 的核心代码如下:

```java
@RestController
public class MyUserController {
    @Autowired
    private MyUserService myUserService;
    @GetMapping("/findAll")
    public List<MyUser> findAll(){
        return myUserService.findAll();
    }
}
```

❽ 在应用程序的主类中扫描 Mapper 接口

在应用程序的 Ch88Application 主类中使用@MapperScan 注解扫描 MyBatis 的 Mapper 接口,核心代码如下:

```java
@SpringBootApplication
//配置扫描 MyBatis 接口的包路径
@MapperScan(basePackages={"com.ch.ch8_8.repository"})
public class Ch88Application {
    public static void main(String[] args) {
        SpringApplication.run(Ch88Application.class, args);
    }
}
```

❾ 运行

首先运行 Ch88Application 主类,然后访问"http://localhost:8080/ch8_8/findAll",运行效果如图 8.43 所示。

图 8.43 查询所有用户信息

本章小结

本章是本书的重点章节，讲解了 Spring Data JPA、Spring Data REST、Spring Boot 整合 MongoDB、Spring Boot 整合 Redis 以及数据缓存 Cache。通过本章的学习，读者不仅要掌握 Spring Boot 访问关系数据库的解决方案，还要掌握 Spring Boot 访问非关系数据库的解决方案。

习题 8

1. 在 Spring Boot 应用中，数据缓存技术解决了什么问题？
2. 什么是 RESTful 架构？什么是 REST 服务？
3. Spring 框架提供了哪些缓存注解？这些注解如何使用？
4. 在 Spring Data JPA 中如何实现一对一、一对多、多对多关联查询？请举例说明。

第 9 章 电子商务平台的设计与实现（Spring Boot+MyBatis+Thymeleaf）

学习目的与要求

本章通过一个小型的电子商务平台讲述如何使用 Spring Boot＋Thymeleaf＋MyBatis 开发一个 Web 应用，其中主要涉及 Spring 与 Spring MVC 框架技术、MyBatis 持久层技术、Thymeleaf 表现层技术。通过本章的学习，掌握基于 Thymeleaf＋MyBatis 的 Spring Boot Web 应用开发的流程、方法以及技术。

本章主要内容

- 系统设计
- 数据库设计
- 系统管理
- 组件设计
- 系统实现

本章系统使用 Spring Boot＋Thymeleaf＋MyBatis 实现各个模块，Web 服务器使用内嵌的 Servlet 容器，数据库采用的是 MySQL 8，集成开发环境为 IntelliJ IDEA。

9.1 系统设计

电子商务平台分为两个子系统，一个是后台管理子系统，另一个是电子商务子系统。下面分别说明这两个子系统的功能需求与模块划分。

▶9.1.1 系统功能需求

❶ 后台管理子系统

后台管理子系统要求管理员登录成功后才能对商品进行管理，包括添加商品、查询商品、修改商品以及删除商品。除对商品进行管理外，管理员还需要对商品类型、注册用户以及用户的订单等进行管理。

❷ 电子商务子系统

1）非注册用户

非注册用户或未登录用户具有浏览首页、查看商品详情以及搜索商品的权限。

2）用户

成功登录的用户除具有未登录用户具有的权限外，还具有购买商品、查看购物车、收藏商品、查看订单、查看收藏以及查看用户个人信息的权限。

▶9.1.2 系统模块划分

❶ 后台管理子系统

管理员登录成功后，进入后台管理主页面（selectGoods.html），可以对商品、商品类型、注册用户以及用户的订单进行管理。后台管理子系统的模块划分如图 9.1 所示。

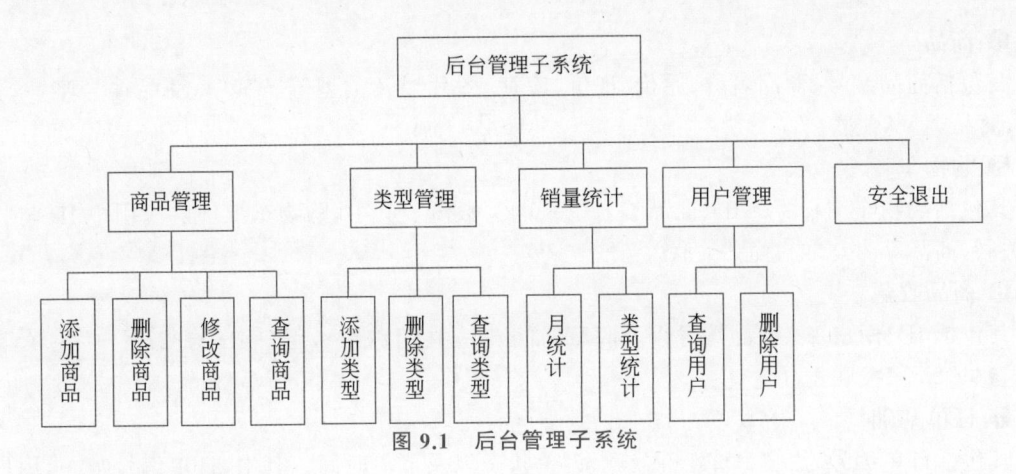

第9章 电子商务平台的设计与实现（Spring Boot+ MyBatis+ Thymeleaf）

图 9.1 后台管理子系统

❷ 电子商务子系统

非注册用户只可以浏览商品、搜索商品，不能购买商品、收藏商品、查看购物车，以及查看用户中心、我的订单和我的收藏。成功登录的用户可以完成电子商务子系统的所有功能，包括购买商品、支付等功能。电子商务子系统的模块划分如图 9.2 所示。

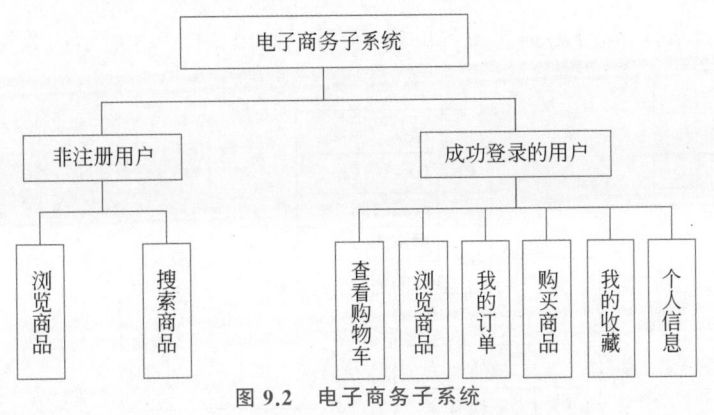

图 9.2 电子商务子系统

9.2 数据库设计

本系统采用加载纯 Java 数据库驱动程序的方式连接 MySQL 8 数据库。在 MySQL 8 中创建数据库 ch9，并在 ch9 中创建 8 张与系统相关的数据表，分别为 ausertable、busertable、carttable、focustable、goodstable、goodstype、orderdetail 和 orderbasetable。

▶9.2.1 数据库概念结构设计

根据系统设计与分析，可以设计出如下数据结构：

❶ 管理员

其包括管理员 ID、用户名和密码。管理员的用户名和密码由数据库管理员预设，不需要注册。

❷ 用户

其包括用户 ID、邮箱和密码。注册用户的邮箱不能相同，用户 ID 唯一。

❸ 商品类型

其包括类型 ID 和类型名称。商品类型由数据库管理员管理，包括新增和删除管理。

❹ 商品

其包括商品编号、商品名称、原价、现价、库存、图片以及类型等。其中,商品编号唯一,类型与"3.商品类型"关联。

❺ 购物车

其包括购物车 ID、用户 ID、商品编号以及购买数量。其中,购物车 ID 唯一,用户 ID 与"2.用户"关联,商品编号与"4.商品"关联。

❻ 商品收藏

其包括 ID、用户 ID、商品编号以及收藏时间。其中,ID 唯一,用户 ID 与"2.用户"关联,商品编号与"4.商品"关联。

❼ 订单基础

其包括订单编号、用户 ID、订单金额、订单状态以及下单时间。其中,订单编号唯一,用户 ID 与"2.用户"关联。

❽ 订单详情

其包括订单编号、商品编号以及购买数量等。其中,订单编号与"7.订单基础"关联,商品编号与"4.商品"关联。

根据以上数据结构,结合数据库设计的特点,可以画出如图 9.3 所示的数据库概念结构图。

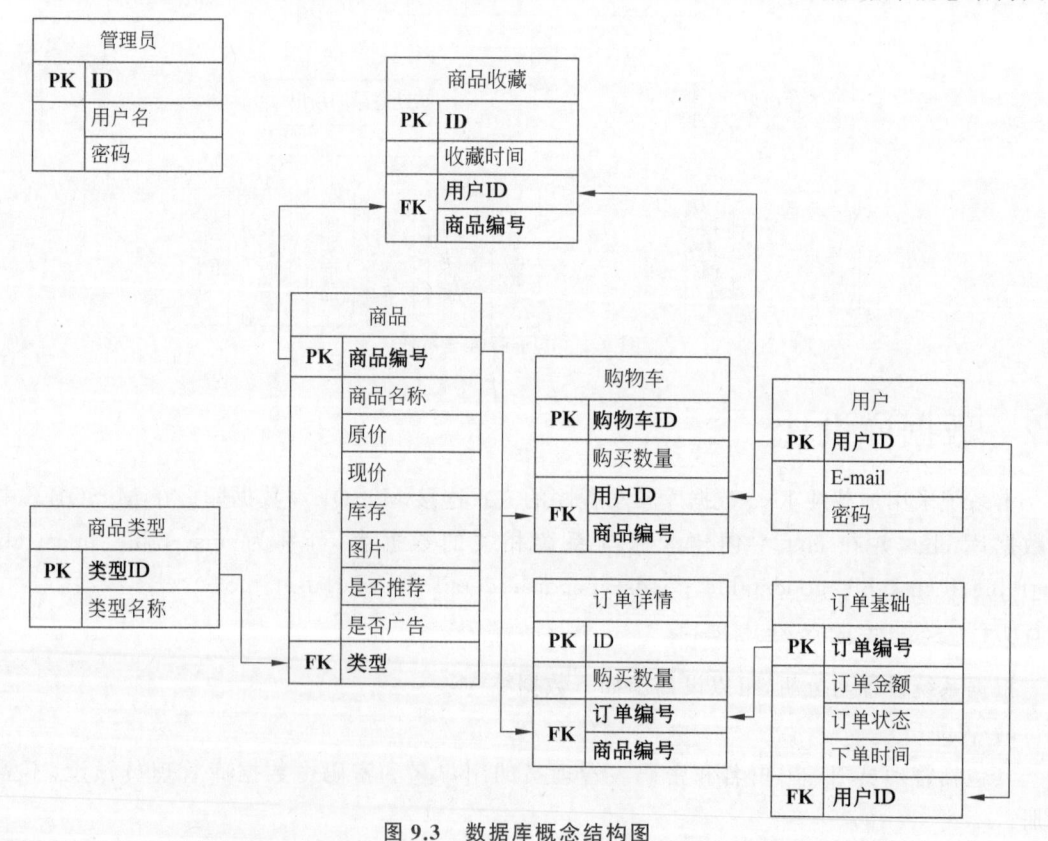

图 9.3 数据库概念结构图

9.2.2 数据逻辑结构设计

将数据库概念结构图转换为 MySQL 数据库所支持的实际数据模型,即数据库的逻辑结构。管理员信息表(ausertable)的设计如表 9.1 所示。

第9章 电子商务平台的设计与实现（Spring Boot+ MyBatis+ Thymeleaf）

表 9.1 管理员信息表

字 段	含 义	类 型	长 度	是否为空
id	管理员 ID（PK 自增）	int	11	no
aname	用户名	varchar	50	no
apwd	密码	varchar	50	no

用户信息表（busertable）的设计如表 9.2 所示。

表 9.2 用户信息表

字 段	含 义	类 型	长 度	是否为空
id	用户 ID（PK 自增）	int	11	no
bemail	E-mail	varchar	50	no
bpwd	密码	varchar	50	no

商品类型表（goodstype）的设计如表 9.3 所示。

表 9.3 商品类型表

字 段	含 义	类 型	长 度	是否为空
id	类型 ID（PK 自增）	int	11	no
typename	类型名称	varchar	50	no

商品信息表（goodstable）的设计如表 9.4 所示。

表 9.4 商品信息表

字 段	含 义	类 型	长 度	是否为空
id	商品编号（PK 自增）	int	11	no
gname	商品名称	varchar	50	no
goprice	原价	double		no
grprice	现价	double		no
gstore	库存	int	11	no
gpicture	图片	varchar	50	no
isRecommend	是否推荐	tinyint	2	no
isAdvertisement	是否广告	tinyint	2	no
goodstype_id	类型（FK）	int	11	no

购物车表（carttable）的设计如表 9.5 所示。

表 9.5 购物车表

字 段	含 义	类 型	长 度	是否为空
id	购物车 ID（PK 自增）	int	11	no
busertable_id	用户 ID（FK）	int	11	no
goodstable_id	商品编号（FK）	int	11	no
shoppingnum	购买数量	int	11	no

商品收藏表（focustable）的设计如表 9.6 所示。

表 9.6 商品收藏表

字 段	含 义	类 型	长 度	是否为空
id	ID(PK 自增)	int	11	no
goodstable_id	商品编号(FK)	int	11	no
busertable_id	用户 ID(FK)	int	11	no
focustime	收藏时间	datetime		no

订单基础表(orderbasetable)的设计如表 9.7 所示。

表 9.7 订单基础表

字 段	含 义	类 型	长 度	是否为空
id	订单编号(PK 自增)	int	11	no
busertable_id	用户 ID(FK)	int	11	no
amount	订单金额	double		no
status	订单状态	tinyint	4	no
orderdate	下单时间	datetime		no

订单详情表(orderdetail)的设计如表 9.8 所示。

表 9.8 订单详情表

字 段	含 义	类 型	长 度	是否为空
id	ID(PK 自增)	int	11	no
orderbasetable_id	订单编号(FK)	int	11	no
goodstable_id	商品编号(FK)	int	11	no
shoppingnum	购买数量	int	11	no

▶9.2.3 创建数据表

根据 9.2.2 节的逻辑结构创建数据表。由于篇幅有限,创建数据表的代码请读者参考本书提供的源程序 ch9.sql。

9.3 系统管理

▶9.3.1 添加相关依赖

首先创建基于 Thymeleaf、MyBatis、Spring Web、Lombok、MySQL Driver 依赖的 Spring Boot Web 应用 ch9;然后向 ch9 应用的 pom.xml 文件中添加表单验证依赖 Hibernate Validator,具体见源程序 ch9 中的 pom.xml 文件。

▶9.3.2 HTML 页面及静态资源管理

本系统由后台管理和电子商务两个子系统组成,为了方便管理,将两个子系统的 HTML 页面分开存放。在 src/main/resources/templates/admin 目录下存放与后台管理子系统相关的 HTML 页面;在 src/main/resources/templates/user 目录下存放与电子商务子系统相关的 HTML 页面;在 src/main/resources/static 目录下存放与整个系统相关的 BootStrap 及 jQuery。由于篇幅受限,本章仅附上部分 HTML 和 Java 文件的核心代码,具体代码请读者参考本书提供的源程序 ch9。

❶ 后台管理子系统

管理员在浏览器的地址栏中输入 http://localhost:8080/ch9/admin/toLogin 访问登录页面,登录成功后,进入后台商品管理主页面(adminGoods.html),adminGoods.html 的运行效果如图 9.4 所示。

图 9.4 后台商品管理主页面

❷ 电子商务子系统

注册用户或游客在浏览器的地址栏中输入 http://localhost:8080/ch9 可以访问电子商务子系统的首页(index.html),index.html 的运行效果如图 9.5 所示。

图 9.5 电子商务子系统的首页

▶9.3.3 应用的包结构

❶ com.ch.ch9 包

该包中包括应用的主程序类 Ch9Application、统一异常处理类 GlobalExceptionHandleController 以及自定义异常类 NoLoginException。

❷ **com.ch.ch9.controller 包**

本系统的控制器类都在该包中，与后台管理相关的控制器类在 admin 子包中，与电子商务相关的控制器类在 before 子包中。

❸ **com.ch.ch9.entity 包**

实体类存放在该包中。

❹ **com.ch.ch9.repository 包**

该包中存放的 Java 接口程序是实现数据库的持久化操作。每个接口方法与 SQL 映射文件中的 id 相同。与后台管理相关的数据库操作在 admin 子包中，与电子商务相关的数据库操作在 before 子包中。

❺ **com.ch.ch9.service 包**

service 包中有 admin 和 before 两个子包，admin 子包中存放与后台管理相关的业务层的接口与实现类；before 子包中存放与电子商务相关的业务层的接口与实现类。

❻ **com.ch.ch9.util 包**

该包中存放的是系统的工具类。

▶ 9.3.4 配置文件

在配置文件 application.properties 中配置了数据源、实体类包别名、映射文件位置、SQL 日志及文件上传等信息，具体如下：

```
server.servlet.context-path=/ch9
spring.datasource.url=jdbc:mysql://localhost:3306/ch9?useUnicode=true&characterEncoding=UTF-8&allowMultiQueries=true&serverTimezone=GMT%2B8
spring.datasource.username=root
spring.datasource.password=root
spring.datasource.driver-class-name=com.mysql.cj.jdbc.Driver
mybatis.type-aliases-package=com.ch.ch9.entity
mybatis.mapperLocations=classpath:mappers/*.xml
logging.level.com.ch.ch9.repository=debug
#在上传文件时，默认单个上传文件的大小是 1MB，max-file-size 设置单个上传文件的大小
spring.servlet.multipart.max-file-size=50MB
#默认总文件大小是 10MB，max-request-size 设置总上传文件的大小
spring.servlet.multipart.max-request-size=500MB
```

9.4 组件设计

本系统的组件包括管理员登录权限验证控制器、前台用户登录权限验证控制器、验证码、统一异常处理以及工具类。

▶ 9.4.1 管理员登录权限验证

从系统分析得知，管理员成功登录后才能管理商品、商品类型、用户、订单等功能模块，因此本系统需要对这些功能模块的操作进行管理员登录权限控制。在 com.ch.ch9.controller.admin 包中创建 AdminBaseController 控制器类，在该类中有一个 @ModelAttribute 注解的方法 isLogin。isLogin 方法的功能是判断管理员是否已成功登录。需要进行管理员登录权限控制的控制器类继承 AdminBaseController 类即可，因为带有 @ModelAttribute 注解的方法首先被控制器执行。AdminBaseController 控制器类的核心代码如下：

```
@Controller
public class AdminBaseController {
    @ModelAttribute
    public void isLogin(HttpSession session) throws NoLoginException {
        if(session.getAttribute("auser") == null){
            throw new NoLoginException("没有登录");
        }
    }
}
```

9.4.2 前台用户登录权限验证

从系统分析得知,用户成功登录后才能购买商品、收藏商品、查看购物车,以及查看我的订单和个人信息。与管理员登录权限验证同理,在 com.ch.ch9.controller.before 包中创建 BeforeBaseController 控制器类,在该类中有一个 @ModelAttribute 注解的方法 isLogin。isLogin 方法的功能是判断前台用户是否已成功登录。需要进行前台用户登录权限控制的控制器类继承 BeforeBaseController 类即可。BeforeBaseController 控制器类的代码与 AdminBaseController 基本一样,为了节省篇幅,这里不再赘述。

9.4.3 验证码

本系统的验证码的使用步骤如下:

❶ 创建产生验证码的控制器类

在 com.ch.ch9.controller.before 包中创建产生验证码的控制器类 ValidateCodeController,具体代码参见本书提供的源程序 ch9。

❷ 使用验证码

在需要验证码的 HTML 页面中调用产生验证码的控制器显示验证码,示例代码片段如下:

```
<img th:src="@{/validateCode}" id="mycode">
```

9.4.4 统一异常处理

本系统对未登录异常、数据库操作异常以及程序未知异常进行了统一异常处理,具体步骤如下:

❶ 创建未登录自定义异常

创建未登录自定义异常 NoLoginException,代码如下:

```
package com.ch.ch9;
public class NoLoginException extends Exception{
    private static final long serialVersionUID = 1L;
    public NoLoginException() {
        super();
    }
    public NoLoginException(String message) {
        super(message);
    }
}
```

❷ 创建统一异常处理类

使用 @ControllerAdvice 和 @ExceptionHandler 注解创建统一异常处理类 GlobalExceptionHandleController。使用@ControllerAdvice 注解的类是一个增强的 Controller 类,在增强的控制器类中使用@ExceptionHandler 注解的方法对所有控制器类进行统一处理异

常。核心代码如下：

```
@ControllerAdvice
public class GlobalExceptionHandleController {
    @ExceptionHandler(value=Exception.class)
    public String exceptionHandler(Exception e, Model model) {
        String message = "";
        //数据库异常
        if (e instanceof SQLException) {
            message = "数据库异常";
        } else if (e instanceof NoLoginException) {
            message = "未登录异常";
        } else {                                         //未知异常
            message ="未知异常";
        }
        model.addAttribute("mymessage",message);
        return "myError";
    }
}
```

▶ 9.4.5 工具类

本系统使用的工具类有 MD5Util 和 MyUtil 两个。MD5Util 工具用来对明文密码加密，在 MyUtil 工具中包含文件重命名和获得用户信息两个功能。MD5Util 和 MyUtil 的代码参见本书提供的源程序 ch9。

9.5 后台管理子系统的实现

管理员登录成功后，可以对商品及商品类型、注册用户以及用户的订单进行管理。本节将详细讲解管理员权限的实现。

▶ 9.5.1 管理员登录

管理员输入用户名和密码后，系统将对管理员的用户名和密码进行验证。如果用户名和密码同时正确，则成功登录，进入后台商品管理主页面（adminGoods.html）；如果用户名或密码有误，则提示错误。实现步骤如下：

❶ 编写视图

login.html 页面提供输入登录信息的页面，效果如图 9.6 所示。

图 9.6　管理员登录页面

在 src/main/resources/templates/admin 目录下创建 login.html。该页面的代码请参见本书提供的源程序 ch9。

❷ 编写控制器层

视图 Action 的请求路径为"admin/login"，系统根据请求路径和@RequestMapping 注解找到对应控制器类 com.ch.ch9.controller.admin 的 login 方法处理登录。在控制器类的 login 方法中调用 com.ch.ch9.service.admin.AdminService 接口的 login 方法处理登录。在登录成功后，首先将登录人的信息存入 session，然后转发到查询商品请求方法。如果登录失败，回到本页面。控制器层的相关代码如下：

```
@Controller
@RequestMapping("/admin")
@SuppressWarnings("all")
public class AdminController {
    @Autowired
    private AdminService adminService;
    @GetMapping("/toLogin")
    public String toLogin(@ModelAttribute("aUser") AUser aUser) {
        return "admin/login";
    }
    @PostMapping("/login")
    public String login(@ModelAttribute("aUser") AUser aUser, HttpSession session, Model model) {
        return adminService.login(aUser, session, model);
    }
}
```

❸ 编写 Service 层

Service 层由接口 com.ch.ch9.service.admin.AdminService 和接口的实现类 com.ch.ch9.service.admin.AdminServiceImpl 组成。Service 层是功能模块实现的核心，Service 层调用数据访问层（Repository）进行数据库操作。管理员登录的业务处理方法 login 的代码如下：

```
public String login(AUser aUser, HttpSession session, Model model) {
    List<AUser> list = adminRepository.login(aUser);
    if(list.size() > 0) {                            //登录成功
        session.setAttribute("auser", aUser);
        return "forward:/goods/selectAllGoodsByPage?currentPage=1&act=select";
    }else {                                          //登录失败
        model.addAttribute("errorMessage", "用户名或密码错误！");
        return "admin/login";
    }
}
```

❹ 编写 SQL 映射文件

数据访问层（Repository）仅由@Repository 注解的接口组成，接口方法与 SQL 映射文件中 SQL 语句的 id 相同。管理员登录的 SQL 映射文件为 src/main/resources/mappers 目录下的 AdminMapper.xml，实现的 SQL 语句如下：

```
<select id="login" parameterType="AUser" resultType="AUser">
    select * from ausertable where aname = #{aname} and apwd = #{apwd}
</select>
```

9.5.2 类型管理

管理员登录成功后能够进行类型管理，类型管理分为添加类型和删除类型，如图 9.7 所示。

❶ 添加类型

添加类型的实现步骤如下。

1）编写视图

单击图 9.7 中的"添加类型"超链接（type/toAddType），打开如图 9.8 所示的添加页面。

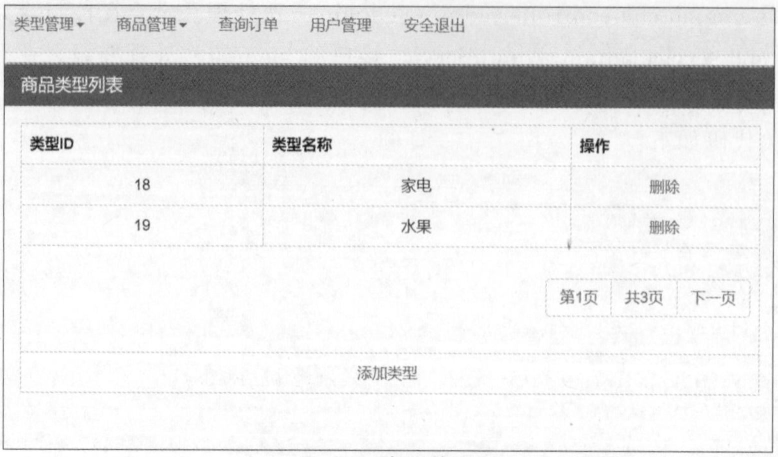

图 9.7 类型管理

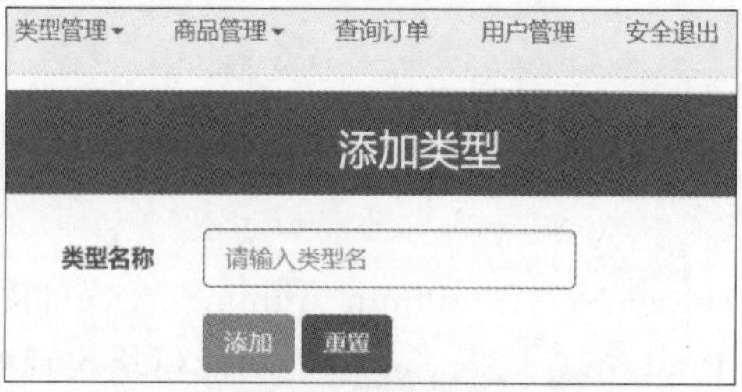

图 9.8 添加类型

在 src/main/resources/templates/admin 目录下创建添加类型页面（addType.html），该页面的代码请参见本书提供的源程序 ch9。

2）编写控制器层

此功能共有两个处理请求，分别为"添加类型"超链接"type/toAddType"与视图 Action 的请求路径"type/addType"。系统根据@RequestMapping 注解找到对应控制器类 com.ch.ch9.controller.admin.TypeController 的 toAddType 和 addType 方法处理请求。在控制器类的处理方法中调用 com.ch.ch9.service.admin.TypeService 接口的 addType 方法处理业务。控制器层的相关代码如下：

```java
@GetMapping("/toAddType")
public String toAddType(@ModelAttribute("goodsType") GoodsType goodsType) {
    return "admin/addType";
}
@PostMapping("/addType")
public String addType(@ModelAttribute("goodsType") GoodsType goodsType) {
    return typeService.addType(goodsType);
}
```

3）编写 Service 层

添加类型"type/addType"的业务处理方法 addType 的代码如下：

```java
@Override
public String addType(GoodsType goodsType) {
    typeRepository.addType(goodsType);
    return "redirect:/type/selectAllTypeByPage?currentPage=1";
}
```

4）编写 SQL 映射文件

实现添加类型"type/addType"的 SQL 语句如下（位于 src/main/resources/mappers/TypeMapper.xml 文件中）：

```xml
<insert id="addType" parameterType="GoodsType">
    insert into goodstype (id, typename) values(null, #{typename})
</insert>
```

❷ 删除类型

删除类型的实现步骤如下：

1）编写视图

单击图 9.7 中的"类型管理"超链接（type/selectAllTypeByPage？currentPage＝1），打开如图 9.7 所示的查询页面。

在 src/main/resources/templates/admin 目录下创建查询类型页面（selectGoodsType.html），该页面的代码参见本书提供的源程序 ch9。

2）编写控制器层

此功能模块共有两个处理请求，分别为"查询类型"超链接"type/selectAllTypeByPage？currentPage＝1"与视图"删除"的请求路径"type/deleteType"。系统根据@RequestMapping 注解找到对应控制器类 com.ch.ch9.controller.admin.TypeController 的 selectAllTypeByPage 和 delete 方法处理请求。在控制器类的处理方法中调用 com.ch.ch9.service.admin.TypeService 接口的 selectAllTypeByPage 和 delete 方法处理业务。控制器层的相关代码如下：

```java
@RequestMapping("/selectAllTypeByPage")
public String selectAllTypeByPage(Model model, int currentPage) {
    return typeService.selectAllTypeByPage(model, currentPage);
}
@RequestMapping("/deleteType")
@ResponseBody                                          //返回字符串数据而不是视图
public String delete(int id) {
    return typeService.delete(id);
}
```

3）编写 Service 层

超链接"type/selectAllTypeByPage？currentPage＝1"的业务处理方法 selectAllTypeByPage 的代码如下：

```java
public String selectAllTypeByPage(Model model, int currentPage) {
    //共有多少个类型
    int totalCount = typeRepository.selectAll();
    //计算共有多少页
    int pageSize = 2;
    int totalPage = (int)Math.ceil(totalCount * 1.0/pageSize);
    List<GoodsType> typeByPage =
    typeRepository.selectAllTypeByPage((currentPage-1) * pageSize, pageSize);
    model.addAttribute("allTypes", typeByPage);
    model.addAttribute("totalPage", totalPage);
    model.addAttribute("currentPage", currentPage);
    return "admin/selectGoodsType";
}
```

删除"type/deleteType"的业务处理方法 delete 的代码如下：

```
public String delete(int id) {
    List<Goods> list = typeRepository.selectGoods(id);
    if(list.size() > 0) {
        //该类型下有商品,不允许删除
        return "no";
    }else {
        typeRepository.deleteType(id);
        //删除后回到查询页面
        return "/type/selectAllTypeByPage?currentPage=1";
    }
}
```

4）编写 SQL 映射文件

实现超链接"type/selectAllTypeByPage？currentPage＝1"的 SQL 语句如下：

```
<select id="selectAll" resultType="integer">
    select count(*) from goodstype
</select>
<!-- 分页查询 -->
<select id="selectAllTypeByPage" resultType="GoodsType">
    select * from goodstype limit #{startIndex}, #{perPageSize}
</select>
```

实现删除"type/deleteType"的 SQL 语句如下：

```
<!-- 删除类型 -->
<delete id="deleteType" parameterType="integer">
    delete from goodstype where id=#{id}
</delete>
<!-- 查询该类型下是否有商品 -->
<select id="selectGoods" parameterType="integer" resultType="Goods">
    select * from goodstable where goodstype_id = #{goodstype_id}
</select>
```

▶9.5.3 添加商品

单击图 9.4 中的"添加商品"超链接打开如图 9.9 所示的添加商品页面。添加商品的实现步骤如下：

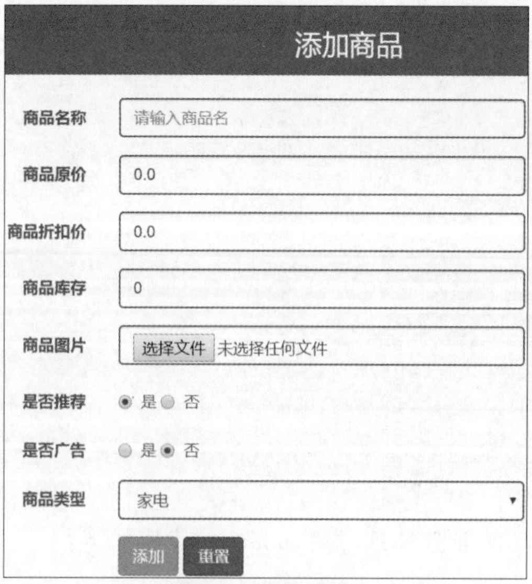

图 9.9 添加商品

第 9 章 电子商务平台的设计与实现（Spring Boot+ MyBatis+ Thymeleaf）

❶ 编写视图

在 src/main/resources/templates/admin 目录下创建添加商品页面（addGoods.html）。该页面的代码请参见本书提供的源程序 ch9。

❷ 编写控制器层

此功能模块共有两个处理请求，分别为"添加商品"超链接"goods/toAddGoods"与视图"添加"的请求路径"goods/addGoods？ act＝add"。系统根据＠RequestMapping 注解找到对应控制器类 com.ch.ch9.controller.admin.GoodsController 的 toAddGoods 和 addGoods 方法处理请求。在控制器类的处理方法中调用 com.ch.ch9.service.admin.GoodsService 接口的 toAddGoods 和 addGoods 方法处理业务。控制器层的相关代码如下：

```
@GetMapping("/toAddGoods")
public String toAddGoods(@ModelAttribute("goods") Goods goods, Model model) {
    goods.setIsAdvertisement(0);
    goods.setIsRecommend(1);
    return goodsService.toAddGoods(goods, model);
}
@PostMapping("/addGoods")
public String addGoods(@ModelAttribute("goods") Goods goods, HttpServletRequest request,
String act) throws IllegalStateException, IOException {
    return goodsService.addGoods(goods, request, act);
}
```

❸ 编写 Service 层

添加商品的 Service 层的相关代码如下：

```
@Override
public String addGoods (Goods goods, HttpServletRequest request, String act) throws
IllegalStateException, IOException {
    MultipartFile myfile = goods.getFileName();
    //如果选择了上传文件,将文件上传到指定的目录 images
    if(!myfile.isEmpty()) {
        //上传文件路径(生产环境)
        //String path = request.getServletContext().getRealPath("/images/");
        //获得上传文件原名
        //上传文件路径(开发环境)
        String path = "C:\\ workspace-spring-tool-suite-4-4.9.0.RELEASE\\ch9\\src\\main\\resources\\static\\images";
        //获得上传文件原名
        String fileName = myfile.getOriginalFilename();
        //对文件重命名
        String fileNewName = MyUtil.getNewFileName(fileName);
        File filePath = new File(path + File.separator + fileNewName);
        //如果文件目录不存在,创建目录
        if(!filePath.getParentFile().exists()) {
            filePath.getParentFile().mkdirs();
        }
        //将上传文件保存到一个目标文件中
        myfile.transferTo(filePath);
        //将重命名后的图片名存到 goods 对象中,在添加时使用
        goods.setGpicture(fileNewName);
    }
    if("add".equals(act)) {
        int n = goodsRepository.addGoods(goods);
        if(n > 0)                                    //成功
            return "redirect:/goods/selectAllGoodsByPage? currentPage=1&act=select";
        //失败
```

```
            return "admin/addGoods";
        }else {                                              //修改
            int n = goodsRepository.updateGoods(goods);
            if(n > 0)                                        //成功
                return "redirect:/goods/selectAllGoodsByPage?currentPage=1&act=updateSelect";
            //失败
            return "admin/UpdateAGoods";
        }
    }
    @Override
    public String toAddGoods(Goods goods, Model model) {
        model.addAttribute("goodsType", goodsRepository.selectAllGoodsType());
        return "admin/addGoods";
    }
```

❹ **编写 SQL 映射文件**

添加商品的 SQL 语句如下：

```xml
<!-- 添加商品 -->
<insert id="addGoods" parameterType="Goods">
    insert into goodstable (id,gname,goprice,grprice,gstore,gpicture,isRecommend,
isAdvertisement,goodstype_id)
     values (null, #{gname}, #{goprice}, #{grprice}, #{gstore}, #{gpicture}, #
{isRecommend}, #{isAdvertisement}, #{goodstype_id})
</insert>
<!-- 查询商品类型 -->
<select id="selectAllGoodsType" resultType="GoodsType">
    select * from goodstype
</select>
```

▶9.5.4 查询商品

管理员登录成功后，进入如图 9.4 所示的后台管理主页面，在该主页面中单击"详情"链接，显示如图 9.10 所示的商品详情页面。

类型管理▼	商品管理▼	查询订单	用户管理	安全退出
商品详情				
商品名称			衣服155	
商品原价			400.0	
商品折扣价			300.0	
商品库存			900	
商品图片			连衣裙	

图 9.10 商品详情

❶ **编写视图**

在 src/main/resources/templates/admin 目录下创建商品管理主页面（adminGoods.html），该页面显示查询商品、修改商品查询以及删除商品查询的结果。其代码参见本书提供的源程序 ch9。

在 src/main/resources/templates/admin 目录下创建商品详情页面（detail.html），该页面的

第9章 电子商务平台的设计与实现（Spring Boot+ MyBatis+ Thymeleaf）

代码参见本书提供的源程序 ch9。

❷ 编写控制器层

此功能模块共有两个处理请求，分别为 goods/selectAllGoodsByPage? currentPage＝1 和 @{goods/detail(id＝ ${gds.id}, act＝detail)}。系统根据@RequestMapping 注解找到对应控制器类 com.ch.ch9.controller.admin.GoodsController 的 selectAllGoodsByPage 和 detail 方法处理请求。在控制器类的处理方法中调用 com.ch.ch9.service.admin.GoodsService 接口的 selectAllGoodsByPage 和 detail 方法处理业务。控制器层的相关代码如下：

```java
@RequestMapping("/selectAllGoodsByPage")
public String selectAllGoodsByPage(Model model, int currentPage) {
    return goodsService.selectAllGoodsByPage(model, currentPage);
}
@GetMapping("/detail")
public String detail(Model model, Integer id, String act) {
    return goodsService.detail(model, id, act);
}
```

❸ 编写 Service 层

查询商品和查看详情的 Service 层的相关代码如下：

```java
@Override
public String selectAllGoodsByPage(Model model, int currentPage) {
    //共有多少个商品
    int totalCount = goodsRepository.selectAllGoods();
    //计算共有多少页
    int pageSize = 5;
    int totalPage = (int)Math.ceil(totalCount * 1.0/pageSize);
    List<Goods> typeByPage = goodsRepository.selectAllGoodsByPage((currentPage - 1) * pageSize, pageSize);
    model.addAttribute("allGoods", typeByPage);
    model.addAttribute("totalPage", totalPage);
    model.addAttribute("currentPage", currentPage);
    return "admin/adminGoods";
}
@Override
public String detail(Model model, Integer id, String act) {
    model.addAttribute("goods", goodsRepository.selectAGoods(id));
    if("detail".equals(act))
        return "admin/detail";
    else {
        model.addAttribute("goodsType", goodsRepository.selectAllGoodsType());
        return "admin/updateAGoods";
    }
}
```

❹ 编写 SQL 映射文件

查询商品和查看详情的 SQL 语句如下：

```xml
<select id="selectAllGoods" resultType="integer">
    select count(*) from goodstable
</select>
<!-- 分页查询 -->
<select id="selectAllGoodsByPage" resultType="Goods">
    select gt.*,gy.typename
    from goodstable gt,goodstype gy
    where gt.goodstype_id = gy.id
    order by id desc limit #{startIndex}, #{perPageSize}
```

```xml
</select>
<!-- 查询商品详情 -->
<select id="selectAGoods" resultType="Goods">
    select gt.*, gy.typename
    from goodstable gt,goodstype gy
    where gt.goodstype_id = gy.id and gt.id = #{id}
</select>
```

9.5.5 修改商品

单击图9.4中的"修改"超链接(goods/detail(id＝${gds.id},act＝update)),打开修改商品信息页面(updateAGoods.html),如图9.11所示。在图9.11中输入要修改的信息,然后单击"修改"按钮,将商品信息提交给 goods/addGoods?act＝update 处理。修改商品的实现步骤如下:

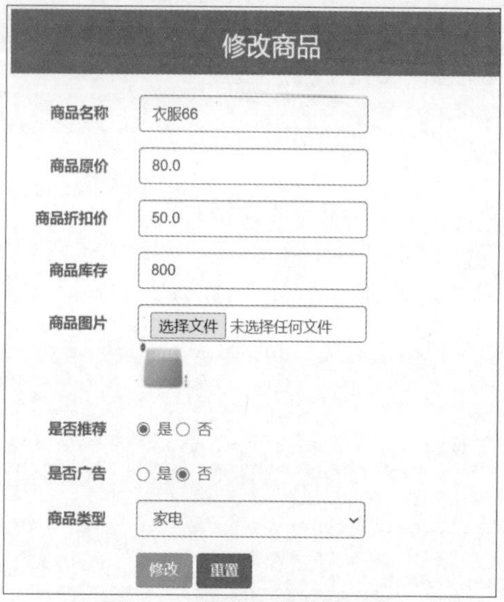

图9.11 修改商品信息页面

❶ 编写视图

在 src/main/resources/templates/admin 目录下创建修改商品信息页面 updateAGoods.html。updateAGoods.html 与添加商品页面的内容基本一样,这里不再赘述。

❷ 编写控制器层

此功能模块共有两个处理请求,分别为 goods/detail(id＝${gds.id},act＝update)和 goods/addGoods?act＝update。goods/detail(id＝${gds.id},act＝update)请求已在9.5.4节介绍,goods/addGoods?act＝update 请求已在9.5.3节介绍。

❸ 编写Service层

同理,Service层请参考9.5.3节与9.5.4节。

❹ 编写SQL映射文件

"修改商品"的SQL语句如下:

```xml
<!-- 修改一个商品 -->
<update id="updateGoods" parameterType="Goods">
update goodstable set
    gname = #{gname},
    goprice = #{goprice},
    grprice = #{grprice},
```

```
        gstore = #{gstore},
        gpicture = #{gpicture},
        isRecommend = #{isRecommend},
        isAdvertisement = #{isAdvertisement},
        goodstype_id = #{goodstype_id}
    where id = #{id}
</update>
```

▶ 9.5.6 删除商品

单击图 9.4 中的"删除"超链接('javascript：deleteGoods(' ＋ ＄{gds.id} ＋')')，可实现单个商品的删除，成功删除（关联商品不允许删除）后返回删除商品管理主页面。

❶ 编写控制器层

此功能模块的处理请求是 goods/delete。相关控制器层代码如下：

```
@RequestMapping("/delete")
@ResponseBody
public String delete(Integer id) {
    return goodsService.delete(id);
}
```

❷ 编写 Service 层

删除商品的相关业务处理代码如下：

```
@Override
public String delete(Integer id) {
    if(goodsRepository.selectCartGoods(id).size() > 0
            || goodsRepository.selectFocusGoods(id).size() > 0
            || goodsRepository.selectOrderGoods(id).size() > 0)
        return "no";
    else {
        goodsRepository.deleteAGoods(id);
        return "/goods/selectAllGoodsByPage?currentPage=1";
    }
}
```

❸ 编写 SQL 映射文件

"删除商品"功能模块的相关 SQL 语句如下：

```
<select id="selectFocusGoods" parameterType="integer" resultType="map">
    select * from focustable where goodstable_id = #{id}
</select>
    <select id="selectCartGoods" parameterType="integer" resultType="map">
    select * from carttable where goodstable_id = #{id}
</select>
<select id="selectOrderGoods" parameterType="integer" resultType="map">
    select * from orderdetail where goodstable_id = #{id}
</select>
<delete id="deleteAGoods" parameterType="Integer">
    delete from goodstable where id=#{id}
</delete>
```

▶ 9.5.7 查询订单

单击后台管理主页面中的"查询订单"超链接（selectOrder？currentPage＝1），打开查询订单页面（allOrder.html），如图 9.12 所示。

❶ 编写视图

在 src/main/resources/templates/admin 目录下创建查询订单页面（allOrder.html）。该页

图 9.12 查询订单页面

面的代码请参见本书提供的源程序 ch9。

❷ 编写控制器层

此功能模块有一个处理请求,即 selectOrder？currentPage=1。系统根据@GetMapping 注解找到对应控制器类 com.ch.ch9.controller.admin.UserAndOrderAndOutController 的 selectOrder 方法处理请求。在控制器类的处理方法中调用 com.ch.ch9.service.admin.UserAndOrderAndOutService 接口的 selectOrder 方法处理业务。相关控制器层代码如下：

```
@GetMapping("/selectOrder")
public String selectOrder(Model model, int currentPage) {
    return userAndOrderAndOutService.selectOrder(model, currentPage);
}
```

❸ 编写 Service 层

"查询订单"功能模块的相关 Service 层代码如下：

```
@Override
public String selectOrder(Model model, int currentPage) {
    //共有多少个订单
    int totalCount = userAndOrderAndOutRepository.selectAllOrder();
    //计算共有多少页
    int pageSize = 5;
    int totalPage = (int)Math.ceil(totalCount * 1.0/pageSize);
    List<Map<String, Object>> orderByPage = userAndOrderAndOutRepository.
selectOrderByPage((currentPage-1) * pageSize, pageSize);
    model.addAttribute("allOrders", orderByPage);
    model.addAttribute("totalPage", totalPage);
    model.addAttribute("currentPage", currentPage);
    return "admin/allOrder";
}
```

❹ 编写 SQL 映射文件

"查询订单"功能模块的相关 SQL 语句如下：

```
<select id="selectAllOrder" resultType="integer">
    select count(*) from orderbasetable
</select>
<!-- 分页查询 -->
<select id="selectOrderByPage" resultType="map">
    select obt.*, bt.bemail from orderbasetable obt, busertable bt where obt.
busertable_id = bt.id limit #{startIndex}, #{perPageSize}
</select>
```

9.5.8 用户管理

单击后台管理主页面中的"用户管理"超链接（selectUser？currentPage=1），打开用户管理页面(allUser.html),如图 9.13 所示。

单击图 9.13 中的"删除"超链接('javascript：deleteUsers('+ ＄{u.id} +')'),可删除未关联的用户。

图 9.13 用户管理页面

"用户管理"与 9.5.7 节中"查询订单"的实现方式基本一样,这里不再赘述。

▶9.5.9 按月统计

单击后台管理主页面中"销量统计"下的"按月统计"超链接(selectOrderByMonth),打开近一年的销量统计,如图 9.14 所示。

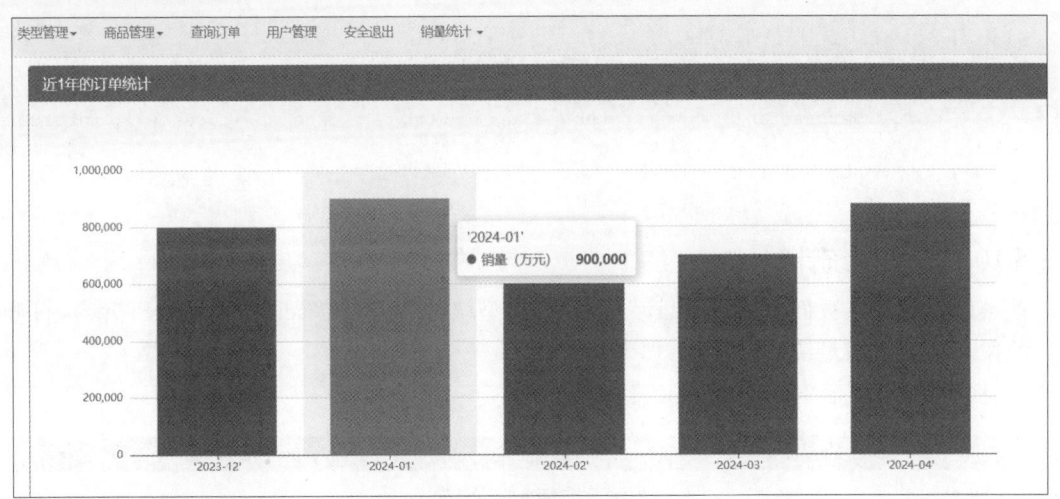

图 9.14 按月统计页面

按月统计的实现步骤如下：

❶ 编写视图

在 src/main/resources/templates/admin 目录下创建按月统计页面 selectOrderByMonth.html。该页面的代码请参见本书提供的源程序 ch9。

❷ 编写控制器层

此功能模块的处理请求为 selectOrderByMonth。系统根据@GetMapping 注解找到对应控制器类 UserAndOrderAndOutController 的 selectOrderByMonth 方法处理请求。在控制器类的处理方法中调用 UserAndOrderAndOutService 接口的 selectOrderByMonth 方法处理业务。相关控制器层代码如下：

```java
@GetMapping("/selectOrderByMonth")
public String selectOrderByMonth(Model model) {
    return adminService.selectOrderByMonth(model);
}
```

❸ 编写 Service 层

"按月统计"功能模块的相关 Service 层代码如下：

```java
@Override
public String selectOrderByMonth(Model model) {
    List<Map<String, Object>>myList = adminMapper.selectOrderByMonth();
    List<String> months = new ArrayList<String>();
    List<Double> totalAmount = new ArrayList<Double>();
    for (Map<String, Object> map : myList) {
        months.add("'" + map.get("months") + "'");
        totalAmount.add((Double)map.get("totalamount"));
    }
    model.addAttribute("months", months);
    model.addAttribute("totalAmount", totalAmount);
    return "admin/selectOrderByMonth";
}
```

❹ 编写 SQL 映射文件

"按月统计"功能模块的相关 SQL 语句如下：

```xml
<!-- 订单销量按月统计(最近一年的) -->
<select id="selectOrderByMonth" resultType="map">
    select sum(amount) totalamount, date_format(orderdate,'%Y-%m') months
        from orderbasetable where status = 1 and orderdate > date_sub(curdate(), interval 1 year)
    group by months order by months
</select>
```

▶9.5.10 按类型统计

单击后台管理主页面中"销量统计"下的"按类型统计"超链接（selectOrderByType），打开近一年的按商品类型销量统计，如图 9.15 所示。

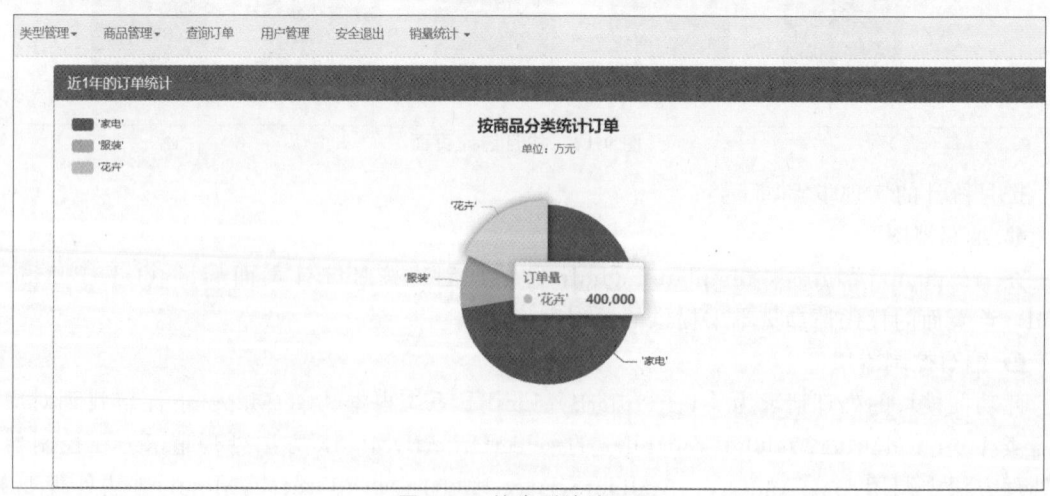

图 9.15 按类型统计页面

按类型统计的实现步骤如下：

❶ 编写视图

在 src/main/resources/templates/admin 目录下创建按类型统计页面 selectOrderByType.html。该页面的代码请参见本书提供的源程序 ch9。

❷ 编写控制器层

此功能模块的处理请求为 selectOrderByType。系统根据@GetMapping 注解找到对应控制器类 UserAndOrderAndOutController 的 selectOrderByType 方法处理请求。在控制器类的处理方法中调用 UserAndOrderAndOutService 接口的 selectOrderByType 方法处理业务。相关控制器层代码如下：

```java
@GetMapping("/selectOrderByType")
public String selectOrderByType(Model model) {
    return adminService.selectOrderByType(model);
}
```

❸ 编写 Service 层

"按类型统计"功能模块的相关 Service 层代码如下：

```java
@Override
public String selectOrderByType(Model model) {
    List<Map<String, Object>>myList = adminMapper.selectOrderByType();
    List<String> typenames = new ArrayList<String>();
    List<Double> totalAmount = new ArrayList<Double>();
    for (Map<String, Object> map : myList) {
        typenames.add("'" + (String)map.get("name") + "'");
        totalAmount.add((Double)map.get("value"));
    }
    model.addAttribute("typenames", typenames);
    model.addAttribute("totalAmount", totalAmount);
    return "admin/selectOrderByType";
}
```

❹ 编写 SQL 映射文件

"按类型统计"功能模块的相关 SQL 语句如下：

```xml
<!-- 按类型统计(最近一年的) -->
<select id="selectOrderByType" resultType="map">
    select sum(gt.grprice * od.shoppingnum) value, gy.typename name
    from orderbasetable ob, orderdetail od, goodstype gy, goodstable gt
        where ob.status = 1 and ob.orderdate > date_sub(curdate(), interval 1 year)
    and od.orderbasetable_id=ob.id
    and gt.id=od.goodstable_id
    and gt.goodstype_id = gy.id
    group by gy.typename
</select>
```

▶9.5.11 安全退出

在后台管理主页面中单击"安全退出"超链接(loginOut)，将返回后台登录页面。系统根据@GetMapping 注解找到对应控制器类 com.ch.ch9.controller.admin.UserAndOrderAndOutController 的 loginOut 方法处理请求。在 loginOut 方法中执行 session.invalidate()，session 将失效，并返回后台登录页面。其具体代码如下：

```java
@GetMapping("/loginOut")
public String loginOut(@ModelAttribute("aUser") AUser aUser, HttpSession session) {
```

```
        session.invalidate();
        return "admin/login";
}
```

9.6 前台电子商务子系统的实现

游客具有浏览首页、查看商品详情和搜索商品等权限。成功登录的用户除具有游客的权限外，还具有购买商品、查看购物车、收藏商品、查看我的订单以及用户信息的权限。本节将详细讲解前台电子商务子系统的实现。

▶9.6.1 导航栏及首页搜索

在前台的每个 HTML 页面中都引入了一个名为 header.html 的页面，引入代码如下：

```
<div th:include="user/header"></div>
```

header.html 中的商品类型以及广告区域的商品信息都从数据库中获取。header.html 页面的运行效果如图 9.16 所示。

图 9.16 导航栏

在导航栏的搜索框中输入信息，单击"搜索"按钮，将搜索信息提交给 search 请求处理，系统根据 @RequestMapping 注解找到 com.ch.ch9.controller.before.IndexController 控制器类的 search 方法处理请求，并将搜索到的商品信息转发给 searchResult.html。searchResult.html 页面的运行效果如图 9.17 所示。

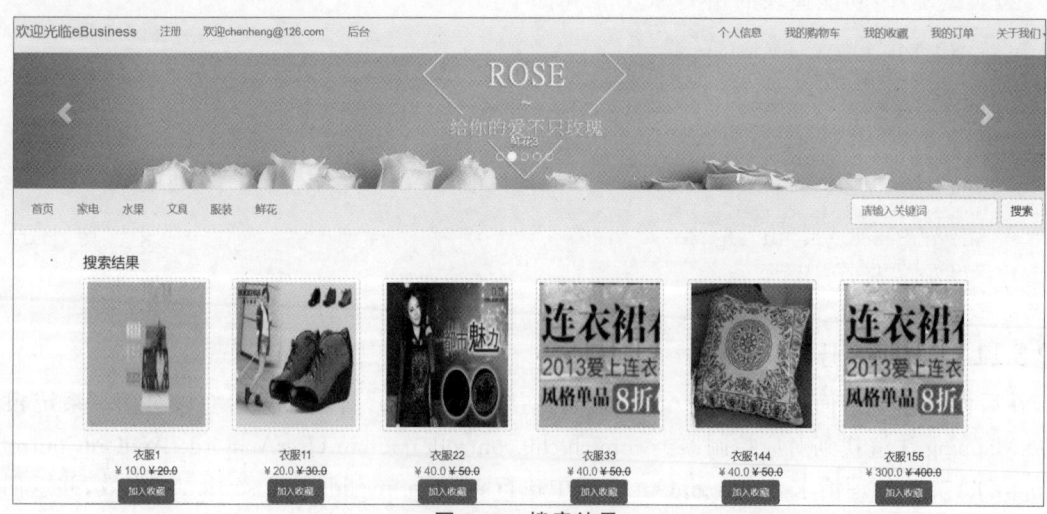

图 9.17 搜索结果

❶ 编写视图

该模块的视图涉及 src/main/resources/templates/user 目录下的两个 HTML 页面，分别为

header.html 和 searchResult.html。header.html 和 searchResult.html 页面的代码请参见本书提供的源程序 ch9。

❷ 编写控制器层

该功能模块的控制器层涉及 com.ch.ch9.controller.before.IndexController 控制器类的处理方法 search，具体代码如下：

```
@RequestMapping("/search")
public String search(Model model, String mykey) {
    return indexService.search(model, mykey);
}
```

❸ 编写 Service 层

该功能模块的 Service 层代码如下：

```
@Override
public String search(Model model, String mykey) {
    //广告区中的商品
    model.addAttribute("advertisementGoods", indexRepository.selectAdvertisementGoods());
    //导航栏中的商品类型
    model.addAttribute("goodsType", indexRepository.selectGoodsType());
    //商品的搜索
    model.addAttribute("searchgoods", indexRepository.search(mykey));
    return "user/searchResult";
}
```

❹ 编写 SQL 映射文件

该功能模块涉及的 SQL 语句如下：

```xml
<!-- 查询广告商品 -->
<select id="selectAdvertisementGoods" resultType="Goods">
    select
        gt.*, gy.typename
    from
        goodstable gt,goodstype gy
    where
        gt.goodstype_id = gy.id
        and gt.isAdvertisement = 1
    order by gt.id desc limit 5
</select>
<!-- 查询商品类型 -->
<select id="selectGoodsType" resultType="GoodsType">
    select * from goodstype
</select>
<!-- 首页搜索 -->
<select id="search" resultType="Goods" parameterType="String">
    select gt.*, gy.typename from GOODSTABLE gt,GOODSTYPE gy where gt.goodstype_id = gy.id
    and gt.gname like concat('%',#{mykey},'%')
</select>
```

▶9.6.2 推荐商品及最新商品

推荐商品是根据商品表中 isRecommend 字段的值判断的。最新商品是以商品 ID 排序的，因为商品 ID 是用 MySQL 自动递增产生的。其具体实现步骤如下：

❶ 编写视图

该模块的视图涉及 src/main/resources/templates/user 目录下的 index.html 页面，其代码请参见本书提供的源程序 ch9。

❷ 编写控制器层

该功能模块的控制器层涉及 com.ch.ch9.controller.before.IndexController 控制器类的处理方法 index,具体代码如下:

```java
@RequestMapping("/")
public String index(Model model, Integer tid) {
    return indexService.index(model, tid);
}
```

❸ 编写 Service 层

该功能模块的 Service 层代码如下:

```java
@Override
public String index(Model model, Integer tid) {
    if(tid == null)
        tid = 0;
    //广告区中的商品
    model.addAttribute("advertisementGoods", indexRepository.selectAdvertisementGoods());
    //导航栏中的商品类型
    model.addAttribute("goodsType", indexRepository.selectGoodsType());
    //推荐商品
    model.addAttribute("recommendGoods", indexRepository.selectRecommendGoods(tid));
    //最新商品
    model.addAttribute("lastedGoods", indexRepository.selectLastedGoods(tid));
    return "user/index";
}
```

❹ 编写 SQL 映射文件

该功能模块涉及的 SQL 语句如下:

```xml
<!-- 查询推荐商品 -->
<select id="selectRecommendGoods" resultType="Goods" parameterType="integer">
    select
        gt.*, gy.typename
    from
        goodstable gt,goodstype gy
    where
        gt.goodstype_id = gy.id
        and gt.isRecommend = 1
        <if test="tid != 0">
            and gy.id = #{tid}
        </if>
    order by gt.id desc limit 6
</select>
<!-- 查询最新商品 -->
<select id="selectLastedGoods" resultType="Goods" parameterType="integer">
    select
        gt.*, gy.typename
    from
        goodstable gt,goodstype gy
    where
        gt.goodstype_id = gy.id
        <if test="tid != 0">
            and gy.id = #{tid}
        </if>
    order by gt.id desc limit 6
</select>
```

▶9.6.3 用户注册

单击导航栏中的"注册"超链接(user/toRegister),将打开注册页面(register.html),如图 9.18 所示。

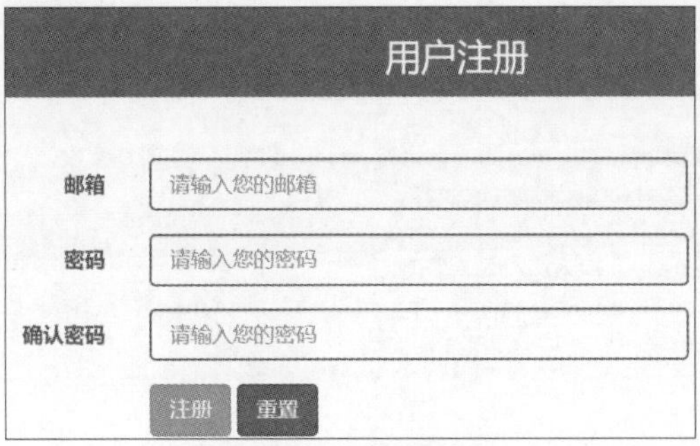

图 9.18 注册页面

输入用户信息,然后单击"注册"按钮,将用户信息提交给 user/register 处理请求,系统根据 @RequestMapping 注解找到 com.ch.ch9.controller.before.UserController 控制器类的 toRegister 和 register 方法处理请求。注册模块的具体实现步骤如下:

❶ 编写视图

该模块的视图涉及 src/main/resources/templates/user 目录下的 register.html,其代码与后台登录页面的代码类似,这里不再赘述。

❷ 编写控制器层

该功能模块涉及 com.ch.ch9.controller.before.UserController 控制器类的 toRegister 和 register 方法,具体代码如下:

```java
@RequestMapping("/toRegister")
public String toRegister(@ModelAttribute("bUser") BUser bUser) {
    return "user/register";
}
@RequestMapping("/register")
public String register(@ModelAttribute("bUser") @Validated BUser bUser,BindingResult rs) {
    if(rs.hasErrors()){                                            //验证失败
        return "user/register";
    }
    return userService.register(bUser);
}
```

❸ 编写 Service 层

该功能模块的 Service 层代码如下:

```java
@Override
public String isUse(BUser bUser) {
    if(userRepository.isUse(bUser).size() > 0 {
        return "no";
    }
    return "ok";
}
@Override
public String register(BUser bUser) {
    //对密码进行 MD5 加密
    bUser.setBpwd(MD5Util.MD5(bUser.getBpwd()));
    if(userRepository.register(bUser) > 0) {
```

```
            return "user/login";
    }
    return "user/register";
}
```

❹ 编写 SQL 映射文件

该功能模块涉及的 SQL 语句如下：

```
<select id="isUse" parameterType="BUser" resultType="BUser">
    select * from busertable where bemail = #{bemail}
</select>
<insert id="register" parameterType="BUser">
    insert into busertable (id, bemail, bpwd) values(null, #{bemail}, #{bpwd})
</insert>
```

▶ 9.6.4 用户登录

用户注册成功后，跳转到登录页面（login.html），如图 9.19 所示。

图 9.19 登录页面

在图 9.19 中输入信息后单击"登录"按钮，将用户输入的 E-mail、密码以及验证码提交给 user/login 请求处理。系统根据 @RequestMapping 注解找到 com.ch.ch9.controller.before. UserController 控制器类的 login 方法处理请求。登录成功后，将用户的登录信息保存到 session 对象中，然后回到网站首页。其具体实现步骤如下：

❶ 编写视图

该模块的视图涉及 src/main/resources/templates/user 目录下的 login.html。其代码与后台登录页面的代码类似，这里不再赘述。

❷ 编写控制器层

该功能模块涉及 com.ch.ch9.controller.before.UserController 控制器类的 login 方法，具体代码如下：

```
@RequestMapping("/login")
public String login(@ModelAttribute("bUser") @Validated BUser bUser,
        BindingResult rs, HttpSession session, Model model) {
    if(rs.hasErrors()){                                              //验证失败
        return "user/login";
    }
    return userService.login(bUser, session, model);
}
```

❸ 编写 Service 层

该功能模块的 Service 层代码如下：

```
@Override
public String login(BUser bUser, HttpSession session, Model model) {
    //对密码进行 MD5 加密
    bUser.setBpwd(MD5Util.MD5(bUser.getBpwd()));
    String rand = (String)session.getAttribute("rand");
    if(!rand.equalsIgnoreCase(bUser.getCode())) {
        model.addAttribute("errorMessage", "验证码错误!");
        return "user/login";
    }
    List<BUser> list = userRepository.login(bUser);
    if(list.size() > 0) {
        session.setAttribute("bUser", list.get(0));
        return "redirect:/";                          //回到首页
    }
    model.addAttribute("errorMessage", "用户名或密码错误!");
    return "user/login";
}
```

❹ 编写 SQL 映射文件

该功能模块的 SQL 语句如下：

```
<select id="login" parameterType="BUser" resultType="BUser">
    select * from busertable where bemail = #{bemail} and bpwd = #{bpwd}
</select>
```

9.6.5 商品详情

用户可以从推荐商品、最新商品、广告商品以及搜索商品结果等位置单击商品图片进入商品详情页面（goodsDetail.html），如图 9.20 所示。

图 9.20 商品详情页面

商品详情页面的具体实现步骤如下。

❶ 编写视图

该模块的视图涉及 src/main/resources/templates/user 目录下的 goodsDetail.html，其代码请参见本书提供的源程序 ch9。

❷ 编写控制器层

该功能模块涉及 com.ch.ch9.controller.before.IndexController 控制器类的 goodsDetail 方法，具体代码如下：

```
@RequestMapping("/goodsDetail")
public String goodsDetail(Model model, Integer id) {
    return indexService.goodsDetail(model, id);
}
```

❸ 编写 Service 层

该功能模块的 Service 层代码如下:

```
@Override
public String goodsDetail(Model model, Integer id) {
    //广告区中的商品
    model.addAttribute("advertisementGoods", indexRepository.selectAdvertisementGoods());
    //导航栏中的商品类型
    model.addAttribute("goodsType", indexRepository.selectGoodsType());
    //商品详情
    model.addAttribute("goods", indexRepository.selectAGoods(id));
    return "user/goodsDetail";
}
```

❹ 编写 SQL 映射文件

该功能模块的 SQL 语句如下:

```xml
<!-- 查询商品详情 -->
<select id="selectAGoods" resultType="Goods">
    select
        gt.*, gy.typename
    from
        goodstable gt,goodstype gy
    where
        gt.goodstype_id = gy.id
        and gt.id = #{id}
</select>
```

▶9.6.6 收藏商品

登录成功的用户可以在商品详情页面、首页以及搜索商品结果页面单击"加入收藏"按钮收藏商品,此时请求路径为 cart/focus(Ajax 实现)。系统根据@RequestMapping 注解找到 com.ch.ch9.controller.before.CartController 控制器类的 focus 方法处理请求。其具体实现步骤如下:

❶ 编写控制器层

该功能模块涉及 com.ch.ch9.controller.before.CartController 控制器类的 focus 方法,具体代码如下:

```
@RequestMapping("/focus")
@ResponseBody
public String focus(@RequestBody Goods goods, Model model, HttpSession session) {
    return cartService.focus(model, session, goods.getId());
}
```

❷ 编写 Service 层

该功能模块的 Service 层代码如下:

```
@Override
public String focus(Model model, HttpSession session, Integer gid) {
    Integer uid = MyUtil.getUser(session).getId();
    List<Map<String,Object>>list = cartRepository.isFocus(uid, gid);
```

```
        //判断是否已收藏
        if(list.size() > 0) {
            return "no";
        }else {
            cartRepository.focus(uid, gid);
            return "ok";
        }
    }
```

❸ 编写 SQL 映射文件

该功能模块的 SQL 语句如下：

```xml
<!-- 处理加入收藏 -->
<select id="isFocus" resultType="map">
    select * from focustable where goodstable_id = #{gid} and busertable_id = #{uid}
</select>
<insert id="focus">
insert into focustable (id, goodstable_id, busertable_id, focustime) values(null, #{gid}, #{uid}, now())
</insert>
```

▶ **9.6.7 购物车**

单击商品详情页面中的"加入购物车"按钮或导航栏中的"我的购物车"超链接，打开购物车页面（cart.html），如图 9.21 所示。

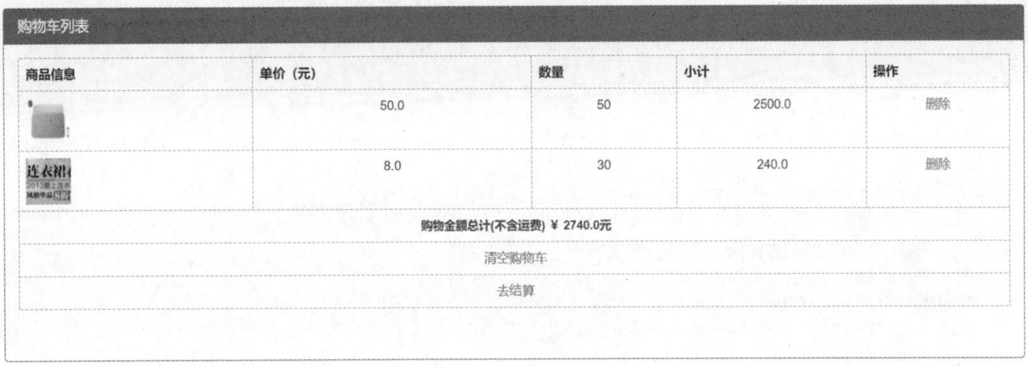

购物车列表					
商品信息	单价（元）		数量	小计	操作
	50.0		50	2500.0	删除
	8.0		30	240.0	删除
	购物金额总计(不含运费)￥2740.0元				
	清空购物车				
	去结算				

图 9.21 购物车

与购物车相关的处理请求有 cart/putCart（加入购物车）、cart/clearCart（清空购物车）、cart/selectCart（查询购物车）和 cart/deleteCart（删除购物车）。系统根据@RequestMapping 注解分别找到 com.ch.ch9.controller.before.CartController 控制器类的 putCart、clearCart、selectCart、deleteCart 等方法处理请求。其具体实现步骤如下：

❶ 编写视图

该模块的视图涉及 src/main/resources/templates/user 目录下的 cart.html，其代码请参见本书提供的源程序 ch9。

❷ 编写控制器层

该功能模块涉及 com.controller.before.CartController 控制器类的 putCart、clearCart、selectCart、deleteCart 等方法，具体代码如下：

```java
@RequestMapping("/putCart")
public String putCart(Goods goods, Model model, HttpSession session) {
    return cartService.putCart(goods, model, session);
```

```java
}
@RequestMapping("/selectCart")
public String selectCart(Model model, HttpSession session, String act) {
    return cartService.selectCart(model, session, act);
}
@RequestMapping("/deleteCart")
public String deleteCart(HttpSession session, Integer gid) {
    return cartService.deleteCart(session, gid);
}
@RequestMapping("/clearCart")
public String clearCart(HttpSession session) {
    return cartService.clearCart(session);
}
```

❸ 编写 Service 层

该功能模块的 Service 层代码如下：

```java
@Override
public String putCart(Goods goods, Model model, HttpSession session) {
    Integer uid = MyUtil.getUser(session).getId();
    //如果商品已经在购物车,那么只更新购买数量
    if(cartRepository.isPutCart(uid, goods.getId()).size() > 0) {
        cartRepository.updateCart(uid, goods.getId(), goods.getBuyNumber());
    }else {                                              //新增到购物车
        cartRepository.putCart(uid, goods.getId(), goods.getBuyNumber());
    }
    //跳转到查询购物车
    return "forward:/cart/selectCart";
}
@Override
public String selectCart(Model model, HttpSession session, String act) {
    List<Map<String, Object>> list = cartRepository.selectCart(MyUtil.getUser(session).getId());
    double sum = 0;
    for (Map<String, Object> map : list) {
        sum = sum + (Double)map.get("smallsum");
    }
    model.addAttribute("total", sum);
    model.addAttribute("cartlist", list);
    //广告区中的商品
    model.addAttribute("advertisementGoods", indexRepository.selectAdvertisementGoods());
    //导航栏中的商品类型
    model.addAttribute("goodsType", indexRepository.selectGoodsType());
    if("toCount".equals(act)) {                         //去结算页面
        return "user/count";
    }
    return "user/cart";
}
@Override
public String deleteCart(HttpSession session, Integer gid) {
    Integer uid = MyUtil.getUser(session).getId();
    cartRepository.deleteAgoods(uid, gid);
    return "forward:/cart/selectCart";
}
@Override
public String clearCart(HttpSession session) {
    cartRepository.clear(MyUtil.getUser(session).getId());
    return "forward:/cart/selectCart";
}
```

❹ 编写 SQL 映射文件

该功能模块的 SQL 语句如下：

```xml
<!-- 是否已添加购物车 -->
<select id="isPutCart" resultType="map">
    select * from carttable where goodstable_id=#{gid} and busertable_id=#{uid}
</select>
<!-- 添加购物车 -->
<insert id="putCart">
    insert into carttable (id, busertable_id, goodstable_id, shoppingnum) values(null,
#{uid},#{gid},#{bnum})
</insert>
<!-- 更新购物车 -->
<update id="updateCart">
    update carttable set shoppingnum=shoppingnum+#{bnum} where busertable_id=#{uid} and
goodstable_id=#{gid}
</update>
<!-- 查询购物车 -->
<select id="selectCart" parameterType="Integer" resultType="map">
    select gt.id, gt.gname, gt.gpicture, gt.grprice, ct.shoppingnum, ct.shoppingnum * gt.
grprice smallsum
    from goodstable gt, carttable ct where gt.id=ct.goodstable_id and ct.busertable_id=
#{uid}
</select>
<!-- 删除购物车 -->
<delete id="deleteAgoods">
    delete from carttable where busertable_id=#{uid} and goodstable_id=#{gid}
</delete>
<!-- 清空购物车 -->
<delete id="clear" parameterType="Integer">
    delete from carttable where busertable_id=#{uid}
</delete>
```

▶9.6.8 下单

在购物车页面中单击"去结算"按钮，进入订单确认页面（count.html），如图 9.22 所示。

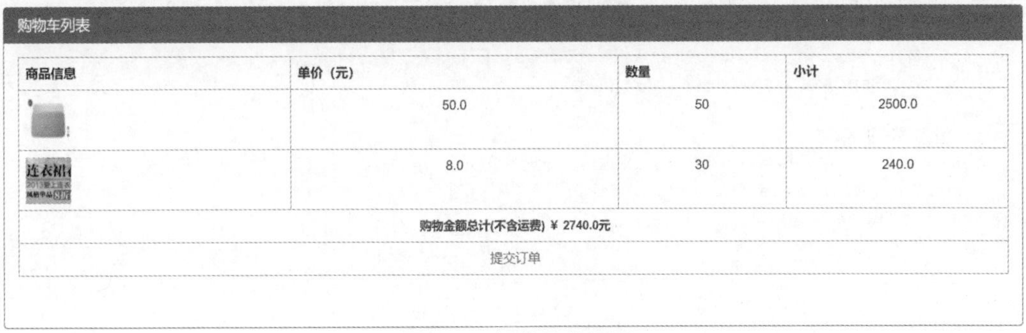

图 9.22 订单确认页面

在订单确认页面中单击"提交订单"按钮，完成订单的提交。在订单完成时，页面效果如图 9.23 所示。

图 9.23 订单提交完成页面

单击图 9.23 中的"去支付"按钮完成订单的支付。
其具体实现步骤如下：

❶ 编写视图

该模块的视图涉及 src/main/resources/templates/user 目录下的 count.html 和 pay.html。count.html 的代码与购物车页面的代码基本一样，这里不再赘述。pay.html 的代码请参见本书提供的源程序 ch9。

❷ 编写控制器层

该功能模块涉及 com.ch.ch9.controller.before.CartController 控制器类的 submitOrder 和 pay 方法，具体代码如下：

```java
@RequestMapping("/submitOrder")
public String submitOrder(Order order, Model model, HttpSession session) {
    return cartService.submitOrder(order, model, session);
}
@RequestMapping("/pay")
@ResponseBody
public String pay(@RequestBody Order order) {
    return cartService.pay(order);
}
```

❸ 编写 Service 层

该功能模块的 Service 层代码如下：

```java
@Override
@Transactional
public String submitOrder(Order order, Model model, HttpSession session) {
    order.setBusertable_id(MyUtil.getUser(session).getId());
    //生成订单
    cartRepository.addOrder(order);
    //生成订单详情
    cartRepository.addOrderDetail(order.getId(), MyUtil.getUser(session).getId());
    //减少商品库存
    List<Map<String, Object>> listGoods = cartRepository.selectGoodsShop(MyUtil.getUser(session).getId());
    for (Map<String, Object> map : listGoods) {
        cartRepository.updateStore(map);
    }
    //清空购物车
    cartRepository.clear(MyUtil.getUser(session).getId());
    model.addAttribute("order", order);
    return "user/pay";
}
@Override
public String pay(Order order) {
    cartRepository.pay(order.getId());
    return "ok";
}
```

❹ 编写 SQL 映射文件

该功能模块涉及的 SQL 语句如下：

```xml
<!-- 添加一个订单,成功后将主键值回填给 id(实体类的属性) -->
<insert id="addOrder" parameterType="Order" keyProperty="id" useGeneratedKeys="true">
    insert into orderbasetable (busertable_id, amount, status, orderdate) values (#{busertable_id}, #{amount}, 0, now())
</insert>
```

```xml
<!-- 生成订单详情 -->
<insert id="addOrderDetail">
    insert into orderdetail (orderbasetable_id, goodstable_id, shoppingnum) select #{ordersn}, goodstable_id, shoppingnum from carttable where busertable_id = #{uid}
</insert>
<!-- 查询商品购买量,以便更新库存时使用 -->
<select id="selectGoodsShop" parameterType="Integer" resultType="map">
    select shoppingnum gshoppingnum, goodstable_id gid from carttable where busertable_id=#{uid}
</select>
<!-- 更新商品库存 -->
<update id="updateStore" parameterType="map">
    update goodstable set gstore= gstore -#{gshoppingnum} where id=#{gid}
</update>
<!-- 支付订单 -->
<update id="pay" parameterType="Integer">
    update orderbasetable set status=1 where id=#{ordersn}
</update>
```

▶9.6.9 个人信息

成功登录的用户在导航栏的上方单击"个人信息"超链接(cart/userInfo),进入用户修改密码页面(userInfo.html),如图 9.24 所示。

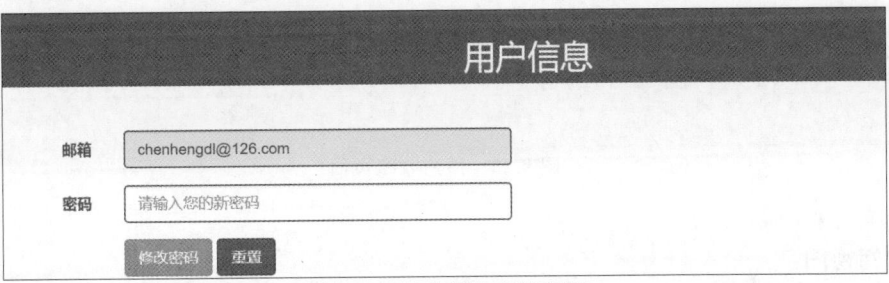

图 9.24 用户修改密码页面

其具体实现步骤如下:

❶ 编写视图

该模块的视图涉及 src/main/resources/templates/user 目录下的 userInfo.html,其代码与登录页面的代码类似,这里不再赘述。

❷ 编写控制器层

该功能模块涉及 com.ch.ch9.controller.before.CartController 控制器类的 userInfo 和 updateUpwd 方法,具体代码如下:

```java
@RequestMapping("/userInfo")
public String userInfo() {
    return "user/userInfo";
}
@RequestMapping("/updateUpwd")
public String updateUpwd(HttpSession session, String bpwd) {
    return cartService.updateUpwd(session, bpwd);
}
```

❸ 编写 Service 层

该功能模块的 Service 层代码如下:

```java
@Override
```

```java
public String updateUpwd(HttpSession session, String bpwd) {
    Integer uid = MyUtil.getUser(session).getId();
    cartRepository.updateUpwd(uid, MD5Util.MD5(bpwd));
    return "forward:/user/toLogin";
}
```

❹ 编写 SQL 映射文件

该功能模块的 SQL 语句如下：

```xml
<!-- 修改密码 -->
<update id="updateUpwd">
    update busertable set bpwd=#{bpwd} where id=#{uid}
</update>
```

▶9.6.10 我的收藏

成功登录的用户在导航栏的上方单击"我的收藏"超链接（cart/myFocus），进入用户收藏页面（myFocus.html），如图 9.25 所示。

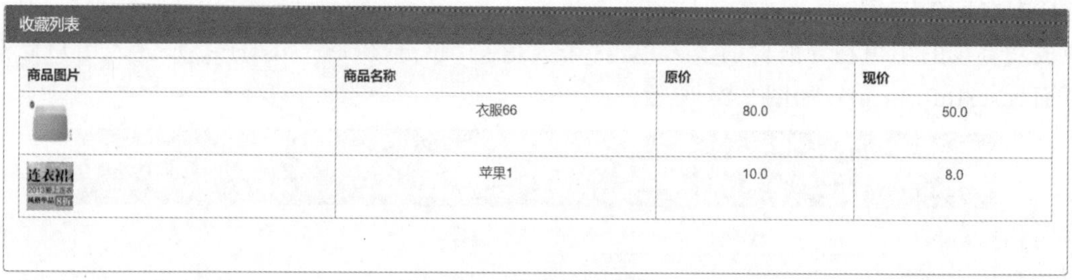

收藏列表			
商品图片	商品名称	原价	现价
	衣服66	80.0	50.0
	苹果1	10.0	8.0

图 9.25　用户收藏页面

其具体实现步骤如下：

❶ 编写视图

该模块的视图涉及 src/main/resources/templates/user 目录下的 myFocus.html，其代码请参见本书提供的源程序 ch9。

❷ 编写控制器层

该功能模块涉及 com.ch.ch9.controller.before.CartController 控制器类的 myFocus 方法，具体代码如下：

```java
@RequestMapping("/myFocus")
public String myFocus(Model model, HttpSession session) {
    return cartService.myFocus(model, session);
}
```

❸ 编写 Service 层

该功能模块的 Service 层代码如下：

```java
@Override
public String myFocus(Model model, HttpSession session) {
    //广告区中的商品
    model.addAttribute("advertisementGoods", indexRepository.selectAdvertisementGoods());
    //导航栏中的商品类型
    model.addAttribute("goodsType", indexRepository.selectGoodsType());
    model.addAttribute("myFocus", cartRepository.myFocus(MyUtil.getUser(session).getId()));
    return "user/myFocus";
}
```

❹ 编写 SQL 映射文件

该功能模块的 SQL 语句如下：

```xml
<!-- 我的收藏 -->
<select id="myFocus" resultType="map" parameterType="Integer">
    select gt.id, gt.gname, gt.goprice, gt.grprice, gt.gpicture from FOCUSTABLE ft, GOODSTABLE gt
    where ft.goodstable_id=gt.id and ft.busertable_id = #{uid}
</select>
```

▶9.6.11 我的订单

成功登录的用户在导航栏的上方单击"我的订单"超链接（cart/myOrder），进入用户订单页面（myOrder.html），如图 9.26 所示。

图 9.26　用户订单页面

单击图 9.26 中的"查看详情"超链接（'cart/orderDetail?id=' + ${order.id}），进入订单详情页面（orderDetail.html），如图 9.27 所示。

图 9.27　订单详情页面

其具体实现步骤如下：

❶ 编写视图

该模块的视图涉及 src/main/resources/templates/user 目录下的 myOrder.html 和 orderDetail.html。myOrder.html 和 orderDetail.html 的代码请参见本书提供的源程序 ch9。

❷ 编写控制器层

该功能模块涉及 com.ch.ch9.controller.before.CartController 控制器类的 myOrder 和 orderDetail 方法，具体代码如下：

```java
@RequestMapping("/myOrder")
public String myOrder(Model model, HttpSession session) {
    return cartService.myOrder(model, session);
}
```

```
@RequestMapping("/orderDetail")
public String orderDetail(Model model, Integer id) {
    return cartService.orderDetail(model, id);
}
```

❸ 编写 Service 层

该功能模块的 Service 层代码如下：

```
@Override
public String myOrder(Model model, HttpSession session) {
    //广告区中的商品
    model.addAttribute("advertisementGoods", indexRepository.selectAdvertisementGoods());
    //导航栏中的商品类型
    model.addAttribute("goodsType", indexRepository.selectGoodsType());
    model.addAttribute("myOrder", cartRepository.myOrder(MyUtil.getUser(session).getId()));
    return "user/myOrder";
}
@Override
public String orderDetail(Model model, Integer id) {
    model.addAttribute("orderDetail", cartRepository.orderDetail(id));
    return "user/orderDetail";
}
```

❹ 编写 SQL 映射文件

该功能模块的 SQL 语句如下：

```
<!-- 我的订单 -->
<select id="myOrder" resultType="map" parameterType="Integer">
select id, amount, busertable_id, status, orderdate from ORDERBASETABLE where busertable_id
= #{uid}
</select>
<!-- 订单详情 -->
<select id="orderDetail" resultType="map" parameterType="Integer">
select gt. id, gt. gname, gt. goprice, gt. grprice, gt. gpicture, odt. shoppingnum from
GOODSTABLE gt, ORDERDETAIL odt
    where odt.orderbasetable_id=#{id} and gt.id=odt.goodstable_id
</select>
```

本章小结

本章讲述了电子商务平台通用功能的设计与实现。通过本章的学习，读者不仅要掌握 Spring Boot 应用开发的流程、方法和技术，还应该熟悉电子商务平台的业务需求、设计以及实现。

习题 9

1. 在本章电子商务平台中是如何控制管理员登录权限的？
2. 在本章电子商务平台中有几对关联数据表？

第 10 章　Spring Boot 的安全控制

学习目的与要求

本章首先重点讲解 Spring Security 安全控制机制，然后介绍 Spring Boot Security 操作实例。通过本章的学习，掌握如何使用 Spring Security 安全控制机制解决企业应用程序的安全问题。

本章主要内容

- Spring Security 快速入门
- Spring Boot Security 操作实例

在 Web 应用开发中，安全毋庸置疑是十分重要的，使用 Spring Security 来保护 Web 应用是一个非常好的选择。Spring Security 是 Spring 框架的一个安全模块，可以非常方便地与 Spring Boot 应用无缝集成。

10.1　Spring Security 快速入门

▶10.1.1　什么是 Spring Security

Spring Security 是一个专门针对 Sping 应用系统的安全框架，充分使用了 Spring 框架的依赖注入和 AOP 功能，为 Spring 应用系统提供安全访问控制解决方案。

在 Spring Security 安全框架中有两个重要概念，即授权（Authorization）和认证（Authentication）。授权即确定用户在当前应用系统下所拥有的功能权限；认证即确认用户访问当前系统的身份。

▶10.1.2　Spring Security 的用户认证

验证用户最常见的方法之一是验证用户名和密码。Spring Security 为使用用户名和密码进行身份验证提供了全面的支持。

在 Spring Security 安全框架中可以通过配置 AuthenticationManager 认证管理器完成用户认证，示例代码如下：

```
@Bean
public AuthenticationManager authenticationManager ( AuthenticationConfiguration
authenticationConfiguration) throws Exception{
    return authenticationConfiguration.getAuthenticationManager();
}
```

❶ 内存中的用户认证

在 Spring Security 安全框架中，InMemoryUserDetailsManager 实现了 UserDetailsService，以支持存储在内存中的用户名/密码的身份验证。InMemoryUserDetailsManager 通过实现 UserDetailsManager 接口提供对 UserDetails 的管理。当 Spring Security 配置为接受用户名和密码进行身份验证时，会使用基于 UserDetails 的身份验证。示例代码如下：

```
@Bean
public UserDetailsService users() {
```

```java
        UserDetails user = User.builder()
            .username("chenheng")
            .password("xxxx")
            .roles("USER")
            .build();
        UserDetails admin = User.builder()
            .username("admin")
            .password("yyyy")
            .roles("DBA", "ADMIN")
            .build();
        return new InMemoryUserDetailsManager(user, admin);
}
```

在上述示例代码中添加了两个用户,一个用户名为"admin",密码为"yyyy",用户权限为"ROLE_ADMIN"和"ROLE_DBA";另一个用户名为"chenheng",密码为"xxxx",用户权限为"ROLE_USER"。"ROLE_"是 Spring Security 保存用户权限的时候默认加上的。

❷ 通用的用户认证

在实际应用中可以查询数据库获取用户和权限,这时需要自定义实现 org.springframework.security.core.userdetails.UserDetailsService 接口的类,并重写 public UserDetails loadUserByUsername(String username)方法查询对应的用户和权限,示例代码如下:

```java
@Service
public class MyUserSecurityService implements UserDetailsService{
    @Autowired
    private MyUserRepository myUserRepository;
    /**
     * 通过重写 loadUserByUsername 方法查询对应的用户
     * UserDetails 是 Spring Security 的一个核心接口
     * UserDetails 定义了可以获取用户名、密码、权限等与认证相关信息的方法
     */
    @Override
    public UserDetails loadUserByUsername(String username) throws UsernameNotFoundException {
        //根据用户名(页面接收的用户名)查询当前用户
        MyUser myUser = myUserRepository.findByUsername(username);
        if(myUser == null) {
            throw new UsernameNotFoundException("用户名不存在");
        }
        //GrantedAuthority 代表赋予当前用户的权限(认证权限)
        List<GrantedAuthority> authorities = new ArrayList<GrantedAuthority>();
        //获得当前用户权限集合
        List<Authority> roles = myUser.getAuthorityList();
        //将当前用户的权限保存为用户的认证权限
        for (Authority authority : roles) {
            GrantedAuthority sg = new SimpleGrantedAuthority(authority.getName());
            authorities.add(sg);
        }
        //org.springframework.security.core.userdetails.User 是 Spring Security 内部的实现,
        //专门用于保存用户名、密码、权限等与认证相关的信息
        User su = new User(myUser.getUsername(), myUser.getPassword(), authorities);
        return su;
    }
}
```

▶10.1.3 Spring Security 的请求授权

在 Spring Security 安全框架中可以通过配置 SecurityFilterChain 完成用户授权。示例代码

如下：

```
@Bean
public SecurityFilterChain filterChain(HttpSecurity http) throws Exception {
    http
    .authorizeHttpRequests(authorize -> authorize
        .requestMatchers("/xxx").permitAll()
        .requestMatchers("/user/**").hasRole("USER")
        //其他所有请求认证授权后才能访问
        .anyRequest().authenticated()
        );
    return http.build();
}
```

在 filterChain(HttpSecurity http)方法中，使用 HttpSecurity 的 authorizeHttpRequests()方法的子节点给指定用户授权访问 URL 模式。可以通过 requestMatchers()方法匹配 URL 路径。在匹配请求路径后，可以针对当前用户对请求进行安全处理。Spring Security 提供了许多安全处理方法，部分方法如表 10.1 所示。

表 10.1 Spring Security 的安全处理的部分方法

方　　法	用　　途
anyRequest()	匹配所有请求路径
access(String attribute)	Spring EL 表达式的结果为 true 时可以访问
anonymous()	匿名可以访问
authenticated()	用户登录后可以访问
denyAll()	用户不能访问
fullyAuthenticated()	用户完全认证可以访问（非 remember-me 下自动登录）
hasAnyAuthority(String...)	参数表示权限，用户权限与其中任一权限相同就可以访问
hasAnyRole(String...)	参数表示角色，用户角色与其中任一角色相同就可以访问
hasAuthority(String authority)	参数表示权限，用户权限与参数相同才可以访问
hasRole(String role)	参数表示角色，用户角色与参数相同才可以访问
permitAll()	任何用户都可以访问
rememberMe()	允许通过 remember-me 登录的用户访问

▶10.1.4　Spring Security 的核心类

Spring Security 的核心类包括 Authentication、SecurityContextHolder、UserDetails、UserDetailsService、GrantedAuthority、DaoAuthenticationProvider 和 PasswordEncoder。

❶ Authentication

Authentication 用来封装用户认证信息的接口，在用户登录认证之前，Spring Security 将相关信息封装为一个 Authentication 具体实现类的对象，在登录认证成功后将生成一个信息更全面、包含用户权限等信息的 Authentication 对象，然后将该对象保存在 SecurityContextHolder 所持有的 SecurityContext 中，以方便后续程序进行调用，如当前用户名、访问权限等。

❷ SecurityContextHolder

SecurityContextHolder 是用来持有 SecurityContext 的类。在 SecurityContext 中包含当前认证用户的详细信息。Spring Security 使用一个 Authentication 对象描述当前用户的相关信息。最常见的是获得当前登录用户的用户名和权限，示例代码如下：

```java
/**
 * 获得当前用户名
 */
private String getUname() {
    return SecurityContextHolder.getContext().getAuthentication().getName();
}
/**
 * 获得当前用户权限
 */
private String getAuthorities() {
    Authentication authentication = SecurityContextHolder.getContext().getAuthentication();
    List<String> roles = new ArrayList<String>();
    for (GrantedAuthority ga : authentication.getAuthorities()) {
        roles.add(ga.getAuthority());
    }
    return roles.toString();
}
```

❸ **UserDetails**

UserDetails 是 Spring Security 的一个核心接口。该接口定义了一些可以获取用户名、密码、权限等与认证相关的信息的方法。通常需要在应用中获取当前用户的其他信息，如 E-mail、电话等。这时只包含与认证相关的 UserDetails 对象可能就不能满足需要了。此时可以实现自己的 UserDetails，在该实现类中定义一些获取用户其他信息的方法，这样就可以直接从当前 SecurityContext 的 Authentication 的 principal 中获取用户的其他信息。

Authentication.getPrincipal() 的返回类型是 Object，但通常返回一个 UserDetails 的实例，通过强制类型转换可以将 Object 转换为 UserDetails 类型。

❹ **UserDetailsService**

UserDetails 是通过 UserDetailsService 的 loadUserByUsername(String username) 方法加载的。UserDetailsService 也是一个接口，也需要实现自己的 UserDetailsService 来加载自定义的 UserDetails 信息。

在登录认证时，Spring Security 将通过 UserDetailsService 的 loadUserByUsername(String username) 方法获取对应的 UserDetails 进行认证，在认证通过后将该 UserDetails 赋给认证通过的 Authentication 的 principal，然后将该 Authentication 保存到 SecurityContext 中。在应用中，如果需要使用用户信息，可以通过 SecurityContextHolder 获取存放在 SecurityContext 中的 Authentication 的 principal，即 UserDetails 实例。

❺ **GrantedAuthority**

Authentication 的 getAuthorities() 方法可以返回当前 Authentication 对象拥有的权限（一个 GrantedAuthority 类型的数组），即当前用户拥有的权限。GrantedAuthority 是一个接口，通常通过 UserDetailsService 进行加载，然后赋给 UserDetails。

❻ **DaoAuthenticationProvider**

在 Spring Security 安全框架中，默认使用 DaoAuthenticationProvider 实现 AuthenticationProvider 接口进行用户认证的处理。在 DaoAuthenticationProvider 进行认证时，需要一个 UserDetailsService 来获取用户信息 UserDetails。当然用户可以实现自己的 AuthenticationProvider，进而改变认证方式。

❼ **PasswordEncoder**

在 Spring Security 安全框架中通过 PasswordEncoder 接口完成对密码的加密。Spring

Security 对 PasswordEncoder 有多种实现，包括 MD5 加密、SHA-256 加密等，开发者直接使用即可。在 Spring Boot 应用中，使用 BCryptPasswordEncoder 加密是较好的选择。BCryptPasswordEncoder 使用 BCrypt 的强散列哈希加密实现，并可以由客户端指定加密强度，强度越高安全性越高。

▶10.1.5　Spring Security 的验证机制

Spring Security 的验证机制是由许多 Filter 实现的，Filter 将在 Spring MVC 前拦截请求，主要包括注销 Filter(LogoutFilter)、用户名密码验证 Filter(UsernamePasswordAuthenticationFilter)等内容。Filter 再交由其他组件完成细分的功能，最常用的 UsernamePasswordAuthenticationFilter 会持有一个 AuthenticationManager 引用，AuthenticationManager 是一个验证管理器，专门负责验证。AuthenticationManager 持有一个 AuthenticationProvider 集合，AuthenticationProvider 是做验证工作的组件，在验证成功或失败之后调用对应的 Handler(处理)。

10.2　Spring Boot 的支持

在 Spring Boot 应用中，只需引入 spring-boot-starter-security 依赖即可使用 Spring Security 安全框架，这是因为 Spring Boot 对 Spring Security 提供了自动配置功能。从 org.springframework.boot.autoconfigure.security.SecurityProperties 类中可以看到使用以"spring.security"为前缀的属性配置了 Spring Security 的相关默认配置。

10.3　实际开发中的 Spring Security 操作实例

本节将讲解一个基于 Spring Data JPA 的 Spring Boot Security 操作实例，演示在 Spring Boot 应用中如何使用基于 Spring Data JPA 的 Spring Security 安全框架。

【例 10-1】　在 Spring Boot 应用中使用基于 Spring Data JPA 的 Spring Security 安全框架。其具体实现步骤如下。

❶ 创建 Spring Boot Web 应用 ch10

创建基于 Lombok、Spring Data JPA、MySQL Driver、Thymeleaf 及 Spring Security 的 Web 应用 ch10，如图 10.1 所示。

❷ 设置 Web 应用 ch10 的上下文路径及数据源配置信息

在应用 ch10 的 application.properties 文件中配置内容与例 8-1 中的相同，这里不再赘述。

❸ 整理脚本、样式等静态文件

JS 脚本、CSS 样式、图片等静态文件默认放置在 src/main/resources/static 目录下，ch10 应用引入的 BootStrap 和 jQuery 与例 7-5 中的一样，这里不再赘述。

图 10.1　Web 应用 ch10 的依赖

❹ 创建用户和权限持久化实体类

在应用 ch10 的 src/main/java 目录下创建名为 com.ch.ch10.entity 的包，并在该包中创建持久化实体类 MyUser 和 Authority。MyUser 类用来保存用户数据，用户名唯一。Authority 用来保存权限信息。用户和权限是多对多的关系。

MyUser 的核心代码如下：

```java
@Entity
@Table(name = "user")
@JsonIgnoreProperties(value = {"hibernateLazyInitializer"})
@Data
public class MyUser implements Serializable{
    private static final long serialVersionUID = 1L;
    @Id
    @GeneratedValue(strategy = GenerationType.IDENTITY)
    private int id;
    private String username;
    private String password;
//这里不能是懒加载 lazy,否则在 MyUserSecurityService 的 loadUserByUsername 方法中获得不到权限
    @ManyToMany(cascade = {CascadeType.REFRESH}, fetch = FetchType.EAGER)
    @JoinTable(name = "user_authority",joinColumns = @JoinColumn(name = "user_id"),
    inverseJoinColumns = @JoinColumn(name = "authority_id"))
    private List<Authority> authorityList;
    //repassword 不映射到数据表
    @Transient
    private String repassword;
    //省略 set 和 get 方法
}
```

需要注意的是,在实际开发中 MyUser 还可以实现 org.springframework.security.core.userdetails.UserDetails 接口,在实现该接口后即可成为 Spring Security 所使用的用户。本例为了区分 Spring Data JPA 的 pojo 和 Spring Security 的用户对象,并没有实现 UserDetails 接口,而是在实现 UserDetailsService 接口的类中进行绑定。

Authority 的核心代码如下：

```java
@Entity
@Table(name = "authority")
@JsonIgnoreProperties(value = {"hibernateLazyInitializer"})
@Data
public class Authority implements Serializable{
    private static final long serialVersionUID = 1L;
    @Id
    @GeneratedValue(strategy = GenerationType.IDENTITY)
    private int id;
    @Column(nullable = false)
    private String name;
    @ManyToMany(mappedBy = "authorityList")
    @JsonIgnore
    private List<MyUser> userList;
}
```

❺ 创建数据访问层接口

在应用 ch10 的 src/main/java 目录下创建名为 com.ch.ch10.repository 的包,并在该包中创建名为 MyUserRepository 的接口,该接口继承了 JpaRepository,核心代码如下：

```java
public interface MyUserRepository extends JpaRepository<MyUser, Integer>{
    //根据用户名查询用户,方法名的命名符合 Spring Data JPA 规范
    MyUser findByUsername(String username);
}
```

在 com.ch.ch10.repository 包中创建名为 AuthorityRepository 的接口,该接口继承了 JpaRepository,具体代码如下：

```java
public interface AuthorityRepository extends JpaRepository<Authority, Integer>{}
```

❻ 创建业务层

在应用 ch10 的 src/main/java 目录下创建名为 com.ch.ch10.service 的包，并在该包中创建 UserService 接口和 UserServiceImpl 实现类。UserService 接口的代码略。

UserServiceImpl 的核心代码如下：

```java
@Service
public class UserServiceImpl implements UserService{
    @Autowired
    private MyUserRepository myUserRepository;
    @Autowired
    private AuthorityRepository authorityRepository;
    /**
     * 实现注册
     */
    @Override
    public String register(MyUser userDomain) {
        String username = userDomain.getUsername();
        List<Authority> authorityList = new ArrayList<Authority>();
        //管理员权限
        if("admin".equals(username)) {
            Authority a1 = new Authority();
            Authority a2 = new Authority();
            a1.setName("ROLE_ADMIN");
            a2.setName("ROLE_DBA");
            authorityList.add(a1);
            authorityList.add(a2);
        }else {                              //用户权限
            Authority a1 = new Authority();
            a1.setName("ROLE_USER");
            authorityList.add(a1);
        }
        //注册权限
        authorityRepository.saveAll(authorityList);
        userDomain.setAuthorityList(authorityList);
        //加密密码
        String secret= new BCryptPasswordEncoder().encode(userDomain.getPassword());
        userDomain.setPassword(secret);
        //注册用户
        MyUser mu = myUserRepository.save(userDomain);
        if(mu != null)                       //注册成功
            return "/login";
        return "/register";                  //注册失败
    }
    /**
     * 用户登录成功
     */
    @Override
    public String loginSuccess(Model model) {
        model.addAttribute("user", getUname());
        model.addAttribute("role", getAuthorities());
        return "/user/loginSuccess";
    }
    /**
     * 管理员登录成功
     */
    @Override
    public String main(Model model) {
        model.addAttribute("user", getUname());
        model.addAttribute("role", getAuthorities());
```

```java
        return "/admin/main";
    }
    /**
     * 注销用户
     */
    @Override
    public String logout(HttpServletRequest request, HttpServletResponse response) {
        //获得用户认证信息
        Authentication authentication = SecurityContextHolder.getContext().getAuthentication();
        if(authentication != null) {
            //注销
            new SecurityContextLogoutHandler().logout(request, response, authentication);
        }
        return "redirect:/login?logout";
    }
    /**
     * 没有权限拒绝访问
     */
    @Override
    public String deniedAccess(Model model) {
        model.addAttribute("user", getUname());
        model.addAttribute("role", getAuthorities());
        return "deniedAccess";
    }
    /**
     * 获得当前用户名
     */
    private String getUname() {
        return SecurityContextHolder.getContext().getAuthentication().getName();
    }
    /**
     * 获得当前用户权限
     */
    private String getAuthorities() {
        Authentication authentication = SecurityContextHolder.getContext().getAuthentication();
        List<String> roles = new ArrayList<String>();
        for (GrantedAuthority ga : authentication.getAuthorities()) {
            roles.add(ga.getAuthority());
        }
        return roles.toString();
    }
}
```

❼ 创建控制器类

在应用 ch10 的 src/main/java 目录下创建名为 com.ch.ch10.controller 的包，并在该包中创建控制器类 TestSecurityController，核心代码如下：

```java
@Controller
public class TestSecurityController {
    @Autowired
    private UserService userService;
    @RequestMapping("/")
    public String index() {
        return "/index";
    }
    @RequestMapping("/toLogin")
    public String toLogin() {
        return "/login";
```

```java
    }
    @RequestMapping("/toRegister")
    public String toRegister(@ModelAttribute("userDomain") MyUser userDomain) {
        return "/register";
    }
    @RequestMapping("/register")
    public String register(@ModelAttribute("userDomain") MyUser userDomain) {
        return userService.register(userDomain);
    }
    @RequestMapping("/login")
    public String login() {
        //这里什么都不做,由 Spring Security 负责登录验证
        return "/login";
    }
    @RequestMapping("/user/loginSuccess")
    public String loginSuccess(Model model) {
        return userService.loginSuccess(model);
    }
    @RequestMapping("/admin/main")
    public String main(Model model) {
        return userService.main(model);
    }
    @RequestMapping("/logout")
    public String logout(HttpServletRequest request, HttpServletResponse response) {
        return userService.logout(request, response);
    }
    @RequestMapping("/deniedAccess")
    public String deniedAccess(Model model) {
        return userService.deniedAccess(model);
    }
}
```

❽ **创建应用的安全控制相关实现**

在应用 ch10 的 src/main/java 目录下创建名为 com.ch.ch10.security 的包,并在该包中创建 MyUserSecurityService、MyAuthenticationSuccessHandler 和 SpringSecurityConfig 类。

MyUserSecurityService 实现了 UserDetailsService 接口,通过重写 loadUserByUsername (String username)方法查询对应的用户,并将用户名、密码、权限等与认证相关的信息封装在 UserDetails 对象中。

MyUserSecurityService 的核心代码如下:

```java
/**
 * 获得对应的 UserDetails,保存与认证相关的信息
 */
@Service
public class MyUserSecurityService implements UserDetailsService {
    @Autowired
    private MyUserRepository myUserRepository;
    /**
     * 通过重写 loadUserByUsername 方法查询对应的用户
     * UserDetails 是 Spring Security 的一个核心接口
     * UserDetails 定义了可以获取用户名、密码、权限等与认证相关的信息的方法
     */
    @Override
    public UserDetails loadUserByUsername(String username) throws UsernameNotFoundException {
        //根据用户名(页面接收的用户名)查询当前用户
        MyUser myUser = myUserRepository.findByUsername(username);
        if(myUser == null) {
```

```java
            throw new UsernameNotFoundException("用户名不存在");
        }
        //GrantedAuthority 代表赋予当前用户的权限(认证权限)
        List<GrantedAuthority> authorities = new ArrayList<GrantedAuthority>();
        //获得当前用户权限集合
        List<Authority> roles = myUser.getAuthorityList();
        //将当前用户的权限保存为用户的认证权限
        for (Authority authority : roles) {
            GrantedAuthority sg = new SimpleGrantedAuthority(authority.getName());
            authorities.add(sg);
        }
        //org.springframework.security.core.userdetails.User 是 Spring Security 的内部实现,
        //专门用于保存用户名、密码、权限等与认证相关的信息
        User su = new User(myUser.getUsername(), myUser.getPassword(), authorities);
        return su;
    }
}
```

MyAuthenticationSuccessHandler 继承了 SimpleUrlAuthenticationSuccessHandler 类, 并重写了 handle(HttpServletRequest request, HttpServletResponse response, Authentication authentication)方法, 根据当前认证用户的角色指定对应的 URL。

MyAuthenticationSuccessHandler 的核心代码如下:

```java
/**
 * 用户授权、认证成功处理类
 */
@Component
public class MyAuthenticationSuccessHandler extends SimpleUrlAuthenticationSuccessHandler {
    //Spring Security 的重定向策略
    private RedirectStrategy redirectStrategy = new DefaultRedirectStrategy();
    /**
     * 重写 handle 方法, 通过 RedirectStrategy 重定向到指定的 URL
     */
    @Override
    protected void handle(HttpServletRequest request, HttpServletResponse response,
Authentication authentication) throws IOException, ServletException {
        //根据当前认证用户的角色返回适当的 URL
        String targetURL = getTargetURL(authentication);
        //重定向到指定的 URL
        redirectStrategy.sendRedirect(request, response, targetURL);
    }
    /**
     * 从 Authentication 对象中提取当前登录用户的角色, 并根据其角色返回适当的 URL
     */
    protected String getTargetURL(Authentication authentication) {
        String url = "";
        //获得当前登录用户的权限(角色)集合
        Collection<? extends GrantedAuthority> authorities =authentication.getAuthorities();
        List<String> roles = new ArrayList<String>();
        //将权限(角色)名称添加到 List 集合
        for (GrantedAuthority au : authorities) {
            roles.add(au.getAuthority());
        }
        //判断不同角色的用户跳转到不同的 URL
        //这里的 URL 是控制器的请求匹配路径
        if(roles.contains("ROLE_USER")) {
            url = "/user/loginSuccess";
        }else if(roles.contains("ROLE_ADMIN")) {
            url = "/admin/main";
```

```
        }else {
            url = "/deniedAccess";
        }
        return url;
    }
}
```

SpringSecurityConfig 类是认证和授权处理的配置类，需要使用 @Configuration 和 @EnableWebSecurity 注解。在该类中配置了加密规则、认证实现方式、认证管理器以及授权操作。MySecurityConfigurerAdapter 的核心代码如下：

```
/**
 * 认证和授权处理类
 */
@Configuration
@EnableWebSecurity
public class SpringSecurityConfig {
    //依赖注入通用的用户服务类
    @Autowired
    private MyUserSecurityService myUserSecurityService;
    @Autowired
    private MyAuthenticationSuccessHandler myAuthenticationSuccessHandler;
    /**
     * BCryptPasswordEncoder 是 PasswordEncoder 的接口实现,实现加密功能
     */
    @Bean
    public PasswordEncoder passwordEncoder() {
        return new BCryptPasswordEncoder();
    }
    /**
     * DaoAuthenticationProvider 是 AuthenticationProvider 的实现,认证的实现方式
     */
    @Bean
    public AuthenticationProvider authenticationProvider() {
        DaoAuthenticationProvider provide = new DaoAuthenticationProvider();
        //不隐藏用户未找到异常
        provide.setHideUserNotFoundExceptions(false);
        //设置自定义认证方式,用户登录认证
        provide.setUserDetailsService(myUserSecurityService);
        //设置密码加密程序认证
        provide.setPasswordEncoder(passwordEncoder());
        return provide;
    }
    /**
     * 获取 AuthenticationManager(认证管理器)
     */
    @Bean
    public AuthenticationManager authenticationManager (AuthenticationConfiguration authenticationConfiguration) throws Exception{
        return authenticationConfiguration.getAuthenticationManager();
    }
    /**
     * 请求授权,用户授权操作
     */
    @Bean
    public SecurityFilterChain filterChain(HttpSecurity http) throws Exception {
        return http
                //设置权限
                .authorizeHttpRequests(authorize -> authorize
                //将首页、登录注册页面、登录注册功能以及静态资源过滤掉,即可任意访问
```

```
                .requestMatchers("/toLogin","/toRegister","/","/login","/register","/css/**","/fonts/**",
"/js/**").permitAll()
                        //这里默认追加 ROLE_,/user/**是控制器的请求匹配路径
                        .requestMatchers("/user/**").hasRole("USER")
                        .requestMatchers("/admin/**").hasAnyRole("ADMIN", "DBA")
                        //其他所有请求登录后才能访问
                        .anyRequest().authenticated()
                )
                //将输入的用户名与密码和授权的内容进行比较
                .formLogin()
                        .loginPage("/login").successHandler(myAuthenticationSuccessHandler)
                        .usernameParameter("username")
                        .passwordParameter("password")
                        //登录失败
                        .failureUrl("/login?error")
                .and()
                //注销行为可任意访问
                .logout().permitAll()
                .and()
                //指定异常处理页面
                .exceptionHandling().accessDeniedPage("/deniedAccess")
                .and().build();
    }
}
```

❾ **创建用于测试的视图页面**

在 src/main/resources/templates 目录下创建应用首页、注册、登录以及拒绝访问页面；在 src/main/resources/templates/admin 目录下创建管理员用户认证成功后访问的页面；在 src/main/resources/templates/user 目录下创建普通用户认证成功后访问的页面。相关视图页面的代码参见本书提供的源程序 ch10。

❿ **测试应用**

运行 Ch10Application 的主方法启动项目。在 Spring Boot 应用启动后，可以通过"http://localhost:8080/ch10"访问首页，如图 10.2 所示；然后单击"去注册"超链接打开注册页面，如图 10.3 所示。在成功注册用户后，打开登录页面进行用户的登录。

图 10.2 首页 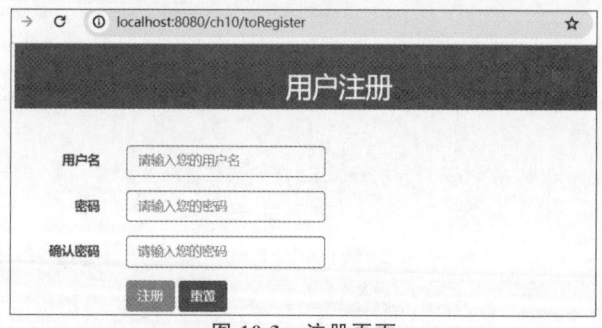　　　　图 10.3 注册页面

如果在图 10.3 中输入的用户名不是 admin，那么就注册了一个普通用户，其权限为"ROLE_USER"；如果在图 10.3 中输入的用户名是 admin，那么就注册了一个管理员用户，其权限为"ROLE_ADMIN"和"ROLE_DBA"。

在登录页面中任意输入用户名和密码，单击"登录"按钮，会提示用户名或密码错误，如图 10.4 所示。

图 10.4　用户名或密码错误

在登录页面中输入管理员用户名和密码,成功登录后打开管理员登录成功页面,如图 10.5 所示。

图 10.5　管理员登录成功页面

单击图 10.5 中的"去访问用户登录成功页面"显示拒绝访问页面,如图 10.6 所示。

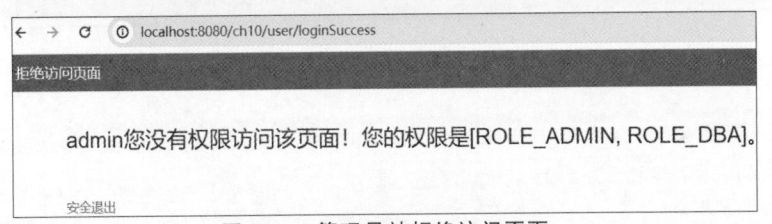

图 10.6　管理员被拒绝访问页面

在登录页面中输入普通用户名和密码,成功登录后打开用户登录成功页面,如图 10.7 所示。

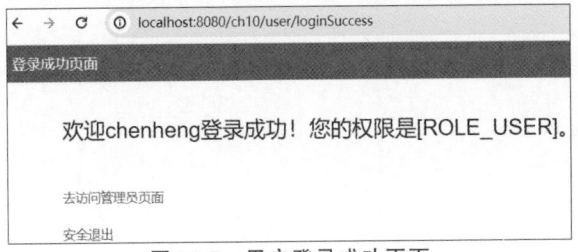

图 10.7　用户登录成功页面

单击图 10.7 中的"去访问管理员页面"显示拒绝访问页面,如图 10.8 所示。

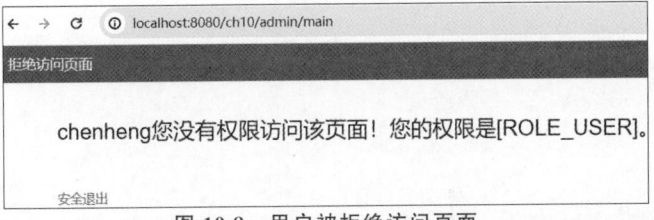

图 10.8　用户被拒绝访问页面

本章小结

本章首先介绍了 Spring Security 快速入门知识,然后详细介绍了实际开发中的 Spring Security 操作实例。通过本章的学习,读者应该了解 Spring Security 安全机制的基本原理,掌握如何在实际应用开发中使用 Spring Security 安全机制提供系统安全解决方案。

习题 10

1. 简述 Spring Security 的验证机制。
2. Spring Security 的用户认证和请求授权是如何实现的?请举例说明。

第 11 章 Spring Boot 的异步消息

学习目的与要求

本章主要讲解了企业级消息代理 JMS 和 AMQP。通过本章的学习，理解异步消息通信原理，掌握异步消息通信技术。

本章主要内容
- JMS
- AMQP

当跨越多个微服务进行通信时，异步消息就显得至关重要了。例如电子商务系统中，订单服务在下单时需要和库存服务进行通信，完成库存的扣减操作，这时就需要基于异步消息和最终一致性的通信方式来进行这样的操作，并且能够在发生故障时正常工作。

11.1 消息模型

异步消息的主要目的是解决跨系统的通信。所谓异步消息，即消息发送者无须等待消息接收者的处理及返回，甚至无须关心消息是否发送与接收成功。在异步消息中有两个极其重要的概念，即消息代理和目的地。当消息发送者发送消息后，消息将由消息代理管理，消息代理保证消息传递到目的地。

异步消息的目的地主要有两种形式，即队列和主题。队列用于点对点式的消息通信，即端到端通信（单接收者）；主题用于发布/订阅式的消息通信，即广播通信（多接收者）。

11.1.1 点对点式

在点对点式的消息通信中，消息代理获得发送者发送的消息后，将消息存入一个队列中，当有消息接收者接收消息时，将从队列中取出消息传递给接收者，这时队列中清除该消息。

在点对点式的消息通信中，确保的是每一条消息只有唯一的发送者和接收者，但并不能说明只有一个接收者可以从队列中接收消息。这是因为队列中有多个消息，点对点式的消息通信只保证每一条消息只有唯一的发送者和接收者。

11.1.2 发布/订阅式

多接收者是消息通信中一种更加灵活的方式，而点对点式的消息通信只保证每一条消息只有唯一的接收者。可以使用发布/订阅式的消息通信解决多接收者的问题。与点对点式不同，发布/订阅式是消息发送者将消息发送到主题，而多个消息接收者监听这个主题。此时的消息发送者称为发布者，接收者称为订阅者。

11.2 企业级消息代理

异步消息传递常用的技术有 JMS 和 AMQP。JMS 是面向基于 Java 的企业应用的异步消息代理。AMQP 是面向所有应用的异步消息代理。

11.2.1 JMS

JMS（Java Messaging Service）即 Java 消息服务，是 Java 平台上有关面向消息中间件的技术规范，它便于消息系统中的 Java 应用程序进行消息交换，并且通过提供标准的产生、发送、接收消息的接口简化企业应用的开发。

❶ **JMS 元素**

JMS 由以下元素组成。

（1）JMS 消息代理实现：连接面向消息中间件，JMS 消息代理接口的一个实现。JMS 的消息代理实现可以是 Java 平台的 JMS 实现，也可以是非 Java 平台的面向消息中间件的适配器。开源的 JMS 实现有 Apache ActiveMQ Artemis、JBoss 社区所研发的 HornetQ、The OpenJMS Group 的 OpenJMS 等实现。

（2）JMS 客户：生产或消费基于消息的 Java 应用程序或对象。

（3）JMS 生产者：创建并发送消息的 JMS 客户。

（4）JMS 消费者：接收消息的 JMS 客户。

（5）JMS 消息：包括可以在 JMS 客户之间传递的数据对象。JMS 定义了 5 种不同的消息正文格式，以及调用的消息类型，允许用户发送并接收一些不同形式的数据，提供现有消息格式的一些级别的兼容性。常见的消息格式有 StreamMessage（指 Java 原始值的数据流消息）、MapMessage（映射消息）、TextMessage（文本消息）、ObjectMessage（一个序列化的 Java 对象消息）、BytesMessage（字节消息）。

（6）JMS 队列：一个容纳被发送的等待阅读的消息区域。与队列名所暗示的意思不同，消息的接收顺序并不一定要与消息的发送顺序相同。一旦一个消息被阅读，该消息将被从队列中移走。

（7）JMS 主题：一种支持发送消息给多个订阅者的机制。

❷ **JMS 的应用接口**

JMS 的应用接口包括以下接口类型：

1）ConnectionFactory 接口（连接工厂）

用户用它来创建到 JMS 消息代理实现的连接的被管对象。JMS 客户通过可移植的接口访问连接，这样当下层的实现改变时代码不需要进行修改。管理员在 JNDI 名字空间中配置连接工厂，这样 JMS 客户才能够查找到它们。根据目的地的不同，用户将使用队列连接工厂，或者使用主题连接工厂。

2）Connection 接口（连接）

连接代表了应用程序和消息服务器之间的通信链路。在获得连接工厂后，就可以创建一个与 JMS 消息代理实现（提供者）的连接。根据不同的连接类型，连接允许用户创建会话，以发送和接收队列或主题到目的地。

3）Destination 接口（目的地）

目的地是一个包装了消息目的地标识符的被管对象，消息目的地是指消息发布和接收的地点，或者是队列，或者是主题。JMS 管理员创建这些对象，然后用户通过 JNDI 发现它们。和连接工厂一样，管理员可以创建两种类型的目的地，即点对点模型的队列以及发布者/订阅者模型的主题。

4）Session 接口（会话）

会话表示一个单线程的上下文，用于发送和接收消息。由于会话是单线程的，所以消息是连续的，也就是说消息是按照发送的顺序一个一个接收的。会话的好处是它支持事务。如果用户选择了事务支持，会话上下文将保存一组消息，直到事务被提交才发送这些消息。在提交事务之前，用户可以使用回滚操作取消这些消息。一个会话允许用户创建消息，生产者来发送消息，消费者来接收消息。

5）MessageConsumer 接口（消息消费者）

消息消费者是由会话创建的对象，用于接收发送到目的地的消息。消费者可以同步地（阻塞模式）或异步地（非阻塞）接收队列和主题类型的消息。

6）MessageProducer 接口（消息生产者）

消息生产者是由会话创建的对象，用于发送消息到目的地。用户可以创建某个目的地的发送者，也可以创建一个通用的发送者，在发送消息时指定目的地。

7）Message 接口（消息）

消息是在消费者和生产者之间传送的对象，也就是说从一个应用程序传送到另一个应用程序。一个消息有三个主要部分。

消息头（必须）：包含用于识别和为消息寻找路由的操作设置。

一组消息属性（可选）：包含额外的属性，支持其他消息代理实现和用户的兼容，可以创建定制的字段和过滤器（消息选择器）。

一个消息体（可选）：允许用户创建 5 种类型的消息（文本消息、映射消息、字节消息、流消息和对象消息）。

JMS 各接口角色间的关系如图 11.1 所示。

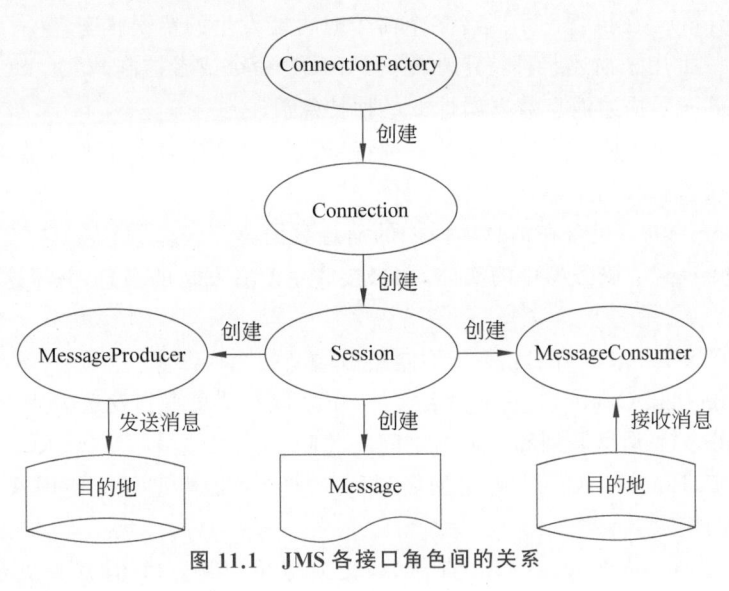

图 11.1　JMS 各接口角色间的关系

11.2.2　AMQP

AMQP（Advanced Message Queuing Protocol）即高级消息队列协议，它是一个提供统一消息服务的应用层标准高级消息队列协议，是应用层协议的一个开放标准，为面向消息的中间件设计。基于此协议的客户端与消息中间件可传递消息，且不受客户端/中间件的不同产品、不同开发语言等条件的限制。AMQP 的技术术语如下。

AMQP 模型（AMQP Model）：一个由关键实体和语义表示的逻辑框架，遵从 AMQP 规范的服务器必须提供这些实体和语义。为了实现本规范中定义的语义，客户端可以发送命令来控制 AMQP 服务器。

连接（Connection）：一个网络连接，例如 TCP/IP 套接字连接。

会话(Session)：端点之间的命名对话。在一个会话上下文中，保证"恰好传递一次"。

信道(Channel)：多路复用连接中的一条独立的双向数据流通道。为会话提供物理传输介质。

客户端(Client)：AMQP 连接或者会话的发起者。AMQP 是非对称的，客户端生产和消费消息，服务器存储和路由这些消息。

服务器(Server)：接受客户端连接，实现 AMQP 消息队列和路由功能的进程，也称为"消息代理"。

端点(Peer)：AMQP 对话的任意一方。一个 AMQP 连接包括两个端点(一个是客户端，另一个是服务器)。

搭档(Partner)：当描述两个端点之间的交互过程时，使用术语"搭档"来表示"另一个"端点的简记法。例如定义端点 A 和端点 B，当它们进行通信时，端点 B 是端点 A 的搭档，端点 A 是端点 B 的搭档。

片段集(Assembly)：段的有序集合，形成一个逻辑工作单元。

段(Segment)：帧的有序集合，形成片段集中的一个完整子单元。

帧(Frame)：AMQP 传输的一个原子单元。一个帧是一个段中的任意分片。

控制(Control)：单向指令，AMQP 规范假设这些指令的传输是不可靠的。

命令(Command)：需要确认的指令，AMQP 规范规定这些指令的传输是可靠的。

异常(Exception)：在执行一个或者多个命令时可能发生的错误状态。

类(Class)：一批用来描述某种特定功能的 AMQP 命令或者控制。

消息头(Header)：描述消息数据属性的一种特殊段。

消息体(Body)：包含应用程序数据的一种特殊段。消息体段对于服务器来说完全透明，即服务器不能查看或者修改消息体。

消息内容(Content)：包含在消息体段中的消息数据。

交换器(Exchange)：服务器中的实体，用来接收生产者发送的消息，并将这些消息路由给服务器中的队列。

交换器类型(Exchange Type)：基于不同路由语义的交换器类。

消息队列(Message Queue)：一个命名实体，用来保存消息直到发送给消费者。

绑定器(Binding)：消息队列和交换器之间的关联。

绑定器关键字(Binding Key)：绑定的名称。一些交换器类型可能使用这个名称作为定义绑定器路由行为的模式。

路由关键字(Routing Key)：一个消息头，交换器可以用这个消息头决定如何路由某条消息。

持久存储(Durable)：一种服务器资源，当服务器重启时保存的消息数据不会丢失。

临时存储(Transient)：一种服务器资源，当服务器重启时保存的消息数据会丢失。

持久化(Persistent)：服务器将消息保存在可靠的磁盘中，当服务器重启时消息不会丢失。

非持久化(Non-Persistent)：服务器将消息保存在内存中，当服务器重启时消息可能丢失。

消费者(Consumer)：一个从消息队列中请求消息的客户端应用程序。

生产者(Producer)：一个向交换器发布消息的客户端应用程序。

虚拟主机(Virtual Host)：一批交换器、消息队列和相关对象。虚拟主机是共享相同的身份认证和加密环境的独立服务器域。客户端应用程序在登录到服务器之后可以选择一个虚拟主机。

11.3 Spring Boot 的支持

11.3.1 JMS 的自动配置

Spring Boot 对 JMS 的自动配置位于 org.springframework.boot.autoconfigure.jms 包下，在 Spring Boot 3.0 中支持 JMS 的实现有 ActiveMQ Artemis（下一代 ActiveMQ）。Spring Boot 为用户定义了 ArtemisConnectionFactoryFactory 的 Bean 作为连接，并通过以"spring.artemis"为前缀的属性配置 ActiveMQ Artemis 的连接属性，主要包含以下内容：

```
spring.artemis.broker-url= tcp://localhost:61616    #消息代理路径
spring.artemis.user=
spring.artemis.password=
spring.artemis.mode=
```

另外，Spring Boot 在 JmsAutoConfiguration 自动配置类中为用户配置了 JmsTemplate；并且在 JmsAnnotationDrivenConfiguration 配置类中为用户开启了注解式消息监听的支持，即自动开启@EnableJms。

11.3.2 AMQP 的自动配置

Spring Boot 对 AMQP 的自动配置位于 org.springframework.boot.autoconfigure.amqp 包下，RabbitMQ 是 AMQP 的主要实现。在 RabbitAutoConfiguration 自动配置类中为用户配置了连接的 RabbitConnectionFactoryBean 和 RabbitTemplate，并且在 RabbitAnnotationDrivenConfiguration 配置类中开启了@EnableRabbit。从 RabbitProperties 类中可以看出 RabbitMQ 的配置可通过以"spring.rabbitmq"为前缀的属性进行配置，主要包含以下内容：

```
spring.rabbitmq.host=localhost              #RabbitMQ 服务器地址，默认为 localhost
spring.rabbitmq.port=5672                   #RabbitMQ 端口，默认为 5672
spring.rabbitmq.username=guest              #默认用户名
spring.rabbitmq.password=guest              #默认密码
```

11.4 异步消息通信实例

本节通过两个实例讲解异步消息通信的实现过程。

11.4.1 JMS 实例

下面使用 JMS 的一种实现 ActiveMQ Artemis 讲解 JMS 实例，因此需要事先安装 ActiveMQ Artemis（注意需要安装 Java 11+）。读者可以访问 http://activemq.apache.org/下载适合自己的 ActiveMQ Artemis。在编写本书时，作者下载了 apache-artemis-2.27.1-bin.zip。该版本的 ActiveMQ Artemis，解压缩即可完成安装。

解压缩后，需要创建一个代理服务器，以 Windows 10 为例，具体步骤如下：

（1）以管理员身份运行 cmd，进入 ActiveMQ Artemis 解压缩后的 bin 目录，例如执行"D:\>cd D:\soft\Java EE\apache-artemis-2.27.1\bin"。

（2）执行"artemis.cmd create 代理所在目录"命令，例如执行"D:\soft\Java EE\apache-artemis-2.27.1\bin>artemis.cmd create D:\soft\ActiveMQArtemis"。

（3）输入默认用户名和密码（admin），并输入"y"允许匿名访问。在代理服务器创建成功后，

cmd 窗口中的信息显示如图 11.2 所示。

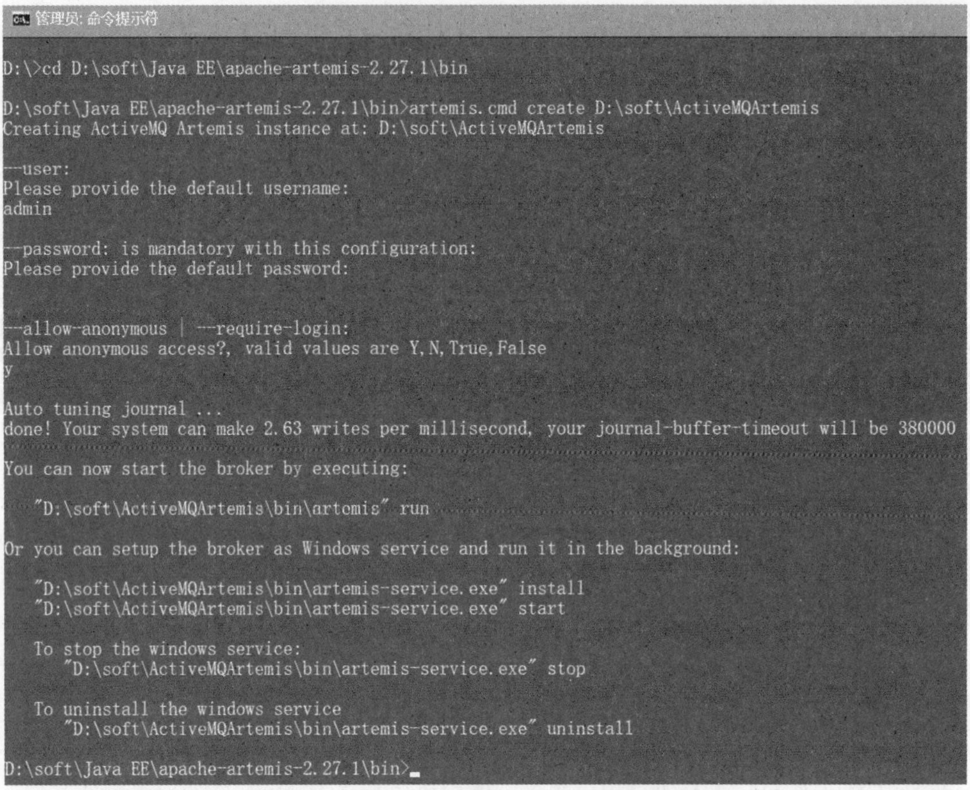

```
管理员: 命令提示符

D:\>cd D:\soft\Java EE\apache-artemis-2.27.1\bin

D:\soft\Java EE\apache-artemis-2.27.1\bin>artemis.cmd create D:\soft\ActiveMQArtemis
Creating ActiveMQ Artemis instance at: D:\soft\ActiveMQArtemis

--user:
Please provide the default username:
admin

--password: is mandatory with this configuration:
Please provide the default password:


--allow-anonymous | --require-login:
Allow anonymous access?, valid values are Y,N,True,False
y

Auto tuning journal ...
done! Your system can make 2.63 writes per millisecond, your journal-buffer-timeout will be 380000

You can now start the broker by executing:

   "D:\soft\ActiveMQArtemis\bin\artemis" run

Or you can setup the broker as Windows service and run it in the background:

   "D:\soft\ActiveMQArtemis\bin\artemis-service.exe" install
   "D:\soft\ActiveMQArtemis\bin\artemis-service.exe" start

   To stop the windows service:
      "D:\soft\ActiveMQArtemis\bin\artemis-service.exe" stop

   To uninstall the windows service
      "D:\soft\ActiveMQArtemis\bin\artemis-service.exe" uninstall

D:\soft\Java EE\apache-artemis-2.27.1\bin>
```

图 11.2　创建 ActiveMQ Artemis 的代理服务器

在代理服务器创建成功后，继续执行"D:\soft\ActiveMQArtemis\bin\artemis-service.exe" install 命令安装 Apache ActiveMQ Artemis 服务，然后执行"D:\soft\ActiveMQArtemis\bin\artemis-service.exe" start 命令启动 Apache ActiveMQ Artemis 服务(也可以在系统服务中启动)，如图 11.3 所示。

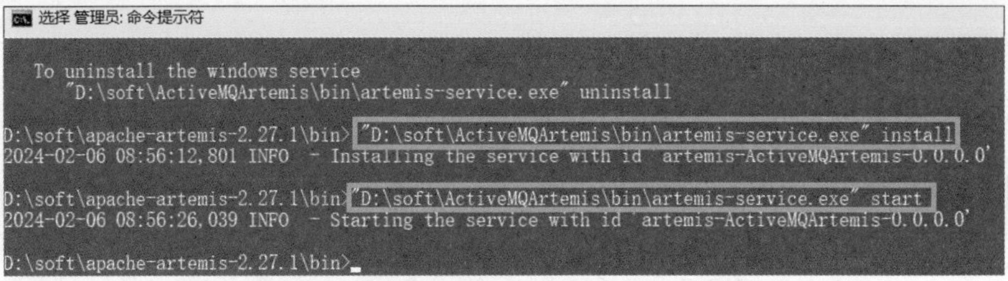

```
选择 管理员: 命令提示符

   To uninstall the windows service
      "D:\soft\ActiveMQArtemis\bin\artemis-service.exe" uninstall

D:\soft\apache-artemis-2.27.1\bin>"D:\soft\ActiveMQArtemis\bin\artemis-service.exe" install
2024-02-06 08:56:12,801 INFO  - Installing the service with id 'artemis-ActiveMQArtemis-0.0.0.0'

D:\soft\apache-artemis-2.27.1\bin>"D:\soft\ActiveMQArtemis\bin\artemis-service.exe" start
2024-02-06 08:56:26,039 INFO  - Starting the service with id 'artemis-ActiveMQArtemis-0.0.0.0'

D:\soft\apache-artemis-2.27.1\bin>
```

图 11.3　启动 Apache ActiveMQ Artemis 服务

接下来通过"http://localhost:8161"进入 ActiveMQ Artemis 的管理界面，管理员账号和密码默认为 admin/admin，如图 11.4 所示。

在启动 Apache ActiveMQ Artemis 服务后，通过实例讲解如何使用 JMS 的实现 ActiveMQ Artemis 进行两个应用系统间的点对点式通信。

【例 11-1】 使用 JMS 的实现 ActiveMQ Artemis 进行两个应用系统间的点对点式通信。

其具体实现步骤如下。

第 11 章　Spring Boot 的异步消息

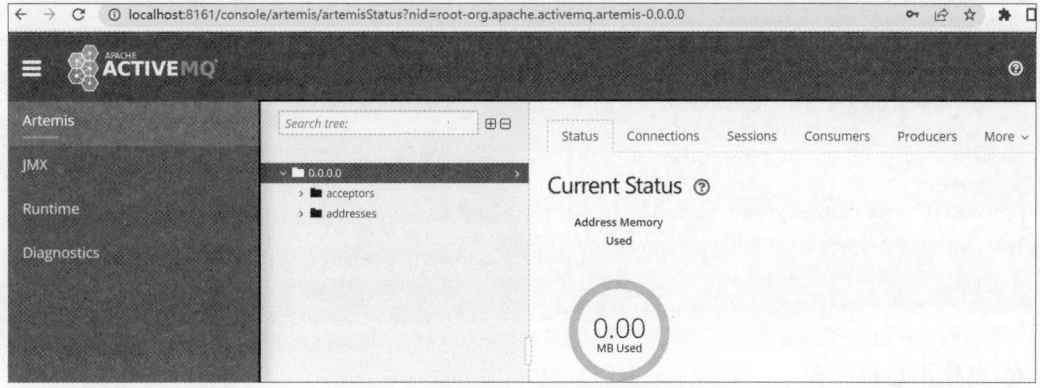

图 11.4　ActiveMQ Artemis 的管理界面

❶ 创建基于 Spring for Apache ActiveMQ Artemis 的 Spring Boot 应用 ch11_1Sender（消息发送者）

创建基于 Spring for Apache ActiveMQ Artemis 的 Spring Boot 应用 ch11_1Sender，该应用作为消息发送者。

❷ 配置 ActiveMQ Artemis 的消息代理地址

在应用 ch11_1Sender 的配置文件 application.properties 中配置 ActiveMQ Artemis 的消息代理地址，具体如下：

```
spring.artemis.broker-url= tcp://localhost:61616
```

❸ 定义消息

在应用 ch11_1Sender 的 com.ch.ch11_1sender 包下创建消息定义类 MyMessage，该类需要实现 MessageCreator 接口，并重写接口方法 createMessage 进行消息定义。核心代码如下：

```
public class MyMessage implements MessageCreator{
    @Override
    public Message createMessage(Session session) throws JMSException {
        MapMessage mapm = session.createMapMessage();
        ArrayList<String> arrayList = new ArrayList<String>();
        arrayList.add("陈恒 1");
        arrayList.add("陈恒 2");
        mapm.setString("mesg1", arrayList.toString());          //只能存储 Java 的基本对象
        mapm.setString("mesg2", "测试消息 2");
        return mapm;
    }
}
```

❹ 发送消息

在应用 ch11_1Sender 的主类 Ch111SenderApplication 中实现 Spring Boot 的 CommandLineRunner 接口，并重写 run 方法，用于程序启动后执行的代码。在该 run 方法中，使用 JmsTemplate 的 send 方法向目的地 mydestination 发送 MyMessage 的消息，也相当于在消息代理上定义了一个叫 mydestination 的目的地。核心代码如下：

```
@SpringBootApplication
public class Ch111SenderApplication implements CommandLineRunner{
    @Autowired
    private JmsTemplate jmsTemplate;
    public static void main(String[] args) {
```

303

```
        SpringApplication.run(Ch111SenderApplication.class, args);
    }
    /**
     * 这里为了方便操作,使用 run 方法发送消息,
     * 当然完全可以使用控制器通过 Web 访问
     */
    @Override
    public void run(String... args) throws Exception {
        //new MyMessage()回调接口方法 createMessage 产生消息
        jmsTemplate.send("mydestination", new MyMessage());
    }
}
```

❺ **创建消息接收者**

按照步骤 1 创建 Spring Boot 应用 ch11_1Receive,该应用作为消息接收者;并按照步骤 2 配置 ch11_1Receive 的 ActiveMQ Artemis 的消息代理地址。

❻ **定义消息监听器接收消息**

在应用 ch11_1Receive 的 com.ch.ch11_1receive 包中创建消息监听器类 ReceiverMsg。在该类中使用@JmsListener 注解不停地监听目的地 mydestination 是否有消息发送过来,如果有就获取消息。核心代码如下:

```
@Component
public class ReceiverMsg {
    @JmsListener(destination="mydestination")
    public void receiverMessage(MapMessage mapm) throws JMSException {
        System.out.println(mapm.getString("mesg1"));
        System.out.println(mapm.getString("mesg2"));
    }
}
```

❼ **运行测试**

启动消息接收者应用 ch11_1Receive,然后单击 JMX,可以看到如图 11.5 所示的界面。

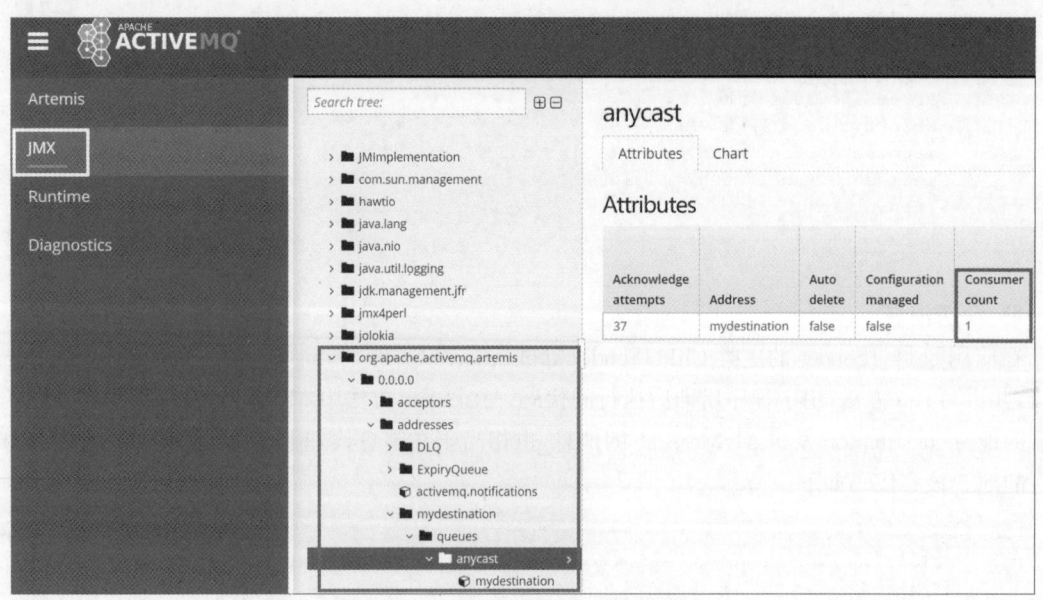

图 11.5 ActiveMQ Artemis 的 Queues

从图 11.5 可以看出目的地 mydestination 有一个消费者,正在等待接收消息。此时启动消

息发送者应用 ch11_1Sender，可以在消息接收者应用 ch11_1Receive 的控制台上看到有消息打印，如图 11.6 所示。刷新图 11.5，可以看到如图 11.7 所示的界面。

图 11.6　消息接收者接收的消息

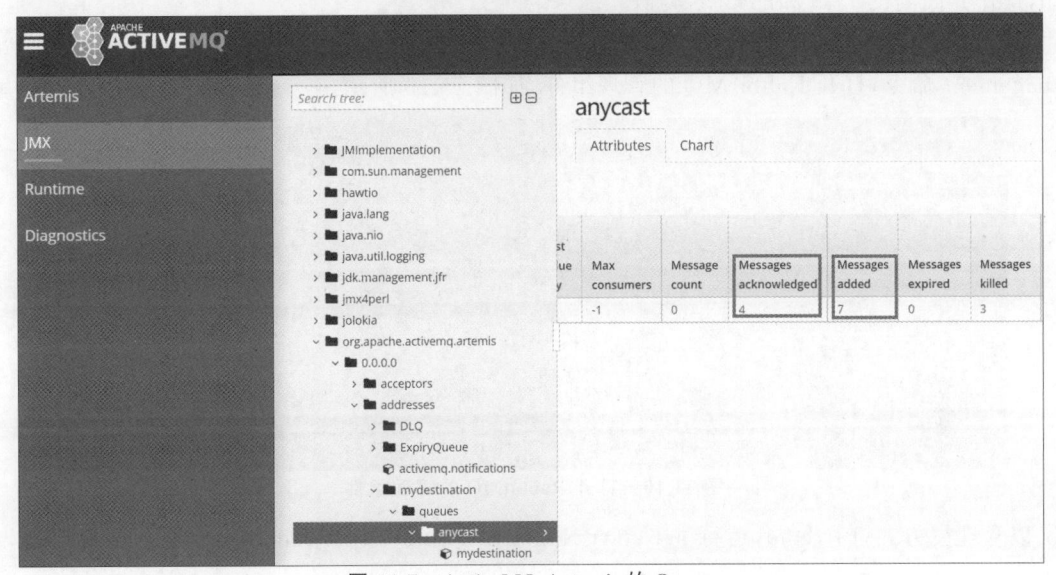

图 11.7　ActiveMQ Artemis 的 Queues

从图 11.7 可以看出目的地 mydestination 的 Messages added 有数据增加（表示发送成功），同时 Messages acknowledged 有数据增加（表示接收成功）。

11.4.2　AMQP 实例

下面使用 AMQP 的主要实现 RabbitMQ 讲解 AMQP 实例，因此需要事先安装 RabbitMQ。又因为 RabbitMQ 是基于 erlang 语言开发的，所以在安装 RabbitMQ 之前，先下载安装 erlang。erlang 语言的下载地址为 https://www.erlang.org/downloads；RabbitMQ 的下载地址为 https://www.rabbitmq.com/download.html。在编写本书时，作者下载的 erlang 语言版本是"otp_win64_25.2.2.exe"，下载的 RabbitMQ 版本是"rabbitmq-server-3.11.8.exe"。

运行 erlang 语言安装包"otp_win64_25.2.2.exe"，一直单击 Next 按钮即可安装 erlang。在安装 erlang 后需要配置环境变量 ERLANG_HOME 以及 path 中新增的 %ERLANG_HOME%\bin，如图 11.8 和图 11.9 所示。

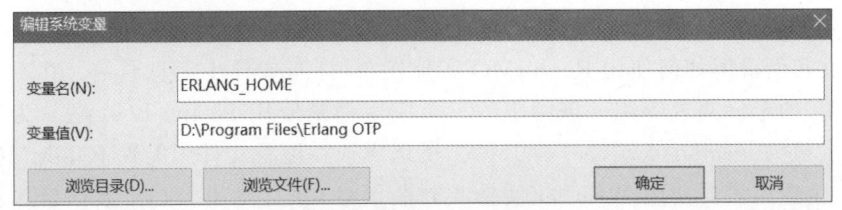

图 11.8　ERLANG_HOME

运行 RabbitMQ 安装包"rabbitmq-server-3.11.8.exe"，一直单击 Next 按钮即可安装

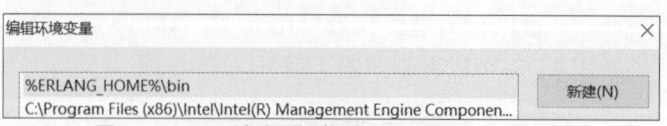

图 11.9　path 中新增%ERLANG_HOME%\bin

RabbitMQ。在安装 RabbitMQ 后需要配置环境变量 RABBITMQ_SERVER=安装目录\RabbitMQ Server\rabbitmq_server-3.11.8 以及 path 中新增的%RABBITMQ_SERVER%\sbin,操作界面与图 11.8 和图 11.9 类似。

在 cmd 窗口中进入 RabbitMQ 的 sbin 目录下,运行 rabbitmq-plugins.bat enable rabbitmq_management 命令,打开 RabbitMQ 的管理组件,如图 11.10 所示。

图 11.10　打开 RabbitMQ 的管理组件

以管理员方式打开 cmd,运行 net start RabbitMQ 命令,提示 RabbitMQ 服务已经启动,如图 11.11 所示。

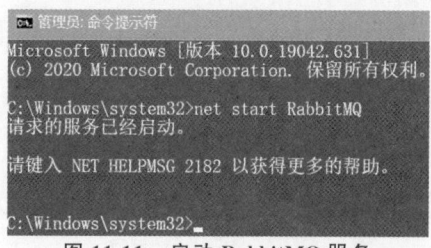

图 11.11　启动 RabbitMQ 服务

在浏览器的地址栏中输入"http://localhost：15672",账号和密码默认为 guest/guest,进入 RabbitMQ 的管理界面(如果进不去界面,请重启计算机服务中的 RabbitMQ),如图 11.12 所示。

至此完成了 RabbitMQ 服务器的搭建。

在例 11-1 中,不管是消息发送者(生产者)还是消息接收者(消费者),都必须知道一个指定的目的地(队列)才能发送、获取消息。如果同一个消息,要求每个消费者都处理,则需要发布/订阅式的消息分发模式。

这里通过实例讲解如何使用 RabbitMQ 实现发布/订阅式异步消息通信。在本例中创建一个发布者应用、两个订阅者应用。该例中的三个应用都是使用 Spring Boot 默认为用户配置的 RabbitMQ,主机为 localhost、端口号为 15672,所以无须在配置文件中配置 RabbitMQ 的连接信息。另外,三个应用需要使用 Weather 实体类封装消息,并且使用 JSON 数据格式发布和订阅消息。

【例 11-2】　使用 RabbitMQ 实现发布/订阅式异步消息通信(天气预报的发布与接收)。

第 11 章　Spring Boot 的异步消息

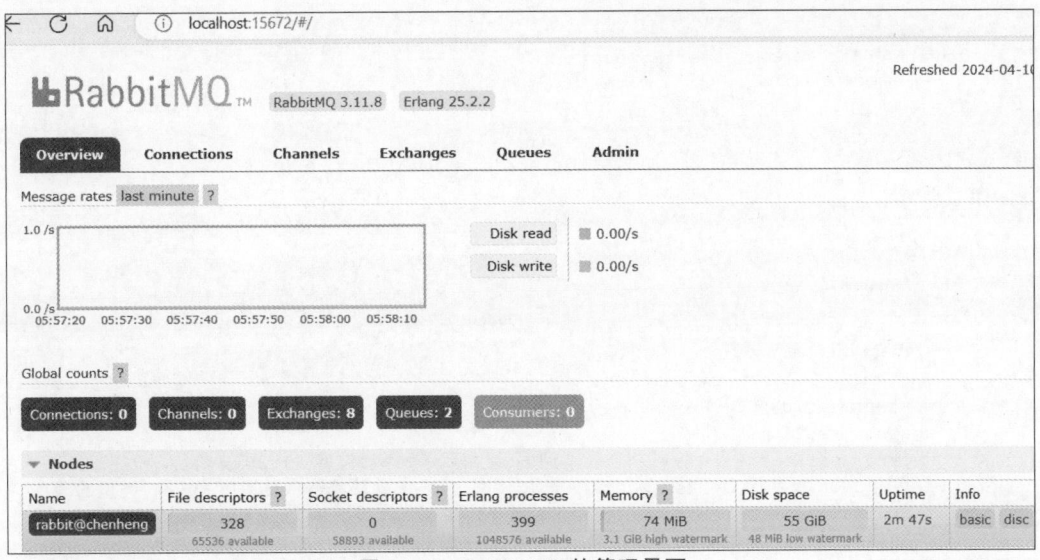

图 11.12　RabbitMQ 的管理界面

其具体实现步骤如下。

❶ 创建发布者应用 ch11_2Sender

创建发布者应用 ch11_2Sender，包括以下步骤。

（1）创建基于 Lombok、Spring for RabbitMQ 的 Spring Boot 应用 ch11_2Sender。

（2）在 ch11_2Sender 应用的 pom.xml 中添加 spring-boot-starter-json 依赖，代码如下：

```xml
<dependency>
    <groupId>org.springframework.boot</groupId>
    <artifactId>spring-boot-starter-json</artifactId>
</dependency>
```

（3）在 ch11_2Sender 应用中创建名为 com.ch.ch11_2sender.entity 的包，并在该包中创建 Weather 实体类，核心代码如下：

```java
@Data
public class Weather implements Serializable{
    private String id;
    private String city;
    private String weatherDetail;
    @Override
    public String toString() {
        return "Weather [id=" + id + ", city=" + city + ", weatherDetail=" + weatherDetail
+ "]";
    }
}
```

（4）在应用 ch11_2Sender 的主类 Ch112SenderApplication 中实现 Spring Boot 的 CommandLineRunner 接口，并重写 run 方法，用于程序启动后执行的代码。在该 run 方法中，使用 RabbitTemplate 的 convertAndSend 方法通过特定的路由"weather.message"发送 Weather 消息对象到指定的交换机"weather-exchange"。在发布消息前需要使用 ObjectMapper 将 Weather 对象转换成 byte[]类型的 JSON 数据。核心代码如下：

```java
@SpringBootApplication
public class Ch112SenderApplication implements CommandLineRunner{
```

```java
    @Autowired
    private ObjectMapper objectMapper;
    @Autowired
    RabbitTemplate rabbitTemplate;
    public static void main(String[] args) {
        SpringApplication.run(Ch112SenderApplication.class, args);
    }
    /**
     * 定义发布者
     */
    @Override
    public void run(String... args) throws Exception {
        //定义消息对象
        Weather weather = new Weather();
        weather.setId("010");
        weather.setCity("北京");
        weather.setWeatherDetail("今天晴到多云,南风 5-6 级,温度 19-26℃");
        //指定 JSON 转换器,Jackson2JsonMessageConverter 默认将消息转换成 byte[]类型的消息
        rabbitTemplate.setMessageConverter(new Jackson2JsonMessageConverter());
        //objectMapper 将 weather 对象转换为 JSON 字节数组
        Message msg=MessageBuilder.withBody(objectMapper.writeValueAsBytes(weather))
                .setDeliveryMode(MessageDeliveryMode.NON_PERSISTENT)
                .build();
        //消息唯一 ID
        CorrelationData correlationData = new CorrelationData(weather.getId());
        //通过特定的路由 Key 发送消息到指定的交换机
        rabbitTemplate.convertAndSend(
                "weather-exchange",         //分发消息的交换机的名称
                "weather.message",          //用来匹配消息的路由 Key
                msg,                        //消息体
                correlationData);
    }
}
```

❷ **创建订阅者应用 ch11_2Receiver-1**

创建订阅者应用 ch11_2Receiver-1,包括以下步骤。

(1) 创建基于 Lombok、Spring for RabbitMQ 的 Spring Boot 应用 ch11_2Receiver-1。

(2) 在 ch11_2Receiver-1 应用的 pom.xml 中添加 spring-boot-starter-json 依赖。

(3) 将 ch11_2Sender 中的 Weather 实体类复制到 com.ch.ch11_2Receiver1 包中。

(4) 在 com.ch.ch11_2receiver1 包中创建订阅者类 Receiver1,在该类中使用@RabbitListener 和 @RabbitHandler 注解监听发布者并接收消息,核心代码如下:

```java
/**
 * 定义订阅者 Receiver1
 */
@Component
public class Receiver1 {
    @Autowired
    private ObjectMapper objectMapper;
    @RabbitListener(
            bindings =
            @QueueBinding(
                    //队列名 weather-queue1 保证和其他订阅者不一样,可以随机起名
                    value = @Queue(value = "weather-queue1",durable = "true"),
                    //weather-exchange 与发布者的交换机名相同
                    exchange = @Exchange(value = "weather-exchange",durable = "true",type = "topic"),
                    //weather.message 与发布者的消息的路由 Key 相同
```

```
                key = "weather.message"
            )
    )
    @RabbitHandler
    public void receiveWeather(@Payload byte[] weatherMessage)throws Exception{
        System.out.println("-----------订阅者 Receiver1 接收到消息--------");
        //将 JSON 字节数组转换为 Weather 对象
        Weather w=objectMapper.readValue(weatherMessage, Weather.class);
        System.out.println("Receiver1 收到的消息内容:"+w);
    }
}
```

❸ **创建订阅者应用 ch11_2Receiver-2**

与创建订阅者应用 ch11_2Receiver-1 的步骤一样,这里不再赘述,但需要注意两个订阅者的队列名不同。

❹ **测试运行**

首先运行发布者应用 ch11_2Sender 的主类 Ch112SenderApplication。

然后运行订阅者应用 ch11_2Receiver-1 的主类 Ch112Receiver1Application,此时接收到的消息如图 11.13 所示。

```
2024-04-10T05:59:07.360+08:00  INFO 17512 --- [ch11_2Receiver-1] [           main] c.c.c.Ch112R
-----------订阅者Receiver1接收到消息--------
Receiver1收到的消息内容:Weather [id=010, city=北京, weatherDetail=今天晴到多云，南风5-6级，温度19-26°C]
```

图 11.13　订阅者 ch11_2Receiver-1 接收到的消息

最后运行订阅者应用 ch11_2Receiver-2 的主类 Ch112Receiver2Application,此时接收到的消息如图 11.14 所示。现在再看图 11.12 中的 Connections 为 3。

```
2024-04-10T06:04:37.919+08:00  INFO 11028 --- [ch11_2Receiver-2] [           main] c.c.c.Ch112R
-----------订阅者Receiver2接收到消息--------
Receiver2收到的消息内容:Weather [id=010, city=北京, weatherDetail=今天晴到多云，南风5-6级，温度19-26°C]
```

图 11.14　订阅者 ch11_2Receiver-2 接收到的消息

从例 11-2 可以看出,一个发布者发布的消息可以被多个订阅者订阅,这就是所谓的发布/订阅式异步消息通信。

本章小结

本章主要介绍了多个应用系统间的异步消息。通过本章的学习,读者应该了解 Spring Boot 对 JMS 和 AMQP 的支持,掌握如何在实际应用开发中使用 JMS 或 AMQP 提供异步通信解决方案。

习题 11

1. 在多个应用系统间的异步消息中有哪些消息模型?
2. JMS 和 AMQP 有什么区别?

第三阶段　Vue.js 3 前端框架开发

第 12 章 　Vue.js 基础

学习目的与要求

本章主要讲解 Vue.js 的基础知识，包括 Vue.js 的安装方法、集成开发环境、生命周期、插值与表达式、计算属性、监听器、内置指令、组件、自定义指令以及 setup 组件选项等内容。通过本章的学习，掌握 Vue.js 的基础知识，为学习 Vue.js 的进阶知识做准备。

本章主要内容

- Vue.js 是什么
- 如何安装 Vue.js
- 如何安装 Visual Studio Code 及其插件
- Vue.js 的生命周期
- 插值与表达式
- 计算属性和监听器
- 内置指令
- 组件
- 自定义指令
- 响应性
- setup 组件选项

目前，广泛应用的 Web 前端三大主流框架是 Angular.js、React.js 和 Vue.js。

Angular.js 由 Google 公司开发，诞生于 2009 年，之前多用 jQuery 开发，它最大的特点是把后端的一些开发模式移植到前端实现，如 MVC、依赖注入等。

React.js 由 Facebook 公司开发，正式版推出于 2013 年，比 Angular.js 晚 4 年，采用函数式编程，门槛稍高，但也更灵活，开发具有更多可能性。

Vue.js 作为后起之秀(2014 年)，借鉴了前辈 Angular.js 和 React.js 的特点，并做了相关优化，使其更加方便，更容易上手，比较适合初学者。

不管读者学习哪种前端框架，建议事先了解 HTML、CSS 和 JavaScript 知识。在学习本章前，假设读者已了解有关 HTML、CSS 和 JavaScript 的知识。

12.1　网站交互方式

Web 网站有单页应用程序(Single-page Application，SPA)和多页应用程序(Multi-page Application，MPA)两种交互方式。

▶12.1.1　多页应用程序

多页应用程序，顾名思义是由多个页面组成的站点。在多页应用程序中，每个网页在每次收到相应的请求时都会重新加载。多页应用程序很大，由于不同页面的数量和层数，有时甚至可以认为很麻烦，大家可以在大多数电子商务网站上找到 MPA 的示例。

多页应用程序以服务端为主导，前后端混合开发，例如.php、.aspx、.jsp。技术堆栈包括

HTML、CSS、JavaScript、jQuery，有时还包括 Ajax。

❶ 多页应用程序的优点

多页应用程序的优点如下。

（1）搜索引擎优化效果好：搜索引擎在做网页排名时，需要根据网页内容给网页添加权重，进行网页排名。搜索引擎可以识别 HTML 内容，而多页应用程序的每个页面的所有内容都放在 HTML 中，所以排名效果较好。

（2）更容易扩展：添加到现有应用程序的页面数几乎没有限制。如果需要显示很多信息，建议使用 MPA，因为它可以确保将来能够更轻松地扩展。

（3）深入的数据分析：有许多数据分析工具可以为 MPA 提供有关客户行为、系统功能和其他重要内容的深刻解析，可以分析每个功能的性能、每个网页的受欢迎程度、每个功能所花费的时间、每日和每月的用户数量以及按年龄、城市、国家/地区划分的受众群体细分等。与 MPA 相比，单页应用程序通常只提供有关访问者数量及其会话持续时间的信息。

❷ 多页应用程序的缺点

多页应用程序的缺点如下。

（1）开发及维护更加困难且昂贵：与单页解决方案相比，由于多页应用程序具有更多功能，所以创建它们需要更多的精力和资源。开发时间与要构建的页面数和实现的功能成比例地增加。

Web 应用程序的开发工具更新快。同时，市场上还引入了其他库、框架、编程语言或者至少发布了新版本。但是，多页应用程序通常需要在开发过程中使用多种技术。因此，当维护 Web 系统时变得更加困难且昂贵。

（2）页面切换慢：多页应用程序每次跳转时都发出一个 HTTP 请求。如果网速较慢，在页面之间来回跳转时将发生明显的卡顿现象。

（3）较低的绩效指标：多页应用程序中的内容会不断重新加载，增加了服务器的负载，对网页速度和整体系统的性能产生负面影响。

▶12.1.2 单页应用程序

单页应用程序，就是只有一张 Web 页面的应用。单页应用程序是加载单个 HTML 页面并在用户与应用程序交互时动态更新该页面的 Web 应用程序。浏览器一开始会加载必需的 HTML、CSS 和 JavaScript，所有的操作都在这张页面上完成，都由 JavaScript 来控制。因此，对单页应用程序来说模块化的开发和设计显得相当重要。单页应用程序的开发技术复杂，所以诞生了许多前端开发框架，如 Angular.js、React.js、Vue.js 等。

在选择单页应用程序开发时，软件工程师通常采用 HTML5、Angular.js、React.js、Vue.js、Ember.js、Ajax 等技术堆栈。

❶ 单页应用程序的优点

单页应用程序的优点如下。

（1）用户体验好：单页应用程序就像一个原生的客户端软件一样使用，在切换过程中不会频繁有被"打断"的感觉。

（2）前后端分离：开发效率高，可维护性强。服务端不关心页面，只关心数据；客户端不关心数据及数据操作，只关心通过接口拿到数据和服务端交互、处理页面。

（3）局部刷新：只需要加载渲染局部视图即可，不需要整页刷新。

（4）完全的前端组件化：前端开发不再以页面为主，更多采用组件化的思想，代码结构和组织方式更加规范化，便于修改和调整。

（5）API共享：如果服务是多端的（浏览器端、Android、iOS、微信等），单页面应用的模式便于在多端共用API，可以显著地减少服务端的工作量。

（6）组件共享：在某些对性能体验要求不高的场景或者产品处于快速试错阶段，借助于技术Hybrid、React Native在多端共享组件，便于产品的快速迭代，节约资源。

❷ 单页应用程序的缺点

单页应用程序的缺点如下。

（1）首次加载大量资源：需要在一个页面上为用户提供产品的所有功能，在加载该页面时首先加载大量的静态资源，加载时间相对比较长。

（2）对搜索引擎不友好：单页面应用的界面数据绝大部分是异步加载的，很难被搜索引擎搜索到。

（3）安全问题：不幸的是，单页面应用更容易受到所谓的跨站点脚本（XSS）攻击，这意味着黑客可以将各种恶意脚本注入应用程序中。原因是缺乏经验的Web开发人员将某些功能和逻辑移至客户端。

12.2 MVVM模式

MVVM是Model-View-ViewModel的缩写，它是一种基于前端开发的架构模式，其核心是提供对View和ViewModel的双向数据绑定，这使得ViewModel的状态改变可以自动传递给View，即所谓的数据双向绑定。

MVVM由Model、View、ViewModel三部分构成，Model层代表数据模型，也可以在Model中定义数据修改和操作的业务逻辑；View代表UI组件，它负责将数据模型转化成UI展现出来，ViewModel是一个同步View和Model的对象。

在MVVM架构下，View和Model之间并没有直接的联系，而是通过ViewModel进行交互，Model和ViewModel之间的交互是双向的，因此View数据的变化会同步到Model中，而Model数据的变化也会立即反映到View上。Model、View、ViewModel三者之间的关系如图12.1所示。

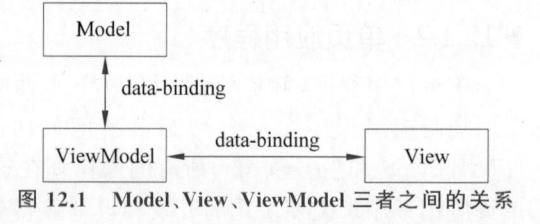

图 12.1 Model、View、ViewModel三者之间的关系

ViewModel通过双向数据绑定把View层和Model层连接了起来，而View和Model之间的同步工作完全是自动的，无须人为干涉，因此开发者只需关注业务逻辑，不需要手动操作DOM，不需要关注数据状态的同步问题，复杂的数据状态维护完全由MVVM来统一管理。

12.3 Vue.js是什么

Vue（读音/vju:/，类似于view）是一套构建用户界面的渐进式框架。与其他重量级框架不同的是，Vue.js采用自底向上增量开发的设计。Vue.js的核心库仅关注视图层，它不仅易于上手，还便于与第三方库或既有项目整合。另外，Vue.js采用单文件组件和Vue.js生态系统支持的库开发复杂的单页应用。

Vue.js本身只是一个JS库，它的目标是通过尽可能简单的API实现响应的数据绑定和组合的视图组件。Vue.js 可以轻松地构建 SPA(Single Web Application)应用程序,通过指令扩展 HTML,通过表达式将数据绑定到 HTML,最大程度地解放 DOM 操作。

Vue.js 具有简单、易用、灵活、高效等特点,在 HTML、CSS、JavaScript 的基础上可快速上手。

12.4 安装 Vue.js

将 Vue.js 添加到项目中主要有 4 种方法,分别为本地独立版本方法、CDN 方法、NPM 方法以及命令行工具(CLI)方法。

▶12.4.1 本地独立版本方法

读者可以通过地址"https://unpkg.com/vue"将最新版本的 Vue.js 库(vue.global.js)下载到本地,然后在界面文件中引入 Vue.js 库,示例代码如下:

```
<script src="js/vue.global.js"></script>
```

▶12.4.2 CDN 方法

读者在学习或开发时,在界面文件中可通过 CDN(Content Delivery Network,内容分发网络)引入最新版本的 Vue.js 库,示例代码如下:

```
<script src="https://unpkg.com/vue@next"></script>
```

对于生产环境,建议使用固定版本,以免因版本不同带来兼容性问题,示例代码如下:

```
<script src="https://unpkg.com/vue@3.4.21/dist/vue.global.js"></script>
```

▶12.4.3 NPM 方法

在使用 Vue.js 构建大型应用时推荐使用 NPM 安装最新稳定版的 Vue.js,因为 NPM 能很好地和 webpack 模块打包器配合使用。示例如下:

```
npm install vue@next
```

▶12.4.4 命令行工具(CLI)方法

Vue.js 提供了一个官方命令行工具(Vue CLI),为单页面应用快速搭建繁杂的脚手架。对于初学者,不建议使用 NPM 和 Vue CLI 方法安装 Vue.js。NPM 和 Vue CLI 方法的安装过程将在本书第 13 章 Vue.js 进阶中介绍。

12.5 第一个 Vue.js 程序

对于前端开发工具,极少数人使用记事本,而大多数程序员使用 JetBrains WebStorm 和 Visual Studio Code(VSCode)。JetBrains WebStorm 是收费的,本书推荐使用 VSCode。

▶12.5.1 安装 Visual Studio Code 及其插件

用户可通过"https://code.visualstudio.com"地址下载 VSCode,本书使用的安装文件是 VSCodeUserSetup-x64-1.52.1.exe(双击即可安装)。在 VSCode 中有许多插件需要用户安装,例如安装 Vue.js 的插件 Vetur。打开 VSCode,单击左侧最下面一个图标,按照图 12.2 所示的步骤安装即可。

图 12.2　VSCode 插件的安装

其他插件的安装方式与图 12.2 类似，这里不再赘述。在 VSCode 中，有关 Vue.js 的部分插件具体描述如下。

（1）Vetur：此插件能够在 .vue 文件中实现语法错误检查、语法高亮显示以及代码自动补全。

（2）ESLint：此插件能够检测代码语法问题与格式问题，对项目代码风格的统一至关重要。

（3）EditorConfig for Visual Studio Code：EditorConfig 是一种被各种编辑器广泛支持的配置，使用此配置有助于项目在整个团队中保持一致的代码风格。

（4）Path Intellisense：此插件能够在编辑器中输入路径时实现自动补全。

（5）View In Browser：在 VSCode 中使用浏览器预览运行静态文件。

（6）Live Server：此插件很有用，在安装后可以打开一个简单的服务器，而且还会自动更新。在安装后，在文件上右击会出现一个名为 Open with Live Server 的选项，自动打开浏览器，默认端口号是 5500。

（7）GitLens：此插件可查看 Git 文件提交的历史。

（8）Document This：此插件生成注释文档。

（9）HTML CSS Support：在编写样式表的时候自动补全，缩减编写时间。

（10）JavaScript Snippet Pack：针对 JS 的插件，包含 JS 的常用语法关键字。

（11）HTML Snippets：此插件包含 HTML 标签。

（12）One Monokai Theme：此插件能够让用户选择自己喜欢的颜色主题编写代码。

（13）vscode-icons：此插件能够让用户选择自己喜欢的图标主题。

12.5.2　创建第一个 Vue.js 应用

每个 Vue.js 应用都是通过用 createApp 函数创建一个新实例开始，具体语法如下：

```
const app = Vue.createApp({ /* 选项 */ })
```

传递给 createApp 的选项用于配置根组件（渲染的起点）。在 Vue.js 应用创建后，调用 mount 方法将 Vue.js 应用挂载到一个 DOM 元素（HTML 元素或 CSS 选择器）中。例如，如果把一个 Vue.js 应用挂载到<div id="app"></div>上，应传递#app。示例代码如下：

```
const HelloVueApp = {}                              //配置根组件
const vueApp = Vue.createApp(HelloVueApp)           //创建 Vue 实例
const vm = vueApp.mount('#app')                     //将 Vue 实例挂载到#app
```

下面使用 VSCode 开发第一个 Vue.js 程序。

【例 12-1】 使用 VSCode 新建一个名为 hellovue.html 的页面，在此页面中使用"＜script src="js/vue.global.js"＞＜/script＞"语句引入 Vue.js。

hellovue.html 的具体代码如下：

```
<div id="hello-vue" class="demo">
    {{ message }}
</div>
<script src="js/vue.global.js"></script>
<script>
    const HelloVueApp = {
        data() {                            //Vue 实例的数据对象,ES 语法,等价于 data: function() {}
            return {
                message: 'Hello Vue!!'
            }
        }
    }
    //每个 Vue.js 应用都是通过用 createApp 函数创建一个新的应用实例开始
    //mount 函数把一个 Vue.js 应用实例挂载到<div id="hello-vue"></div>
    Vue.createApp(HelloVueApp).mount('#hello-vue')
</script>
<style>
.demo {
    font-family: sans-serif;
}
</style>
```

从上述代码中可以看出，hellovue.html 文件内容由 HTML、JavaScript 以及 CSS 三部分组成，所以读者在学习 Vue.js 之前应掌握 HTML、JavaScript 以及 CSS 等内容。

从上述代码中还可以看出，创建一个 Vue.js 应用程序只需要三个步骤，分别为引入 Vue.js 库文件、创建一个 Vue 实例、渲染 Vue 实例。

在 VSCode 中，按照 12.5.1 节的方法安装 Live Server 插件后，在 hellovue.html 代码上右击会出现一个名为 Open with Live Server 的选项，自动打开浏览器，默认端口号是 5500，如图 12.3 所示。

在 VSCode 中，可以设置 Live Server 插件默认打开的浏览器，首先通过 File→Preferences→Settings 菜单项打开 Search settings，然后通过 User→Extensions→Live Server Config→Custom Browser 修改默认打开的浏览器，如图 12.4 所示。

图 12.3 hellovue.html 的运行效果

▶12.5.3 声明式渲染

Vue.js 的核心是采用简洁的模板将数据渲染到 DOM 中，例如在 12.5.2 节的 hellovue.html 文件中，通过模板＜div id="hello-vue" class="demo"＞{{message}}＜/div＞声明将属性变量 message 的值"Hello Vue!!"渲染到页面显示。

Vue.js 框架在声明式渲染时，做的主要工作就是将数据和 DOM 建立关联，一切皆响应。例如例 12-2 中的 counter 属性每秒递增。

【例 12-2】 使用 VSCode 新建一个名为 ch12_2.html 的页面，在该页面中使用时钟函数 setInterval 来演示响应式程序。

ch12_2.html 的具体代码如下：

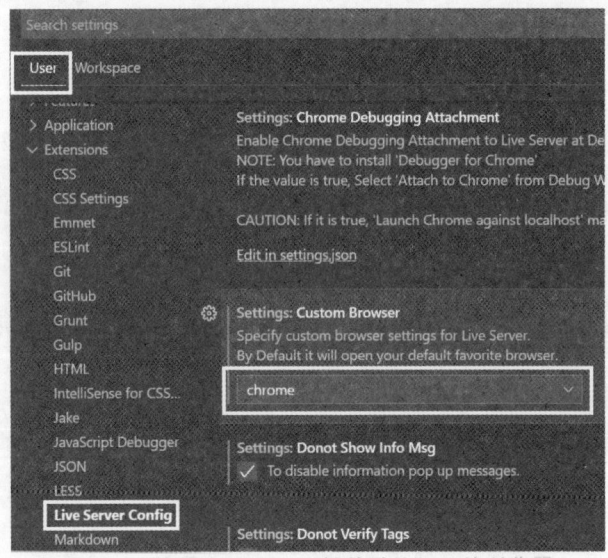

图 12.4 设置 Live Server 插件默认打开的浏览器

```
<div id="counter" class="demo">
    <!--通过模板获取变量 counter 的值-->
    {{ counter }}
</div>
<script src="js/vue.global.js"></script>
<script>
    const CounterApp = {
        data() {
            return {
                counter: 0
            }
        },
        /* mounted 是一个钩子函数,当挂载到实例上后(初始化页面后),调用该函数,一般是第一个业务逻辑在这里开始 */
        mounted() {
            //定时器
            setInterval(() => {
                this.counter++
            }, 1000)
        },
        /* Vue.js 实例销毁前调用(离开页面前调用),这是一个常用的生命周期钩子函数,一般在此时做一些重置的操作,例如清除定时器和监听的 DOM 事件 */
        beforeUnmount(){
            //清除定时器
            clearTimeout(this.counter)
        }
    }
    Vue.createApp(CounterApp).mount('#counter')
</script>
<style>
    .demo {
        font-family: sans-serif;
    }
</style>
```

12.5.4 Vue.js 的生命周期

每个 Vue.js 实例在被创建时都要经过一系列的初始化过程,例如数据监听、编译模板、将实例挂载到 DOM 并在数据变化时更新 DOM 等。同时在这个过程中也会调用一些叫生命周期钩子的函数,在适当的时机执行业务逻辑。

例如,created 钩子函数可以用来在一个 Vue.js 实例被创建后执行代码(Vue.js 实例创建后被立即调用,即 HTML 加载完成前):

```
Vue.createApp({
    data() {
        return {
            message: '测试钩子函数'
        }
    },
    created() {
        //this 指向调用它的 Vue.js 实例
        console.log('message 是: ' + this.message)          //"message 是:测试钩子函数"
    }
})
```

Vue.js 的生命周期共分 8 个阶段(如图 12.5 所示),分别对应 8 个与 created 类似的钩子函数。

beforeCreate(创建前):在 Vue.js 实例初始化后,数据观测和事件配置前调用,此时 el 和 data 并未初始化,因此无法访问 methods、data、computed 等上的方法和数据。

created(创建后):Vue.js 实例创建后被立即调用,即 HTML 加载完成前。此时 Vue.js 实例已完成数据观测、属性和方法的运算、watch/event 事件回调、data 数据的初始化。然而挂载阶段还没有开始,el 属性目前不可见。这是一个常用的生命周期钩子函数,可以调用 methods 中的方法、改变 data 中的数据、获取 computed 中的计算属性等,通常在此钩子函数中对实例进行预处理。

beforeMount(载入前):挂载开始前被调用,Vue.js 实例已完成编译模板、把 data 里面的数据和模板生成 HTML、el 和 data 初始化,注意此时还没有挂载 HTML 到页面上。

mounted(载入后):页面加载后调用该函数,这是一个常用的生命周期钩子函数,一般是第一个业务逻辑在此钩子开始,mounted 只会执行一次。

beforeUpdate(更新前):在数据更新前被调用,发生在虚拟 DOM 重新渲染和打补丁之前,可以在该钩子中进一步地更改状态,不会触发附加地重渲染过程。

updated(更新后):在由数据更改导致虚拟 DOM 重新渲染和打补丁时调用,在调用时 DOM 已经更新,所以可以执行依赖于 DOM 的操作,应该避免在此期间更改状态,这可能会导致更新无限循环。

beforeUnmount(销毁前):Vue.js 实例销毁前调用(离开页面前调用),这是一个常用的生命周期钩子函数,一般在此时做一些重置的操作,例如清除定时器和监听的 DOM 事件。

unmounted(销毁后):在实例销毁后调用,调用后事件监听器被移出,所有子实例也被销毁。

图 12.5 展示了 Vue.js 实例的生命周期,读者现在不需要弄明白所有阶段的钩子函数,随着不断地学习和使用,慢慢理解它们。

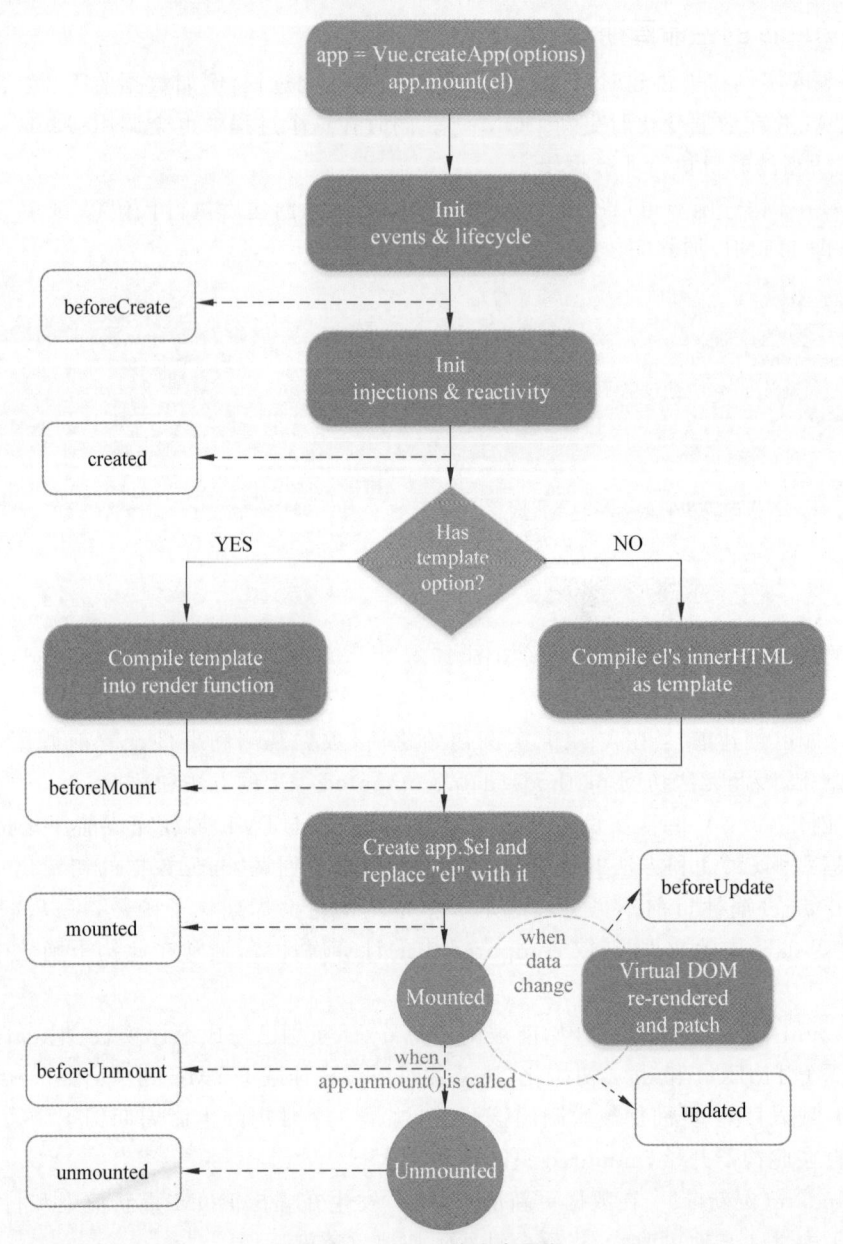

图 12.5　Vue.js 的生命周期

12.6　插值与表达式

　　Vue 的插值表达式"{{}}"的作用是读取 Vue.js 中的 data 数据,显示在视图中,数据更新,视图也随之更新。"{{ }}"中只能放表达式(有返回值),不可以放语句。例如,{{ var a = 1 }}与{{ if（ok）{ return message } }}都是无效的。

▶12.6.1　文本插值

　　数据绑定最常见的形式就是使用"Mustache(小胡子)"语法(双花括号)的文本插值,它将绑定的数据实时显示出来。例如,例 12-2 中的{{ counter }},无论何时,绑定的 Vue.js 实例的counter 属性值发生改变,插值处的内容都将更新。

用户可通过使用 v-once 指令执行一次性插值,即当数据改变时,插值处的内容不会更新。示例代码如下:

```
<span v-once>{{ counter }}</span>
```

12.6.2 原始 HTML 插值

"{{ }}"将数据解释为普通文本,而非 HTML 代码。当需要输出真正的 HTML 代码时,可使用 v-html 指令。动态渲染任意的 HTML 是非常危险的,因为很容易导致 XSS 攻击。最好只对可信内容使用 HTML 插值,绝不可将用户提供的 HTML 作为插值。v-html 指令示例如下。

假如,Vue.js 实例的 data 为:

```
data() {
    return {
        rawHtml: '<hr>'
    }
}
```

则"<p>无法显示 HTML 元素内容:{{rawHtml}}</p>"显示的结果是<hr>;而"<p>可正常显示 HTML 元素内容:<span v-html="rawHtml"></span></p>"显示的结果是一条水平线。

12.6.3 JavaScript 表达式

在前面的学习中,仅用表达式绑定简单的属性值。但实际上,对于所有的数据绑定,Vue.js 都提供了完全的 JavaScript 表达式支持。示例如下:

```
{{ number + 1 }}
{{ isLogin? 'True' : 'False' }}
{{ message.split('').reverse().join('') }}
```

12.7 计算属性和监听器

Vue 模板的插值表达式"{{ }}"用起来非常便利,设计它们的初衷是用于简单运算。但在表达式中放入太多逻辑,将会难以维护。计算属性就是用于解决该类问题的。

12.7.1 计算属性 computed

在 Vue 实例的 computed 选项中定义一些属性(可使用 this 引用),这些属性称作"计算属性"。所有的计算属性都是以方法(函数)的形式定义,但仅当作属性来使用。在一个计算属性中可以完成各种复杂的逻辑,包括运算、方法调用等,但最终必须返回一个结果。计算属性的结果还可以依赖于多个数据,只要其中任一数据发生变化,计算属性都将重新执行,视图也会更新。下面以计算购物车总价为例讲解计算属性的用法。

【例 12-3】 使用计算属性计算购物车总价。

其具体代码如下:

```html
<div id="myCart">
    <!-- v-model 指令在表单元素上实现双向数据绑定,在后续章节讲解 -->
    商品 1 数量:<input type="text" v-model="num1">
    商品 1 价格:<input type="text" v-model="price1"><br>
    商品 2 数量:<input type="text" v-model="num2">
    商品 2 价格:<input type="text" v-model="price2"><br>
```

```
        商品总价:{{ total }}
    </div>
    <script src="js/vue.global.js"></script>
    <script>
        const CounterApp = {
            data() {
                return {
                    num1: 1,
                    price1: 10,
                    num2: 2,
                    price2: 20
                }
            },
            computed: {
                total() {
                    return this.num1 * this.price1 + this.num2 * this.price2
                }
            }
        }
        Vue.createApp(CounterApp).mount('#myCart')
    </script>
```

商品数量或商品价格只要有一个属性发生变化,购物车商品总价属性将会自动更新。例12-3 的运行结果如图 12.6 所示。

图 12.6　计算购物车总价

Vue 的每一个计算属性都包含一个 getter 和一个 setter 方法,在例 12-3 中只是使用了计算属性的默认用法,即仅使用了 getter 方法来读取计算属性。当然,也可以使用计算属性的 setter 方法来修改其值。下面讲解计算属性的 setter 方法。

【例 12-4】　通过单击按钮重新为计算属性赋值。

其具体代码如下:

```
<div id="app">
    <!--调用计算属性的 getter 方法获取计算属性值-->
    姓名:{{fullName}}<br>
    <!-- v-on 指令给 HTML 元素添加一个事件监听器,在后续章节讲解 -->
    <button v-on:click="changeName">修改计算属性</button>
</div>
<script src="js/vue.global.js"></script>
<script>
    const CounterApp = {
        data() {
            return {
                firstName: '陈',
                lastName: '恒'
            }
        },
        methods:{                                  //方法定义
            changeName(){
                //计算属性可使用 this 引用,当计算属性 fullName 值发生变化时调用 setter 方法
                this.fullName = '张 三'
            }
        },
```

```
        computed: {
            fullName: {                             //计算属性
                get(){
                    return this.firstName + this.lastName
                },
                set(newValue){
                    var names = newValue.split(' ')
                    this.firstName = names[0]
                    this.lastName = names[1]
                }
            }
        }
    }
    Vue.createApp(CounterApp).mount('#app')
</script>
```

从例 12-4 的代码可以看出,当单击"修改计算属性"按钮时,执行 changeName()方法;在 changeName()方法中,执行"this.fullName = '张 三'"语句时,调用计算属性 fullName 的 setter 方法;在 setter 方法中,数据 firstName 和 lastName 相继更新,视图同样也会更新。

在大多数情况下,使用默认的 getter 方法来读取计算属性值即可,不必声明 setter,例如例 12-3。

用户可以在表达式中调用方法达到与计算属性同样的效果。下面通过修改例 12-3 的程序,演示 methods 达到与计算属性同样的效果。

【例 12-5】 用 methods 改写例 12-3。

修改后的代码如下:

```
<div id="myCart">
    商品 1 数量:<input type="text" v-model="num1">
    商品 1 价格:<input type="text" v-model="price1"><br>
    商品 2 数量:<input type="text" v-model="num2">
    商品 2 价格:<input type="text" v-model="price2"><br>
    商品总价:{{ total() }}
</div>
<script src="js/vue.global.js"></script>
<script>
    const CounterApp = {
        data() {
            return {
                num1: 1,
                price1: 10,
                num2: 2,
                price2: 20
            }
        },
        methods: {
            total() {
                return this.num1 * this.price1 + this.num2 * this.price2
            }
        }
    }
    Vue.createApp(CounterApp).mount('#myCart')
</script>
```

在例 12-5 中没有使用计算属性,在 methods 选项中定义了一个方法实现同样的效果,方法还可以接收参数,使用起来更方便。那么为什么还使用计算属性呢? 这是因为计算属性是基于它的依赖缓存的。也就是说,一个计算属性所依赖的数据发生变化时,它才会重新取值,所以只要商品数量和价格不发生改变,计算属性 total 也就不更新。下面分别使用计算属性和 methods 显示时间,演示它们的区别。

【例 12-6】 分别使用计算属性和 methods 显示时间,每次显示时间后弹出警告框暂停时间显示。

其具体代码如下:

```
<div id="app">
    时间 1:{{ mytime() }}{{ mystop() }}<br>
    时间 2:{{ mytime() }}{{ mystop() }}<br>
    时间 3:{{ yourtime }}{{ mystop() }}<br>
    时间 4:{{ yourtime }}
</div>
<script src="js/vue.global.js"></script>
<script>
    const CounterApp = {
        methods: {
            mytime() {
                return Date.now()
            },
            mystop() {
                alert('暂停一下')
            }
        },
        computed: {
            yourtime(){
                return Date.now()
            }
        }
    }
    Vue.createApp(CounterApp).mount('#app')
</script>
```

运行例 12-6 的程序,连续三次弹出暂停警告框后,显示结果如图 12.7 所示。

从图 12.7 可以看出,使用 methods 显示时间 1 和时间 2 时执行了两次 Date.now(),说明调用了两次 methods;而使用计算属性显示时间 3 和时间 4 时,即使中间弹出暂停警告框,显示的时间也相同,说明从缓存中获取计算属性(Date.now()不是响应式依赖)。

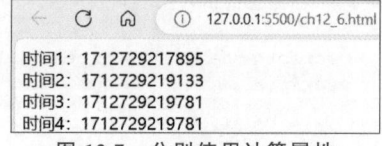

图 12.7 分别使用计算属性和 methods 显示时间

至于使用计算属性还是 methods 取决于是否需要缓存,当遍历大数组或做大量计算时,应该使用计算属性,从缓存中获取计算结果,提高执行效率。

▶12.7.2 监听器属性 watch

虽然计算属性在大多数情况下更合适,但有时也需要一个监听器来响应数据的变化。Vue 通过 watch 选项提供监听数据属性的方法(方法名与属性名相同)来响应数据的变化。当被监视的数据发生变化时,触发 watch 中对应的处理方法。下面通过实例讲解 watch 属性的用法。

【例 12-7】 使用 watch 属性监视 data 中 question 的变化(watch 中需提供与 question 同名的方法)。

其具体代码如下:

```
<div id="watch-example">
    <p>
        请问一个问题,包含英文字符?:
```

```
            <input v-model="question" />
        </p>
        <p>{{ answer }}</p>
    </div>
    <script src="js/vue.global.js"></script>
    <script>
        const watchExampleVM = Vue.createApp({
            data() {
                return {
                    question: '',
                    answer: '这是一个好问题。'
                }
            },
            watch: {                        //watch 选项提供监听数据属性的方法
                //question 方法名与数据属性名 question 一致
                question(newQuestion, oldQuestion) {
                                        //newQuestion 是改变的值,oldQuestion 是没改变的值
                    if (newQuestion.indexOf('?') > -1) {
                        //包含英文字符?时,返回 yes 答案
                        this.answer = 'yes'
                    } else {
                        this.answer = 'no'
                    }
                }
            }
        }).mount('#watch-example')
    </script>
```

从前面的学习可知:

(1) computed 属性所依赖的数据发生变化时将自动重新计算,并把计算结果缓存起来。另外,computed 属性最后一定有一个返回值,而且不带参数。

(2) watch 属性用来监听某些特定数据的变化,从而进行具体的业务逻辑操作。另外,watch 选项中的方法可传入被监听属性的新旧值,通过这两个值可以做一些特定的操作;computed 属性通常是做简单的计算。

那么应该如何选择 computed 属性和 watch 属性呢?如果一个值依赖多个属性,建议使用 computed 属性;如果一个值发生变化后引起一系列业务逻辑操作,或者引起一系列值的变化,建议使用 watch 属性。

12.8 内置指令

指令是 Vue.js 模板中最常用的一项功能,它带有特殊前缀"v-"。指令的主要职责是当其表达式的值改变时相应地将某些行为应用到 DOM 上。Vue.js 内置了许多指令,可以快速完成常见的 DOM 操作,例如循环渲染、条件渲染等,本节将学习这些内置指令。

▶12.8.1 v-bind 指令

在 HTML 元素的属性中不能使用表达式动态更新属性值。幸运的是,Vue.js 提供了 v-bind 指令绑定 HTML 元素的属性,并可动态更新属性值,例如 id、class 等。下面通过实例讲解 v-bind 指令的用法。

【例 12-8】 使用 v-bind 指令绑定超链接的 href 属性、图片的 src 属性以及 class 属性。

其具体代码如下:

```
<div id="app">
    <a v-bind:href="myurl.baiduUrl">去百度</a>
    <img v-bind:src="myurl.imgUrl"/><br>
    <!-- v-bind可缩写为":",这种缩写称为语法糖-->
    <a :href="myurl.baiduUrl">去百度</a>
    <img :src="myurl.imgUrl"/><br>
    <div :class="mycolor">对象语法</div>
    <div class="static" :class="{'active': isActive, 'text-danger': hasError}">在对象中传
入更多字段</div>
    <div :class="[activeClass, errorClass]">数组语法</div>
    <div :class="[isActive ? activeClass : '', errorClass]">数组中使用三元表达式</div>
    <div :class="[{ 'active': isActive }, errorClass]">数组中嵌套对象</div>
</div>
<script src="js/vue.global.js"></script>
<script>
    Vue.createApp({
        data() {
            return {
                myurl: {
                    baiduUrl: 'https://www.baidu.com/',
                    imgUrl:'images/ok.gif'
                },
                mycolor: 'my',
                isActive: true,
                hasError: false,
                activeClass:'your',
                errorClass:'his'
            }
        }
    }).mount('#app')
</script>
<style>
    .my {
        background-color: red
    }
    .your {
        font-size: 20px
    }
    .his {
        background-color: blue
    }
    .static {
        background-color: yellow
    }
    .active {
        font-size: 40px
    }
</style>
```

12.8.2 条件渲染指令 v-if 和 v-show

与 JavaScript 的条件语句 if、else、else if 类似，Vue.js 的条件指令 v-if 也可以根据表达式的值渲染或销毁元素/组件。下面通过具体实例讲解 v-if 指令的具体用法。

【例 12-9】 使用条件渲染指令判断成绩的等级。

其具体代码如下：

```
<div id="if-handling">
    <div v-if="score >= 90">优秀</div>
    <div v-else-if="score >= 80">良好</div>
```

```
        <div v-else-if="score >= 70">中等</div>
        <div v-else-if="score >= 60">及格</div>
        <div v-else>不及格</div>
</div>
<script src="js/vue.global.js"></script>
<script>
    Vue.createApp({
        data() {
            return {
                score: 87
            }
        }
    }).mount('#if-handling')
</script>
```

从上述示例代码可以看出，v-else 元素必须紧跟在 v-if 或者 v-else-if 元素后面；v-else-if 必须紧跟在 v-if 或者 v-else-if 元素后面。

v-if 条件渲染指令必须添加到一个元素上。但是如果想包含多个元素呢？此时可以使用 <template> 元素（模板占位符）帮助用户包裹元素，并在上面使用 v-if。最终的渲染结果将不包含 <template> 元素。示例代码如下：

```
<template v-if="ok">
    <h1>Title</h1>
    <p>Paragraph 1</p>
    <p>Paragraph 2</p>
</template>
```

v-show 指令的用法基本与 v-if 一样，也是根据条件展示元素，例如 <h1 v-show="yes">一级标题</h1>。不同的是，v-if 每次都会重新删除或创建元素，而带有 v-show 的元素始终会被渲染并保留在 DOM 中，只是切换元素的 display:none 样式。所以，v-if 有更高的切换消耗，而 v-show 有更高的初始渲染消耗。因此，如果需要频繁切换，使用 v-show 较好；如果在运行时条件不太可能改变，使用 v-if 较好。另外，v-show 不支持 <template> 元素，也不支持 v-else。

【例 12-10】 演示 v-if 与 v-show 的区别。

其具体代码如下：

```
<div id="event-handling">
    <div v-if="flag">一直显示</div>
    <div v-show="flag">反复无常</div>
    <!-- v-on:click 调用事件方法, v-on:可缩写为"@"-->
    <button @click="flag=!flag">隐藏/显示</button>
</div>
<script src="js/vue.global.js"></script>
<script>
    Vue.createApp({
        data() {
            return {
                flag: true
            }
        }
    }).mount('#event-handling')
</script>
```

使用 Google 浏览器第一次运行程序时，按 F12 键，页面初始化效果如图 12.8 所示。
单击"隐藏/显示"按钮后，页面如图 12.9 所示。

```
[左图]
Elements  Console  Sources  Network
<html>
  <head></head>
  ▼<body> == $0
    ▼<div id="event-handling" data-v-app>
      <div>一直显示</div>
      <div>反复无常</div>
      <button>隐藏/显示</button>
    </div>
    <script src="js/vue.global.js"></script>
    ▶<script>…</script>
  </body>
</html>
```

图 12.8　页面初始化效果

```
[右图]
Elements  Console  Sources  Network
<html>
  <head></head>
  ▼<body> == $0
    ▼<div id="event-handling" data-v-app>
      <!--v-if-->
      <div style="display: none;">反复无常</div>
      <button>隐藏/显示</button>
    </div>
    <script src="js/vue.global.js"></script>
    ▶<script>…</script>
  </body>
</html>
```

图 12.9　单击"隐藏/显示"按钮后的效果

从图 12.9 可以看出，通过 v-if 控制的元素，如果隐藏，从 DOM 中移除，而通过 v-show 控制的元素并没有真正移除，只是给其添加了 CSS 样式"display:none"。

▶12.8.3　列表渲染指令 v-for

可以使用 v-for 指令遍历一个数组或对象，它的表达式需结合 in 来使用，形式为 item in items，其中 items 是源数据，item 是被迭代集合中元素的别名。v-for 还支持一个可选的参数作为当前项的索引。v-for 指令的常用方式如下：

❶ 遍历普通数组

```
<ul>
    <li v-for="(item,index) in items">
        {{index}} - {{ item }}
    </li>
</ul>
```

❷ 遍历对象数组

```
<ul>
<li v-for="user in users">
    {{ user.uname }}
</li>
</ul>
```

❸ 遍历对象属性

```
<li v-for="(value, key, index) in myObject">
    {{ ++index }}. {{ key }}: {{ value }}
</li>
```

❹ 迭代数字

```
<li v-for="i in 100">
    {{ i }}
</li>
```

【例 12-11】　演示 v-for 指令的常用方式。

其具体代码如下：

```
<div id="myfor">
    <ul>
        <li v-for="(book, index) in books">
            {{++index}}. {{book}}
        </li>
    </ul>
    <ul>
```

```
            <li v-for="(author, index) in authors">
                {{++index}}. {{author.name}} - {{author.sex}}
            </li>
    </ul>
    <p v-for="(value, key, index) in userinfo">键是:{{key}}, 值是:{{value}}, 索引是:
{{index}}</p>
    <p v-for="i in 5">这是第{{i}}段。</p>
</div>
<script src="js/vue.global.js"></script>
<script>
    Vue.createApp({
        data() {
            return {
                //普通数组
                books: ['Java Web 开发从入门到实战',
                'Java EE 框架整合开发入门到实战——Spring+Spring MVC+MyBatis',
                'Spring Boot 从入门到实战'],
                //对象数组
                authors:[{name: '陈恒', sex: '男'}, {name: '陈恒 11', sex: '女'},
                        {name: '陈恒 22', sex: '男'}],
                //对象
                userinfo:{
                    uname: '陈恒 3',
                    age: 88
                }
            }
        }
    }).mount('#myfor')
</script>
```

▶12.8.4 事件处理

所有的事件处理都离不开事件监听器,在 Vue.js 中可以使用 v-on 指令给 HTML 元素添加一个事件监听器,通过该指令调用在 Vue 实例中定义的方法。下面使用 v-on 指令监听按钮事件,实现字符串反转。

【例 12-12】 使用 v-on 指令监听按钮事件,实现字符串反转。

其具体代码如下:

```
<div id="event-handling">
    <p>{{ message }}</p>
    <button v-on:click="reverseMessage">反转 Message</button>
    <!-- v-on:可缩写为"@",是一个语法糖-->
    <button @click="reverseMessage">反转 Message</button>
</div>
<script src="js/vue.global.js"></script>
<script>
    const EventHandling = {
        data() {
            return {
                message: 'Hello Vue.js!'
            }
        },
        methods: {                              //方法定义
            reverseMessage() {
                this.message = this.message.split('').reverse().join('')
            }
        }
```

```
    Vue.createApp(EventHandling).mount('#event-handling')
</script>
```

在上述代码中，@click 等同于 v-on：click，是一个语法糖；在 methods 属性中定义了触发事件时调用的方法。@click 调用的方法如果没有参数，在方法名后可以不写括号"()"。

在 Vue.js 中有时需要访问原生的 DOM 事件。Vue.js 提供了一个特殊变量 $event，用于访问原生的 DOM 事件，例如下面的实例阻止打开链接。

【例 12-13】 阻止打开链接。

其具体代码如下：

```
<div id="event-handling">
    <a href="https://www.baidu.com/" @click="warn('考试期间禁止百度!', $event)">去百度</a>
</div>
<script src="js/vue.global.js"></script>
<script>
    Vue.createApp({
        methods: {
            warn(message, event) {
                //event 访问原生的 DOM 事件
                event.preventDefault()
                alert(message)
            }
        }
    }).mount('#event-handling')
</script>
```

在事件处理中调用 event.preventDefault() 或 event.stopPropagation() 是非常常见的需求。尽管可以在方法中轻松实现这类需求，但方法最好只有纯粹的数据逻辑，而不是去处理 DOM 事件细节。

为了解决该问题，Vue.js 为 v-on 提供了事件修饰符。修饰符是由点开头的指令后缀表示。Vue.js 支持的修饰符有 .stop、.prevent、.capture、.self、.once 以及 .passive。其用法是在 @绑定的事件后加小圆点"."，再跟修饰符，具体如下：

```
<!-- 阻止单击事件-->
<a @click.stop="doThis"></a>
<!-- 提交事件不再重载页面 -->
<form @submit.prevent="onSubmit"></form>
<!-- 修饰符可以串联 -->
<a @click.stop.prevent="doThat"></a>
<!-- 只有修饰符 -->
<form @submit.prevent></form>
<!-- 添加事件监听器时使用事件捕获模式，即内部元素触发的事件先在此处理，然后才交由内部元素进行处理 -->
<div @click.capture="doThis">...</div>
<!-- 当事件在该元素自身触发时触发回调，即事件不是从内部元素触发的-->
<div @click.self="doThat">...</div>
<!--只触发一次 -->
<a @click.once="doThis"></a>
<!-- 滚动事件的默认行为(即滚动行为)将会立即触发，而不会等待"onScroll"完成 -->
<div @scroll.passive="onScroll">...</div>
```

▶12.8.5 表单与 v-model

表单控件是网页数据交互的必备手段，用于向服务器传输数据，较为常见的表单控件有单选或多选下拉列表框、输入框等，用表单控件可以完成数据的输入、校验、提交等。Vue.js 用

v-model 指令在表单＜input＞、＜textarea＞及＜select＞元素上创建双向数据绑定（Model 到 View 以及 View 到 Model）。使用 v-model 指令的表单元素将忽略该元素的 value、checked、selected 等属性初始值，而是将当前活动的 Vue 实例的数据作为数据来源。所以，在使用 v-model 指令时，应通过 JavaScript 在 Vue 实例的 data 选项中声明初始值。

从 Model 到 View 的数据绑定，即 ViewModel 驱动将数据渲染到视图；从 View 到 Model 的数据绑定，即 View 中元素上的事件被触发后导致数据变更，将通过 ViewModel 驱动修改数据层。下面通过一个实例演示 v-model 指令在表单元素上实现双向数据绑定。

【例 12-14】 v-model 指令在表单元素上实现双向数据绑定。

其具体代码如下：

```html
<div id="vmodel-databinding">
    用户名:<input v-model="uname" />
    <p>输入的用户名是：{{ uname }}</p>
    <textarea v-model="introduction"></textarea>
    <p>输入的个人简介是：{{ introduction }}</p>
    <p>
        备选歌手：
        <input type="checkbox" id="zhangsan" value="张三" v-model="singers" />
        <label for="zhangsan">张三</label>
        <input type="checkbox" id="lisi" value="李四" v-model="singers" />
        <label for="lisi">李四</label>
        <input type="checkbox" id="wangwu" value="王五" v-model="singers" />
        <label for="wangwu">王五</label>
        <input type="checkbox" id="chenheng" value="陈恒" v-model="singers" />
        <label for="chenheng">陈恒</label>
        <br />
        <span>你喜欢的歌手：{{ singers }}</span>
    </p>
    <p>
        性别：
        <input type="radio" id="male" value="男" v-model="sex" />
        <label for="male">男</label>
        <input type="radio" id="female" value="女" v-model="sex" />
        <label for="female">女</label>
        <br />
        <span>你的性别：{{ sex }}</span>
    </p>
    <p>
        <!--单选下拉列表框-->
        备选国籍：
        <select v-model="single">
            <option v-for="option in options" :value="option.value">
                {{ option.text }}
            </option>
        </select><br>
        <span>你的国籍：{{ single }}</span>
    </p>
    <p>
        <!--多选下拉列表框-->
        备选国家：
        <select v-model="moreselect" multiple>
            <option v-for="option in moreoptions" :value="option.value">
                {{ option.text }}
            </option>
        </select><br>
        <span>你去过的国家：{{ moreselect }}</span>
    </p>
```

```
</div>
<script src="js/vue.global.js"></script>
<script>
    Vue.createApp({
        data() {
            return {
                uname: '陈恒',
                introduction: '我是一个好少年',
                //多个复选框,绑定到同一个数组,歌手默认选择'陈恒'
                singers: ['陈恒'],
                sex: '女',                    //默认性别为女
                //单选下拉列表绑定变量,默认国籍为'中国'
                single: '中国',
                options: [
                    { text: '中国', value: '中国' },
                    { text: '日本', value: '日本' },
                    { text: '美国', value: '美国' }
                ],
                moreselect: ['中国'],         //多选下拉列表绑定一个数组,默认去过中国
                moreoptions: [
                    { text: '中国', value: '中国' },
                    { text: '英国', value: '英国' },
                    { text: '日本', value: '日本' },
                    { text: '美国', value: '美国' }
                ],
            }
        }
    }).mount('#vmodel-databinding')
</script>
```

例 12-14 的运行效果如图 12.10 所示。

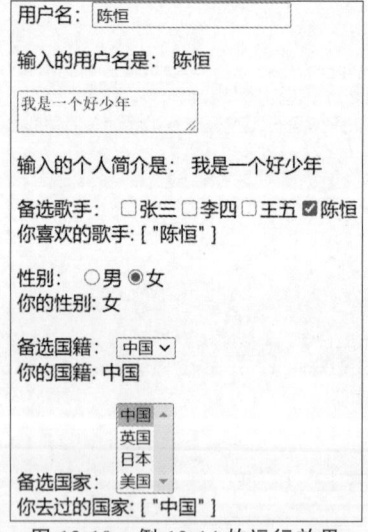

图 12.10 例 12-14 的运行效果

在默认情况下,v-model 在每次 input 事件触发后将输入框的值与数据进行同步。如果不想在每次 input 事件触发后同步,可以添加 lazy 修饰符,从而转为在 change 事件后进行同步。示例代码如下:

```
<!-- 在"change"时更新-->
<input v-model.lazy="msg"/>
```

如果需要将用户的输入值自动转换为数值类型，可以给 v-model 添加 number 修饰符，示例代码如下：

```
<input v-model.number="age" type="number" />
```

如果需要将用户输入的首尾空格自动去除，可以给 v-model 添加 trim 修饰符，示例代码如下：

```
<input v-model.trim="msg" />
```

▶12.8.6 实战：购物车实例

本节以 Vue.js 的计算属性、内置指令、方法等技术为基础，完成一个在电商平台中具有代表性的小功能——购物车。购物车的具体需求如下：

(1) 展示已加入购物车的商品列表，包括商品名称、商品单价、购买数量以及实时购买的总价。

(2) 购买数量可以增加或减少，每类商品还可以从购物车中删除。最终实现的效果如图 12.11 所示。

图 12.11 购物车效果图

因购物车的代码实现稍微复杂，这里将 JavaScript 从 HTML 中分离出来，以便于阅读和维护。具体代码文件为 cart.html（引入资源、模板及 CSS）和 cart.js（Vue 实例及业务代码）。

在 cart.html 中引入 Vue.js 和 cart.js，创建一个根元素（<div id="cart">）来挂载 Vue 实例。cart.html 的具体代码如下：

```
<div id="cart">
    <template v-if="cartList.length">
        <table>
            <thead>
                <tr>
                    <th></th>
                    <th>商品名称</th>
                    <th>商品单价</th>
                    <th>购买数量</th>
                    <th>操作</th>
                </tr>
                <tbody>
                    <tr v-for="(item, index) in cartList">
                        <td>{{ index + 1 }}</td>
                        <td>{{ item.gname }}</td>
                        <td>{{ item.gprice }}</td>
                        <td>
                            <button @click="reduce(index)" :disabled="item.count === 1">-</button>
                            {{ item.count }}
                            <button @click="add(index)">+</button>
```

```html
                    </td>
                    <td>
                        <button @click="remove(index)">删除</button>
                    </td>
                </tr>
            </tbody>
        </thead>
    </table>
    <div>总价:¥{{ totalPrice }} <button @click="removeAll">清空购物车</button></div>
    </template>
    <div v-else>购物车为空</div>
</div>
<script src="js/vue.global.js"></script>
<script src="cart.js"></script>
<style>
    table{
    border: 1px solid #e9e9e9;
    border-collapse:collapse;
    }
    th, td{
        padding: 8px 16px;
        border: 1px solid #e9e9e9;
    }
    th{
        background: #f7f7f7;
    }
</style>
```

在 cart.js 中,首先初始化 Vue 实例,然后在 data 选项中初始化商品列表 cartList,再使用计算属性 totalPrice 计算购物车总价,最后在 methods 选项中定义事件处理方法。cart.js 的具体代码如下:

```js
Vue.createApp({
    data() {
        return {
            cartList:[
                {
                    id: 1,
                    gname: 'Spring Boot 从入门到实战',
                    gprice: 79.8,
                    count: 5
                },
                {
                    id: 2,
                    gname: 'Java Web 开发从入门到实战',
                    gprice: 69.8,
                    count: 10
                },
                {
                    id: 3,
                    gname: 'Java EE 框架整合开发入门到实战',
                    gprice: 69.8,
                    count: 100
                }
            ]
        }
    },
    computed: {
        totalPrice() {
            var total = 0
```

第 12 章　Vue.js 基础

```
        for (var i = 0; i < this.cartList.length; i++) {
            var item = this.cartList[i]
            total = total + item.gprice * item.count
        }
        return total
    }
},
methods: {
    reduce(index) {
        if(this.cartList[index].count === 1)
            return
        this.cartList[index].count--
    },
    add(index) {
        this.cartList[index].count++
    },
    remove(index) {
        this.cartList.splice(index, 1)
    },
    removeAll() {
        this.cartList.splice(0, this.cartList.length)
    }
}
}).mount('#cart')
```

12.9　组件

组件（Component）是 Vue.js 最核心的功能，是可扩展的 HTML 元素（可看作自定义的 HTML 元素），能够封装可重用的代码，同时也是 Vue 实例，可以接受与 Vue 相同的选项对象并提供相同的生命周期钩子。

组件系统是 Vue.js 中一个重要的概念，它提供了一种抽象，让人们可以使用独立可复用的小组件来构建大型应用，任意类型的应用界面都可以抽象为一个组件树。这种前端组件化方便 UI 组件的重用。

本节将和读者一起由浅入深地学习组件的相关内容。

▶12.9.1　组件的注册

为了能在 UI 模板中使用组件，必须先注册组件，以便 Vue 识别。通常有两种组件的注册类型，即全局注册和局部注册。

❶ 全局注册

组件可通过 component 方法实现全局注册，全局注册的示例代码如下：

```
const app = Vue.createApp({})
app.component('component-a', {
    //选项
})
app.component('component-b', {
    //选项
})
app.component('component-c', {
    //选项
})
app.mount('#app')
```

app.component 的第一个参数是 component-a 组件的名称（自定义标签），组件的名称推荐全部小写，包含连字符（即有多个单词），以避免与 HTML 元素相冲突。

在注册后任何 Vue 实例都可以使用这些组件，示例代码如下：

```html
<div id="app">
    <component-a></component-a>
    <component-b></component-b>
    <component-c></component-c>
</div>
```

❷ 局部注册

全局注册往往是不够理想的。例如，在使用打包工具构建系统时，全局注册的组件即使不再使用，仍然被包含在最终的构建结果中，造成用户无意义地下载 JavaScript。在这些情况下，可以通过一个普通的 JavaScript 对象来定义组件：

```javascript
const ComponentA = {
    /* ... */
}
const ComponentB = {
    /* ... */
}
const ComponentC = {
    /* ... */
}
```

然后，使用 Vue 实例的 components 选项局部注册组件：

```javascript
const app = Vue.createApp({
    components: {
        'component-a': ComponentA,              //component-a 为局部组件名称
        'component-b': ComponentB
    }
})
```

局部注册的组件只在该组件的作用域下有效。例如，希望 ComponentA 在 ComponentB 中可用，需要在 ComponentB 中使用 components 选项局部注册 ComponentA：

```javascript
const ComponentA = {
    /* ... */
}
const ComponentB = {
    components: {
        'component-a': ComponentA
    }
    //…
}
```

【例 12-15】 全局组件和局部组件示例。本例定义一个名为 button-counter 的全局组件，以及两个局部组件 ComponentA 和 ComponentB。

其具体代码如下：

```html
<template id="button-counter">
    <button @click="count++">You clicked me {{ count }} times.</button>
</template>
<div id="components-demo">
    <!--在模板中任意使用组件-->
    <!--每个组件各自独立维护它的 count。因为每用一次组件，就会有一个它的新实例被创建-->
    <button-counter></button-counter><br><br>
```

```
        <button-counter></button-counter><br><br>
        <button-counter></button-counter><br><br>
        <component-b></component-b>
        <!--component-a 失效-->
        <component-a></component-a>
    </div>
    <script src="js/vue.global.js"></script>
    <script>
        const ComponentA = {
            template: '<span>这是私有组件A!</span>'
        }
        const ComponentB = {
            //使用 components 选项局部注册 ComponentA
            components: {
                'component-a': ComponentA
            },
            template: '<span>这是私有组件B!</span>'
        }
        //创建一个 Vue 应用
        const app = Vue.createApp({
            //使用 Vue 实例的 components 选项局部注册组件
            components: {
                'component-b': ComponentB
            }
        })
        //定义一个名为 button-counter 的全局组件(注册)
        app.component('button-counter', {
            data() {
                return {
                    count: 0
                }
            },
            //组件显示的内容
            template: '#button-counter'
        })
        app.mount('#components-demo')
    </script>
```

该例的运行结果如图 12.12 所示。

从图 12.12 可以看出，component-a 组件没有显示，这是因为它不是 Vue 实例 app 的组件，而是 ComponentB 的私有组件。

在例 12-15 组件定义的代码中，template 定义组件显示的内容，必须被一个 HTML 元素包含（如<button>），否则无法渲染。除了 template 选项外，组件可以像 Vue 实例一样使用 data、computed、methods 选项，例如上述 button-counter 组件。

图 12.12　例 12-15 的运行结果

▶12.9.2　使用 props 传递数据

组件除了把模板内容复用外，更重要的是向组件传递数据。在组件中，使用 props 选项来声明从父级组件接收的数据，props 的值可以是字符串数组，也可以是对象。

使用 props 实现的数据传递都是单向的，即父组件数据变化时，子组件中所有的 prop 将刷新为最新的值，但是反过来不行。这样设计的原因是尽可能将父子组件解耦，避免子组件无意中修改父组件的状态。如果在业务中需要改变 prop，一种是父组件传递初始值进来，子组件将它作为初始值保存起来，在子组件自己的作用域下随意修改；另一种是使用计算属性修改。

【例 12-16】 使用 props 传递数据示例。本例构造两个数组 props，一个数组接收来自父组件的数据 message（实现静态传递），另一个数组接收来自父组件的数据 id 和 title（实现动态传递），并将它们在组件模板中渲染。在子组件中声明数据 count 保存来自父组件的 mycount，count 的改变不影响 mycount。

其具体代码如下：

```html
<!--父组件显示-->
<template id="parent">
    <h4>{{ message }}</h4>
    <!--使用 v-bind 将父组件 parent 的 data(posts)动态传递给 props,children 组件只能在 parent 中-->
    <children v-for="post in posts" :id="post.id" :title="post.title"></children>
    <!--将一个对象的所有属性都作为 prop 传入，与上面一句等价-->
    <children v-for="post in posts" v-bind="post" ></children>
</template>
<!--子组件显示-->
<template id="children">
    <h4>{{id}} : {{ title }}</h4>
</template>
<template id="child-app">
    <button @click="count++">You clicked me {{ count }} times.</button>
</template>
<div id="message-post-demo">
    <!--静态传递字符串,父组件就是 Vue 当前的实例-->
    <parent message="来自父组件的消息"></parent>
    父组件的计数器:{{mycount}}<br>
    <child-counter :parent-count="mycount"></child-counter>
</div>
<script src="js/vue.global.js"></script>
<script>
    const messageApp = Vue.createApp({
        data() {
            return {
                mycount: 10
            }
        }
    })
    messageApp.component('parent', {
        data() {
            return {
                //posts 是对象数组
                posts: [
                    { id: 1, title: 'My journey with Vue' },
                    { id: 2, title: 'Blogging with Vue' },
                    { id: 3, title: 'Why Vue is so fun' }
                ]
            }
        },
        props: ['message'],                     //接收父组件 messageApp 传递的数据
        components: {                           //创建子组件 children
            'children':{
                props: ['id','title'],          //接收父组件 parent 传递的数据
                template: '#children'
            }
        },
        template: '#parent'
    })
```

```
    messageApp.component('child-counter', {
        props: ['parentCount'],
        data() {
            return {
                count: this.parentCount
            }
        },
        //组件显示的内容
        template: '#child-app'
    })
    messageApp.mount('#message-post-demo')
</script>
```

例 12-16 的运行效果如图 12.13 所示。

在例 12-16 中，因为 HTML 不区分大小写，当使用 DOM 模板时，以驼峰式命名法命名的 props 名称 parentCount 要转换为短横线命名 parent-count。在图 12.13 中，单击按钮改变子组件计数器 count 的值，而父组件计数器 mycount（通过 props 传递给子组件）不受影响。

▶12.9.3　组件的通信

通过学习 12.9.2 节，大家知道 props 可以实现父组件向子组件传递数据，即通信。Vue 组件通信的场景有多种，包括父子组件通信、兄弟组件通信、组件链通信。下面介绍各种组件间通信的方法。

图 12.13　例 12-16 的运行效果

❶ 使用自定义事件通信

用户可通过 props 从父组件向子组件传递数据，并且这种传递是单向的。当需要从子组件向父组件传递数据时，可以首先给子组件自定义事件，并使用 $emit（事件名，要传递的数据）方法触发事件，然后父组件使用 v-on 或 @ 监听子组件的事件。下面通过一个实例讲解通过自定义事件实现通信的方法。

【例 12-17】　使用自定义事件通信示例。在本例中子组件触发两个事件，分别实现字体变大、变小。

其具体代码如下：

```
<template id="blog">
    <!--0.1是传递给父组件 blogApp 的数据,可以不填。当在父组件监听这个事件时,可以通过$event访
问该数据。如果事件处理函数是一个方法,那么该数据将会作为第一个参数传入该方法(如 onEnlargeText)
-->
    <h4>{{id}} : {{ title }}</h4>
    <button @click="$emit('enlarge-text', 0.1)">变大</button>
    <button @click="$emit('ensmall-text', 0.1)">变小</button>
</template>
<div id="blog-post-demo">
    <div v-bind:style="{ fontSize: postFontSize + 'em' }">
    <!--将一个对象的所有属性作为 prop 传给子组件,@父组件监听事件并更新 postFontSize 值-->
        <!--$event 接收子组件传递过来的数据 0.1-->
        <blog-post v-for="post in posts" v-bind:post" @ensmall-text="postFontSize -=
$event"  @enlarge-text="onEnlargeText"></blog-post>
    </div>
</div>
<script src="js/vue.global.js"></script>
```

```
<script>
    const blogApp = Vue.createApp({
        data() {
            return {
                //posts是对象数组
                posts: [
                    { id: 1, title: 'My journey with Vue' },
                    { id: 2, title: 'Blogging with Vue' },
                    { id: 3, title: 'Why Vue is so fun' }
                ],
                postFontSize: 1
            }
        },
        methods: {
            onEnlargeText(enlargeAmount) {
                this.postFontSize += enlargeAmount
            }
        }
    })
    blogApp.component('blog-post', {          //定义子组件
        props: ['id', 'title'],               //接收父组件 blogApp 的两个参数 id 和 title
        template: '#blog'
    })
    blogApp.mount('#blog-post-demo')
</script>
```

注意：在上述代码中，事件名推荐使用短横线命名（例如 enlarge-text），这是因为 HTML 是不区分大小写的。如果事件名为 enlargeText，@enlargeText 将变成 @enlargetext，enlargeText 事件不可能被父组件监听到。

❷ 使用 v-model 通信

除了自定义事件实现子组件向父组件传值外，还可以在子组件上使用 v-model 向父组件传值，实现双向绑定。

下面通过一个实例讲解如何在子组件上使用 v-model 向父组件传值。

【例 12-18】 使用 v-model 通信示例。在本例中使用 v-model 实现子组件向父组件传值，并实现双向绑定。

其具体代码如下：

```
<template id="custom">
<!--为了让子组件正常工作,子组件内的 <input> 必须将其 value 属性绑定到一个名为 modelValue 的
props 上,在其 input 事件被触发时,将新的值通过自定义的 update:modelValue 事件传递-->
    <input :value="modelValue" @input="$emit('update:modelValue', $event.target.value)">
</template>
<div id="vmodel-post-demo">
    {{searchText}}<br>
    <custom-input v-model="searchText"></custom-input><br>
    <!--这两个子组件等价-->
    <custom-input :model-value="searchText" @update:model-value="searchText = $event">
    </custom-input>
</div>
<script src="js/vue.global.js"></script>
<script>
    const blogApp = Vue.createApp({
        data() {
            return {
                searchText: '陈恒'
```

```
            }
        }
    })
    blogApp.component('custom-input', {
        props: ['modelValue'],
        template: '#custom'
    })
    blogApp.mount('#vmodel-post-demo')
</script>
```

例 12-18 的运行效果如图 12.14 所示。

在例 12-18 中实现了一个具有双向绑定的 v-model 组件，需要满足以下两个要求：

(1) 接收一个 value 属性。

(2) 在有新的 value 时触发 input 事件。

图 12.14　例 12-18 的运行效果

❸ 使用 mitt 实现非父子组件通信

在 Vue.js 中，推荐使用一个空的 Vue 实例作为媒介(中央事件总线)实现父子组件、兄弟组件及组件链通信。例如在人们的生活中，买房卖房中介帮忙，买卖双方通过房产中介(中央事件总线)实现需求对接。

在 Vue 2.x 中，Vue 实例可通过事件触发 API（＄on、＄off 和 ＄once）实现中央事件总线，但是在 Vue 3.x 中移除了 ＄on、＄off 和 ＄once 实例方法，推荐使用外部库 mitt 来代替 ＄on、＄emit 和 ＄off 实例方法。使用 TypeScript 编写的 mitt 源代码具体如下：

```
/**
 * @param 入参为 EventHandlerMap 对象
 * @returns 返回一个对象,对象包含属性 all,以及方法 on、off、emit
 */
function mitt(all) {
    all = all || new Map();
    return {
        //事件键值对映射对象
        all,
        /**
         * 注册一个命名的事件处理
         * @param type 事件名
         * @param handler 事件处理函数
         */
        on(type, handler) {
            //根据 type 去查找事件
            const handlers = all.get(type);
            //如果找到有相同的事件,则继续添加,Array.prototype.push 返回值为添加后的新长度
            const added = handlers && handlers.push(handler);
            //如果已添加了 type 事件,则不再执行 set 操作
            if (!added) {
                all.set(type, [handler]);          //注意此处值是数组类型,可以添加多个相同的事件
            }
        },
        /**
         * 移除指定的事件处理
         * @param type 事件名,和第二个参数一起用来移除指定的事件
         * @param handler 事件处理函数
         */
        off(type, handler) {
            //根据 type 去查找事件
            const handlers = all.get(type);
```

```
            //如果找到则进行删除操作
            if (handlers) {
                handlers.splice(handlers.indexOf(handler) >>> 0, 1);
            }
        },
        /**
         * 触发所有 type 事件,如果有 type 为 * 的事件,则最后执行
         * @param type 事件名
         * @param evt 传递给处理函数的参数
         */
        emit(type, evt) {
            //找到 type 的事件循环执行
            (all.get(type) || []).slice().map((handler) => { handler(evt); });
            //然后找到所有为 * 的事件,循环执行
            (all.get('*') || []).slice().map((handler) => { handler(type, evt); });
        }
    };
}
```

下面通过一个实例讲解如何使用 mitt 实现非父子组件通信。

【例 12-19】 使用 mitt 实现非父子组件通信示例。在本例中首先使用 mitt 新建一个中央事件总线 bus,然后分别创建两个 Vue 实例 buyer(买方)和 seller(卖方),买卖双方互相通信。其具体代码如下:

```
<div id="buyer">
    <h1>{{ message1 }}</h1>
    <button @click="transferb">我是买方,向卖方传递信息</button>
</div>
<div id="seller">
    <h1>{{ message2 }}</h1>
    <button @click="transfers">我是卖方,向买方传递信息</button>
</div>
<script src="js/vue.global.js"></script>
<script src="js/mitt.js"></script>
<script>
    const bus = mitt()
    //买方
    const buyer = Vue.createApp({
        data() {
            return {
                message1: ''
            }
        },
        methods: {
            transferb() {
                //用 emit 触发事件传值
                bus.emit('on-message1', '来自买方的信息')
            }
        },
        mounted(){
            //监听
            bus.on('on-message2', (msg) => {        //(msg)相当于 function(msg)
                this.message1 = msg
            })
        }
    })
    buyer.mount('#buyer')
    //卖方
    const seller = Vue.createApp({
```

```
        data() {
            return {
                message2: ''
            }
        },
        methods: {
            transfers() {
                //用 emit 触发事件传值
                bus.emit('on-message2', '来自卖方的信息')
            }
        },
        mounted(){
            //监听
            bus.on('on-message1', (msg) => {      //(msg)相当于 function(msg)
                this.message2 = msg
            })
        }
    })
    seller.mount('#seller')
</script>
```

❹ 提供/注入（组件链传值）

当需要将数据从父组件传递到子组件时，可以使用 props 实现。但有时一些子组件是深嵌套的，如果将 props 传递到整个组件链中将很麻烦，更不可取。对于这种情况，可以使用 provide 和 inject 实现组件链传值。父组件可以作为其所有子组件的依赖项提供程序，而不管组件的层次结构有多深，父组件有一个 provide 选项来提供数据，子组件有一个 inject 项来使用这个数据。下面通过一个实例演示组件链传值的用法。

【例 12-20】 组件链传值示例。在本例中创建 Vue 实例为祖先组件，并用 provide 提供一个数据供其子孙组件 inject 使用。

其具体代码如下：

```
<template id="son">
    <div>{{ todos.length }}</div>
    <!--todo-son 是 todo-list 的私有组件-->
    <todo-son></todo-son>
</template>
<template id="grandson">
    <div>
        <!--使用注入的数据-->
        {{ todoLength }}
    </div>
</template>
<div id="vmodel-post-demo">
    <!--父组件 Vue 实例传递数据 todos 给子组件 todo-list-->
    <todo-list :todos="todos"></todo-list>
</div>
<script src="js/vue.global.js"></script>
<script>
    const app = Vue.createApp({
        data() {
            return {
                todos: ['Feed a cat', 'Buy tickets']
            }
        },
        provide() {                    //祖先组件 App 提供一个数据 todoLength
            return {
```

```
                    todoLength: this.todos.length
                }
            }
        })
        app.component('todo-list', {
            props: ['todos'],
            components:{                        //在父组件todo-list中定义子组件todo-son
                'todo-son': {
                    inject: ['todoLength'],     //子孙组件注入数据todoLength供自己使用
                    template: '#grandson'
                }
            },
            template: '#son'
        })
        app.mount('#vmodel-post-demo')
</script>
```

▶12.9.4 动态组件与异步组件

组件间切换或异步加载是常见的应用场景。本节将介绍动态组件与异步组件的实现方法。

❶ 动态组件

在不同组件之间进行动态切换是常见的场景，例如在一个多标签的页面中进行内容的收纳和展现。Vue 可通过＜component＞元素动态挂载不同的组件，进行组件的切换。示例代码如下：

```
<!-- is属性选择挂载的组件,currentView是已注册组件的名称或一个组件的选项对象-->
<component :is="currentView"></component>
```

下面通过一个实例讲解动态组件的用法。

【例 12-21】 通过＜component＞元素动态切换组件，在该例中有三个按钮代表标签，单击不同按钮展示不同组件的信息。

其具体代码如下：

```
<div id="app">
    <button @click="changeCom('add')">添加信息</button> 
    <button @click="changeCom('update')">修改信息</button> 
    <button @click="changeCom('delete')">删除信息</button>
    <component :is="currentCom"></component>
</div>
<script src="js/vue.global.js"></script>
<script>
    const blogApp = Vue.createApp({
        data() {
            return {
                currentCom: 'add'
            }
        },
        components:{                        //在组件选项中定义三个局部组件供切换
            'add': {
                template: '<div>添加信息展示界面</div>'
            },
            'update': {
                template: '<div>修改信息展示界面</div>'
            },
            'delete': {
                template: '<div>删除信息展示界面</div>'
            }
```

第 12 章　Vue.js 基础

```
        },
        methods:{
            changeCom(com) {
                this.currentCom = com
            }
        }
    })
    blogApp.mount('#app')
</script>
```

❷ 异步组件

在大型应用中，可能需要将应用分割成许多小的代码块，并且只在需要时从服务器异步加载一个模块。这样可以避免一开始就把所有组件加载，浪费不必要的开销。Vue 使用 defineAsyncComponent 方法将组件定义为一个工厂函数，动态地解析组件。Vue 只在组件需要渲染时触发工厂函数，并把结果缓存起来，以备再次渲染。下面通过一个实例讲解异步组件的用法。

【例 12-22】 实现 5 秒钟后加载组件信息。

其具体代码如下：

```
<div id="app">
    <!--5秒钟后才下载组件并展示-->
    <async-example></async-example>
</div>
<script src="js/vue.global.js"></script>
<script>
    const blogApp = Vue.createApp({})
    //定义异步组件
    const AsyncComp = Vue.defineAsyncComponent(() =>
                                    //定义 defineAsyncComponent 方法的返回值
        new Promise((resolve, reject) => {//返回 Promise 的工厂函数
            window.setTimeout(() => {    //window.setTimeout 只是演示异步
                resolve({/*从服务器收到加载组件定义后，调用 Promise 的 resolve 方法异步下载组件，也可以调用 reject(reason)指示加载失败 */
                    template: '<div>5秒钟后才展示我!</div>'
                })
            }, 5000)
        })
    )
    blogApp.component('async-example', AsyncComp)
    blogApp.mount('#app')
</script>
```

上述工厂函数返回 Promise 对象，当从服务器收到加载组件定义后，调用 Promise 的 resolve 方法异步下载组件，也可以调用 reject(reason)指示加载失败。这里 window.setTimeout 只是演示异步，具体的异步下载逻辑由开发者自己决定。

12.9.5　实战: 正整数数字输入框组件

本节将普通输入框扩展成正整数数字输入框，用来快捷地输入一个标准的数字，如图 12.15 所示。

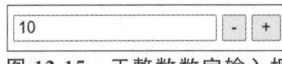

图 12.15　正整数数字输入框

其具体实现过程如下：

❶ 定义正整数数字输入框组件

在 ch12 的 js 目录中创建 input-num.js 文件。在 input-num.js 文件中定义正整数数字输入框组件 inputNumber，并在该组件中定义三个方法 handleDown()、handleUp()和 handleChange()，分别

实现减一、加一和数值判断的功能。input-num.js 的具体代码如下:

```javascript
function isValueNumber (value){
    return (/^[1-9]\d*$/).test(value+'')
}
//定义正整数数字输入组件
const inputNumber = {
    //组件显示的内容
    template: '\
<div>\
    <input type="text" :value="currentValue" @change="handleChange">\
    <button @click="handleDown" :disabled="currentValue <= 1">-</button>\
    <button @click="handleUp">+</button>\
</div>',
    data() {
        return {
            currentValue: 1
        }
    },
    methods: {
        handleDown() {
            if (this.currentValue <= 1)
                return
            this.currentValue -= 1
        },
        handleUp() {
            this.currentValue += 1
        },
        handleChange(event) {
            var val = event.target.value.trim()
            if(isValueNumber(val)) {
                this.currentValue = Number(val)
            } else {
                event.target.value = this.currentValue
            }
        }
    }
}
```

❷ 引用正整数数字输入框组件

在 ch12 目录中创建 inputNumber.html 文件。在 inputNumber.html 文件中创建 Vue 实例 App,并引入正整数数字输入框组件 inputNumber。inputNumber.html 的具体代码如下:

```html
<head>
    <meta charset="utf-8">
    <title>数字输入框</title>
</head>
<body>
    <div id="app">
        <input-number></input-number>
    </div>
    <script src="js/vue.global.js"></script>
    <script src="js/input-num.js"></script>
    <script>
        const app = Vue.createApp({
            //引入数字输入组件
            components: {
                'input-number': inputNumber
            }
        })
```

```
        app.mount('#app')
    </script>
</body>
```

12.10 自定义指令

Vue 为用户提供了功能丰富的内置指令，例如 v-model、v-show 等。这些内置指令可以满足大部分业务需求，但有时用户需要一些特殊功能，例如对普通 DOM 元素进行底层操作。幸运的是，Vue 允许用户自定义指令，实现特殊功能。

▶12.10.1 自定义指令的注册

与组件类似，自定义指令的注册也分为全局注册和局部注册，例如注册一个名为 v-focus 的指令，用于输入元素（<input>和<textarea>）在初始化时自动获得焦点。两种注册的示例代码如下：

```
const app = Vue.createApp({})
//注册一个全局自定义指令 v-focus
app.directive('focus', {
    //指令选项
})
//注册一个局部自定义指令 v-focus
const app = Vue.createApp({
    directives: {
        focus: {
            //指令选项
        }
    }
})
```

一个自定义指令的选项是由几个钩子函数（均为可选）组成的。

（1）beforeMount：只调用一次，当指令第一次绑定到元素时调用，并进行初始化设置。

（2）mounted：在挂载绑定元素的父组件时调用。

（3）beforeUpdate：在元素本身更新前调用。

（4）updated：在元素本身更新后调用。

（5）beforeUnmount：在卸载绑定元素的父组件前调用。

（6）unmounted：只调用一次，当指令与元素解除绑定时调用。

用户可以根据业务需求，在不同的钩子函数中完成业务逻辑代码，例如下面的实例。

【例 12-23】 自定义名为 v-focus 的指令，用于输入元素<input>（挂载绑定父组件调用 mounted 函数）在初始化时自动获得焦点。

其具体代码如下：

```
<div id="app">
    <input v-focus type="text" />
</div>
<script src="js/vue.global.js"></script>
<script>
    const app = Vue.createApp({})
    //自定义指令 focus,在模板中使用 v-focus
    app.directive('focus', {
        mounted(el) {
            el.focus()
```

```
      }
   })
   app.mount('#app')
</script>
```

在例 12-23 的代码中,el 为钩子函数的参数,除了 el 参数外,还有 binding、vnode 和 prevNnode 参数。它们的含义具体如下。

- el:指令所绑定的元素,可以用来直接操作 DOM。
- binding:一个对象,包含以下常用属性。
  - value:指令的绑定值,例如在 v-my-directive="1 + 1"中,绑定值为 2。
  - oldValue:指令绑定的前一个值,仅在 updated 钩子函数中可用。无论值是否改变都可用。
  - arg:传给指令的参数,可选。例如在 v-my-directive:foo 中,参数为"foo"。
  - modifiers:一个包含修饰符的对象。例如在 v-my-directive.foo.bar 中,修饰符对象为 { foo: true, bar: true }。
- vnode:Vue 编译生成的虚拟节点。
- prevNnode:上一个虚拟节点,仅在 updated 钩子中可用。

▶ **12.10.2　实战: 实时时间转换指令**

例如,大家发布的朋友圈会有一个相对本机时间转换后的时间,如图 12.16 中方框内的时间。

图 12.16　某朋友圈发文

为了显示实时性,在一些社交类软件中,经常将 Unix 时间戳转换为可读的时间格式,如几分钟前、几小时前、几天前等不同的格式。本实战将实现这样一个自定义指令 v-time,该指令将传入的时间戳实时转换为需要的时间格式。其具体实现步骤如下:

❶ **为统一使用时间戳进行逻辑判断**

在编写指令 v-time 前,事先编写一系列与时间相关的函数,将这些函数封装在 Time 对象中。time.js 的代码如下:

```
var Time = {
    //获得当前时间戳
    getUnix: function() {
        var date = new Date();
        return date.getTime();
    },
    //获得今天 0 点 0 分 0 秒的时间戳
    getTodayUnix: function() {
        var date = new Date();
        date.setHours(0);
        date.setMinutes(0);
        date.setSeconds(0);
```

```
        date.setMilliseconds(0);
        return date.getTime();
    },
    //获取标准年月日
    getNormalDate: function(time) {
        var date = new Date(time);
        var month = date.getMonth() + 1;
        var monthFormate = month < 10 ?('0' + month) : month;
        var day = date.getDate() < 10 ?('0' + date.getDate()) : date.getDate();
        return date.getFullYear() + '-' + monthFormate + '-' + day;
    },
    //自定义指令需要调用的函数,参数为毫秒级时间戳
    getFormatTime: function(timeStamp) {
        //获得当前时间戳
        var now = this.getUnix();
        //获得今天 0 点 0 分 0 秒的时间戳
        var today = this.getTodayUnix();
        //转换秒级时间
        var timer = (now - timeStamp) / 1000;
        var timeFormat = '';
        if (timer <= 0) {
            timeFormat = '刚刚';
        } else if (Math.floor(timer / 60) <= 0) {   //一分钟以前,显示刚刚
            timeFormat = '刚刚';
        } else if (timer < 3600) {                   //一分钟到一小时,显示××分钟前
            timeFormat = Math.floor(timer / 60) + '分钟前';
        } else if (timer >= 3600 && (timeStamp - today) >= 0) {
                                                     //一小时到一天,显示××小时前
            timeFormat = Math.floor(timer / 3600) + '小时前';
        } else if (timer / 86400 <= 31) {            //一天到一个月,显示××天前
            timeFormat = Math.ceil(timer / 86400) + '天前';
        } else {                                     //大于一个月,显示××年××月××日
            timeFormat = this.getNormalDate(timeStamp);
        }
        return timeFormat;
    }
}
```

❷ **在 HTML 文件 v-time.html 中注册一个全局指令 v-time**

在 v-time 的钩子函数 beforeMount 中,将指令表达式的值 binding.value 作为参数传入 Time.getFormatTime()方法得到格式化时间,再通过 el.innerHTML 写入指令元素,并且每分钟触发一次定时器 el.timeout 更新时间。同时,在 v-time 的钩子函数 unmounted 中清除定时器。v-time.html 的具体代码如下:

```
<div id="app">
    <div v-time="nowTime"></div>
    <div v-time="beforeTime"></div>
</div>
<script src="js/vue.global.js"></script>
<script src="js/time.js"></script>
<script>
    const app = Vue.createApp({
        data() {
            return {
                //nowTime 是目前的时间
                nowTime: (new Date()).getTime(),
                //beforeTime 是 2021-08-08(固定时间)
                beforeTime: 1628407242588
            }
```

```
        }
    })
    app.directive('time', {
        //绑定一次性事件等初始化操作
        beforeMount(el, binding) {
            el.innerHTML = Time.getFormatTime(binding.value)
            //定时器一分钟更新一次
            el.timeout = setInterval(function () {
                el.innerHTML = Time.getFormatTime(binding.value)
            }, 60000)
        },
        //解除相关绑定
        unmounted(el) {
            clearInterval(el.timeout)
            delete el.timeout
        }
    })
    app.mount('#app')
</script>
```

12.11 响应性

非侵入性的响应性系统是 Vue.js 最独特的特性之一。在本节将学习 Vue.js 响应性系统的实现细节。

▶12.11.1 什么是响应性

响应性是一种允许用户以声明式的方式去适应变化的编程范例。例如,在某个 Excel 电子表格中,将数字 x 放在第一个单元格中,数字 y 放在第二个单元格中,并要求自动计算 x + y 的值放在第三个单元格中。同时,如果更新数字 x 或 y,第三个单元格中的值也会自动更新。

Vue.js 如何追踪数据的变化呢？在生成 Vue.js 实例时,使用带有 getter 和 setter 的处理程序遍历传入的 data,将其所有 property 转换为 Proxy 对象,见例 12-24。Proxy 代理对象,顾名思义,在访问对象前增加一个中间层,通过中间层做中转,通过操作代理对象实现目标对象的修改。Proxy 对象对于用户来说是不可见的,但在内部,它使 Vue.js 能够在 property 值被访问或修改的情况下进行依赖跟踪和变更通知。

【例 12-24】 将 property 转换为 Proxy 对象。

其具体代码如下：

```
<script>
    const data = {
        uname: 'chenheng',
        age: 90
    }
    const handler = {
        get(target, name, receiver) {
            alert('执行 get 方法')
            //Reflect.get 方法查找并返回 target 对象的 name 属性,如果没有该属性,则返回 undefined
            return Reflect.get(...arguments)
        },
        set(target, name, value, receiver) {
            alert('执行 set 方法')
            //Reflect.set 方法设置 target 对象的 name 属性等于 value
            return Reflect.set(...arguments)
        }
```

```
            }
            const proxy = new Proxy(data, handler)
            alert(proxy.uname)                      //执行 get 方法
            proxy.uname = 'hhhhh'                   //执行 set 方法
            alert(proxy.uname)                      //执行 get 方法
        </script>
```

▶12.11.2 响应性原理

reactive()方法和 watchEffect()方法是 Vue3 中响应式的两个核心方法,reactive()方法负责将数据变成响应式代理对象,watchEffect()方法的作用是监听数据变化去更新视图或调用函数。reactive()方法和 watchEffect()方法的应用示例见例 12-25。

【例 12-25】 reactive()方法和 watchEffect()方法的应用。

其具体代码如下。

```
        <script src="js/vue.global.js"></script>
        <script>
            //reactive()方法接收一个普通对象,然后返回该对象的响应式代理
            let book = Vue.reactive({
                title: 'SSM + Spring Boot + Vue.js 3全栈开发从入门到实战(微课视频版)',
                author: '陈恒'
            })
            //watchEffect()方法监听数据变化去更新视图或调用函数
            Vue.watchEffect(() => {
                alert(book.title)
            })
            book.title = 'Vue.js 3从入门到实战(微课视频版)'
        </script>
```

12.12 setup 组件选项

Vue 组件提供 setup 选项,供开发者使用组合 API。setup 选项在创建组件前执行,一旦 props 被解析,便充当组合式 API 的入口点。由于在执行 setup 时尚未创建组件实例,所以在 setup 选项中没有 this。这意味着除了 props 之外,无法访问组件中声明的任何属性,包括本地状态、计算属性或方法。

setup 选项是一个接受 props 和 context 参数的函数。此外,从 setup 返回的所有内容都将暴露给组件的其余部分(计算属性、方法、生命周期钩子、模板等)。

▶12.12.1 setup 函数的参数

❶ setup 函数中的第一个参数(props)

setup 函数中的 props 是响应式的,当传入新的属性时它将被更新,见例 12-26。

【例 12-26】 在 setup 函数中参数 props 是响应式的。

其具体代码如下。

```
        <template id="stesting">
            {{mybook}}
        </template>
        <div id="app">
            <setup-testing :abook="book"></setup-testing>
        </div>
        <script src="js/vue.global.js"></script>
        <script>
            const app = Vue.createApp({
```

```
        data() {
            return {
                book: {
                    id: 1,
                    title: 'My journey with Vue'
                }
            }
        }
    })
    app.component('setup-testing', {
        props: ['abook'],
        setup(props) {                                      //props 是响应式的
            console.log(props.abook.id)
            mybook = props.abook
            //暴露给 template
            return {
                mybook
            }
        },
        template: '#stesting'
    })
    app.mount('#app')
</script>
```

但是，因为 props 是响应式的，不能使用 ES6 解构，将会消除 props 的响应性。如果需要解构 props，可以在 setup 函数中使用 toRefs 函数来完成，见例 12-27。

【例 12-27】 在 setup 函数中使用 toRefs 函数创建 props 属性的响应式引用。

其具体代码如下。

```
<template id="stesting">
    {{mybook}}
</template>
<div id="app">
    <setup-testing :mytitle="book.title"></setup-testing>
</div>
<script src="js/vue.global.js"></script>
<script>
    const app = Vue.createApp({
        data() {
            return {
                book: {
                    id: 1,
                    title: 'My journey with Vue'
                }
            }
        }
    })
    app.component('setup-testing', {
        props: ['mytitle'],
        setup(props) {
            //使用 toRefs 函数创建 props 属性的响应式引用
            title =Vue.toRefs(props)
            //使用 ES6 解构
            console.log(title.mytitle.value)
            mybook = title
            //暴露给 template
            return {
                mybook
            }
```

```
        },
        template: '#stesting'
    })
    app.mount('#app')
</script>
```

❷ setup 函数中的第二个参数（context）

context 上下文是一个普通的 JavaScript 对象，它暴露组件的 4 个属性，分别为 attrs、slots、emit 以及 expose。示例代码如下：

```
setup(props, context) {
    //Attribute (非响应式对象,等同于 $attrs)
    console.log(context.attrs)
    //插槽 (非响应式对象,等同于 $slots)
    console.log(context.slots)
    //触发事件 (方法,等同于 $emit)
    console.log(context.emit)
    //暴露公共 property (函数)
    console.log(context.expose)
}
```

context 是一个普通的 JavaScript 对象，也就是说它不是响应式的，这意味着可以安全地对 context 使用 ES6 解构。示例代码如下：

```
setup(props, { attrs, slots, emit, expose }) {
    ...
}
```

attrs 和 slots 是有状态的对象，它们随组件本身的更新而更新。这意味着应该避免对它们进行解构，并始终以 attrs.x 或 slots.x 的方式引用属性。

▶12.12.2 setup 函数的返回值

❶ 对象

如果 setup 返回一个对象，则可以在组件的模板中访问该对象的属性。下面通过一个实例讲解 setup 函数的使用方法。

【例 12-28】 setup 函数的使用。在该例中 setup 函数返回一个对象。

其具体代码如下。

```
<template id="stesting">
    <!-- 在模板中使用 readersNumber 对象会被自动开箱,所以不需要.value-->
    <div>{{ readersNumber }} {{ book.title }}</div>
</template>
<div id="app">
    <setup-testing></setup-testing>
</div>
<script src="js/vue.global.js"></script>
<script>
    const app = Vue.createApp({})
    app.component('setup-testing', {
        setup() {
            //使用 ref 函数,对值创建一个响应式引用,并返回一个具有 value 属性的对象
            const readersNumber = Vue.ref(1000)
            //reactive()接收一个普通对象,然后返回该对象的响应式代理
            const book = Vue.reactive({ title: '好书' })
            //暴露给 template
            return {
```

```
                readersNumber,
                book
            }
        },
        template:'#stesting'
    })
    app.mount('#app')
</script>
```

❷ 渲染函数

setup 还可以返回一个渲染函数,该函数可以直接使用在同一作用域中声明的响应式状态。下面通过一个实例讲解 setup 返回渲染函数。

【例 12-29】 实现例 12-28 的功能,要求 setup 返回渲染函数。

其具体代码如下:

```
<div id="app">
    <setup-testing></setup-testing>
</div>
<script src="js/vue.global.js"></script>
<script>
    const app = Vue.createApp({})
    app.component('setup-testing', {
        setup() {
            //使用 ref 函数,对值创建一个响应式引用,并返回一个具有 value 属性的对象
            const readersNumber = Vue.ref(1000)
            //reactive()接收一个普通对象,然后返回该对象的响应式代理
            const book = Vue.reactive({ title: '好书' })
            //返回渲染函数
            return () => Vue.h('div', [readersNumber.value, book.title])
        }
    })
    app.mount('#app')
</script>
```

▶12.12.3　使用 ref 创建响应式引用

❶ 声明响应式状态

如果要为 JavaScript 对象创建响应式状态,可以使用 reactive()方法。reactive()方法接收一个普通对象,然后返回该对象的响应式代理。示例代码如下:

```
const book = Vue.reactive({ title: '好书' })
```

reactive()方法响应式转换是"深层的",即影响对象内部嵌套的所有属性。基于 ES 的 Proxy 实现,返回的代理对象不等于原始对象。建议大家使用代理对象,避免依赖原始对象。例如,在例 12-28 中使用代理对象 book。

❷ 使用 ref 创建独立的响应式值对象

ref 接收一个参数值,并返回一个响应式且可改变的 ref 对象。ref 对象拥有一个指向内部值的单一属性.value。示例代码如下:

```
//const readersNumber = Vue.ref(1000)
//console.log(readersNumber.value)            //1000
//readersNumber.value++
console.log(readersNumber.value)              //1001
```

当 ref 作为渲染上下文的属性返回(即在 setup()返回的对象中)并在模板中使用时,它会自

动开箱,无须在模板内额外书写.value。例如,在例12-28中使用{{ readersNumber }}。

当嵌套在响应式对象中时,ref才会自动开箱。从Array或者Map等原生集合类中访问ref时不会自动开箱。示例代码如下:

```
const map = reactive(new Map([['foo', ref(0)]]))
//这里需要.value
console.log(map.get('foo').value)
```

在例12-28和例12-29中,使用ref将值封装在一个对象中,看似没有必要。但为了保持JavaScript中不同数据类型的行为统一,这是必需的。也就是说,任何数据类型的值都有一个封装对象,这样就可以在整个应用中安全地传递它,而不必担心在某个地方失去它的响应性。

▶12.12.4 在setup内部调用生命周期钩子函数

在setup内部,可以通过在生命周期钩子函数前面加上"on"来访问组件的生命周期钩子函数。因为setup是围绕beforeCreate和created生命周期钩子函数运行的,所以不需要显式地定义它们。换句话说,在这些钩子函数中编写的任何代码都应该直接在setup函数中编写。这些on函数接受一个回调函数,当钩子函数被组件调用时将会被执行。示例代码如下:

```
setup() {
    //mounted时执行
    onMounted(() => {
        console.log('Component is mounted!')
    })
}
```

本章小结

本章主要介绍了Vue.js的基础知识,在学习本章后,读者对Vue.js的插值与表达式、生命周期、计算属性与监听器、指令、组件系统等基础知识应该有一个初步了解。组件是Vue最核心的功能,是非常抽象的概念,也是本章的重点。在以后的学习中,需要读者慢慢理解组件的内在机制。

习题12

1. 下列选项中能够定义Vue.js根实例对象的元素是( )。
    A. template B. script C. style D. title
2. 定义Vue.js根实例,需要调用的方法是( )。
    A. createApp() B. mount() C. createVue() D. create()
3. MVVM模式中的VM是指( )。
    A. Model B. View C. ViewModel D. VueModel
4. 下列选项中插值不正确的是( )。
    A. {{myValue}} B. {{one.join(two)}}
    C. {{const a = 1}} D. {{x + y}}
5. 下列选项中能够动态渲染数据属性的是( )。
    A. methods B. watch C. computed D. data
6. 下列有关computed属性的描述错误的是( )。

A. computed 属性默认只有 getter 方法，但可以为其提供一个 setter 方法

B. computed 属性所依赖的数据发生变化时将自动重新计算，并把计算结果缓存起来

C. computed 属性一定有返回值，而且不带参数

D. computed 属性只有 getter 方法

7. 下列选项中能够实现绑定属性的指令是（　　）。

  A. v-once　　　B. v-bind　　　C. v-model　　　D. v-on

8. 下列选项中能够实现表单数据的双向绑定的指令是（　　）。

  A. v-once　　　B. v-if　　　C. v-model　　　D. v-on

9. 下列指令中能够捕获事件的是（　　）。

  A. v-for　　　B. v-bind　　　C. v-if　　　D. v-on

10. 下列事件修饰符中能够阻止事件冒泡的是（　　）。

  A. .stop　　　B. .prevent　　　C. .capture　　　D. .self

11. 下列事件修饰符中能够实现事件只被触发一次的是（　　）。

  A. .prevent　　　B. .once　　　C. .capture　　　D. .self

12. 下列 v-model 修饰符中能够将用户的输入值自动转为数值类型的是（　　）。

  A. .trim　　　B. .lazy　　　C. .number　　　D. 所有选项都不是

13. 下列指令中不属于条件渲染指令的是（　　）。

  A. v-if　　　B. v-show　　　C. v-else-if　　　D. v-for

14. 父组件中绑定 myData="[1,2,3,5]" 传递给子组件，在子组件中显示 myData.length 的值为（　　）。

  A. 5　　　B. 3　　　C. 4　　　D. 9

15. 子组件可以通过（　　）属性声明使用父组件的变量。

  A. data　　　B. message　　　C. parent　　　D. props

16. 父组件可以作为其所有子组件的依赖项提供程序，而不管组件的层次结构有多深，父组件有一个（　　）选项来提供数据，子组件有一个（　　）项来使用这个数据。

  A. provide　inject　B. set　get　C. push　get　D. provide　props

17. 什么是计算属性？为什么要使用计算属性？

18. 简述计算属性和监听器属性的区别。

19. 参考例 12-3，编写一个 HTML 页面 practice12_1.html，在该页面中输入三个数字。使用计算属性判断这三个数字是否构成三角形，运行结果如图 12.17 所示。

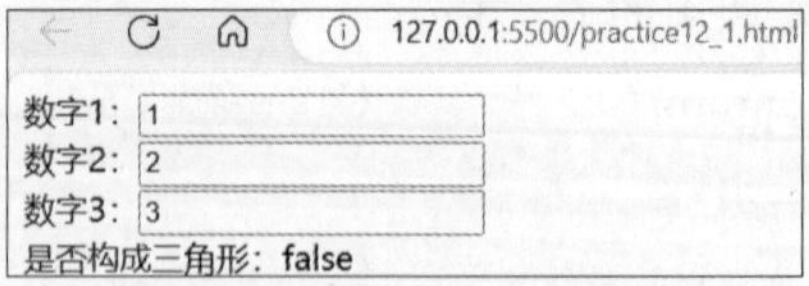

图 12.17　是否构成三角形

20. 简述 Vue.js 的生命周期。

21. 什么是 MVVM 模式？简述 Model、View、ViewModel 三者之间的关系。

22. 如何创建一个 Vue.js 实例？又如何将 Vue.js 实例挂载到一个 DOM 元素上？

23. 参考 12.8.6 节，将以下二维数组的数据显示在购物车界面 practice12_2.html（如图 12.18 所示）。

```
cartList:[
    {
        type:'图书',
        items:[
            {
                id: 1,
                gname: 'Spring Boot 从入门到实战',
                gprice: 79.8,
                count: 5
            },
            {
                id: 2,
                gname: 'Java Web 开发从入门到实战',
                gprice: 69.8,
                count: 10
            },
            {
                id: 3,
                gname: 'Java EE 框架整合开发入门到实战',
                gprice: 69.8,
                count: 100
            }
        ]
    },
    {
        type:'家电',
        items:[
            {
                id: 1,
                gname: '电视机 X',
                gprice: 888,
                count: 5
            },
            {
                id: 2,
                gname: '冰箱 Y',
                gprice: 999,
                count: 10
            }
        ]
    }
]
```

图 12.18　practice12_2.html 的运行效果

24. 如何注册组件？如何区分父子组件？你了解的组件传值有哪几种？它们是如何实现的？

25. 在 12.9.5 节实战内容的基础上自定义一个奇偶数判定输入框组件,运行效果如图 12.19 所示。

图 12.19　奇偶数判定输入框

26. 开发一个自定义指令 v-birthdayformat,接收一个出生日期(YYYY-MM-DD),将它转换为具体年龄,例如 8 岁 8 个月 8 天,运行效果如图 12.20 所示。

图 12.20　年龄计算自定义指令

第 13 章　Vue.js 进阶

学习目的与要求

本章主要讲解 Vue.js 的进阶知识，包括 Vue CLI、Vue Router、setup 语法糖以及 Element Plus UI 组件库。通过本章的学习，将 Vue.js 的进阶知识灵活应用于单页面综合应用中。

本章主要内容

- Vue Router 的基本用法与高级应用
- setup 语法糖
- Element Plus UI 组件库

高效的开发离不开高效的项目构建工具以及实用的插件，本章将学习 Vue CLI 脚手架、Vue Router 插件以及 Element Plus UI 组件库。

13.1　单文件组件与 webpack

Vue.js 是一个渐进式的 JavaScript 框架，在使用 webpack 构建 Vue 应用时可以使用一种新的构建模式——.vue 单文件组件。

.vue 是 Vue.js 自定义的一种文件格式，一个 .vue 文件就是一个单独的组件，在文件内封装了组件的相关代码（HTML、CSS、JS）。

.vue 文件由三部分组成，即 <template>、<style>、<script>，示例如下：

```
<template>
    HTML
</template>
<style>
    CSS
</style>
<script>
    JS
</script>
```

但是浏览器本身并不识别 .vue 文件，因此必须对 .vue 文件进行加载解析，此时需要 webpack 的 vue-loader 加载器。

webpack 是一个用于 JavaScript 应用程序的静态模块打包工具。当 webpack 处理应用程序时，它会在内部从一个或多个入口点（即入口文件）构建一个依赖图（Dependency Graph），然后将项目中所需的每一个模块组合成一个或多个 bundles，它们均为静态资源，用于展示项目内容。

webpack 根据模块的依赖关系进行静态分析，然后将这些模块按照指定的规则生成对应的静态资源。图 13.1 是来自 webpack 官方网站（https://webpack.js.org/）的模块化示意图。

图 13.1 的左边是在业务中编写的各种类型文件，例如 TypeScript、JPG、Less、CSS，以及后续将要学习的 .vue 格式的文件。这些类型的文件通过特定的加载器（Loader）编译后，最终统一生成 .js、.css、.jpg、.png 等静态资源文件。在 webpack 中，一张图片、一个 CSS 文件等都被称为模块，它们彼此存在依赖关系。使用 webpack 的目的就是处理模块间的依赖关系，并将它们进行打包。

图 13.1　webpack 模块化示意图

webpack 的主要适用场景是单页面应用（SPA），SPA 通常由一个 HTML 文件和一堆按需加载的 JS 文件组成。

在用 webpack 搭建单页面应用程序时，需要安装许多插件并编写复杂的项目配置，大大降低了开发效率。为了提高单页面应用程序的开发效率，本章将使用 Vue CLI（Vue 脚手架）搭建 Vue.js 项目。

13.2　安装 Node.js 和 NPM

本章将使用 NPM 安装 Vue CLI，而 NPM 是 Node.js 的包管理器，集成在 Node.js 中，所以需要首先安装 Node.js。

▶ 13.2.1　安装 Node.js

通过访问官网 https://nodejs.org/en/ 即可下载对应版本的 Node.js，本书下载的是"16.15.1 LTS"。

在下载完成后运行安装包 node-v16.15.1-x64.msi，一直单击"下一步"按钮即可完成安装。然后在命令行窗口中输入命令 node -v，检查是否安装成功，如图 13.2 所示。

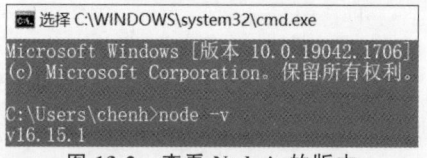

图 13.2　查看 Node.js 的版本

如图 13.2 所示，出现了 Node.js 的版本号，说明 Node.js 已经安装成功。同时，NPM 包管理器也已经安装成功，可以输入 npm -v 查看版本号。最后输入 npm -g install npm 命令，可以将 NPM 更新至最新版本。

▶ 13.2.2　NPM 的常用命令

NPM（Node Package Manager）是随 Node.js 一起安装的包管理工具，方便下载、安装、上传及管理包，解决了 Node 代码部署上的很多问题，常用于以下三种场景。

- 允许用户从 NPM 服务器下载别人编写的第三方包到本地使用。
- 允许用户从 NPM 服务器下载并安装别人编写的命令行程序到本地使用。
- 允许用户将自己编写的包或命令行程序上传到 NPM 服务器供别人使用。

NPM 的背后有一个开源的面向文档的数据库管理系统支撑,详细地记录了每个包的信息,包括作者、版本、依赖、授权等信息。它的作用是将开发者从烦琐的包管理工作中解放出来,从而更加专注于功能业务的开发。

下面讲解 NPM 的常用命令。

❶ **检测是否安装及版本**

```
npm -v                                          #显示版本号说明已经安装相应的版本
```

❷ **生成 package.json 文件**

package.json 用来描述项目中用到的模块和其他信息,可使用如下命令生成该文件。

```
npm init
```

❸ **安装模块**

安装 package.json 定义好的模块,简写为 npm i。

```
npm install
```

安装包指定模块,具体命令如下。

```
npm i <ModuleName>
```

全局安装,具体命令如下。

```
npm i <ModuleName> -g
```

在安装包的同时将信息写入 package.json 中的 dependencies 配置,具体命令如下。

```
npm i <ModuleName> --save
```

在安装包的同时将信息写入 package.json 中的 devDependencies 配置,具体命令如下。

```
npm i <ModuleName> --save-dev
```

安装多模块,具体命令如下。

```
npm i <ModuleName1> <ModuleName2>
```

安装方式参数,具体如下。

```
-save                     #简写为-S,加入生产依赖中
-save-dev                 #简写为-D,加入开发依赖中
-g                        #全局安装,将安装包放在 /usr/local 下或者 Node 的安装目录
```

❹ **查看**

查看所有全局安装的包,具体命令如下。

```
npm ls -g
```

查看本地项目中安装的包,具体命令如下。

```
npm ls
```

查看包的 package.json 文件,具体命令如下。

```
npm view <ModuleName>
```

查看包的依赖关系,具体命令如下。

```
npm view <ModuleName> dependencies
```

查看包的源文件地址,具体命令如下。

```
npm view <ModuleName> repository.url
```

查看包所依赖的 Node 版本,具体命令如下。

```
npm view <ModuleName> engines
```

查看帮助,具体命令如下。

```
npm help
```

❺ 更新模块

更新本地模块,具体命令如下。

```
npm update <ModuleName>
```

更新全局模块,具体命令如下。

```
npm update -g <ModuleName>              #更新全局软件包
npm update -g                           #更新所有的全局软件包
npm outdated -g --depth=0               #找出需要更新的包
```

❻ 卸载模块

卸载本地模块,具体命令如下。

```
npm uninstall <ModuleName>
```

卸载全局模块,具体命令如下。

```
npm uninstall -g <ModuleName>           #卸载全局软件包
```

❼ 清空缓存

清空 NPM 缓存,具体命令如下。

```
npm cache clear
```

❽ 使用淘宝镜像

使用淘宝镜像,具体命令如下。

```
npm install -g cnpm --registry=https://registry.npm.taobao.org
```

❾ 其他

更改包内容后进行重建,具体命令如下。

```
npm rebuild <ModuleName>
```

检查包是否已经过时,此命令会列出所有已经过时的包,可以及时进行包的更新,具体命令如下。

```
npm outdated
```

访问 NPM 的 JSON 文件,此命令将会打开一个网页,具体命令如下。

```
npm help json
```

在发布一个包的时候需要检验某个包名是否存在,具体命令如下。

```
npm search <ModuleName>
```

撤销自己发布过的某个版本代码,具体命令如下。

```
npm unpublish <package> <version>
```

❿ 使用技巧

多次安装不成功尝试先清除缓存,具体命令如下。

```
npm cache clean -f
```

查看已安装的依赖包的版本号,具体命令如下。

```
npm ls <ModuleName>
```

注意:用此方法才能准确地知道项目使用的版本号,在查看 package.json 时,有"^"符号表示高于此版本。

13.3 Vue Router

Vue Router 是 Vue.js 官方的路由管理器,它和 Vue.js 的核心深度集成,使构建单页面应用变得更加容易。Vue Router 包含的功能如下。

(1) 嵌套路由映射。
(2) 动态路由选择。
(3) 模块化、基于组件的路由配置。
(4) 路由参数、查询、通配符。
(5) 展示由 Vue.js 的过渡系统提供的过渡效果。
(6) 细粒度的导航控制。
(7) 自动激活 CSS 类的链接。
(8) HTML5 history 模式或 hash 模式。
(9) 可定制的滚动行为。
(10) URL 的正确编码。

▶13.3.1 Vue Router 的安装

将 Vue Router 添加到项目中主要有 4 种方法,分别为本地独立版本方法、CDN 方法、NPM 方法以及命令行工具(Vue CLI)方法。

❶ 本地独立版本方法

用户可通过地址"https://unpkg.com/vue-router@next"将最新版本的 Vue Router 库(vue-router.global.js)下载到本地(在页面上右击,在弹出的快捷菜单中选择"另存为"),在编写本书时,其最新版本是 4.0.13。然后在界面文件中引入 vue-router.global.js 库,示例代码如下。

```
<script src="js/vue-router.global.js"></script>
```

❷ CDN 方法

读者在学习或开发时,可在界面文件中通过 CDN(Content Delivery Network,内容分发网络)引入最新版本的 Vue Router 库,示例代码如下。

```
<script src="https://unpkg.com/vue-router@next"></script>
```

对于生产环境,建议使用固定版本,以免因版本不同带来兼容性问题,示例代码如下。

```
<script src="https://unpkg.com/vue-router@4.0.13/dist/vue-router.global.js"></script>
```

❸ **NPM 方法**

在使用 Vue.js 构建大型应用时推荐使用 NPM 安装最新稳定版的 Vue Router,因为 NPM 能很好地和 webpack 模块打包器配合使用,示例代码如下。

```
npm install vue-router@next
```

❹ **命令行工具(Vue CLI)方法**

为了提高单页面应用程序的开发效率,这里使用 Vue CLI(Vue 脚手架)搭建 Vue.js 项目。Vue CLI 是一个基于 Vue.js 进行快速开发的完整系统,提供如下功能。

- 通过 @vue/cli 实现交互式项目脚手架。
- 通过 @vue/cli + @vue/cli-service-global 实现零配置原型开发。
- 一个运行时依赖 @vue/cli-service,该依赖可升级,基于 webpack 构建,并带有合理的默认配置;可通过项目的配置文件进行配置;可通过插件进行扩展。
- 一个丰富的官方插件集合,集成了前端生态工具。
- 提供了一套创建和管理 Vue.js 项目的用户界面。

Vue CLI 致力于将 Vue.js 生态工具基础标准化,确保各种构建工具平稳衔接,让开发者专注于应用的撰写上,而不必纠结配置的问题。下面讲解如何安装 Vue CLI 以及如何使用 Vue CLI 创建 Vue.js 项目,具体步骤如下。

1) 全局安装 Vue CLI

打开 cmd 命令行窗口中输入命令 npm install -g @vue/cli 全局安装 Vue 脚手架,输入命令 vue --version 查看版本(测试是否安装成功)。如果需要升级全局的 Vue CLI,在 cmd 命令行窗口中运行 npm update -g @vue/cli 命令即可。

2) 打开图形化界面

安装成功后,在命令行窗口中继续输入命令 vue ui 打开一个浏览器窗口,并以图形化界面引导至项目创建的流程,如图 13.3 所示。

图 13.3 Vue CLI 图形化界面

3) 创建项目

在图 13.3 中单击"创建"按钮进入创建项目界面,如图 13.4 所示。

在图 13.4 中输入并选择项目位置信息,然后单击"＋ 在此创建新项目"进入项目详情界面,如图 13.5 所示。

图 13.4　创建项目界面

图 13.5　项目详情界面

在图 13.5 中输入并选择项目的相关信息,然后单击"下一步"按钮进入项目预设界面,选择手动,单击"下一步"按钮进入功能界面,在功能界面中激活 Router,安装 vue-router 插件,为本节后续内容做准备,如图 13.6 所示。

在图 13.6 中单击"下一步"按钮进入项目配置界面,在配置后单击"创建项目"按钮即可完成项目 router-demo 的创建(可能需要一定的创建时间),如图 13.7 所示。

4)使用 VS Code 打开项目

使用 VS Code 打开(File→Open Folder,选择项目目录)第 3 步创建的项目 router-demo。打开后,在 Terminal 终端输入 npm run serve 命令启动服务,如图 13.8 所示。

5)运行项目

在浏览器的地址栏中访问 http://localhost:8080/即可运行项目 router-demo,如图 13.9 所示。

当通过 http://localhost:8080/访问时,打开的页面是 public 目录下的 index.html。index.html 是一个普通的 HTML 文件,让它与众不同的是"<div id="app"></div>"这句代码,下

图 13.6　项目功能界面

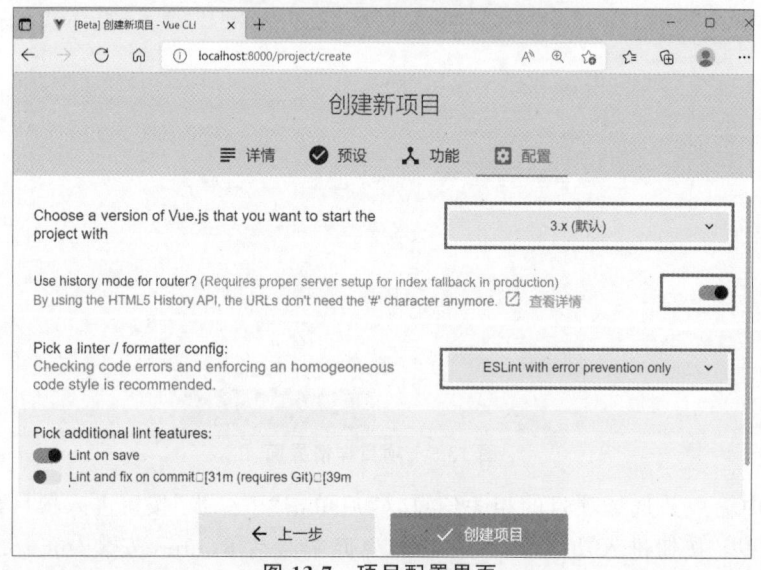

图 13.7　项目配置界面

面有一行注释,构建的文件将会被自动注入,也就是说用户编写的其他内容将都在这个 div 中展示。另外,整个项目只有这一个 HTML 文件,所以这是一个单页面应用,当打开这个应用时,表面上可以看到很多页面,实际上它们都在这一个 div 中显示。

在 main.js 中创建了一个 Vue 对象。该 Vue 对象的挂载目标是"#app"(与 index.html 中的 id="app"对应);router 代表该对象包含 Vue Router,并使用项目中定义的路由(在 src/router 目录下的 index.js 文件中定义)。

综上所述,main.js 与 index.html 是项目启动的首加载页面资源与 JS 资源,App.vue 则是 Vue 页面资源的首加载项,称为根组件。Vue 项目的具体执行过程是,首先启动项目,找到 index.html 与 main.js,执行 main.js(入口程序),根据 import 加载 App.vue 根组件;然后将组件内容渲染到 index.html 中 id="app"的 DOM 元素上。

第 13 章　Vue.js 进阶

图 13.8　启动服务

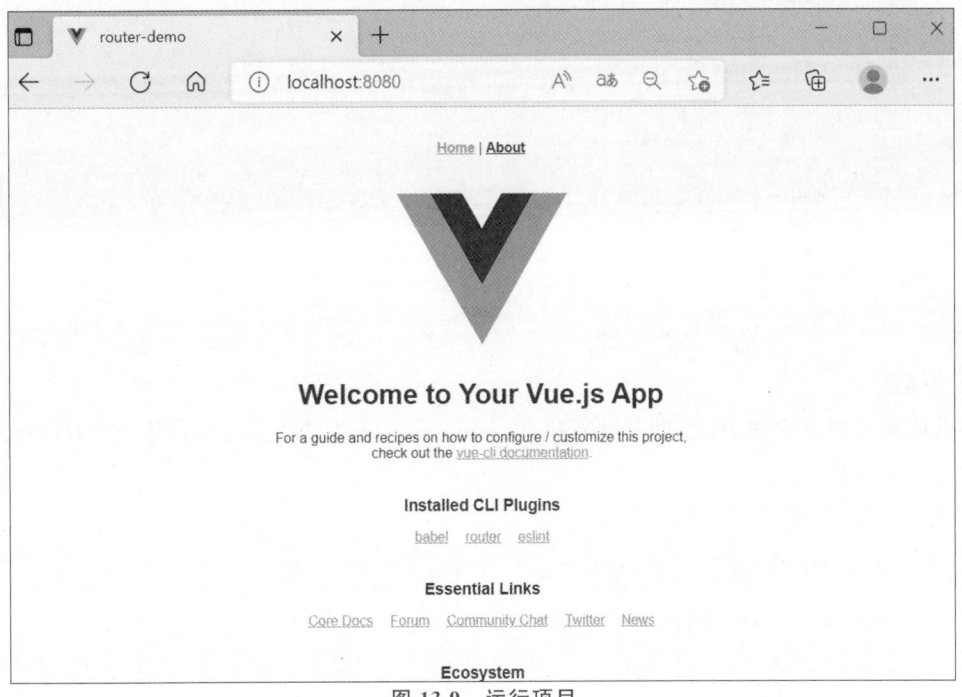

图 13.9　运行项目

▶13.3.2　Vue Router 的基本用法

在使用 Vue Router 动态加载不同组件时，需要将组件（Components）映射到路由（Routers），然后告诉 Vue Router 在哪里显示它们。

❶ 跳转

Vue Router 有两种跳转，第一种是使用内置的＜router-link＞组件，默认渲染一个＜a＞标签，示例代码如下。

```
<div id="nav">
    <router-link to="/">第一个页面</router-link> |
    <router-link to="/MView2">第二个页面</router-link>
</div>
```

<router-link>组件和一般组件一样，to 是一个 prop，指定跳转的路径。使用<router-link>组件，在 HTML5 的 history 模式下将拦截点击，避免浏览器重新加载页面。<router-link>组件还有如下常用属性。

（1）tag 属性：指定渲染的标签，例如<router-link to="/" tag="li">，渲染的结果是<li>而不是<a>。

（2）replace 属性：使用 replace 不会留下 history 记录，所以在导航后不能用后退键返回上一个页面，例如<router-link to="/" replace>。

Vue Router 的第二种跳转方式需要在 JavaScript 中进行，类似于 window.location.href。这种方式需要使用 router 实例方法 push 或 replace。例如，在 MView1.vue 中通过点击事件跳转，示例代码如下。

```
<template>
  <div>第一个页面</div>
  <button @click="goto">去第二个页面</button>
</template>
<script>
export default {
  methods: {
    goto() {
      //也可以使用 replace 方法，与 replace 属性一样不会向 history 中添加新记录
      this.$router.push('/MView2')
    }
  }
}
</script>
```

❷ 传参

路由传参一般有两种方式，即 query 和 params，不管哪种方式都是通过修改 URL 来实现。

1）query 传参

query 传递参数的示例代码如下：

```
<router-link to="/?id=888&pwd=999">
```

通过 $route.query 获取路由中的参数，示例代码如下：

```
<h4>id:{{$route.query.id}}</h4>
<h4>pwd:{{$route.query.pwd}}</h4>
```

2）params 传参

在路由规则中定义参数，修改路由规则的 path 属性（动态匹配），示例代码如下。

```
{
path: '/:id/:pwd',
name: 'MView1',
component: MView1
}
<router-link to="/888/999">
```

通过 $route.params 获取路由中的参数，示例代码如下：

```
<h4>id:{{$route.params.id}}</h4>
<h4>pwd:{{$route.params.pwd}}</h4>
```

❸ 配置路由

路由配置通常在前端工程项目的 src/router/index.js 文件中进行。首先需要在前端工程项目的 src/main.js 和 src/router/index.js 文件中分别导入 vue 和 vue-router 模块，然后在 main.js 中执行 use 方法注册路由。

main.js 的示例代码如下。

```
import { createApp } from 'vue'
import App from './App.vue'
import router from './router'          //导入router目录中的index.js,路由的创建与配置在该文件中
//将router注册到根实例App
createApp(App).use(router).mount('#app')
```

index.js 的示例代码如下。

```
import { createRouter, createWebHistory } from 'vue-router'
//导入组件
import HomeView from '../views/HomeView.vue'
//定义路由
const routes = [
  {
    path: '/',
    name: 'home',
    component: HomeView
  },
  {
    path: '/about',
    name: 'about',
    //导入组件
    component: () => import('../views/AboutView.vue')
  }
]
//创建路由实例router(管理路由),传入routes配置
const router = createRouter({
  //用createWebHistory()创建HTML5模式,推荐用户使用这个模式
  history: createWebHistory(process.env.BASE_URL),
  routes
})
export default router
```

下面通过实例讲解路由的跳转、传参以及配置过程。

【例 13-1】 Vue Router 实战：三个组件间的跳转与传参。

本例在 13.3.1 节中 Vue Router 项目 router-demo 的基础上进行，其他步骤具体如下。

1）创建共通组件 CommonView.vue

在 router-demo/src/components 目录下创建信息显示共通组件 CommonView.vue，具体代码如下。

```
<template>
  <div>
    <img alt="Vue logo" src="../assets/logo.png">
    <h1>{{ msg }}</h1>
  </div>
</template>
<script>
```

```
export default {
  name: 'CommonView',
  props: {
    msg: String
  }
}
</script>
```

2）创建视图组件 FirstView.vue

在 router-demo/src/views 目录下创建视图组件 FirstView.vue。在该视图组件中使用 CommonView.vue 组件显示信息，并通过 $route.query 获取路由中的参数，具体代码如下：

```
<template>
  <div>
    <CommonView msg="欢迎来访第一个 View"/>
    <br>
    <h4>uname:{{$route.query.uname}}</h4>
    <h4>pwd:{{$route.query.pwd}}</h4>
  </div>
</template>
<script>
import CommonView from '@/components/CommonView.vue'
export default {
  name: 'FirstView',
  components: {
    CommonView
  }
}
</script>
```

3）创建视图组件 SecondView.vue

在 router-demo/src/views 目录下创建视图组件 SecondView.vue。在该视图组件中使用 CommonView.vue 组件显示信息，并通过 $route.params 获取路由中的参数，具体代码如下：

```
<template>
  <div>
    <CommonView msg="欢迎来访第二个 View"/>
    <br>
    <h4>uname:{{$route.params.uname}}</h4>
    <h4>pwd:{{$route.params.pwd}}</h4>
  </div>
</template>
<script>
import CommonView from '@/components/CommonView.vue'
export default {
  name: 'SecondView',
  components: {
    CommonView
  }
}
</script>
```

4）修改根组件 App.vue

在根组件 App.vue 中通过路由跳转组件＜router-link＞跳转到视图组件 FirstView.vue 和 SecondView.vue，并传递参数，具体代码如下：

```
<template>
  <nav>
```

```
        <router-link to="/first?uname=chenheng&pwd=123456">第一个页面</router-link> |
        <router-link to="/second/chenheng1/654321">第二个页面</router-link>
    </nav>
    <!-- router-view 表示路由出口,将匹配到的组件(相当于链接的页面)渲染在这里 -->
    <router-view/>
</template>
```

5)配置路由

在 router-demo/src/router/index.js 文件中配置根组件 App.vue 中的路由,具体代码如下。

```
import { createRouter, createWebHistory } from 'vue-router'
//导入组件
import SecondView from '../views/SecondView.vue'
//定义路由
const routes = [
  {
    path: '/first',
    name: 'first',
    //导入组件
    component: () => import('../views/FirstView.vue')
  },
  {
    path: '/second/:uname/:pwd',
    name: 'second',
    //导入组件
    component: SecondView
  }
]
//创建路由实例 router(管理路由),传入 routes 配置
const router = createRouter({
  history: createWebHistory(process.env.BASE_URL),
  routes
})
export default router
```

6)运行测试

首先在项目 router-demo 的 Terminal 终端输入 npm run serve 命令启动服务,然后在浏览器的地址栏中输入 http://localhost:8080/,运行效果如图 13.10 所示。

图 13.10 例 13-1 的首页

单击图 13.10 中的"第一个页面"超链接,打开第一个视图组件,如图 13.11 所示。

单击图 13.10 中的"第二个页面"超链接,打开第二个视图组件,如图 13.12 所示。

注意:在本书提供的源代码中,读者可使用 VS Code 打开本章对应的代码目录,并在 Terminal 终端执行 npm install 命令自动安装所有的依赖,然后执行 npm run serve 命令启动服务,即可运行本章实例。

▶13.3.3 Vue Router 的高级应用

❶ 动态路由的匹配

当需要将符合某种匹配模式的所有路由映射到同一个组件时,可以在路由路径中使用动态

图 13.11 例 13-1 的第一个视图组件

图 13.12 例 13-1 的第二个视图组件

路径参数(如 path：'/user/:uname/:pwd')来实现。示例代码如下。

```
//定义路由
const routes = [
  {
    path: '/user/:uname/:pwd',
    name: 'user',
    component: UserView
  }
]
```

在上述示例代码中定义路由后，/user/zhangsan/123456、/user/lisi/654321 等用户都将映射到相同的路由。

每一个动态路径参数使用冒号(：)标记，冒号后面是参数名。当匹配到一个路由时，参数值将被设置到 $ route.params 中，并可以在路由对应的组件内使用"$ route.params.参数名"获得参数值。

第 13 章　Vue.js 进阶

如果有多个参数,即有多个冒号,则 \$route.params 中保存为对象。例如,路由路径 path 为/user/:uname/:pwd,则对应的访问路径为/user/zhangsan/123456,\$route.params 中的对象为{ uname: 'zhangsan', pwd: '123456'}。另外,也可以使用 post 进行多个动态参数传递,例如路由路径 path 为/user/:uname/post/:pwd/post/:age,则对应的访问路径为/user/:lisi/post/:654321/post/:18,\$route.params 中的对象为{ uname: 'lisi', pwd: '123456', age: '18'}。

\$route 路由信息对象表示当前激活的路由状态信息,每次成功导航后都将产生一个新的对象。除了 \$route.params 外,\$route 对象还提供其他许多有用的信息,如表 13.1 所示。

表 13.1　\$route 路由信息对象的属性

序号	属性名称	说　　明
1	\$route.path	对应当前路由的路径,如/third/:张三/post/:654321/post/:18
2	\$route.params	一个 key:value 对象,包含了所有动态参数,如果没有参数,则是一个空对象,如{ "uname": ": 张三", "pwd": ": 654321", "age": ": 18" }
3	\$route.query	一个 key:value 对象,表示 URL 查询参数。例如,/first? uname＝chenheng&pwd＝123456,则 \$route.query.uname 为 chenheng。如果没有查询参数,则是空对象
4	\$route.hash	当前路由的哈希值(不带♯),如果没有哈希值,则为空字符串
5	\$route.fullPath	完成解析后的 URL,包含查询参数和哈希的完整路径
6	\$route.matched	返回数组,包含当前匹配的路径中包含的所有片段所对应的配置
7	\$route.name	当前路径名称
8	\$route.meta	路由元信息

【例 13-2】　Vue Router 实战:\$route 对象的属性。

本例在例 13-1 的基础上进行,其他步骤具体如下。

1) 修改根组件 App.vue

在根组件 App.vue 中使用＜router-link＞添加路由链接,并在该路由链接中使用 post 进行多个动态参数传递。修改后的 App.vue 的具体代码如下。

```
<template>
  <nav>
    <router-link to="/first?uname=chenheng&pwd=123456">第一个页面</router-link> |
    <router-link to="/second/chenheng1/654321">第二个页面</router-link> |
    <router-link to="/third/:张三/post/:654321/post/:18">第三个页面</router-link>
  </nav>
  <router-view/>
</template>
```

2) 创建视图组件 ThirdView.vue

在 router-demo/src/views 目录下创建视图组件 ThirdView.vue。在该视图组件中使用 CommonView.vue 组件显示信息,并显示 \$route 对象的各种属性值,具体代码如下。

```
<template>
  <div>
    <CommonView msg="欢迎来访第三个 View"/>
    <br>
    <h4>uname:{{$route.params.uname}}</h4>
    <h4>pwd:{{$route.params.pwd}}</h4>
    <h4>age: {{$route.params.age}}</h4>
    <h4>route 的 path: {{$route.path}}</h4>
    <h4>route 的 params: {{$route.params}}</h4>
```

```
        <h4>route 的 query: {{$route.query}}</h4>
        <h4>route 的 hash: {{$route.hash}}</h4>
        <h4>route 的 fullPath: {{$route.fullPath}}</h4>
        <h4>route 的 matched: {{$route.matched}}</h4>
        <h4>route 的 name: {{$route.name}}</h4>
        <h4>route 的 meta: {{$route.meta}}</h4>
    </div>
</template>
<script>
import CommonView from '@/components/CommonView.vue'
export default {
  name: 'ThirdView',
  components: {
    CommonView
  }
}
</script>
```

3）添加路由配置

在 router-demo/src/router/index.js 文件中添加路径/third/:张三/post/:654321/post/:18 对应的路由配置，具体代码如下。

```
import { createRouter, createWebHistory } from 'vue-router'
//导入组件
import SecondView from '../views/SecondView.vue'
import ThirdView from '../views/ThirdView.vue'
//定义路由
const routes = [
  {
    path: '/first',
    name: 'first',
    //导入组件
    component: () => import('../views/FirstView.vue')
  },
  {
    path: '/second/:uname/:pwd',
    name: 'second',
    //导入组件
    component: SecondView
  },
  {
    path: '/third/:uname/post/:pwd/post/:age',
    name: 'third',
    //导入组件
    component: ThirdView
  }
]
//创建路由实例 router(管理路由)，传入 routes 配置
const router = createRouter({
  history: createWebHistory(process.env.BASE_URL),
  routes
})
export default router
```

4）运行测试

首先在项目 router-demo 的 Terminal 终端输入 npm run serve 命令启动服务，然后在浏览器的地址栏中输入 http://localhost:8080/，运行效果如图 13.13 所示。

单击图 13.13 中的"第三个页面"超链接，打开第三个视图组件，如图 13.14 所示。

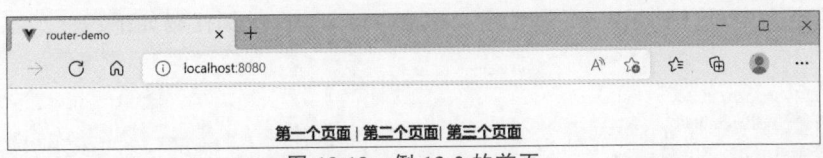

图 13.13　例 13-2 的首页

图 13.14　例 13-2 的第三个视图组件

❷ 嵌套路由

嵌套路由,即路由的多层嵌套,也称为子路由。在实际应用中,嵌套路由相当于多级菜单,一级菜单下有二级菜单,二级菜单下有三级菜单,等等。

创建嵌套路由的步骤一般如下。

首先在根组件 App.vue 中定义基础路由(相当于一级菜单)导航,示例代码如下。

```
<nav>
  <router-link to="/">首页</router-link> |
  <router-link to="/about">关于我们</router-link> |
  <router-link to="/product">产品介绍</router-link>
</nav>
```

然后定义基础路由 product 对应的组件(ProductView.vue),示例代码如下。

```
<template>
    <div>
        <p>
            <!-- 为"产品介绍"定义了嵌套路由-->
            <router-link to="/product/alldev">全栈开发</router-link> |
            <router-link to="/product/JavaEE">Java EE 整合开发</router-link> |
            <router-link to="/product/SpringBoot">Spring Boot 开发</router-link>
        </p>
        <router-view/>
    </div>
</template>
```

最后完成所有嵌套路由组件的定义,并在 router/index.js 文件中定义嵌套路由。在基础路由 product 的定义中，使用 children 属性定义嵌套的子路由，示例代码如下。

```
{
    path: '/product',
    name: 'product',
    component: ProductView,
    children:[                              //子路由
      {
        path: '',                           //空子路由为基础路由的默认显示
        component: () => import('../views/AlldevView.vue')
      },
      {
        path: 'alldev',                     //注意这里没有'/'
        component: () => import('../views/AlldevView.vue')
      },
      {
        path: 'JavaEE',
        component: () => import('../views/JavaEEView.vue')
      },
      {
        path: 'SpringBoot',
        component: () => import('../views/SpringBoot.vue')
      }
    ]
}
```

下面通过一个实例讲解嵌套路由的实现过程。

【例 13-3】 Vue Router 实战：嵌套路由的实现过程。

本例在参考 13.3.1 节创建 Vue Router 项目 nested-routes 的基础上进行，其他步骤具体如下。

1）修改根组件 App.vue

在根组件 App.vue 中使用＜router-link＞添加基础路由链接，修改后的 App.vue 的具体代码如下。

```
<template>
  <h1>嵌套路由</h1>
  <nav>
    <router-link to="/">首页</router-link> |
    <router-link to="/about">关于我们</router-link> |
    <router-link to="/product">产品介绍</router-link>
  </nav>
  <router-view class="my-view"> </router-view>
</template>
```

2）创建视图组件 ProductView.vue

在 nested-routes/src/views 目录下创建视图组件 ProductView.vue。在该视图组件中定义嵌套路由，具体代码如下。

```
<template>
    <div>
      <p>
          <!--定义嵌套路由-->
          <router-link to="/product/alldev">全栈开发</router-link> |
          <router-link to="/product/JavaEE">Java EE 整合开发</router-link> |
          <router-link to="/product/SpringBoot">Spring Boot 开发</router-link>
      </p>
        <router-view/>
```

```
        </div>
</template>
<style scoped>
    p a {
        text-decoration: none;
    }
</style>
```

3）创建视图组件 AlldevView.vue、JavaEEView.vue 和 SpringBoot.vue

在 nested-routes/src/views 目录下分别创建子路由 alldev、JavaEE 和 SpringBoot 对应的视图组件 AlldevView.vue、JavaEEView.vue 和 SpringBoot.vue。

AlldevView.vue 的代码如下。

```
<template>
    <div>
        <img alt="alldev" src="../images/091883-all.jpg" width="200" height="300">
    </div>
</template>
```

JavaEEView.vue 的代码如下。

```
<template>
    <div>
        <img alt="javaee" src="../images/079720-javaee.jpg" width="200" height="300">
    </div>
</template>
```

SpringBoot.vue 的代码如下。

```
<template>
    <div>
        <img alt="springboot" src="../images/083960-springboot.jpg" width="200" height="300">
    </div>
</template>
```

4）添加路由配置

在 nested-routes/src/router/index.js 文件中定义基础路由 product。在基础路由 product 的定义中使用 children 属性定义嵌套的子路由，具体代码如下。

```
import { createRouter, createWebHistory } from 'vue-router'
import HomeView from '../views/HomeView.vue'
import ProductView from '../views/ProductView.vue'
const routes = [
  {
    path: '/',
    name: 'home',
    component: HomeView
  },
  {
    path: '/about',
    name: 'about',
    component: () => import('../views/AboutView.vue')
  },
  {
    path: '/product',
    name: 'product',
    component: ProductView,
    children:[                                              //子路由
```

```
        {
          path: '',                         //空子路由为基础路由的默认显示
          component: () => import('../views/AlldevView.vue')
        },
        {
          path: 'alldev',                   //注意这里没有'/'
          component: () => import('../views/AlldevView.vue')
        },
        {
          path: 'JavaEE',
          component: () => import('../views/JavaEEView.vue')
        },
        {
          path: 'SpringBoot',
          component: () => import('../views/SpringBoot.vue')
        }
      ]
    }
]
const router = createRouter({
  history: createWebHistory(process.env.BASE_URL),
  routes
})
export default router
```

5）运行测试

首先在项目 nested-routes 的 Terminal 终端输入 npm run serve 命令启动服务，然后在浏览器的地址栏中输入 http://localhost:8080/，运行效果如图 13.15 所示。

图 13.15　例 13-3 的首页

单击图 13.15 中的"产品介绍"超链接，打开空子路由对应的 AlldevView.vue 视图组件，如图 13.16 所示。

单击图 13.16 中的"Java EE 整合开发"超链接，在子路由上切换导航，导航到"Java EE 整合开发"视图组件，如图 13.17 所示。

单击图 13.16 中的"Spring Boot 开发"超链接，在子路由上切换导航，导航到"Spring Boot 开发"视图组件，如图 13.18 所示。

第 13 章　Vue.js 进阶

图 13.16　默认显示空子路由对应的视图组件

图 13.17　"Java EE 整合开发"视图组件

❸ 编程式导航

除了使用内置的＜router-link＞组件渲染一个＜a＞标签定义导航链接外，还可以通过编程调用路由（router 或 this.$router）的实例方法实现导航链接。导航常用的路由实例方法如表 13.2 所示。

表 13.2　导航常用的路由实例方法

序号	方法名称	功　能　说　明
1	push()	跳转到由参数指定的新路由地址，在历史记录中添加一条新记录
2	replace()	跳转到由参数指定的新路由地址，替换当前的历史记录

续表

序号	方法名称	功能说明
3	go(n)	n为整数,在历史记录中前进或后退n步
4	forward()	在历史记录中前进一步,相当于this.$router.go(1)
5	back()	在历史记录中后退一步,相当于this.$router.go(−1)

图13.18 "Spring Boot开发"视图组件

　　push()方法和replace()方法的用法相似,唯一不同的是push()方法在历史记录中添加一条新记录,replace()方法不会添加新记录,而是替换当前记录。下面介绍push()方法和replace()方法的参数的意义。

　　push()方法和replace()方法的参数可以是字符串、对象、命名路由、带查询参数等多种形式,示例如下。

```
//字符串路由path
this.$router.push('/')
//对象
this.$router.push({path: '/product'})
//命名路由及params传参,params更像post,是隐性传参
this.$router.push({name: 'home', params:{uname:'123', pwd:'abc'} })
//带查询参数,/product? uname=123&pwd=abc,更像get传参,是显性传参
this.$router.push({path: '/product', query:{uname:'123', pwd:'abc'} })
```

　　注意:在Vue Router更新后,弃用了params传参,可以使用History API方式传递和接收,在跳转前的页面使用state参数,示例代码如下。

```
this.$router.push({name: 'home', state:{uname:'123', pwd:'abc'} })
```

　　在home对应的页面中可以使用history.state接收数据。
　　下面通过一个实例讲解编程式导航的应用。
　　【例13-4】 Vue Router实战:编程式导航的应用。
　　本例在例13-3的基础上修改根组件App.vue即可(本例的项目名称为programming-

navigation），其他与例 13-3 相同，具体实现步骤如下。

1）修改根组件 App.vue

App.vue 的代码如下。

```vue
<template>
  <h1>编程式导航及嵌套路由</h1>
  <!--编程式导航-->
  <button @click="go1">前进一步</button>
  <button @click="back1">后退一步</button>
  <button @click="goHome">回首页</button>
  <button @click="goProduct">看产品介绍</button>
  <button @click="repAbout">代替关于我们</button>
  <nav>
    <router-link to="/">首页</router-link> |
    <router-link to="/about">关于我们</router-link> |
    <router-link to="/product">产品介绍</router-link>
  </nav>
  <router-view class="my-view"> </router-view>
</template>
<script>
export default {
  name: 'App',
  methods: {
    go1(){
      this.$router.forward()
    },
    back1(){
      this.$router.back()
    },
    goHome(){
      this.$router.push('/')                //字符串路由 path
    },
    goProduct(){
      this.$router.push({                   //对象
        path: '/product'
      })
    },
    repAbout(){
      this.$router.replace({
        name: 'home'                        //命名路由
      })
    }
  }
}
</script>
```

2）运行测试

首先在项目 programming-navigation 的 Terminal 终端输入 npm run serve 命令启动服务，然后在浏览器的地址栏中输入 http://localhost:8080/，运行效果如图 13.19 所示。

❹ 命名路由

在链接一个路由或执行跳转时，对路由定义一个名称（name）将显得方便一些。命名路由的示例代码如下。

```
{
  path: '/',
  name: 'home',
  component: HomeView
}
```

图 13.19 例 13-4 的首页

如果需要链接到一个命名路由,可以给<router-link>的：to 属性传递一个对象,示例代码如下(注意 to 前面的冒号)。

```
<router-link :to="{name: 'home', params: {uname: '123', pwd: 'abc'}}">首页</router-link>
```

上述示例代码与编程式导航 this.$router.push({name: 'home', params: {uname: '123', pwd: 'abc'} })的功能相同。

❺ 重定向

通过路由配置可完成路由的重定向。例如,实现从/first 重定向到/second,路由配置代码如下。

```
{
    path: '/first',
    redirect: '/second',
    name: 'first',
    component: FirstView
}
```

重定向的目标也可以是一个命名路由,路由配置代码如下。

```
{
    path: '/first',
    redirect: {name: 'second'},
    name: 'first',
    component: FirstView
}
```

❻ 路由组件 props 传参

在组件中使用$route,将使路由与组件形成高度耦合,从而使组件只能在某些特定的 URL 上使用,限制了组件的灵活性。在配置路由时使用 props 传参,可降低路由与组件的耦合度。路由组件传参的具体示例如下。

1) 导航组件

假设导航组件中有如下链接。

```
<router-link to="/">首页</router-link>
```

2) 使用 props 传参

可以使用 props 配置 URL"/"对应的路由,并传递参数给组件 HomeView,具体代码如下。

```
{
    path: '/',
    name: 'home',
    component: HomeView,
    props: {uname: '张三', upwd: '123456'}
}
```

3）通过 props 接收参数

最后在目标视图组件 HomeView 中通过 props 接收参数，具体代码如下。

```
<template>
  <div class="home">
    <h1>{{uname}}</h1>
    <h1>{{upwd}}</h1>
  </div>
</template>
<script>
export default {
  name: 'HomeView',
  props: {
    uname: {type: String, default: 'lisi'},
    upwd: {type: String, default: '000000'}
  }
}
</script>
```

▶13.3.4 路由钩子函数

在路由跳转时，可能需要一些权限判断或者其他操作，这时需要使用路由钩子函数。路由钩子函数主要是给使用者在路由发生变化时进行一些特殊的处理所定义的函数，又称为路由守卫或导航守卫。

❶ 全局前置钩子函数

在 Vue Router 中，使用 router.beforeEach 注册一个全局前置钩子函数（在路由跳转前执行），注册示例代码如下。

```
const router = new createRouter({ ... })
router.beforeEach((to, from) => {
    //...
    //返回 false,以取消导航
    return false
})
```

当一个导航触发时，全局前置钩子函数按照创建顺序调用。beforeEach 函数接收两个参数，具体如下。

- to：Route：即将要进入的目标路由对象。
- from：Route：当前导航正要离开的路由。

beforeEach 函数可以返回 false 值或一个路由地址，具体如下。

- false：取消当前导航。如果浏览器的 URL 改变（用户手动或者单击浏览器的后退按钮），那么 URL 地址将重置到 from 路由对应的地址。
- 一个路由地址：跳转到该路由地址，即当前的导航被中断，进行一个新的导航。

例如，使用 beforeEach 函数检查用户是否登录，示例代码如下。

```
router.beforeEach(async (to, from) => {
    //在 ES7 标准中新增了 async 和 await 关键字，作为处理异步请求的一种解决方案
```

```
    if (
        //检查用户是否已登录
        !isAuthenticated &&
        //避免无限重定向
        to.name !== 'Login'
    ) {
        //将用户重定向到登录页面
        return { name: 'Login' }
    }
})
```

在之前的 Vue Router 版本中，beforeEach 需要使用第三个参数 next，现在是一个可选的参数。next 参数的相关说明如下。

- next()：进行管道中的下一个钩子函数。如果全部钩子函数执行完，则导航的状态就是 confirmed（确认的）。
- next(false)：中断当前的导航。如果浏览器的 URL 改变（可能是用户手动或者单击浏览器的后退按钮），那么 URL 地址会重置到 from 路由对应的地址。
- next('/') 或者 next({ path: '/' })：跳转到一个不同的地址。当前的导航被中断，然后进行一个新的导航。可以向 next 传递任意位置对象，且允许设置诸如 replace：true、name：'home'之类的选项以及任何用在 router-link 的 to 属性或 router.push 中的选项。
- next(error)：如果传入 next 的参数是一个 Error 实例，则导航被终止，且该错误被传递给 router.onError() 注册过的回调。

确保 next() 函数在任何给定的前置钩子中被严格调用一次。它可以出现多次，但是只能在所有逻辑路径都不重叠的情况下，否则钩子永远都不会被解析或报错。例如，在用户未能验证身份时重定向到 /login 的示例。

```
router.beforeEach((to, from, next) => {
    if (to.name !== 'Login' && !isAuthenticated)
        next({ name: 'Login' })
    else
        next()
})
```

❷ 全局解析钩子函数

在 Vue Router 中，使用 router.beforeResolve 注册一个全局解析钩子函数。与 router.beforeEach 类似，它在每次导航时都会触发，但是要确保在导航被确认之前，同时在所有组件内钩子函数和异步路由组件被解析之后，解析钩子函数被正确调用。例如，确保用户可以访问自定义的路由元信息（meta 属性）requiresCamera 的路由，具体代码如下。

```
router.beforeResolve(async to => {
    if (to.meta.requiresCamera) {
        try {
            await askForCameraPermission()
        } catch (error) {
            if (error instanceof NotAllowedError) {
                //... 处理错误，然后取消导航
                return false
            } else {
                //意料之外的错误，取消导航并把错误传给全局处理器
                throw error
            }
        }
    }
})
```

```
    }
  }
})
```

router.beforeResolve 是获取数据或执行任何其他操作（例如，如果用户无法进入页面，希望避免执行的操作）的理想位置。

❸ **全局后置钩子函数**

在 Vue Router 中也可以使用 router.afterEach 注册全局后置钩子函数，该钩子函数不接受 next 参数，也不会改变导航本身，在跳转之后判断。对于分析、更改页面标题、声明页面等辅助功能都很有用。其示例代码如下。

```
router.afterEach((to, from) => {
    //...
})
```

❹ **某个路由的钩子函数**

顾名思义，它是写在某个路由中的函数，本质上跟组件内的函数没有区别。其示例代码如下。

```
const routes = [
  {
    path: '/users/:id',
    component: UserDetails,
    beforeEnter: (to, from) => {
      //取消导航
      return false
    },
  },
]
```

路由的 beforeEnter 钩子函数，只在进入路由时触发，不会在 params、query 或 hash 改变时触发。例如，从 /users/2 进入 /users/3 或者从 /users/2#info 进入 /users/2#projects。它们只有在从一个不同的路由导航时才会被触发。

另外，也可以将一个函数数组传递给路由的 beforeEnter 钩子函数，这在为不同的路由重用钩子函数时很有用，示例代码如下。

```
function removeQueryParams(to) {
  if (Object.keys(to.query).length)
    return { path: to.path, query: {}, hash: to.hash }
}
function removeHash(to) {
  if (to.hash) return { path: to.path, query: to.query, hash: '' }
}
const routes = [
  {
    path: '/users/:id',
    component: UserDetails,
    beforeEnter: [removeQueryParams, removeHash],
  },
  {
    path: '/about',
    component: UserDetails,
    beforeEnter: [removeQueryParams],
  },
]
```

❺ 组件内的钩子函数

可以在路由组件内直接定义路由导航钩子函数 beforeRouteEnter、beforeRouteUpdate、beforeRouteLeave，具体示例代码如下。

```
const UserDetails = {
  template: '...',
  beforeRouteEnter(to, from) {
    //在渲染该组件的对应路由被验证前调用
    //不能获取组件实例 'this' !
    //因为当该钩子函数执行时,组件实例还没被创建!
  },
  beforeRouteUpdate(to, from) {
    //在当前路由改变,但是该组件被复用时调用
    //举例来说,对于一个带有动态参数的路径'/users/:id',在'/users/1'和'/users/2'之间跳转的时候,
    //由于渲染同样的'UserDetails'组件,所以组件实例会被复用,该钩子函数在此情况下也被调用。
    //因为在这种情况发生的时候,组件已经挂载好了,该钩子函数可以访问组件实例'this'
  },
  beforeRouteLeave(to, from) {
    //在导航离开渲染该组件的对应路由时调用
    //与'beforeRouteUpdate'一样,它可以访问组件实例'this'
  },
}
```

beforeRouteEnter 钩子函数不能访问 this，因为 beforeRouteEnter 在导航确认前被调用，所以即将登场的新组件还没有被创建。用户可以通过传一个回调给 next()函数来访问组件实例。在导航被确认的时候执行回调，并且把组件实例作为回调方法的参数，具体示例代码如下。

```
beforeRouteEnter(to, from, next) {
  next(vm => {
    //通过 'vm' 访问组件实例
  })
}
```

beforeRouteEnter 是支持给 next()函数传递回调的唯一钩子函数。对于 beforeRouteUpdate 和 beforeRouteLeave 来说，this 已经可以使用了，所以不支持传递回调，也没有必要。其示例代码如下。

```
beforeRouteUpdate (to, from) {
  //使用 'this'
  this.name = to.params.name
}
```

beforeRouteLeave 钩子函数通常用来预防用户在还未保存修改前突然离开，导航可以通过返回 false 来取消离开操作。其具体示例代码如下。

```
beforeRouteLeave(to, from) {
  const answer = window.confirm('真的离开? 你还没保存修改!')
  if (!answer)
    return false
}
```

本节只是简单地介绍钩子函数的分类与定义，具体应用在 13.3.6 节中。

▶13.3.5 路由元信息

有时，用户可能希望将任意信息附加到路由上，如过渡名称、访问路由权限等。这些工作可以通过接收属性对象的 meta 属性来实现，并且它可以在路由地址和导航守卫（路由钩子函数）

中被访问到。路由的 meta 属性配置示例如下。

```
const routes = [
  {
    path: '/posts',
    component: PostsLayout,
    children: [
      {
        path: 'new',
        component: PostsNew,
        //只有经过身份验证的用户才能创建帖子
        meta: { requiresAuth: true }
      },
      {
        path: ':id',
        component: PostsDetail
        //任何人都可以阅读帖子
        meta: { requiresAuth: false }
      }
    ]
  }
]
```

那么如何访问路由的 meta 属性呢？

将 routes 配置中的每个路由对象称为路由记录，路由记录可以是嵌套的，因此当一个路由匹配成功后，它可能匹配多个路由记录。

例如，根据上面的路由配置，/posts/new 这个 URL 将会匹配父路由记录（path：'/posts'）以及子路由记录（path：'new'）。

一个路由匹配到的所有路由记录被暴露为 $route 对象的 $route.matched 数组。用户需要遍历这个数组来检查路由记录中的 meta 字段，令人兴奋的是，Vue Router 还提供了一个 $route.meta 方法，它是一个非递归合并所有 meta 字段（从父字段到子字段）的方法。因此，用户可以通过 $route.meta 方法简单地获取路由的 meta 属性值，具体示例代码如下。

```
router.beforeEach((to, from) => {
  //不是去检查每条路由记录
  //to.matched.some(record => record.meta.requiresAuth)
  if (to.meta.requiresAuth && !auth.isLoggedIn()) {
    //此路由需要授权，请检查是否已登录
    //如果没有，则重定向到登录页面
    return {
      path: '/login',
      //保存当前路由所在的位置，以便再回来
      query: { redirect: to.fullPath },
    }
  }
})
```

▶13.3.6 登录权限验证实例

登录权限验证实例的具体要求如下。

- 在 App.vue 根组件中通过＜router-link＞访问登录页面组件 Login.vue、主页面组件 Main.vue 以及 Home.vue 组件。
- 在登录成功后才能访问主页面组件 Main.vue 和 Home.vue 组件。
- 在 main.js 中使用路由钩子函数 beforeEach(to,from) 实现登录权限验证。

【例 13-5】 登录权限验证实例。

其具体实现步骤如下。

❶ **使用 Vue CLI 搭建基于 Router 功能的项目**

参考 13.3.1 节使用 Vue CLI 搭建基于 Router 功能（见图 13.6）的项目 login-validate。

❷ **完善 App.vue**

完善项目 login-validate 的根组件 App.vue 的模板代码，具体如下。

```
<template>
  <nav>
    <router-link to="/login">Login</router-link> |
    <router-link to="/main">Main</router-link> |
    <router-link to="/home">Home</router-link>
  </nav>
  <router-view/>
</template>
```

❸ **配置路由**

在 src/router 目录的 index.js 文件中配置路由，需要登录验证的路由使用 meta 元信息标注。配置路由的具体代码如下。

```
import { createRouter, createWebHistory } from 'vue-router'
import Login from '../views/LoginView.vue'
import Main from '../views/MainView.vue'
import Home from '../views/HomeView.vue'
const routes = [
  {
    path: '/login',
    name: 'Login',
    component: Login
  },
  {
    path: '/home',
    name: 'Home',
    component: Home,
    meta:{auth:true}
  },
  {
    path: '/main',
    name: 'Main',
    component: Main,
    meta:{auth:true}                                           //需要验证登录权限
  }
]
const router = createRouter({
  history: createWebHistory(process.env.BASE_URL),
  routes
})
export default router
```

❹ **登录权限验证**

在配置文件 main.js 中，使用路由钩子函数 beforeEach(to,from) 实现登录权限验证，具体代码如下。

```
import { createApp } from 'vue'
import App from './App.vue'
import router from './router'           //导入 router 目录中的 index.js,路由的创建与配置在该文件中
createApp(App).use(router).mount('#app')
//eslint-disable-next-line no-unused-vars
```

```
router.beforeEach((to,from)=>{
        //提示未使用,ESLint 规则 no-unused vars 关闭为 eslint-disable-next-line
    //如果路由器需要验证
    if(to.meta.auth){
      //对路由进行验证
      if (window.sessionStorage.getItem('user') == null) {
        alert("您没有登录,无权访问!")
        /* 未登录则跳转到登录页面,
        query:{ redirect: to.fullPath}表示把当前路由信息传递过去,方便登录后跳转回来 */
        return {
          path: 'login',
          query: {redirect: to.fullPath}
        }
      }
    }
})
```

❺ 新建登录组件 LoginView.vue

在 views 目录中新建登录组件 LoginView.vue,在该组件中使用 window.sessionStorage.setItem()保存登录状态,LoginView.vue 的代码具体如下。

```
<template>
  <div>
    <h2>登录页面</h2>
    <form>
      用户名:<input type="text" v-model="uname" placeholder="请输入用户名"/><br><br>
      密码: <input type="password" v-model="upwd" placeholder="请输入密码"/><br><br>
      <button type="button" @click="login" :disabled="isDisable">登录</button>
      <button type="reset">重置</button>
    </form>
  </div>
</template>
<script>
export default {
  data() {
    return {
      isDisable:false,
      uname: '',
      upwd: ''
    }
  },
  methods: {
    login() {
      this.isDisable = true
        if (this.uname === 'zhangsan' && this.upwd == '123456') {
          alert('登录成功')
          //将成功登录的用户信息保存到 session
          window.sessionStorage.setItem('user', this.uname)
          //到达成功之后的页面
          let path = this.$route.query.redirect
          this.$router.replace({path: path === '/' || path === undefined ? '/main': path})
        }else {
          alert("用户名或密码错误!")
          this.isDisable = false
        }
    }
  }
}
</script>
```

❻ **新建主页面组件 MainView.vue**

在 views 目录中新建主页面组件 MainView.vue，MainView.vue 的代码具体如下：

```
<template>
  <div>欢迎{{uname}}登录成功</div>
</template>
<script>
export default {
  data() {
    return {
      uname : window.sessionStorage.getItem('user')
    }
  }
}
</script>
```

❼ **测试运行**

在 Terminal 终端输入 npm run serve 命令启动服务，然后在浏览器的地址栏中访问 http://localhost:8080/即可运行项目 login-validate。在登录页面中输入用户名 zhangsan、密码 123456，即可成功登录。登录成功后打开主页面组件，如图 13.20 所示。

在 login-validate 下执行 npm install 命令，将自动安装所有的依赖，然后执行 npm run serve 命令启动服务，即可运行项目。

图 13.20 主页面

13.4 setup 语法糖

在 setup 选项中，属性、方法等必须通过 return 返回暴露出来，然后在 template 中才能使用，很不友好。Vue.js 3.2 增加了 setup 语法糖，即在＜script＞标签中添加 setup。

在使用 setup 语法糖时，组件只需要引入不需要注册，属性和方法不需要 return 返回，不用再写 setup 函数，也不用写 export default 默认输出。

setup 语法糖是在单文件组件中使用组合式 API 的编译时语法糖。与普通的＜script＞语法相比，它具有更多优势：

（1）更少的 template 内容，更简洁的代码。

（2）能够使用纯 TypeScript 声明 props 和抛出事件。

（3）更好的运行时性能（其 template 将被编译成与其同一作用域的渲染函数，没有任何中间代理）。

（4）更好的 IDE 类型推断性能（减少语言服务器从代码中抽离类型的工作）。

许多基于 Vue.js 的开源 UI 组件库的官方示例都是使用 setup 语法糖编写的。为了方便学习 Vue UI 组件库，先学习 setup 语法糖知识。

▶13.4.1 属性与方法的绑定

在使用 setup 语法糖时，属性与方法不需要 return 返回暴露出来，可以直接在 template 中使用。下面通过一个实例讲解在使用 setup 语法糖时如何进行属性与方法的绑定。因为 setup 语法糖是在单文件组件中使用组合式 API 的编译时语法糖，所以本节实例在使用 Vue CLI 搭建的 Vue.js 项目的单文件组件中演示。

【例 13-6】 在使用 setup 语法糖时进行属性与方法的绑定。

其具体步骤如下。

❶ **使用 Vue CLI 搭建基于 Router 功能的 Vue.js 项目**

参考 13.3.1 节，使用 Vue CLI 搭建基于 Router 功能的 Vue.js 项目 setup-sugar。

❷ **修改根组件 App.vue**

在 setup-sugar 的根组件 App.vue 中定义一个非响应式基本数据属性 firstNumber、一个响应式基本数据属性 secondNumber、一个响应式对象（复杂数据）属性 thirdNumber 和一个改变属性的方法 changeNumber，并使用 setup 语法糖进行属性与方法的绑定。App.vue 的代码具体如下。

```
<template>
  <div>firstNumber是一个基本类型数据:{{firstNumber}}</div>
  <div>secondNumber是一个响应式且可改变的基本数据对象:{{secondNumber}}</div>
  <div>thirdNumber是一个响应式且可改变的对象数据:{{thirdNumber}}</div>
  <div><button @click="changeNumber">改变数据</button></div>
</template>
<!-- 只需要添加 setup 属性 -->
<script setup>
//注意,与setup函数一样,在使用setup语法糖时没有this
import { ref,reactive } from 'vue'
//不使用 ref 定义的变量
let firstNumber = 100
//使用ref定义的变量,是一个响应式且可改变的基本数据对象(也可以是对象)
let secondNumber = ref(100)
//使用reactive定义的变量,是一个响应式且可改变的对象
let thirdNumber = reactive({
  uname: '陈恒',
  age: '58'
})
//定义一个无参数的方法,在使用setup语法糖时,方法也被认为属性
let changeNumber = ()=> {
  firstNumber++
  //在模板中使用时,它会自动开箱,无须在模板内额外书写.value
  secondNumber.value++
  thirdNumber.age++
}
</script>
```

❸ **测试运行**

首先在 setup-sugar 项目的 Terminal 终端输入 npm run serve 命令启动服务，然后在浏览器的地址栏中访问 http://localhost:8080/，即可运行项目 setup-sugar 的根组件 App.vue，页面显示效果如图 13.21 所示。

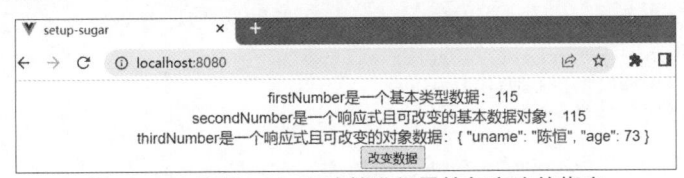

图 13.21　使用 setup 语法糖进行属性与方法的绑定

13.4.2　路由

大家知道可以使用 this.$router.push 进行编程式导航，但使用 setup 语法糖时并没有 this，如何进行编程式导航呢？具体做法是，首先使用 import 语句引入 useRouter，即 import {useRouter} from 'vue-router';然后使用 useRouter 创建路由对象，即 const router = useRouter();

最后使用路由对象 router 进行编程式导航。

下面通过一个实例讲解在使用 setup 语法糖时如何进行编程式导航。

【例 13-7】 在使用 setup 语法糖时如何进行编程式导航。

该例在例 13-6 的基础上完成，其他具体步骤如下。

❶ **修改根组件 App.vue**

在 setup-sugar 的根组件 App.vue 中，首先使用 import 语句引入 useRouter，然后使用 useRouter 创建路由对象 router，最后使用路由对象 router 进行编程式导航。App.vue 的代码如下。

```
<template>
    <div><button @click="goToNewPage">去 Vue 主页 home</button></div>
    <router-view/>
</template>
<script setup>
//记住是在 vue-router 中引入 useRouter
import {useRouter} from 'vue-router'
const router = useRouter()
let goToNewPage = ()=>{
    let params = {uname:'123', pwd:'abc'}
    router.push({name: 'home', state: params})
}
</script>
```

❷ **修改路由配置**

修改 setup-sugar 的路由配置文件 /router/index.js，实现编程式导航。index.js 的代码具体如下。

```
...
const routes = [
  {
    path: '/home',
    name: 'home',
    component: HomeView
  },
  ...
```

❸ **修改 HomeView.vue**

在 /views/HomeView.vue 文件的 <template> 中使用 setup 语法糖简化程序。HomeView.vue 的代码具体如下。

```
<template>
  <div class="home">
    <img alt="Vue logo" src="../assets/logo.png">
    <br>uname:{{historyParams.uname}}
    <br>uname:{{historyParams.pwd}}
    <br>
    <HelloWorld msg="Welcome to Your Vue.js App"/>
  </div>
</template>
<script setup>
//组件只需要引入不需要注册
import HelloWorld from '@/components/HelloWorld.vue'
//接收传递过来的数据
const historyParams = history.state
</script>
```

第 13 章　Vue.js 进阶

❹ 测试运行

首先在 setup-sugar 项目的 Terminal 终端输入 npm run serve 命令启动服务，然后在浏览器的地址栏中访问 http://localhost:8080/，即可运行项目 setup-sugar 的根组件 App.vue，再单击"去 Vue 主页 home"按钮，即可通过编程式导航打开 HomeView.vue，HomeView.vue 的显示效果如图 13.22 所示。

图 13.22　HomeView.vue 的显示效果

在例 13-7 中，使用 setup 语法糖时，父组件 HomeView 只需要引入子组件 HelloWorld，并没有注册子组件 HelloWorld。另外，在子组件 HelloWorld 中，使用 export default｛name：'HelloWorld'，props：｛msg：String｝｝接收父组件 HomeView 传递过来的值。但是此时无法在子组件 HelloWorld 中使用 setup 语法糖，即无法获取 props、emit 等。幸运的是，setup 语法糖提供了新的 API（defineProps、defineEmits、defineExpose 等）供用户使用。下面讲解 defineProps、defineEmits、defineExpose 等 API 在组件传值中的具体应用。

▶13.4.3　组件传值

在子组件中，defineProps 用来接收父组件传来的 props；defineEmits 用来声明触发的事件；defineExpose 用来导出数据，暴露于父组件（在父组件中通过 ref='xxx' 的方法来获取子组件实例）。下面通过一个具体实例讲解 defineProps、defineEmits、defineExpose 等 API 在组件传值中的具体应用。

【例 13-8】　defineProps、defineEmits、defineExpose 等 API 在组件传值中的具体应用。

该例在例 13-7 的基础上完成，其他具体步骤如下。

❶ 修改 HomeView.vue

在 /views/HomeView.vue 父组件中引入子组件 /components/HelloWorld.vue，并传值给子组件，同时接收子组件传递的数据与获得子组件的属性值。HomeView.vue 的代码具体如下。

```
<template>
  <div>
    <!-- myAdd 和 myDel 与子组件中使用 defineEmits 声明向父组件抛出的事件名称相同 -->
    <HelloWorld :info="msg" constv="88" @myAdd="myAddAction" @myDel="myDelAction" ref="comRef"/>
    <button @click="getSon">获取子组件的属性值</button>
  </div>
</template>
<script setup>
//组件只需要引入不需要注册
import HelloWorld from '@/components/HelloWorld.vue'
import { ref } from 'vue'
const msg = "传给子组件"
```

```
//接收子组件传递过来的数据
const myAddAction= (msg) => {
  console.log('单击子组件的新增按钮传值:', msg)
}
//接收子组件传递过来的数据
const myDelAction= (msg) => {
  console.log('单击子组件的删除按钮传值:', msg)
}
const comRef = ref()
//获取子组件的属性值
const getSon = () => {
  console.log('获得子组件中的性别:', comRef.value.prop1)
  console.log('获得子组件中的其他信息:', comRef.value.prop2)
}
</script>
```

❷ 修改 HelloWorld.vue

在/components/HelloWorld.vue 子组件中，使用 defineProps 接收父组件 HomeView.vue 的传值，使用 defineEmits 声明向父组件抛出的自定义事件，使用 defineExpose 将数据暴露于父组件。HelloWorld.vue 的代码具体如下。

```
<template>
  <div>
    <h4>接收父组件传值</h4>
    <h4>info:{{ info }}</h4>
    <h4>constv:{{ constv }}</h4>
  </div>
  <div>
    <button @click="add">新增</button>  
    <button @click="del">删除</button>
  </div>
  <div>
    <h4>性别:{{prop1}}</h4>
    <h4>其他信息:{{prop2}}</h4>
  </div>
</template>
<script setup>
import {ref, reactive, defineProps, defineEmits, defineExpose } from 'vue'
const prop1 = ref('男')
const prop2 = reactive({
  uname: '陈恒',
  age: 88
})
//使用 defineExpose 将数据暴露于父组件
defineExpose({
  prop1,
  prop2
})
//接收父组件传值
defineProps({
  info: {
    type: String,
    default: '-----'
  },
  constv: {
    type: String,
    default: '0'
  }
})
```

```
//使用defineEmits声明向父组件抛出的自定义事件
const myemits = defineEmits(['myAdd', 'myDel'])
const add = () => {
  //通过抛出myAdd事件向父组件传值
  myemits('myAdd', '传向父组件的新增数据')
}
const del = () => {
  myemits('myDel', '传向父组件的删除数据')
}
</script>
<style scoped>
h3 {
  margin: 40px 0 0;
}
</style>
```

❸ 测试运行

首先在 setup-sugar 项目的 Terminal 终端输入 npm run serve 命令启动服务，然后在浏览器的地址栏中访问 http://localhost:8080/，即可运行项目 setup-sugar 的根组件 App.vue，再单击"去 Vue 主页 home"按钮，即可打开 HomeView.vue，HomeView.vue 的显示效果如图 13.23 所示。

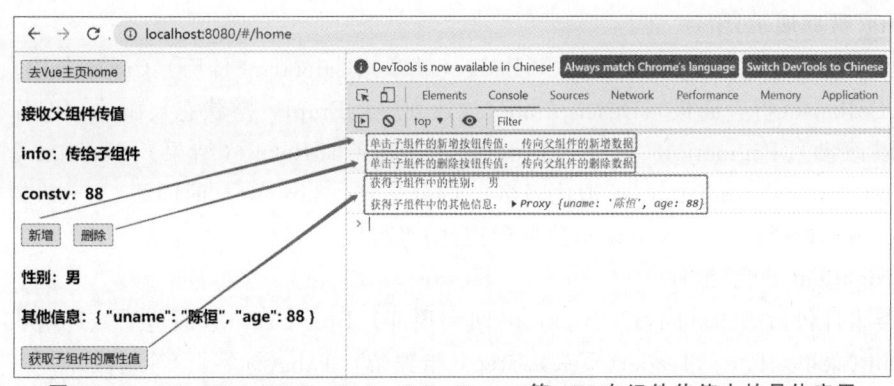

图 13.23　defineProps、defineEmits、defineExpose 等 API 在组件传值中的具体应用

13.5　Element Plus UI 组件库

　　Element Plus 是一套为开发者、设计师和产品经理准备的基于 Vue 3 的组件库，提供了配套设计资源，简化了常用组件的封装，大大降低了开发难度，帮助使用者的网站快速成型。Element Plus 目前还处于快速开发迭代中，其官方文档可以参见官网（https://element-plus.gitee.io/zh-CN/）。

▶13.5.1　Element Plus 的安装

　　建议开发者使用包管理器（如 NPM、YARN、PNPM）安装 Element Plus，这样能更好地和 Vite、WebPack 等打包工具配合使用。根据实际需要选择一个自己喜欢的包管理器，具体安装命令如下。

```
#NPM
$npm install element-plus --save
#YARN
$yarn add element-plus
```

```
# PNPM
$pnpm install element-plus
```

13.5.2 Element Plus 组件的介绍

Element Plus 组件主要包括 Basic（基础）、Form（表单）、Data（数据展示）、Navigation（导航）、Feedback（反馈）五大类组件。每类组件又包含很多组件。下面简单介绍每类组件中所包含的组件。

❶ **Basic（基础）组件**

基础类组件包括 Button（按钮）、Border（边框）、Color（色彩）、Container（布局容器）、Icon（图标）、Layout（布局）、Link（链接）、Scrollbar（滚动条）、Space（间距）、Typography（排版）等组件。

❷ **Form（表单）组件**

表单类组件包括 Cascader（级联选择器）、Checkbox（多选框）、ColorPicker（颜色选择器）、DatePicker（日期选择器）、DateTimePicker（日期时间选择器）、Form（表单）、Input（输入框）、Input Number（数字输入框）、Radio（单选框）、Rate（评分）、Select（选择器）、Select V2（虚拟列表选择器）、Slider（滑块）、Switch（开关）、TimePicker（时间选择器）、TimeSelect（时间选择）、Transfer（穿梭框）、Upload（上传）等组件。

❸ **Data（数据展示）组件**

数据展示类组件包括 Avatar（头像）、Badge（徽章）、Calendar（日历）、Card（卡片）、Carousel（走马灯）、Collapse（折叠面板）、Descriptions（描述列表）、Empty（空状态）、Image（图片）、Infinite Scroll（无限滚动）、Pagination（分页）、Progress（进度条）、Result（结果）、Skeleton（骨架屏）、Table（表格）、Virtualized Table（虚拟化表格）、Tag（标签）、Timeline（时间线）、Tree（树形控件）、TreeSelect（树形选择）、Tree V2（虚拟化树形控件）等组件。

❹ **Navigation（导航）组件**

导航类组件包括 Affix（固钉）、Backtop（回到顶部）、Breadcrumb（面包屑）、Dropdown（下拉菜单）、Menu（菜单）、Page Header（页头）、Steps（步骤条）、Tabs（标签页）等组件。

❺ **Feedback（反馈）组件**

反馈类组件包括 Alert（提示）、Dialog（对话框）、Drawer（抽屉）、Loading（加载）、Message（消息提示）、MessageBox（消息弹框）、Notification（通知）、Popconfirm（气泡确认框）、Popover（气泡卡片）、Tooltip（文字提示）等组件。

13.5.3 Element Plus 组件的应用

Element Plus 提供了一套常用的图标集合，但这些图标默认不在组件中，需要另外安装才能使用，安装方式与 Element Plus 的安装方式相同。

下面通过实例讲解如何在使用 Vue CLI（Vue 脚手架）搭建的 Vue.js 项目中应用 Element Plus 组件。

【例 13-9】 在使用 Vue CLI 搭建的 Vue.js 项目中应用 Element Plus 组件。

其具体步骤如下。

❶ **使用 Vue CLI 搭建 Vue.js 项目**

参考 13.3.1 节，使用 Vue CLI 搭建 Vue.js 项目 elementplus-vue。

❷ **安装 Element Plus 和 @element-plus/icons-vue**

使用 VS Code 打开项目 elementplus-vue，并进入 Terminal 终端，依次执行"npm install element-plus --save"和"npm install @element-plus/icons-vue"命令，进行 Element Plus 和 @element-plus/icons-vue 的安装。

❸ 引入 Element Plus 组件并注册图标组件 ElementPlusIconsVue

在 elementplus-vue 的 main.js 文件中完整引入 Element Plus，并注册图标组件 ElementPlusIconsVue，main.js 的代码具体如下。

```
import { createApp } from 'vue'
import ElementPlus from 'element-plus'
import 'element-plus/dist/index.css'
import * as ElementPlusIconsVue from '@element-plus/icons-vue'
import App from './App.vue'
const app = createApp(App)
//注册所有图标
for (const [key, component] of Object.entries(ElementPlusIconsVue)) {
    app.component(key, component)
}
//使用 ElementPlus
app.use(ElementPlus).mount('#app')
```

❹ 在根组件 App.vue 中使用 Element Plus 组件

在 elementplus-vue 的根组件 App.vue 中使用 Element Plus 组件实现如图 13.24 所示的功能。

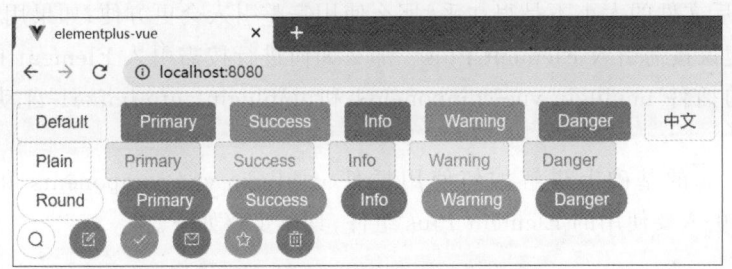

图 13.24 Element Plus 组件的应用

App.vue 的代码具体如下。

```
<template>
  <el-row class="mb-4">
    <el-button>Default</el-button>
    <el-button type="primary">Primary</el-button>
    <el-button type="success">Success</el-button>
    <el-button type="info">Info</el-button>
    <el-button type="warning">Warning</el-button>
    <el-button type="danger">Danger</el-button>
    <el-button>中文</el-button>
  </el-row>
  <el-row class="mb-4">
    <el-button plain>Plain</el-button>
    <el-button type="primary" plain>Primary</el-button>
    <el-button type="success" plain>Success</el-button>
    <el-button type="info" plain>Info</el-button>
    <el-button type="warning" plain>Warning</el-button>
    <el-button type="danger" plain>Danger</el-button>
  </el-row>
  <el-row class="mb-4">
    <el-button round>Round</el-button>
    <el-button type="primary" round>Primary</el-button>
    <el-button type="success" round>Success</el-button>
```

```
        <el-button type="info" round>Info</el-button>
        <el-button type="warning" round>Warning</el-button>
        <el-button type="danger" round>Danger</el-button>
    </el-row>
    <el-row>
        <el-button :icon="Search" circle />
        <el-button type="primary" :icon="Edit" circle />
        <el-button type="success" :icon="Check" circle />
        <el-button type="info" :icon="Message" circle />
        <el-button type="warning" :icon="Star" circle />
        <el-button type="danger" :icon="Delete" circle />
    </el-row>
</template>
<script setup>
import {Check,Delete,Edit,Message,Search,Star,} from '@element-plus/icons-vue'
</script>
```

❺ 运行测试

首先在 elementplus-vue 项目的 Terminal 终端输入 npm run serve 命令启动服务，然后在浏览器的地址栏中访问 http://localhost:8080/，即可运行项目 elementplus-vue。

▶13.5.4 按需引入 Element Plus

在例 13-9 中完整引入 Element Plus，将造成打包后的文件很大，并且首页加载稍慢的问题。如果用户对打包后文件的大小不是很在乎，那么使用完整引入会更方便；如果用户对打包后文件的大小很在乎，建议按需引入 Element Plus。那么如何进行按需引入 Element Plus 呢？推荐开发者使用额外的插件 unplugin-vue-components 和 unplugin-auto-import 自动引入要使用的组件。

下面在例 13-9 的基础上讲解如何使用插件 unplugin-vue-components 和 unplugin-auto-import 自动按需引入要使用的 Element Plus 组件，具体步骤如下。

❶ 安装插件

打开项目 elementplus-vue 的 Terminal 终端，执行"npm install -D unplugin-vue-components unplugin-auto-import"命令，安装 unplugin-vue-components 和 unplugin-auto-import 插件。

❷ 配置插件

打开项目 elementplus-vue 的配置文件 vue.config.js，并配置 unplugin-vue-components 和 unplugin-auto-import 插件，具体配置代码如下。

```
const { defineConfig } = require("@vue/cli-service");
const AutoImport = require("unplugin-auto-import/webpack");
const Components = require("unplugin-vue-components/webpack");
const { ElementPlusResolver } = require("unplugin-vue-components/resolvers");
module.exports = defineConfig({
    transpileDependencies: true,
    configureWebpack: {
        plugins: [
            AutoImport({
                resolvers: [ElementPlusResolver()],
            }),
            Components({
                resolvers: [ElementPlusResolver()],
            }),
        ],
    },
});
```

❸ 修改 main.js 文件，去除完整引入

修改项目 elementplus-vue 的配置文件 main.js，去除完整引入。main.js 的具体代码如下。

```
import { createApp } from 'vue'
//import ElementPlus from 'element-plus'           //完整引入
//import 'element-plus/dist/index.css'
//引入图标
import * as ElementPlusIconsVue from '@element-plus/icons-vue'
import App from './App.vue'
const app = createApp(App)
//注册所有图标
for (const [key, component] of Object.entries(ElementPlusIconsVue)) {
    app.component(key, component)
}
//使用 ElementPlus
//app.use(ElementPlus).mount('#app')               //完整引入
app.mount('#app')
```

经过上述三个步骤，即可按需自动引入 Element Plus 组件，使用起来也极其方便。因此，推荐开发者使用此方式按需自动引入 Element Plus 组件。

❹ 运行测试

首先在 elementplus-vue 项目的 Terminal 终端输入 npm run serve 命令启动服务，然后在浏览器的地址栏中访问 http://localhost:8080/，即可运行项目 elementplus-vue，运行效果如图 13.25 所示。

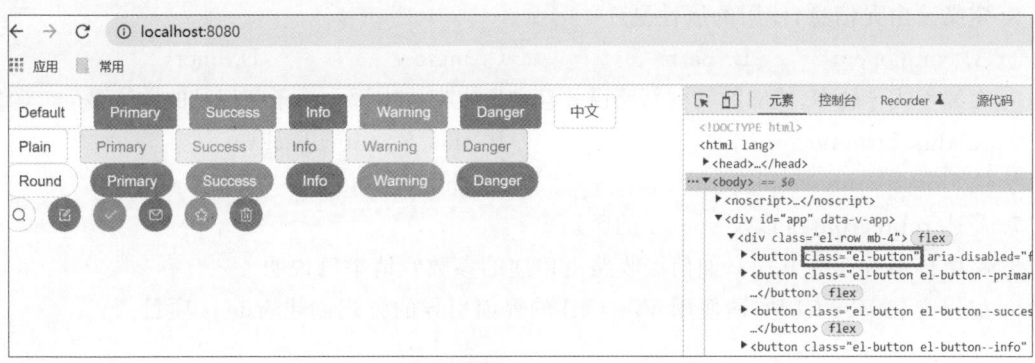

图 13.25　按需引入 Element Plus 组件

从图 13.25 可以看出，在按需自动引入 Element Plus 组件时自动引入了 ElButton 组件，并引入了组件样式。不过，当需要使用命令的方式创建 Element Plus 组件时，还需要 import 方式引入。其示例代码如下。

```
<template>
    <el-button v-on:click="gogo">Round</el-button>
</template>
<script setup>
import { ElMessage } from 'element-plus'
const gogo = () => {
    ElMessage.warning('注意注意!')
}
</script>
```

本章小结

本章主要介绍了 Vue CLI、Vue Router 插件以及 Element Plus UI 组件库等知识，希望读者重点学习 Vue CLI 与 Vue Router 插件的用法，为进行项目实战夯实基础。

习题 13

1. 在 Vue Router 中，下列选项中能够设置页面导航的是（　　）。
   A. <router-link>　　B. <router-view>　　C. <router-a>　　D. <router-nav>
2. 下列选项中能够显示或渲染匹配到的路由信息的标记是（　　）。
   A. <router-link>　　B. <router-view>　　C. <router-v>　　D. <router-vue>
3. 下列选项中能够正确表示跳转到 user/chenheng 的路由是（　　）。
   A. {path：'/user', name：'user', component：UserView}
   B. {path：'/user/:uname', name：'user', component：UserView}
   C. {path：'/user/uname', name：'user', component：UserView}
   D. {path：'/user/name', name：'user', component：UserView}
4. 定义命名路由，使用的属性是（　　）。
   A. component　　B. path　　C. meta　　D. name
5. 定义路由元信息，使用的属性是（　　）。
   A. component　　B. path　　C. meta　　D. query
6. 在编程式导航中，能够跳转到新路由并且在历史记录中添加一条新记录的方法是（　　）。
   A. this.$router.push()　　B. this.$router.back()
   C. this.$router.replace()　　D. this.$router.go()
7. 简述 route、routes 以及 router 的区别。
8. 路由传参有几种方式？如何接收路由传递的参数？请举例说明。
9. 如何安装 Vue CLI？请使用 Vue CLI 的界面引导的方式创建 Vue.js 项目。

第 14 章　MyBatis-Plus

学习目的与要求

本章将重点介绍 MyBatis-Plus 的基础知识，包括 MyBatis-Plus 常用注解、CRUD 接口以及条件构造器等内容。通过本章的学习，掌握 MyBatis-Plus 的基础知识，掌握 Spring Boot 整合 MyBatis-Plus 的基本步骤。

本章主要内容

- Spring Boot 整合 MyBatis-Plus
- MyBatis-Plus 注解
- CRUD 接口
- 条件构造器

MyBatis-Plus 是增强版的 MyBatis，本章将学习 MyBatis-Plus 的基础知识以及在 Spring Boot 应用中如何整合 MyBatis-Plus。

14.1　MyBatis-Plus 简介

MyBatis-Plus 是 MyBatis 的一个增强工具，在 MyBatis 的基础上只做增强，不做改变，为简化开发、提高效率而生。MyBatis-Plus 的特性具体如下。

（1）无侵入：只做增强，不做改变，引入它不会对现有工程产生影响。

（2）损耗小：启动即会自动注入基本 CRUD，性能基本无损耗，直接面向对象操作。

（3）强大的 CRUD 操作：内置通用 Mapper、通用 Service，仅通过少量配置即可实现单表的大部分 CRUD 操作，更有强大的条件构造器，满足各类使用需求。

（4）支持 Lambda 形式调用：通过 Lambda 表达式，方便编写各类查询条件，开发者无须再担心写错字段。

（5）支持主键自动生成：支持多种主键策略，可自由配置，完美解决主键问题。

（6）支持 ActiveRecord 模式：支持 ActiveRecord 形式调用，实体类只需继承 Model 类即可进行强大的 CRUD 操作。

（7）支持自定义全局通用操作：支持全局通用方法注入。

（8）内置代码生成器：采用代码或者 Maven 插件可快速生成 Mapper、Model、Service、Controller 层代码，支持模板引擎，提供更多自定义配置。

（9）内置分页插件：基于 MyBatis 物理分页，开发者无须关心具体操作，配置好插件之后，实现分页等同于普通 List 遍历。

14.2　Spring Boot 整合 MyBatis-Plus

在 Spring Boot 应用中添加 mybatis-plus-boot-starter 依赖即可整合 MyBatis-Plus，具体如下：

```
<dependency>
    <groupId>com.baomidou</groupId>
```

```xml
        <artifactId>mybatis-plus-boot-starter</artifactId>
        <version>3.x.y.z</version>
</dependency>
```

在 Spring Boot 应用中,通过 mybatis-plus-boot-starter 引入 MyBatis-Plus 依赖后,将自动引入 MyBatis、MyBatis-Spring 等相关依赖,所以不再需要引入这些依赖,以避免因版本差异导致问题。

下面通过实例讲解如何在 Spring Boot 应用中使用 MyBatis-Plus 框架操作数据库。

【例 14-1】 在 Spring Boot 应用中使用 MyBatis-Plus 框架操作数据库。

其具体实现步骤如下。

❶ 创建 Spring Boot Web 应用

创建基于 Lombok、MySQL Driver、Spring Web 依赖的 Spring Boot Web 应用 ch14_1。在该应用中操作的数据库是 springtest,操作的数据表是 user 表。

❷ 修改 pom.xml 文件

在编写本书时,MyBatis-Plus 的最新版本是 3.5.5,但是 mybatis-plus-boot-starter 3.5.5 依赖的 mybatis-spring 是 2.1.2(与 Spring Boot 3.2.4 不兼容),因此在 pom.xml 中添加如下依赖:

```xml
<dependency>
    <groupId>com.baomidou</groupId>
    <artifactId>mybatis-plus-boot-starter</artifactId>
    <version>3.5.5</version>
    <exclusions>
        <exclusion>
            <groupId>org.mybatis</groupId>
            <artifactId>mybatis-spring</artifactId>
        </exclusion>
    </exclusions>
</dependency>
<dependency>
    <groupId>org.mybatis</groupId>
    <artifactId>mybatis-spring</artifactId>
    <version>3.0.3</version>
</dependency>
```

❸ 设置 Web 应用 ch14_1 的上下文路径及数据源配置信息

在应用 ch14_1 的 application.properties 文件中配置如下内容:

```
server.servlet.context-path=/ch14_1
#数据库地址
spring.datasource.url=jdbc:mysql://localhost:3306/springtest?useUnicode=true&characterEncoding=UTF-8&allowMultiQueries=true&serverTimezone=GMT%2B8
#数据库用户名
spring.datasource.username=root
#数据库密码
spring.datasource.password=root
#数据库驱动
spring.datasource.driver-class-name=com.mysql.cj.jdbc.Driver
#设置包的别名(在 Mapper 映射文件中直接使用实体类名)
mybatis-plus.type-aliases-package=com.ch.ch14_1.entity
#告诉系统到哪里去找 mapper.xml 文件(映射文件)
mybatis-plus.mapper-locations=classpath:mappers/*.xml
#在控制台中输出 SQL 语句日志
logging.level.com.ch.ch14_1.mapper=debug
#让控制器输出的 JSON 字符串格式更美观
spring.jackson.serialization.indent-output=true
```

❹ 创建实体类

创建名为 com.ch.ch14_1.entity 的包,并在该包中创建 MyUser 实体类,核心代码如下。

```
@Data
@TableName("user")
public class MyUser {
    @TableId(value = "uid", type = IdType.AUTO)
    private Integer uid;
    private String uname;
    private String usex;
}
```

❺ 创建数据访问接口

创建名为 com.ch.ch14_1.mapper 的包,并在该包中创建 UserMapper 接口。UserMapper 接口通过继承 BaseMapper<MyUser> 接口(14.4 节将讲解该接口)对实体类 MyUser 对应的数据表 user 进行 CRUD 操作。UserMapper 接口的核心代码如下:

```
@Repository
public interface UserMapper extends BaseMapper<MyUser> {
    List<MyUser> myFindAll();
}
```

❻ 创建 Mapper 映射文件

在 src/main/resources 目录下创建名为 mappers 的包,并在该包中创建 SQL 映射文件 MyUserMapper.xml(当 Mapper 接口中没有自定义方法时,可以不创建此文件),具体代码如下:

```xml
<?xml version="1.0" encoding="UTF-8"?>
<!DOCTYPE mapper
PUBLIC "-//mybatis.org//DTD Mapper 3.0//EN"
"http://mybatis.org/dtd/mybatis-3-mapper.dtd">
<mapper namespace="com.ch.ch14_1.mapper.UserMapper">
    <select id="myFindAll" resultType="MyUser">
        select * from user
    </select>
</mapper>
```

❼ 创建控制器类 MyUserController

创建名为 com.ch.ch14_1.controller 的包,并在该包中创建控制器类 MyUserController。MyUserController 的核心代码如下:

```
@RestController
public class MyUserController {
    @Autowired
    private UserMapper userMapper;
    @GetMapping("/findAll")
    public List<MyUser> findAll(){
        //通过 BaseMapper 接口方法 selectList 查询
        return userMapper.selectList(null);
    }
    @GetMapping("/myFindAll")
    public List<MyUser> myFindAll(){
        //通过自定义方法 myFindAll 查询
        return userMapper.myFindAll();
    }
}
```

❽ 在应用程序的主类中扫描 Mapper 接口

在应用程序的 Ch141Application 主类中使用@MapperScan 注解扫描 MyBatis 的 Mapper

接口,核心代码如下:

```
@SpringBootApplication
@MapperScan(basePackages={"com.ch.ch14_1.mapper"})
public class Ch141Application {
    public static void main(String[] args) {
        SpringApplication.run(Ch141Application.class, args);
    }
}
```

❾ 运行

首先运行 Ch141Application 主类,然后访问 "http://localhost:8080/ch14_1findAll" 和 "http://localhost:8080/ch14_1/myFindAll" 进行测试。

14.3 MyBatis-Plus 注解

本节将详细介绍 MyBatis-Plus 的相关注解类,具体如下。

❶ @TableName

当实体类的类名与要操作表的表名不一致时,需要使用@TableName 注解标识实体类对应的表。示例代码如下:

```
@TableName("user")
public class MyUser {}
```

@TableName 注解的所有属性都不是必须指定的,如表 14.1 所示。

表 14.1 @TableName 的属性

属性	类型	默认值	描述
value	String	""	表名
schema	String	""	指定模式名称。如果是 MySQL 数据库,则指定数据库名称;如果是 Oracle,则为 schema。例如,schema="scott",scott 就是 Oracle 中的 schema
keepGlobalPrefix	boolean	false	是否保持使用全局的 tablePrefix 值(当设置全局 tablePrefix 时)
resultMap	String	""	XML 中 resultMap 的 ID(用于满足特定类型的实体类对象的绑定)
autoResultMap	boolean	false	是否自动构建 resultMap 并使用(如果设置 resultMap,则不会进行 resultMap 的自动构建与注入)
excludeProperty	String[]	{}	需要排除的属性名

❷ @TableId

@TableId 注解为主键注解,指定实体类中的某属性为主键字段。示例代码如下:

```
@TableName("user")
public class MyUser {
    @TableId(type=IdType.AUTO)
    private Integer uid;
}
```

@TableId 注解有 value 和 type 两个属性,value 属性表示主键字段名,默认值为"";type 属性表示主键类型,默认值为 IdType.NONE。对于 type 属性值,IdType.AUTO 表示数据表 ID 自增;IdType.NONE 表示无状态,该类型为未设置主键类型;IdType.INPUT 表示 insert 前自行 set 主键值;IdType.ASSIGN_ID 表示分配 ID(类型为 Long、Integer 或 String),使用接口

IdentifierGenerator 的 nextId 方法（默认实现类为 DefaultIdentifierGenerator 雪花算法）；IdType.ASSIGN_UUID 表示分配 UUID，类型为 String，使用接口 IdentifierGenerator 的 nextUUID 方法。

❸ @TableField

@TableField 注解为非主键字段注解。若实体类中的属性使用的是驼峰命名风格，而表中的字段使用的是下画线命名风格，例如实体类属性为 userName，表中字段为 user_name，此时 MyBatis-Plus 会自动将下画线命名风格转化为驼峰命名风格。若实体类中的属性和表中的字段不满足上述条件，例如实体类属性为 name，表中字段为 username，此时需要在实体类属性上使用 @TableField("username") 设置属性所对应的字段名。

@TableField 注解的所有属性都不是必须指定的，如表 14.2 所示。

表 14.2 @TableField 的属性

属 性	类 型	默 认 值	描 述
value	String	""	数据库字段名
exist	boolean	true	是否为数据库表字段
condition	String	""	字段 where 实体查询比较条件，有值设置则按设置的值为准，没有则为默认全局的 %s=#{%s}
update	String	""	字段 update set 部分注入，例如，当 version 字段上注解 update="%s+1" 表示更新时会 set version=version+1
insertStrategy	Enum	FieldStrategy.DEFAULT	IGNORED 忽略判断；NOT_NULL 为非 NULL 判断；NOT_EMPTY 为非空判断（只对字符串类型字段，其他类型字段依然为 NULL 判断）；DEFAULT 追随全局配置；NEVER 不加入 SQL。举例：NOT_NULL insert into table_a(<if test="columnProperty != null">column</if>) values (<if test="columnProperty != null">#{columnProperty}</if>)
updateStrategy	Enum	FieldStrategy.DEFAULT	举例：IGNORED update table_a set column=#{columnProperty}
whereStrategy	Enum	FieldStrategy.DEFAULT	举例：NOT_EMPTY where <if test="columnProperty != null and columnProperty!=''">column=#{columnProperty}</if>
fill	Enum	FieldFill.DEFAULT	字段自动填充策略。DEFAULT 默认不处理；INSERT 为插入时填充字段；UPDATE 为更新时填充字段；INSERT_UPDATE 为插入和更新时填充字段
select	boolean	true	是否进行 select 查询
keepGlobalFormat	boolean	false	是否保持使用全局的 format 进行处理
jdbcType	JdbcType	JdbcType.UNDEFINED	JDBC 类型（该默认值不代表会按照该值生效）

续表

属性	类型	默认值	描述
typeHandler	Class<? extends TypeHandler>	UnknownTypeHandler.class	类型处理器（该默认值不代表会按照该值生效）
numericScale	String	""	指定小数点后保留的位数

❹ @Version

乐观锁注解@Version标记在字段上。当要更新一条记录时，希望这条记录没有被别人更新，可以考虑使用乐观锁。乐观锁的实现方式为，当取出记录时，获取当前version；更新时，带上这个version；执行更新时，set version = newVersion where version = oldVersion，如果version不对，则更新失败。具体示例如下：

```
@Data
@TableName("t_product")
public class Product {
    private Long id;
    private String name;
    private Integer price;
    @Version
    private Integer version;
}
@Configuration
public class MybatisPlusConfig {
    @Bean
    public MybatisPlusInterceptor mybatisPlusInterceptor() {
        MybatisPlusInterceptor interceptor =
                new MybatisPlusInterceptor();
        //乐观锁插件
        interceptor.addInnerInterceptor(new OptimisticLockerInnerInterceptor());
        return interceptor;
    }
}
```

这个@Version注解就是实现乐观锁的重要注解，当更新数据库中的数据时，例如价格，version就会加1，如果where语句中的version版本不对，则更新失败。

❺ @EnumValue

普通枚举类注解，注解在枚举字段上。示例代码如下：

```
@Getter                                     //类中属性都生成getter方法
public enum SexEnum {
    MALE(1, "男"),
    FEMALE(2, "女");
    @EnumValue                              //标记数据库中存的值是sex
    private Integer sex;
    private String sexName;
    SexEnum(Integer sex, String sexName) {
        this.sex = sex;
        this.sexName = sexName;
    }
}
```

在实体类中添加枚举类型的字段，示例代码如下：

```
@Data
public class User {
    private Integer id;
    private String name;
    private SexEnum sex;
}
```

❻ @TableLogic

@TableLogic 注解的使用代表着实体类中的属性是逻辑删除的属性。逻辑删除即假删除，将对应数据中代表是否被删除字段的状态修改为"被删除状态"，之后在数据库中仍然能看到此条数据记录。

14.4　CRUD 接口

MyBatis-Plus 使用 MyBatis 接口编程实现机制，默认提供了一系列增、删、改、查基础方法，并且开发人员对于这些基础操作方法不需要编写 SQL 语句即可进行处理操作。

❶ Mapper CRUD 接口

MyBatis-Plus 内置了可以实现对单表 CRUD 的 BaseMapper＜T＞接口，泛型 T 为任意实体对象。BaseMapper＜T＞接口是针对 Dao 层的 CRUD 方法进行封装。

在自定义数据访问接口时继承 BaseMapper＜T＞接口，即可使用 BaseMapper＜T＞接口方法进行单表的 CRUD，例如 public interface UserMapper extends BaseMapper＜MyUser＞ {}。BaseMapper＜T＞接口方法具体如下。

1）insert

BaseMapper＜T＞接口提供了一个实现插入一条记录的方法，具体如下。

```
//插入一条记录
int insert(T entity);
```

2）delete

BaseMapper＜T＞接口提供了许多删除方法，具体如下。

```
//根据 entity 条件删除记录
int delete(@Param(Constants.WRAPPER) Wrapper<T> wrapper);
//删除(根据 ID 批量删除)
int deleteBatchIds (@ Param (Constants. COLLECTION) Collection <? extends Serializable >
idList);
//根据 ID 删除
int deleteById(Serializable id);
//根据 columnMap 条件删除记录
int deleteByMap(@Param(Constants.COLUMN_MAP) Map<String, Object> columnMap);
```

在上述方法中，wrapper（Wrapper＜T＞）为实体对象封装操作类（即条件构造器，可以为null）；idList（Collection＜? extends Serializable＞）为主键 ID 列表（不能为 null 以及 empty）；id（Serializable）为主键 ID；columnMap（Map＜String，Object＞）为表字段 map 对象。

3）update

BaseMapper＜T＞接口提供了两个更新方法，具体如下。

```
//根据 whereWrapper 条件更新记录
int update(@Param(Constants.ENTITY) T updateEntity,@Param(Constants.WRAPPER) Wrapper<T>
whereWrapper);
//根据 ID 修改
int updateById(@Param(Constants.ENTITY) T entity);
```

4）select

BaseMapper<T>接口提供了许多查询方法，具体如下。

```
//根据 ID 查询
T selectById(Serializable id);
//根据 entity 条件查询一条记录
T selectOne(@Param(Constants.WRAPPER) Wrapper<T> queryWrapper);
//查询(根据 ID 批量查询)
List<T> selectBatchIds(@Param(Constants.COLLECTION) Collection<? extends Serializable> idList);
//根据 entity 条件查询全部记录
List<T> selectList(@Param(Constants.WRAPPER) Wrapper<T> queryWrapper);
//查询(根据 columnMap 条件)
List<T> selectByMap(@Param(Constants.COLUMN_MAP) Map<String, Object> columnMap);
//根据 Wrapper 条件查询全部记录
List<Map<String, Object>> selectMaps(@Param(Constants.WRAPPER) Wrapper<T> queryWrapper);
//根据 Wrapper 条件查询全部记录。注意，只返回第一个字段的值
List<Object> selectObjs(@Param(Constants.WRAPPER) Wrapper<T> queryWrapper);
//根据 entity 条件查询全部记录(并翻页)
IPage<T> selectPage(IPage<T> page, @Param(Constants.WRAPPER) Wrapper<T> queryWrapper);
//根据 Wrapper 条件查询全部记录(并翻页)
IPage<Map<String, Object>> selectMapsPage(IPage<T> page, @Param(Constants.WRAPPER) Wrapper<T> queryWrapper);
//根据 Wrapper 条件查询总记录数
Integer selectCount(@Param(Constants.WRAPPER) Wrapper<T> queryWrapper);
```

5）ActiveRecord 模式

所谓 ActiveRecord 模式，在 Spring Boot 应用中，如果已注入对应实体的 BaseMapper，例如 public interface UserMapper extends BaseMapper<MyUser>{}，那么实体类 MyUser 只需继承 Model 类即可具有强大的 CRUD 功能，例如 public class MyUser extends Model<MyUser>{}。

❷ Service CRUD 接口

通用 Service CRUD 封装 IService<T>接口，进一步封装 CRUD，采用 get 查询单行、remove 删除、list 查询集合、page 分页等前缀命名方式区分 Mapper 层，以避免混淆。创建 Service 接口及其实现类的示例代码如下。

```
public interface UserService extends IService<MyUser> { }
/* ServiceImpl 实现了 IService,提供了 IService 中基础功能的实现。若 ServiceImpl 无法满足业务需求，可以使用自定义的 UserService 方法，并在实现类中实现 */
@Service
public class UserServiceImpl extends ServiceImpl<UserMapper, MyUser> implements UserService { }
```

IService<T>接口针对业务逻辑层的封装，需要指定 Dao 层接口和对应的实体类，是在 BaseMapper<T>基础上的加强，ServiceImpl<M, T>是针对业务逻辑层的实现。

1）save

通用 Service CRUD 接口提供如下 save 方法：

```
//插入一条记录(选择字段,策略插入)
boolean save(T entity);
//插入(批量)
boolean saveBatch(Collection<T> entityList);
//插入(批量)
boolean saveBatch(Collection<T> entityList, int batchSize);
//@TableId 注解存在更新记录,否则插入一条记录
boolean saveOrUpdate(T entity);
```

```
//根据 updateWrapper 尝试更新,否则继续执行 saveOrUpdate(T)方法
boolean saveOrUpdate(T entity, Wrapper<T> updateWrapper);
//批量修改插入
boolean saveOrUpdateBatch(Collection<T> entityList);
//批量修改插入
boolean saveOrUpdateBatch(Collection<T> entityList, int batchSize);
```

2）remove

通用 Service CRUD 接口提供如下 remove 方法：

```
//根据 entity 条件删除记录
boolean remove(Wrapper<T> queryWrapper);
//根据 ID 删除
boolean removeById(Serializable id);
//根据 columnMap 条件删除记录
boolean removeByMap(Map<String, Object> columnMap);
//删除(根据 ID 批量删除)
boolean removeByIds(Collection<? extends Serializable> idList);
```

3）update

通用 Service CRUD 接口提供如下 update 方法：

```
//根据 UpdateWrapper 条件更新记录,需要设置 sqlset
boolean update(Wrapper<T> updateWrapper);
//根据 whereWrapper 条件更新记录
boolean update(T updateEntity, Wrapper<T> whereWrapper);
//根据 ID 选择修改
boolean updateById(T entity);
//根据 ID 批量更新
boolean updateBatchById(Collection<T> entityList);
//根据 ID 批量更新
boolean updateBatchById(Collection<T> entityList, int batchSize);
```

4）get、list、page 及 count

通用 Service CRUD 接口提供如下 get、list、page 及 count 查询方法：

```
//根据 ID 查询
T getById(Serializable id);
//根据 Wrapper 查询一条记录。结果集如果是多个会抛出异常,随机取一条加上限制条件 wrapper.last
("LIMIT 1")
T getOne(Wrapper<T> queryWrapper);
//根据 Wrapper 查询一条记录
T getOne(Wrapper<T> queryWrapper, boolean throwEx);
//根据 Wrapper 查询一条记录
Map<String, Object> getMap(Wrapper<T> queryWrapper);
//根据 Wrapper 查询一条记录
<V> V getObj(Wrapper<T> queryWrapper, Function<? super Object, V> mapper);
//查询所有
List<T> list();
//查询列表
List<T> list(Wrapper<T> queryWrapper);
//查询(根据 ID 批量查询)
Collection<T> listByIds(Collection<? extends Serializable> idList);
//查询(根据 columnMap 条件)
Collection<T> listByMap(Map<String, Object> columnMap);
//查询所有列表
List<Map<String, Object>> listMaps();
//查询列表
List<Map<String, Object>> listMaps(Wrapper<T> queryWrapper);
```

```
//查询全部记录
List<Object> listObjs();
//查询全部记录
<V> List<V> listObjs(Function<? super Object, V> mapper);
//根据 Wrapper 条件查询全部记录
List<Object> listObjs(Wrapper<T> queryWrapper);
//根据 Wrapper 条件查询全部记录
<V> List<V> listObjs(Wrapper<T> queryWrapper, Function<? super Object, V> mapper);
//无条件分页查询
IPage<T> page(IPage<T> page);
//条件分页查询
IPage<T> page(IPage<T> page, Wrapper<T> queryWrapper);
//无条件分页查询
IPage<Map<String, Object>> pageMaps(IPage<T> page);
//条件分页查询
IPage<Map<String, Object>> pageMaps(IPage<T> page, Wrapper<T> queryWrapper);
//查询总记录数
int count();
//根据 Wrapper 条件查询总记录数
int count(Wrapper<T> queryWrapper);
```

5）链式 query 及 update

通用 Service CRUD 接口提供如下 query 及 update 链式方法：

```
//链式查询,普通
QueryChainWrapper<T> query();
//链式查询,Lambda 式
LambdaQueryChainWrapper<T> lambdaQuery();
//示例
query().eq("column", value).one();
lambdaQuery().eq(Entity::getId, value).list();
//链式更改,普通
UpdateChainWrapper<T> update();
//链式更改,Lambda 式
LambdaUpdateChainWrapper<T> lambdaUpdate();
//示例
update().eq("column", value).remove();
lambdaUpdate().eq(Entity::getId, value).update(entity);
```

下面通过一个实例演示 Mapper CRUD 接口和 Service CRUD 接口的使用方法。

【例 14-2】 演示 Mapper CRUD 接口和 Service CRUD 接口的使用方法。

其具体实现步骤如下。

❶ 创建 Spring Boot Web 应用

创建基于 Lombok、MySQL Driver、Spring Web 依赖的 Spring Boot Web 应用 ch14_2。在该应用中操作的数据库与 14.2 节一样，都是 springtest，操作的数据表是 user 表。

❷ 修改 pom.xml 文件

在 pom.xml 文件中添加 MyBatis-Plus 依赖，与例 14-1 相同，这里不再赘述。

❸ 设置 Web 应用 ch14_2 的上下文路径及数据源配置信息

在应用 ch14_2 的 application.properties 文件中配置上下文路径及数据源配置信息，配置内容与例 14-1 基本相同，这里不再赘述。

❹ 创建实体类

创建名为 com.ch.ch14_2.entity 的包，并在该包中创建 MyUser 实体类，核心代码如下。

```
@Data
@TableName("user")
public class MyUser extends Model<MyUser> {
    @TableId(value = "uid", type = IdType.AUTO)
    private Integer uid;
    private String uname;
    private String usex;
}
```

❺ 创建数据访问接口

创建名为 com.ch.ch14_2.mapper 的包，并在该包中创建 UserMapper 接口。UserMapper 接口通过继承 BaseMapper<MyUser>接口对实体类 MyUser 对应的数据表 user 进行 CRUD 操作。UserMapper 接口的核心代码如下：

```
@Repository
public interface UserMapper extends BaseMapper<MyUser> {}
```

❻ 创建 Service 接口及实现类

创建名为 com.ch.ch14_2.service 的包，并在该包中创建 UserService 接口及实现类 UserServiceImpl。

UserService 接口继承 IService<MyUser>接口，核心代码如下：

```
public interface UserService extends IService<MyUser> {}
```

实现类 UserServiceImpl 继承 ServiceImpl<UserMapper，MyUser>类，核心代码如下：

```
@Service
public class UserServiceImpl extends ServiceImpl < UserMapper, MyUser > implements UserService {}
```

❼ 配置分页插件 PaginationInnerInterceptor

MyBatis-Plus 基于 PaginationInnerInterceptor 拦截器，实现分页查询功能，所以需要事先配置该拦截器才能实现分页查询功能。

创建名为 com.ch.ch14_2.config 的包，并在该包中创建 MybatisPlusConfig 配置类，核心代码如下：

```
@Configuration
public class MybatisPlusConfig{
    @Bean
    public MybatisPlusInterceptor mybatisPlusInterceptor() {
        MybatisPlusInterceptor interceptor = new MybatisPlusInterceptor();
        interceptor.addInnerInterceptor(new PaginationInnerInterceptor(DbType.MYSQL));
        return interceptor;
    }
}
```

❽ 创建控制器类 MyUserController

创建名为 com.ch.ch14_2.controller 的包，并在该包中创建控制器类 MyUserController。MyUserController 的核心代码如下：

```
@RestController
public class MyUserController {
    @Autowired
    private UserMapper userMapper;
    @Autowired
```

```java
    private UserService userService;
    @GetMapping("/testMapperSave")
    public MyUser testMapperSave(){
        MyUser mu = new MyUser();
        mu.setUname("testMapperSave 陈恒 1");
        mu.setUsex("女");
        int result = userMapper.insert(mu);
        //实体类的主键属性使用@TableId注解后,主键自动回填
        return mu;
    }
    @GetMapping("/testMapperDelete")
    public int testMapperDelete(){
        List<Long> list = Arrays.asList(17L, 7L, 19L);
        int result = userMapper.deleteBatchIds(list);
        return result;
    }
    @GetMapping("/testMapperUpdate")
    public MyUser testMapperUpdate(){
        MyUser mu = new MyUser();
        mu.setUid(1);
        mu.setUname("李四");
        mu.setUsex("男");
        int result = userMapper.updateById(mu);
        return mu;
    }
    @GetMapping("/testMapperSelect")
    public List<MyUser> testMapperSelect(){
        return userMapper.selectList(null);
    }
    @GetMapping("/testModelSave")
    public MyUser testModelSave(){
        MyUser mu = new MyUser();
        mu.setUname("testModelSave 陈恒 2");
        mu.setUsex("男");
        mu.insert();
        return mu;
    }
    @GetMapping("/testServiceSave")
    public List<MyUser> testServiceSave(){
        MyUser mu1 = new MyUser();
        mu1.setUname("testServiceSave 陈恒 1");
        mu1.setUsex("女");
        MyUser mu2 = new MyUser();
        mu2.setUname("testServiceSave 陈恒 2");
        mu2.setUsex("男");
        List<MyUser> list = Arrays.asList(mu1, mu2);
        boolean result = userService.saveBatch(list);
        return list;
    }
    @GetMapping("/testServiceUpdate")
    public List<MyUser> testServiceUpdate(){
        MyUser mu1 = new MyUser();
        mu1.setUid(23);
        mu1.setUname("testServiceSave 陈恒 11");
        mu1.setUsex("女");
        MyUser mu2 = new MyUser();
        mu2.setUid(24);
        mu2.setUname("testServiceSave 陈恒 22");
        mu2.setUsex("男");
        List<MyUser> list = Arrays.asList(mu1, mu2);
        boolean result = userService.updateBatchById(list);
```

```
            return list;
        }
        @GetMapping("/testServicePage")
        public List<MyUser> testServicePage(){
            //1为当前页,5为页面大小
            IPage<MyUser> iPage = new Page<>(1, 5);
            IPage<MyUser> page = userService.page(iPage);
            //条件构造器
            //QueryWrapper<MyUser> wrapper = new QueryWrapper<>();
            //wrapper.like("uname", "陈");
            //IPage<MyUser> page = userService.page(iPage, wrapper);
            System.out.println(page.getPages());
            //返回当前页的记录
            return page.getRecords();
        }
        @GetMapping("/testServiceQuery")
        public List<MyUser> testServiceQuery(){
            //链式查询
            return userService.query().like("uname", "陈").list();
        }
}
```

❾ **在应用程序的主类中扫描 Mapper 接口**

在应用程序的 Ch142Application 主类中使用@MapperScan 注解扫描 MyBatis 的 Mapper 接口,核心代码如下:

```
@SpringBootApplication
@MapperScan(basePackages={"com.ch.ch14_2.mapper"})
public class Ch142Application {
    public static void main(String[] args) {
        SpringApplication.run(Ch142Application.class, args);
    }
}
```

❿ **运行**

首先运行 Ch142Application 主类,然后访问"http://localhost:8080/ch14_2/testMapperSave"进行测试。

14.5 条件构造器

MyBatis-Plus 提供了构造条件的类 Wrapper,开发者使用它可以定义自己需要的条件。Wrapper 是一个抽象类,在一般情况下用它的子类 QueryWrapper 来实现自定义条件查询。在查询前首先创建条件构造器 QueryWrapper wrapper = new QueryWrapper<>(),然后调用构造器中的方法实现按条件查询。

AbstractWrapper 是 Wrapper 的子类,也是 QueryWrapper(LambdaQueryWrapper)和 UpdateWrapper(LambdaUpdateWrapper)的父类,用于生成 SQL 的 where 条件,entity 属性也用于生成 SQL 的 where 条件。条件构造器中的方法具体如下。

❶ **allEq:全部 eq(或个别 isNull)**

```
allEq(Map<R, V> params)
allEq(Map<R, V> params, boolean null2IsNull)
allEq(boolean condition, Map<R, V> params, boolean null2IsNull)
```

其中,params 的 key 为数据库字段名,value 为字段值;null2IsNull 为 true 则在 map 的 value 为

null 时调用 isNull 方法，为 false 则忽略 value 为 null 的条件。

示例 1：map.put("id"，1)；map.put("name"，"老王")；map.put("age"，null)；wrapper.allEq(map)，等价 SQL 为 id = 1 and name = '老王' and age is null。

示例 2：wrapper.allEq(map, false)，等价 SQL 为 id = 1 and name = '老王'。

```
allEq(BiPredicate<R, V> filter, Map<R, V> params)
allEq(BiPredicate<R, V> filter, Map<R, V> params, boolean null2IsNull)
allEq(boolean condition, BiPredicate<R, V> filter, Map<R, V> params, boolean null2IsNull)
```

其中，filter 为过滤函数，表示是否允许字段传入比对条件中。

示例 1：wrapper.allEq((k,v) -> k.contains("a")，map)，等价 SQL 为 name = '老王' and age is null。

示例 2：wrapper.allEq((k,v) -> k.contains("a")，map, false)，等价 SQL 为 name = '老王'。

❷ eq：等于，=

```
eq(R column, Object val)
eq(boolean condition, R column, Object val)
```

示例：eq("name"，"老王")，等价于 name = '老王'。

❸ ne：不等于，<>

```
ne(R column, Object val)
ne(boolean condition, R column, Object val)
```

示例：ne("name"，"老王")，等价于 name <> '老王'。

❹ gt：大于，>

```
gt(R column, Object val)
gt(boolean condition, R column, Object val)
```

示例：gt("age"，7)，等价于 age > 7。

❺ ge：大于或等于，>=

```
ge(R column, Object val)
ge(boolean condition, R column, Object val)
```

示例：ge("age"，7)，等价于 age >= 7。

❻ lt：小于，<

```
lt(R column, Object val)
lt(boolean condition, R column, Object val)
```

示例：lt("age"，7)，等价于 age < 7。

❼ le：小于或等于，<=

```
le(R column, Object val)
le(boolean condition, R column, Object val)
```

示例：le("age"，7)，等价于 age <= 7。

❽ between：between 值 1 and 值 2

```
between(R column, Object val1, Object val2)
between(boolean condition, R column, Object val1, Object val2)
```

示例：between("age"，7，30)，等价于 age between 7 and 30。

❾ notBetween：not between 值1 and 值2

```
notBetween(R column, Object val1, Object val2)
notBetween(boolean condition, R column, Object val1, Object val2)
```

示例：notBetween("age"，7，30)，等价于 age not between 7 and 30。

❿ like：like '%值%'

```
like(R column, Object val)
like(boolean condition, R column, Object val)
```

示例：like("name"，"王")，等价于 name like '%王%'。

⓫ notLike：not like '%值%'

```
notLike(R column, Object val)
notLike(boolean condition, R column, Object val)
```

示例：notLike("name"，"王")，等价于 name not like '%王%'。

⓬ likeLeft：like '%值'

```
likeLeft(R column, Object val)
likeLeft(boolean condition, R column, Object val)
```

示例：likeLeft("name"，"王")，等价于 name like '%王'。

⓭ likeRight：like '值%'

```
likeRight(R column, Object val)
likeRight(boolean condition, R column, Object val)
```

示例：likeRight("name"，"王")，等价于 name like '王%'。

⓮ isNull 及 isNotNull

```
isNull(R column)
isNull(boolean condition, R column)
isNotNull(R column)
isNotNull(boolean condition, R column)
```

示例1：isNull("name")，等价于 name is null。

示例2：isNotNull("name")，等价于 name is not null。

⓯ in 及 notIn

```
in(R column, Collection<?> value)
in(boolean condition, R column, Collection<?> value)
notIn(R column, Collection<?> value)
notIn(boolean condition, R column, Collection<?> value)
```

示例1：in("age",{1,2,3})，等价于 age in (1,2,3)。

示例2：notIn("age",{1,2,3})，等价于 age not in (1,2,3)。

⓰ inSql 及 notInSql

```
inSql(R column, String inValue)
inSql(boolean condition, R column, String inValue)
notInSql(R column, String inValue)
notInSql(boolean condition, R column, String inValue)
```

示例1：inSql("age"，"1,2,3,4,5,6")，等价于 age in (1,2,3,4,5,6)。

示例2：inSql("id", "select id from table where id < 3")，等价于 id in (select id from table where id < 3)。

示例 3：notInSql("age", "1,2,3,4,5,6")，等价于 age not in (1,2,3,4,5,6)。

示例 4：notInSql("id", "select id from table where id < 3")，等价于 id not in (select id from table where id < 3)。

❶⑦ groupBy

```
groupBy(R... columns)
groupBy(boolean condition, R... columns)
```

示例：groupBy("id", "name")，等价于 group by id,name。

❶⑧ orderByAsc、orderByDesc 及 orderBy

```
orderByAsc(R... columns)
orderByAsc(boolean condition, R... columns)
orderByDesc(R... columns)
orderByDesc(boolean condition, R... columns)
orderBy(boolean condition, boolean isAsc, R... columns)
```

示例 1：orderByAsc("id", "name")，等价于 order by id ASC,name ASC。

示例 2：orderByDesc("id", "name")，等价于 order by id DESC,name DESC。

示例 3：orderBy(true, true, "id", "name")，等价于 order by id ASC,name ASC。

❶⑨ having

```
having(String sqlHaving, Object... params)
having(boolean condition, String sqlHaving, Object... params)
```

示例 1：having("sum(age) > 10")，等价于 having sum(age) > 10。

示例 2：having("sum(age) > {0}", 11)，等价于 having sum(age) > 11。

❷⓿ func

```
func(Consumer<Children> consumer)
func(boolean condition, Consumer<Children> consumer)
```

示例：func(i -> if(true) {i.eq("id", 1)} else {i.ne("id", 1)})。

❷① or 及 and

```
or()
or(boolean condition)
or(Consumer<Param> consumer)
or(boolean condition, Consumer<Param> consumer)
and(Consumer<Param> consumer)
and(boolean condition, Consumer<Param> consumer)
```

示例 1：eq("id",1).or().eq("name","老王")，等价于 id = 1 or name = '老王'。

示例 2：or(i -> i.eq("name", "李白").ne("status", "活着"))，等价于 or (name = '李白' and status <> '活着')。

示例 3：and(i -> i.eq("name", "李白").ne("status", "活着"))，等价于 and (name = '李白' and status <> '活着')。

❷② nested：正常嵌套，不带 and 或者 or

```
nested(Consumer<Param> consumer)
nested(boolean condition, Consumer<Param> consumer)
```

示例：nested(i -> i.eq("name", "李白").ne("status", "活着"))，等价于(name = '李白' and status <> '活着')。

❷ **apply**：拼接 SQL

```
apply(String applySql, Object... params)
apply(boolean condition, String applySql, Object... params)
```

示例 1：apply("id = 1")，等价于 id = 1。

示例 2：apply("date_format(dateColumn,'%Y-%m-%d') = '2008-08-08'")，等价于 date_format(dateColumn,'%Y-%m-%d') = '2008-08-08'"。

示例 3：apply("date_format(dateColumn,'%Y-%m-%d') = {0}", "2008-08-08")，等价于 date_format(dateColumn,'%Y-%m-%d') = '2008-08-08'"。

❷ **last**：无视优化规则直接拼接到 SQL 的最后

```
last(String lastSql)
last(boolean condition, String lastSql)
```

示例：last("limit 1")。

❷ **exists 及 notExists**

```
exists(String existsSql)
exists(boolean condition, String existsSql)
notExists(String notExistsSql)
notExists(boolean condition, String notExistsSql)
```

示例 1：exists("select id from table where age = 1")，等价于 exists (select id from table where age = 1)。

示例 2：notExists("select id from table where age = 1")，等价于 not exists (select id from table where age = 1)。

❷ **QueryWrapper 之 select**：设置查询字段

```
select(String... sqlSelect)
select(Predicate<TableFieldInfo> predicate)
select(Class<T> entityClass, Predicate<TableFieldInfo> predicate)
```

示例 1：select("id", "name", "age")。

示例 2：select(i -> i.getProperty().startsWith("test"))。

❷ **UpdateWrapper 之 set 及 setSql**

```
set(String column, Object val)
set(boolean condition, String column, Object val)
setSql(String sql)
```

示例 1：set("name", "老李头")。

示例 2：set("name", "")，数据库字段值变为空字符串。

示例 3：set("name", null)，数据库字段值变为 null。

示例 4：setSql("name = '老李头'")。

本章小结

MyBatis-Plus 是增强版的 MyBatis，对 MyBatis 只做增强，不做改变，因此灵活使用 MyBatis-Plus 的前提是掌握 MyBatis 基础知识。

本章详细介绍了 Spring Boot 如何整合 MyBatis 及 MyBatis-Plus，希望读者掌握 MyBatis 及 MyBatis-Plus 在 Spring Boot 应用中的整合开发过程。

习题 14

1. 简述 MyBatis 与 MyBatis-Plus 的关系。
2. 简述 MyBatis-Plus 的特性。
3. 在 MyBatis-Plus 中，当实体类的类名与要操作表的表名不一致时，需要使用（　　）注解标识实体类对应的表。

  A. @TableName B. @TableId C. @TableField D. @TableLogic

第 15 章　Spring Boot 单元测试

学习目的与要求

本章重点讲解 Spring Boot 单元测试的相关内容，包括 JUnit 5 的注解、断言以及单元测试用例。通过本章的学习，掌握 JUnit 5 的注解与断言机制的用法，掌握单元测试用例的编写。

本章主要内容

- JUnit 5 注解
- JUnit 5 断言
- 单元测试用例
- 使用 Postman 测试 Controller

单元测试（Unit Testing）是指对软件中的最小可测试单元进行检查和验证，是对开发人员所编写的代码进行测试。本章将重点讲解 JUnit 单元测试框架的应用过程。

15.1　JUnit 5

15.1.1　JUnit 5 简介

JUnit 是 Java 语言的一个单元测试框架，是由 Erich Gamma 和 Kent Beck 编写的一个回归测试框架（Regression Testing Framework）。JUnit 测试是程序员测试，即所谓的白盒测试，因为程序员知道被测试的软件如何（How）完成功能和完成什么样（What）的功能。多数 Java 开发环境（如 Eclipse、IntelliJ IDEA）都已经集成了 JUnit 作为单元测试工具。

JUnit 5 由 JUnit Platform、JUnit Jupiter 以及 JUnit Vintage 三部分组成，Java 运行环境的最低版本是 Java 8。

JUnit Platform：JUnit 提供的平台功能模块，通过 JUnit Platform，其他的测试引擎都可以接入 JUnit 实现接口和执行。

JUnit Jupiter：JUnit 5 的核心，是一个基于 JUnit Platform 的引擎实现，JUnit Jupiter 包含许多丰富的新特性，使得自动化测试更加方便和强大。

JUnit Vintage：兼容 JUnit 3、JUnit 4 版本的测试引擎，使旧版本的自动化测试也可以在 JUnit 5 下正常运行。

JUnit 5 使用了 Java 8 或更高版本的特性，例如 Lambda 函数，使测试更强大，更容易维护。JUnit 5 可以同时使用多个扩展，可以轻松地将 Spring 扩展与其他扩展（如自定义扩展）结合起来，包容性强，可以接入其他的测试引擎。其功能更强大，提供了新的断言机制、参数化测试、重复性测试等功能。

15.1.2　JUnit 5 注解

在单元测试中，JUnit 5 有以下应用在方法上的常用注解。

❶ @Test

@Test 注解表示方法是单元测试方法（返回值都是 void）。但是它与 JUnit 4 的 @Test 不同，其职责非常单一，不能声明任何属性，拓展的测试将会由 Jupiter 提供额外测试。示例代码

如下：

```
@Test
void testSelectAllUser() {}
```

❷ @RepeatedTest

@RepeatedTest 注解表示单元测试方法可重复执行，示例代码如下：

```
@Test
@RepeatedTest(value = 5)
void firstTest() {                               //该测试方法重复执行 5 次
    System.out.println(55555);
}
```

❸ @DisplayName

@DisplayName 注解为单元测试方法设置展示名称（默认为方法名），示例代码如下：

```
@Test
@DisplayName("测试用户名查询方法")
void findByUname() {}
```

❹ @BeforeEach

@BeforeEach 注解表示在每个单元测试方法之前执行，示例代码如下：

```
@BeforeEach
void setUp() {}
```

❺ @AfterEach

@AfterEach 注解表示在每个单元测试方法之后执行，示例代码如下：

```
@AfterEach
void tearDown() {}
```

❻ @BeforeAll

@BeforeAll 注解表示在所有单元测试方法之前执行。被@BeforeAll 注解的方法必须为静态方法，该静态方法将在当前测试类的所有@Test 方法前执行一次。示例代码如下：

```
@BeforeAll
static void superBefore(){
    System.out.println("在最前面执行");
}
```

❼ @AfterAll

@AfterAll 注解表示在所有单元测试方法之后执行。被@AfterAll 注解的方法必须为静态方法，该静态方法将在当前测试类的所有@Test 方法后执行一次。示例代码如下：

```
@AfterAll
static void superAfter(){
    System.out.println("在最后面执行");
}
```

❽ @Disabled

@Disabled 注解表示单元测试方法不执行，类似于 JUnit 4 中的@Ignore。

❾ @Timeout

@Timeout 注解表示单元测试方法运行时如果超过了指定时间将会返回错误。示例代码

如下：

```
@Test
@Timeout(value = 500, unit = TimeUnit.MILLISECONDS)
void testTimeout() throws InterruptedException {
    Thread.sleep(600);
}
```

▶15.1.3　JUnit 5 断言

断言，简单理解就是用来判断的语句，判断待测试代码结果和期望结果是否一致，如果不一致，说明单元测试失败。JUnit 5 的断言方法都是 org.junit.jupiter.api.Assertions 的静态方法（返回值为 void）。JUnit 5 有以下常用的断言方法。

❶ **assertEquals 和 assertNotEquals**

Assertions.assertEquals(Object expected，Object actual，String message)方法的第一个参数是期望值，第二个参数是待测试方法的实际返回值，第三个参数 message 是可选的，表示判断失败的提示信息。assertEquals 判断两者的值是否相等，不判断类型是否相同。示例代码如下：

```
@Test
void myTest(){
    int a = 1;
    long b = 1L;
    //虽然 a 和 b 类型不同，但依旧判断是成功的，当 a 和 b 不相等时，测试不通过
    assertEquals(a, b, "a 与 b 不相等");
    MyUser au = new MyUser();
    MyUser bu = new MyUser();
    //虽然 au 和 bu 指向不同的对象，但它们的值相等，依旧判断是成功的
    assertEquals(au, bu, "au 与 bu 的对象属性值不相等");
    bu.setUname("ch");
    assertEquals(au, bu, "au 与 bu 的对象属性值不相等");
}
```

❷ **assertSame 和 assertNotSame**

assertSame 与 assertEquals 有所区别，assertSame 不仅判断值是否相等，还判断类型是否相同。对于对象，assertSame 判断两者的引用是否为同一个。示例代码如下：

```
@Test
void yourTest(){
    int a = 1;
    long b = 1L;
    long c = 1L;
    //b 和 c 比较，判断成功，因为它们的类型也相同
    assertSame(b, c, "测试失败");
    //a 和 b 比较，判断失败，因为它们的类型不相同
    assertSame(a, b, "测试失败");
    MyUser au = new MyUser();
    MyUser bu = new MyUser();
    MyUser cu = bu;
    //bu 和 cu 比较，判断成功，因为它们的引用是同一个
    assertSame(bu, cu, "测试失败");
    //au 和 bu 比较，判断失败，因为它们的引用不相同
    assertSame(au, bu, "测试失败");
}
```

❸ **assertNull 和 assertNotNull**

Assertions.assertNull(Object actual)的实际测试值是 null，则单元测试成功。

❹ assertTrue 和 assertFalse

Assertions.assertTrue(boolean condition)的实际测试值是 true,则单元测试成功。

❺ assertThrows

Assertions.assertThrows（Class $<$ T $>$ expectedType，Executable executable，String message）判断 executable 方法在执行过程中是否抛出指定异常 expectedType,如果没有抛出异常,或者抛出的异常类型不对,则单元测试失败。示例代码如下:

```
@Test
void testAssertThrows() {
    assertThrows(ArithmeticException.class, () -> errorMethod());
}
private void errorMethod() {
    int a[] = {1,2,3,4,5};
    for(int i = 0; i <= 5; i++){
        System.out.println(a[i]);
    }
}
```

❻ assertDoesNotThrow

assertDoesNotThrow(Executable executable)判断测试方法是否抛出异常,如果没有抛出任何异常,则单元测试成功。示例代码如下:

```
@Test
void testAssertDoesNotThrow() {
    assertDoesNotThrow(() -> rightMethod());
}
private void rightMethod() {
    int a = 1/1;
}
```

❼ assertAll

assertAll(Executable... executables)判断一组断言是否都成功,如果都成功,则整个单元测试成功。示例代码如下:

```
@Test
void testAll(){
    assertAll(
            () -> assertEquals(1, 1),
            () -> assertNotEquals(1, 2),
            () -> assertNull(null)
    );
}
```

15.2 单元测试用例

本节将讲解如何使用 JUnit 5 的注解与断言机制编写单元测试用例。在编写单元测试用例前,首先进行测试环境的构建。

▶ 15.2.1 测试环境的构建

在 Spring Boot Web 应用中已经集成了 JUnit 5 和 JSON 相关的 JAR 包,所以可以直接进行单元测试。为了方便本章后续的单元测试,下面构建一个 Spring Boot Web 应用,具体步骤如下。

第 15 章 Spring Boot 单元测试

❶ 创建 Spring Boot Web 应用

创建基于 Lombok、MySQL Driver、Spring Web 依赖的 Spring Boot Web 应用 ch15。

❷ 修改 pom.xml 文件

在 pom.xml 文件中添加 MyBatis-Plus 依赖，与例 14-2 相同，这里不再赘述。

❸ 设置 Web 应用 ch15 的上下文路径及数据源配置信息

在应用 ch15 的 application.properties 文件中配置上下文路径及数据源配置信息，配置内容与例 14-2 的基本相同，这里不再赘述。

❹ 创建实体类

创建名为 com.ch.ch15.entity 的包，并在该包中创建 MyUser 实体类，MyUser 的代码与例 14-2 的相同，这里不再赘述。

❺ 创建数据访问接口

创建名为 com.ch.ch15.mapper 的包，并在该包中创建 UserMapper 接口。UserMapper 接口通过继承 BaseMapper＜MyUser＞接口对实体类 MyUser 对应的数据表 user 进行 CRUD 操作。UserMapper 接口的代码与例 14-2 的相同，这里不再赘述。

❻ 创建 Service 接口及实现类

创建名为 com.ch.ch15.service 的包，并在该包中创建 UserService 接口及实现类 UserServiceImpl。UserService 接口及实现类 UserServiceImpl 的代码与例 14-2 的相同，这里不再赘述。

❼ 创建控制器类 MyUserController

创建名为 com.ch.ch15.controller 的包，并在该包中创建控制器类 MyUserController。MyUserController 的核心代码如下：

```java
@RestController
public class MyUserController {
    @Autowired
    private UserMapper userMapper;
    @Autowired
    private UserService userService;
    @GetMapping("/selectAllUsers")
    public List<MyUser> selectAllUsers(){
        return userMapper.selectList(null);
    }
    @PostMapping("/addAUser")
    public MyUser addAUser(MyUser mu){
        //实体类的主键属性使用@TableId注解后，主键自动回填
        int result = userMapper.insert(mu);
        return mu;
    }
    @PutMapping("/updateAUser")
    public boolean updateAUser(MyUser mu){
        return userService.updateById(mu);
    }
    @DeleteMapping("/deleteAUser")
    public boolean  deleteAUser(MyUser mu){
        return userService.removeById(mu);
    }
    @GetMapping("/getOne")
    public MyUser getOne(int id){
        return userService.getById(id);
    }
}
```

❽ 在应用程序的主类中扫描 Mapper 接口

在应用程序的 Ch15Application 主类中使用@MapperScan 注解扫描 MyBatis 的 Mapper 接口,核心代码如下:

```
@SpringBootApplication
@MapperScan(basePackages={"com.ch.ch15.mapper"})
public class Ch15Application {
    public static void main(String[] args) {
        SpringApplication.run(Ch15Application.class, args);
    }
}
```

经过以上步骤,完成测试环境的构建。下面使用 JUnit 5 测试框架对 ch15 应用中的 Mapper 接口、Service 层进行单元测试。

▶15.2.2 测试 Mapper 接口

在 IntelliJ IDEA 中选中类或接口的名字,按快捷键 Ctrl+Shift+T 创建测试类,此时生成的测试类在 test 文件夹中,测试方法都是 void 方法。图 15.1 所示为 Create Test 窗口。

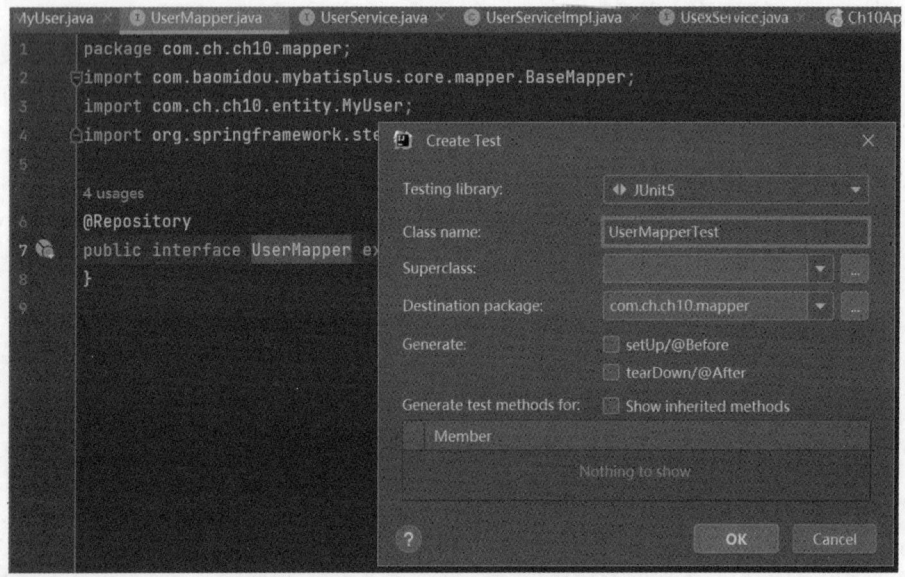

图 15.1 创建 Mapper 接口的测试类

单击图 15.1 中的 OK 按钮,完成 Mapper 接口 UserMapper 的单元测试类 UserMapperTest 的创建。

@SpringBootTest 用于 Spring Boot 应用测试,它默认根据包名逐级往上找,一直找到 Spring Boot 主程序(包含@SpringBootApplication 注解的类),并在单元测试时启动该主程序来创建 Spring 上下文环境,所以需要在单元测试类上使用@SpringBootTest 注解标注后才能进行单元测试。

在测试类 UserMapperTest 中使用 JUnit 5 的注解与断言进行 Mapper 接口方法的测试。UserMapperTest 的代码具体如下:

```
package com.ch.ch15.mapper;
import com.ch.ch15.entity.MyUser;
import org.junit.jupiter.api.Test;
import org.springframework.beans.factory.annotation.Autowired;
```

```
import org.springframework.boot.test.context.SpringBootTest;
import static org.junit.jupiter.api.Assertions.*;
@SpringBootTest
class UserMapperTest {
    @Autowired
    private UserMapper userMapper;
    @Test
    void getOne(){
        MyUser mu = userMapper.selectById(1);
        assertEquals(mu.getUid(), 1, "a与b不相等");
    }
}
```

运行上述测试类 UserMapperTest 中的测试方法即可进行 Mapper 接口方法的测试。例如，右击 getOne 方法名，选择 Run 'getOne()'运行测试方法 getOne，这里测试成功，如图 15.2 所示。

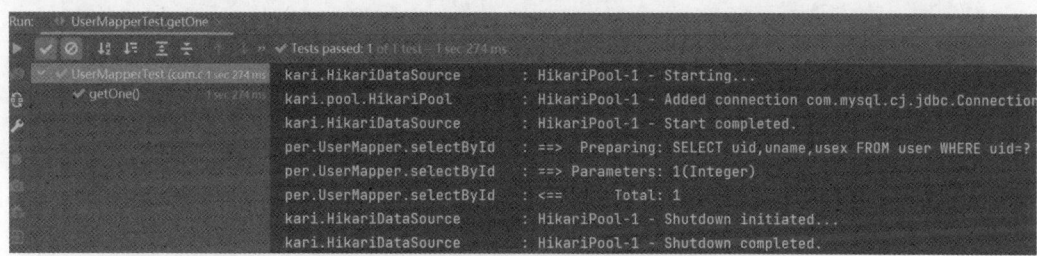

图 15.2　测试方法 getOne 的运行效果

15.2.3　测试 Service 层

单元测试 Service 层与测试 Mapper 接口类似，需要特别考虑 Service 是否依赖其他还未开发完毕的 Service（第三方接口）。如果依赖其他还未开发完毕的 Service，开发者需要使用 Mockito（Java Mock 测试框架，用于模拟任何 Spring 管理的 Bean）来模拟未完成的 Service。

假设应用 ch15 的 UserServiceImpl 类依赖一个还未开发完毕的第三方接口 UsexService。在接口 UsexService 中有一个获得用户性别的接口方法 getUsex，具体代码如下：

```
package com.ch.ch15.service;
public interface UsexService {
    String getUsex(int id);
}
```

现在创建 UserServiceImpl 的测试类 UserServiceImplTest，在 UserServiceImplTest 类中使用 Mockito.mock 方法模拟第三方接口 UsexService 的对象，并进行测试。UserServiceImplTest 类的代码具体如下：

```
package com.ch.ch15.service;
import org.junit.jupiter.api.Test;
import org.mockito.BDDMockito;
import org.mockito.Mockito;
import org.springframework.beans.factory.annotation.Autowired;
import org.springframework.boot.test.context.SpringBootTest;
import static org.junit.jupiter.api.Assertions.*;
import static org.mockito.ArgumentMatchers.anyInt;
@SpringBootTest
class UserServiceImplTest {
    @Autowired
    private UserService userService;
    //模拟第三方接口 UsexService 的对象
```

```
        private UsexService usexService = Mockito.mock(UsexService.class);
        @Test
        void testGetOne() {
            int uid = 1;
            String expectedUsex = "男";
            /* given 是 BDDMockito 的一个静态方法,用来模拟一个 Service 方法调用返回,anyInt()表示可
以传入任何参数,willReturn 方法说明这个调用将返回男。 */
            BDDMockito.given(usexService.getUsex(anyInt())).willReturn(expectedUsex);
            assertEquals(usexService.getUsex(1), userService.getById(uid).getUsex(),"测试失
败,与期望值不一致");
        }
    }
```

运行上述 testGetOne 测试方法,运行效果如图 15.3 所示。

图 15.3　testGetOne 测试方法的运行效果

15.3　使用 Postman 测试 Controller 层

在 Spring 框架中可以使用 org.springframework.test.web.servlet.MockMvc 类进行 Controller 层的测试。MockMvc 类的核心方法如下。

```
public ResultActions perform(RequestBuilder requestBuilder)
```

通过 perform 方法调用 MockMvcRequestBuilders 的 get、post、multipart 等方法来模拟 Controller 请求,进而测试 Controller 层,模拟请求示例如下。

模拟一个 get 请求:

```
mvc.peform(get("/getCredit/{id}", uid));
```

模拟一个 post 请求:

```
mvc.peform(post("/getCredit/{id}", uid));
```

模拟文件上传:

```
mvc.peform(multipart("/upload").file("file", "文件内容".getBytes("UTF-8")));
```

模拟请求参数:

```
//模拟提交 errorMessage 参数
mvc.peform(get("/getCredit/{id}/{uname}", uid, uname).param("errorMessage", "用户名或密码
错误"));
//模拟提交 check
mvc.peform(get("/getCredit/{id}/{uname}", uid, uname).param("job", "收银员", "IT"));
```

综上所知,在通过 MockMvc 类的 perform 方法测试 Controller 时,需要编写复杂的单元测试程序。为了提高 Controller 层的测试效率,本节将讲解一个针对 Controller 层测试的接口测试工具 Postman。

第 15 章　Spring Boot 单元测试

　　Postman 是一个接口测试工具，在做接口测试时，Postman 相当于一个客户端，它可以模拟用户发起的各类 HTTP 请求，将请求数据发送至服务端，获取对应的响应结果，从而验证响应中的结果数据是否和预期值相匹配。

　　Postman 主要用来模拟各种 HTTP 请求（如 get、post、delete、put 等），Postman 与浏览器的区别在于有的浏览器不能输出 JSON 格式，而 Postman 可以更直观地看到接口返回的结果。

　　开发者可以从官网 https://www.postman.com/ 下载对应的 Postman 安装程序。在安装成功后，不需要创建账号即可使用。

　　下面讲解如何使用 Postman 测试 ch15 应用的 Controller 类 MyUserController。首先将 ch15 应用的主类启动，然后打开 Postman 客户端，即可测试 Controller 类中的请求方法。

❶ 测试 selectAllUsers 方法

　　在 Postman 客户端中输入 Controller 请求方法 selectAllUsers 对应的 URL，并选择对应的请求方式 GET，然后单击 Send 按钮，即可完成查询所有用户的测试，如图 15.4 所示。

图 15.4　selectAllUsers 方法的测试结果

❷ 测试 addAUser 方法

　　在 Postman 客户端中输入 Controller 请求方法 addAUser 对应的 URL 与表单数据，并选择对应的请求方式 POST，然后单击 Send 按钮，即可完成添加用户的测试，如图 15.5 所示。

图 15.5　addAUser 方法的测试结果

❸ 测试 updateAUser 方法

在 Postman 客户端中输入 Controller 请求方法 updateAUser 对应的 URL 与表单数据，并选择对应的请求方式 PUT，然后单击 Send 按钮，即可完成修改用户的测试，如图 15.6 所示。

图 15.6　updateAUser 方法的测试结果

❹ 测试 deleteAUser 方法

在 Postman 客户端中输入 Controller 请求方法 deleteAUser 对应的 URL 与表单数据，并选择对应的请求方式 DELETE，然后单击 Send 按钮，即可完成删除用户的测试，如图 15.7 所示。

图 15.7　deleteAUser 方法的测试结果

❺ 测试 getOne 方法

在 Postman 客户端中输入 Controller 请求方法 getOne 对应的 URL 与表单数据，并选择对应的请求方式 GET，然后单击 Send 按钮，即可完成查询一个用户的测试，如图 15.8 所示。

图 15.8　getOne 方法的测试结果

本章小结

本章首先重点讲解了 JUnit 5 的常用注解与断言,然后详细讲解了在 IntelliJ IDEA 中如何使用 JUnit 5 进行 Mapper 接口与 Service 层的单元测试,最后介绍了一个针对 Controller 层测试的接口测试工具 Postman,该测试工具在前后端分离开发中广泛应用于 RESTful 接口测试。

习题 15

1. 在 JUnit 5 测试框架中,测试方法的返回值是( )类型。
   A. boolean　　　　B. int　　　　　　C. void　　　　　　D. 以上都不是
2. 在 JUnit 5 测试框架中,使用( )注解标注的方法是测试方法。
   A. @Test　　　　B. @RepeatedTest　　C. @BeforeEach　　D. @AfterEach
3. 简述断言方法 assertSame 与 assertEquals 的区别。

第 16 章　电子商务平台的设计与实现（Spring Boot+Vue.js 3+MyBatis-Plus）

学习目的与要求

本章以电子商务平台的设计与实现为综合案例，讲述如何使用 Spring Boot ＋ Vue.js 3 ＋ MyBatis-Plus 开发一个前后端分离的应用程序。通过本章的学习，掌握基于 Spring Boot ＋ Vue.js 3 ＋ MyBatis-Plus 的前后端分离的应用程序的开发流程、方法以及技术。

本章主要内容

- 使用 IntelliJ IDEA 构建电子商务平台后端系统
- 使用 Vue CLI 构建电子商务平台前端系统

前后端分离的核心思想是前端页面通过 Ajax 调用后端的 RESTful API 进行数据交互。本章将使用 Spring Boot ＋ MyBatis-Plus 实现后端系统，使用 Vue.js 3 实现前端系统，数据库采用的是 MySQL 8，后端集成开发环境为 IntelliJ IDEA，前端集成开发环境为 Visual Studio Code。

本章电子商务平台的系统设计与数据库设计在本书第 9 章已经介绍，在此不再赘述。本章将重点介绍如何使用 Spring Boot ＋ Vue.js 3 ＋ MyBatis-Plus 开发一个前后端分离的电子商务平台。

16.1　使用 IntelliJ IDEA 构建后端系统

本节将讲解如何使用 IntelliJ IDEA 构建电子商务平台的后端应用 ch16，具体步骤如下。

▶16.1.1　创建 Spring Boot Web 应用

使用 IntelliJ IDEA 创建基于 Lombok、Spring Data Redis、MySQL Driver、Spring Cache Abstraction 以及 Spring Web 依赖的 Spring Boot Web 应用 ch16。

▶16.1.2　修改 pom.xml

修改后端应用 ch16 的 pom.xml 文件，添加 MyBatis-Plus 以及 Java 工具类 Hutool 的依赖，具体代码如下：

```xml
<dependency>
    <groupId>com.baomidou</groupId>
    <artifactId>mybatis-plus-boot-starter</artifactId>
    <version>3.5.5</version>
    <exclusions>
        <exclusion>
            <groupId>org.mybatis</groupId>
            <artifactId>mybatis-spring</artifactId>
        </exclusion>
    </exclusions>
</dependency>
<dependency>
    <groupId>org.mybatis</groupId>
    <artifactId>mybatis-spring</artifactId>
```

```xml
        <version>3.0.3</version>
    </dependency>
    <dependency>
        <groupId>cn.hutool</groupId>
        <artifactId>hutool-all</artifactId>
        <version>5.8.11</version>
    </dependency>
```

▶16.1.3 配置数据源等信息

在后端系统 ch16 的配置文件 application.properties 中配置端口号、数据源以及文件上传等信息，具体内容如下：

```
server.servlet.context-path=/eBusiness
spring.servlet.multipart.location=D:/data/apps/temp
server.port=8443
#数据库地址
spring.datasource.url=jdbc:mysql://localhost:3306/ch16?useUnicode=true&characterEncoding=UTF-8&allowMultiQueries=true&serverTimezone=GMT%2B8
#数据库用户名
spring.datasource.username=root
#数据库密码
spring.datasource.password=root
#数据库驱动
spring.datasource.driver-class-name=com.mysql.cj.jdbc.Driver
#设置包的别名(在 Mapper 映射文件中直接使用实体类名)
mybatis-plus.type-aliases-package=com.ch.ch16.entity
#告诉系统到哪里去找 mapper.xml 文件(映射文件)
mybatis-plus.mapper-locations=classpath:mappers/*.xml
#在控制台中输出 SQL 语句日志
logging.level.com.ch.ch16.mapper=debug
#让控制器输出的 JSON 字符串格式更美观
spring.jackson.serialization.indent-output=true
```

▶16.1.4 创建持久化实体类

根据电子商务平台的功能可知，该系统共有 8 个实体，即管理员、用户、商品类型、商品、购物车、收藏商品、订单基础信息以及订单详情，因此需要在应用 ch16 的 src/main/java 目录下创建 com.ch.ch16.entity 包，并在该包中创建这 8 个实体对应的持久化类。对于持久化实体类的代码，请读者参见本书提供的源程序 ch16。

▶16.1.5 创建 Mapper 接口

在后端应用 ch16 中使用 MyBatis-Plus 进行数据访问，因此在应用 ch16 的 src/main/java 目录下创建 com.ch.ch16.mapper 包，在该包中针对 8 个实体类创建数据访问接口，并分别继承 BaseMapper 接口。对于接口的代码，读者可参见本书提供的源程序 ch16。

在后端应用 ch16 中尽量使用 MyBatis-Plus 的 BaseMapper 接口和 IService 接口操作数据库，但个别功能需要开发者自己编写 SQL 语句操作数据库才能实现。

GoodsMapper 接口对应的 Mapper 映射文件的代码如下：

```xml
<?xml version="1.0" encoding="UTF-8" ?>
<!DOCTYPE mapper
    PUBLIC "-//mybatis.org//DTD Mapper 3.0//EN"
    "http://mybatis.org/dtd/mybatis-3-mapper.dtd">
<mapper namespace="com.ch.ch16.mapper.admin.GoodsMapper">
    <!--${ew.customSqlSegment}是一个动态 SQL 语句占位符，可以用于构建复杂的 SQL 查询条件。
```

其中，ew 是一个 Wrapper 对象，用于构建查询条件。在 ew 对象中，可以使用各种方法来构建查询条件，例如，eq 构建相等条件，ne 构建不等条件，like 构建模糊查询条件，in 构建包含条件，between 构建范围查询条件等。在使用${ew.customSqlSegment}时，MyBatis-Plus 会将 ew 对象中构建的查询条件转化为 SQL 语句，并将其插入${ew.customSqlSegment}占位符位置。这样就可以在 SQL 语句中动态构建查询条件，实现更加灵活的查询功能。-->

```xml
    <select id="iPageGoods" resultType="GoodsEntity">
        select gt.*, gy.typename
        from goodstable as gt
            join goodstype as gy on gy.id = gt.goodstype_id
            ${ew.customSqlSegment}
    </select>
    <!--我的收藏-->
    <select id="iPageMyFocusGoods" resultType="GoodsEntity">
        select gt.*, gy.typename
        from goodstable as gt
             join goodstype as gy on gy.id = gt.goodstype_id
             join focustable as ft on ft.goodstable_id = gt.id
            ${ew.customSqlSegment}
    </select>
    <!--我的购物车-->
    <select id="myCartGoods" resultType="map">
        select gt.*, ct.shoppingnum, ct.id cid
        from goodstable as gt join carttable as ct on ct.goodstable_id = gt.id
        where ct.busertable_id = #{uid}
    </select>
</mapper>
```

OrdersMapper 接口对应的 Mapper 映射文件的代码如下：

```xml
<?xml version="1.0" encoding="UTF-8"?>
<!DOCTYPE mapper
      PUBLIC "-//mybatis.org//DTD Mapper 3.0//EN"
      "http://mybatis.org/dtd/mybatis-3-mapper.dtd">
<mapper namespace="com.ch.ch16.mapper.admin.OrdersMapper">
    <select id="getOrdersDetail" resultType="map" parameterType="OrdersEntity">
        select gt.id, gt.gname, gt.grprice, od.shoppingnum, gt.grprice * od.shoppingnum smallTotalJP
        from orderdetail as od
            join goodstable as gt on od.goodstable_id = gt.id
            where od.orderbasetable_id = #{id}
    </select>
    <!-- 按类型统计(最近一年的) -->
    <select id="selectOrderByType" resultType="map">
        SELECT
            SUM(gt.grprice * od.shoppingnum) AS value,
            gy.typename AS name
        FROM
            orderbasetable ob
            JOIN orderdetail od ON ob.id = od.orderbasetable_id
            JOIN goodstable gt ON od.goodstable_id = gt.id
            JOIN goodstype gy ON gt.goodstype_id = gy.id
        WHERE
            ob.status = 1
            AND ob.orderdate > DATE_SUB(CURDATE(), INTERVAL 1 YEAR)
        GROUP BY
            gy.typename
    </select>
</mapper>
```

16.1.6 创建业务层

在 Spring 框架中提倡使用接口,因此在后端应用 ch16 的业务层中涉及 Service 接口和 Service 实现类。Service 接口继承 IService 接口,Service 实现类继承 ServiceImpl 类。

在应用 ch16 的 src/main/java 目录下创建 com.ch.ch16.service 包,并在该包中创建管理员功能和用户功能对应的 Service 接口与实现类。Service 接口的代码略,本节仅提供 Service 实现类的代码。

与管理员登录相关的 Service 实现类 AUserServiceImpl 的核心代码如下:

```
@Service
@SuppressWarnings("all")
public class AUserServiceImpl extends ServiceImpl<AUserMapper, AUserEntity> implements AUserService{
    @Autowired
    private JwtTokenUtil jwtUtil;
    @Autowired
    private RedisUtil redisUtil;
    @Autowired
    private ConfigrationBean config;
    @Override
    public ResponseResult<Map<String, String>> login(AUserEntity aUserEntity) {
        //链式 query
        long res = lambdaQuery().eq(AUserEntity::getAname, aUserEntity.getAname()).count();
        if(res == 0) {
            //用户名不存在
            return ResponseResult.getMessageResult(null, "A001");
        }
        Map<String, String> mapparam = new HashMap<>();
        mapparam.put("aname", aUserEntity.getAname());
        mapparam.put("apwd", aUserEntity.getApwd());
        List<AUserEntity> mu = query().allEq(mapparam).list();
        if(mu.size() > 0){                    //登录成功
            String token = jwtUtil.createToken(aUserEntity.getAname());
            //在签名时验证是否过期
            redisUtil.set("login_" + aUserEntity.getAname(), aUserEntity.getAname(), config.getRedisExpiration());
            Map<String, String> myres = new HashMap<>();
            myres.put("authtoken", token);
            myres.put("aname", aUserEntity.getAname());
            myres.put("aid", mu.get(0).getId()+"");
            return ResponseResult.getSuccessResult(myres);
        }else{                                //密码错误
            return ResponseResult.getMessageResult(null, "A002");
        }
    }
}
```

与商品相关的 Service 实现类 GoodsServiceImpl 的核心代码如下:

```
@Service
@Service
@SuppressWarnings("all")
public class GoodsServiceImpl extends ServiceImpl< GoodsMapper, GoodsEntity > implements GoodsService{
    @Autowired
    private GoodsMapper goodsMapper;
    @Autowired
    private CartService cartService;
```

```java
    @Autowired
    private FocusService focusService;
    @Autowired
    private OrderdetailService orderdetailService;
    /**
     * 管理员查询商品
     */
    @Override
    public ResponseResult<Map<String, Object>> getGoods(GoodsEntity goodsEntity) {
        //2 为每页大小
        IPage<GoodsEntity> iPage = new Page<>(goodsEntity.getCurrentPage(), 2);
        //构造条件
        LambdaQueryWrapper<GoodsEntity> lambdaQueryWrapper = new LambdaQueryWrapper<>();
        if (goodsEntity.getGoodstypeId() != null && goodsEntity.getGoodstypeId() > 0)
            lambdaQueryWrapper.eq(GoodsEntity::getGoodstypeId, goodsEntity.getGoodstypeId());
        if(goodsEntity.getGname() != null && goodsEntity.getGname().length() >0)
            lambdaQueryWrapper.like(GoodsEntity::getGname, goodsEntity.getGname());
        //分页查询(商品管理首页进来时的初始查询及条件查询)
        IPage<GoodsEntity> page = goodsMapper.iPageGoods(iPage, lambdaQueryWrapper);
        Map<String, Object> myres = new HashMap<>();
        myres.put("allGoods", page.getRecords());
        myres.put("totalPage", page.getPages());
        return ResponseResult.getSuccessResult(myres);
    }
    /**
     * 查询用户的关注商品
     */
    @Override
    public ResponseResult<Map<String, Object>> iPageMyFocusGoods(GoodsEntity goodsEntity) {
        //2 为每页大小
        IPage<GoodsEntity> iPage = new Page<>(goodsEntity.getCurrentPage(), 2);
        //构造条件
        QueryWrapper<FocusEntity> queryWrapper = new QueryWrapper<>();
        queryWrapper.eq("ft.busertable_id", goodsEntity.getBusertableId());
        //分页查询
        IPage<GoodsEntity> page = goodsMapper.iPageMyFocusGoods(iPage, queryWrapper);
        Map<String, Object> myres = new HashMap<>();
        myres.put("allGoods", page.getRecords());
        myres.put("totalPage", page.getPages());
        return ResponseResult.getSuccessResult(myres);
    }
    /**
     * 查询用户的购物车
     */
    @Override
    public ResponseResult<List<Map<String, Object>>> myCartGoods(CartEntity cartEntity) {
        List<Map<String, Object>> cartGoods = goodsMapper.myCartGoods(cartEntity.getBusertableId());
        return ResponseResult.getSuccessResult(cartGoods);
    }
    /**
     * 获得轮播广告商品
     */
    @Override
    public ResponseResult<List<GoodsEntity>> getAdvGoods() {
        List<GoodsEntity> listAdv = lambdaQuery().eq(GoodsEntity::getIsAdvertisement, 1)
                .orderByDesc(GoodsEntity::getId).last("limit 5").list();
        return ResponseResult.getSuccessResult(listAdv);
    }
```

第16章　电子商务平台的设计与实现（Spring Boot+ Vue.js 3+ MyBatis-Plus）

```java
    @Override
    public ResponseResult<Map<String, String>> add(GoodsEntity goodsEntity) {
        byte[] myfile = goodsEntity.getLogoFile();
        //如果选择了上传文件
        if(myfile != null && myfile.length > 0) {
            String path = "D:\\idea-workspace\\ebusiness-vue\\src\\assets";
            //获得上传文件的原名
            String fileName = goodsEntity.getFileName();
            //对文件重命名
            String fileNewName = MyUtil.getNewFileName(fileName);
            File filePath = new File(path + File.separator + fileNewName);
            //如果文件目录不存在,创建目录
            if(!filePath.getParentFile().exists()) {
                filePath.getParentFile().mkdirs();
            }
            FileOutputStream out = null;
            try {
                out = new FileOutputStream(filePath);
                out.write(myfile);
            } catch (IOException e) {
                throw new RuntimeException(e);
            }
            //将重命名后的图片名存到goodsEntity对象中,在添加时使用
            goodsEntity.setGpicture(fileNewName);
        }
        if("add".equals(goodsEntity.getAct())) {
            boolean result = save(goodsEntity);
            if(result)                              //成功
                return ResponseResult.getMessageResult(null, "A001");
            else
                return ResponseResult.getMessageResult(null, "A002");
        }else {                                     //修改
            boolean result = updateById(goodsEntity);
            if(result)
                return ResponseResult.getMessageResult(null, "A001");
            else
                return ResponseResult.getMessageResult(null, "A002");
        }
    }
    @Override
    public ResponseResult<Map<String, String>> delete(GoodsEntity goodsEntity) {
        //先查询商品是否关联
        long res1 = cartService.lambdaQuery().eq(CartEntity::getGoodstableId,
goodsEntity.getId()).count();
        long res2 = focusService.lambdaQuery().eq(FocusEntity::getGoodstableId,
goodsEntity.getId()).count();
        long res3 = orderdetailService.lambdaQuery().eq(OrderdetailEntity::
getGoodstableId, goodsEntity.getId()).count();
        if (res1 > 0 || res2 > 0 || res3 > 0)       //有关联,不能删除
            return ResponseResult.getMessageResult(null, "A001");
        if (removeById(goodsEntity))
            return ResponseResult.getMessageResult(null, "A002");
        return ResponseResult.getMessageResult(null, "A003");
    }
    @Override
    public ResponseResult<List<GoodsEntity>> getGoodsIndex(GoodsEntity goodsEntity) {
        //构造条件
        LambdaQueryWrapper<GoodsEntity> lambdaQueryWrapper = new LambdaQueryWrapper<>();
        if(goodsEntity.getGoodstypeId() != null && goodsEntity.getGoodstypeId() != 0)
            lambdaQueryWrapper.eq(GoodsEntity::getGoodstypeId,goodsEntity.getGoodstypeId());
        if(goodsEntity.getGname() != null && goodsEntity.getGname().trim().length() > 0)
```

```
            lambdaQueryWrapper.like(GoodsEntity::getGname, goodsEntity.getGname());
            lambdaQueryWrapper.orderByDesc(GoodsEntity::getId).last("limit 15");
            //执行查询
            List<GoodsEntity> listIndex = list(lambdaQueryWrapper);
            return ResponseResult.getSuccessResult(listIndex);
        }
        @Override
        public ResponseResult<GoodsEntity> getGoodsById(GoodsEntity goodsEntity) {
            return ResponseResult.getSuccessResult(getById(goodsEntity.getId()));
        }
    }
```

与商品类型相关的 Service 实现类 GoodsTypeServiceImpl 的核心代码如下：

```
@Service
@SuppressWarnings("all")
public class GoodsTypeServiceImpl extends ServiceImpl<GoodsTypeMapper, GoodsTypeEntity>
implements GoodsTypeService {
    @Autowired
    private GoodsService goodsService;
    @Override
    public ResponseResult<Map<String, Object>> getGoodsType(GoodsTypeEntity
goodsTypeEntity) {
        //2 为每页大小
        IPage<GoodsTypeEntity> iPage = new Page<>(goodsTypeEntity.getCurrentPage(), 2);
        //分页查询
        IPage<GoodsTypeEntity> page = page(iPage);
        Map<String, Object> myres = new HashMap<>();
        myres.put("allGoodsTypes", page.getRecords());
        myres.put("totalPage", page.getPages());
        return ResponseResult.getSuccessResult(myres);
    }
    @Override
    public ResponseResult<Map<String, Object>> add(GoodsTypeEntity goodsTypeEntity) {
        boolean b = save(goodsTypeEntity);
        if (b)
            return ResponseResult.getMessageResult(null, "A001");
        return ResponseResult.getMessageResult(null, "A002");
    }
    @Override
    public ResponseResult<Map<String, Object>> update(GoodsTypeEntity goodsTypeEntity) {
        boolean b = updateById(goodsTypeEntity);
        if (b)
            return ResponseResult.getMessageResult(null, "A001");
        return ResponseResult.getMessageResult(null, "A002");
    }
    @Override
    public ResponseResult<Map<String, Object>> delete(GoodsTypeEntity goodsTypeEntity) {
        //先查询类型是否关联
        long res = goodsService.query().eq("goodstype_id", goodsTypeEntity.getId()).count();
        if (res > 0)                                              //有关联,不能删除
            return ResponseResult.getMessageResult(null, "A001");
        if (removeById(goodsTypeEntity))
            return ResponseResult.getMessageResult(null, "A002");
        return ResponseResult.getMessageResult(null, "A003");
    }
}
```

与订单相关的 Service 实现类 OrdersServiceImpl 的核心代码如下：

```
@Service
```

第 16 章 电子商务平台的设计与实现（Spring Boot+ Vue.js 3+ MyBatis-Plus）

```java
@SuppressWarnings("all")
public class OrdersServiceImpl extends ServiceImpl<OrdersMapper, OrdersEntity> implements OrdersService {
    @Autowired
    private OrdersMapper ordersMapper;
    @Autowired
    private OrderdetailService orderdetailService;
    @Autowired
    private GoodsService goodsService;
    @Autowired
    private CartService cartService;
    @Override
    public ResponseResult<Map<String, Object>> getAllOrders(OrdersEntity ordersEntity) {
        //2 为每页大小
        IPage<OrdersEntity> iPage = new Page<>(ordersEntity.getCurrentPage(), 2);
        //构造条件
        QueryWrapper<OrdersEntity> queryWrapper = new QueryWrapper<>();
        if (ordersEntity.getId() != null && ordersEntity.getId() > 0)
            queryWrapper.eq("id", ordersEntity.getId());
        //分页查询(商品管理首页进来时的初始查询及条件查询)
        IPage<OrdersEntity> page = page(iPage,queryWrapper);
        Map<String, Object> myres = new HashMap<>();
        myres.put("allOrders", page.getRecords());
        myres.put("totalPage", page.getPages());
        return ResponseResult.getSuccessResult(myres);
    }
    @Override
    public ResponseResult<Map<String, Object>> getOrdersByUid(OrdersEntity ordersEntity) {
        //2 为每页大小
        IPage<OrdersEntity> iPage = new Page<>(ordersEntity.getCurrentPage(), 2);
        //构造条件
        QueryWrapper<OrdersEntity> queryWrapper = new QueryWrapper<>();
        queryWrapper.eq("busertable_id", ordersEntity.getBusertableId());
        //分页查询(商品管理首页进来时的初始查询及条件查询)
        IPage<OrdersEntity> page = page(iPage,queryWrapper);
        Map<String, Object> myres = new HashMap<>();
        myres.put("ordersByUid", page.getRecords());
        myres.put("totalPage", page.getPages());
        return ResponseResult.getSuccessResult(myres);
    }
    @Override
    public ResponseResult<List<Map<String, Object>>> getOrdersDetail(OrdersEntity ordersEntity) {
        return ResponseResult.getSuccessResult(ordersMapper.getOrdersDetail(ordersEntity));
    }

    @Override
    public ResponseResult<Map<String, String>> goPay(OrdersEntity ordersEntity) {
        ordersEntity.setStatus(1);
        if(updateById(ordersEntity))
            return ResponseResult.getMessageResult(null, "A001");
        return ResponseResult.getMessageResult(null, "A002");
    }
    @Override
    public ResponseResult<List<OrdersEntity>> selectOrderByMonth(OrdersEntity ordersEntity) {
        //默认查询所有订单
        if(ordersEntity.getStartDate().length() == 0)
            ordersEntity.setStartDate("1900-01");
```

```java
            if(ordersEntity.getEndDate().length() == 0)
                ordersEntity.setEndDate("9000-12");
            List<OrdersEntity> ordersEntityList = query()
                    .select("sum(amount) totalamount","date_format(orderdate,'%Y-%m') months")
                    .eq("status", 1)
                    .between("date_format(orderdate,'%Y-%m')", ordersEntity.getStartDate(),
                        ordersEntity.getEndDate())
                    .groupBy("months")
                    .orderByAsc("months")
                    .list();
        return ResponseResult.getSuccessResult(ordersEntityList);
    }
    @Override
    public ResponseResult<List<Map<String, Object>>> selectOrderByType() {
        return ResponseResult.getSuccessResult(ordersMapper.selectOrderByType());
    }
    @Override
    @Transactional                              //事务管理,submitOrder方法是原子性的
    public ResponseResult<OrdersEntity> submitOrder(OrdersEntity ordersEntity) {
        ordersEntity.setStatus(0);
        ordersEntity.setOrderdate(new Date());
        //生成订单
        save(ordersEntity);
        //生成订单详情
        List<Integer> bgid = ordersEntity.getBgid();
        List<Integer> bshoppingnum = ordersEntity.getBshoppingnum();
        List<OrderdetailEntity> bods = new ArrayList<OrderdetailEntity>();
        List<GoodsEntity> ges = new ArrayList<GoodsEntity>();
        for (int i = 0; i < bgid.size(); i++) {
            OrderdetailEntity ot = new OrderdetailEntity();
            GoodsEntity ge = new GoodsEntity();
            ot.setOrderbasetableId(ordersEntity.getId());
            ot.setGoodstableId(bgid.get(i));
            ot.setShoppingnum(bshoppingnum.get(i));
            bods.add(ot);
            ge.setId(bgid.get(i));
            //减少库存
            ge.setGstore(goodsService.getById(bgid.get(i)).getGstore() - bshoppingnum.get(i));
            ges.add(ge);
        }
        orderdetailService.saveBatch(bods);
        //批量更新库存
        goodsService.updateBatchById(ges);
        //清空购物车
        CartEntity cartEntity = new CartEntity();
        cartEntity.setBusertableId(ordersEntity.getBusertableId());
        cartService.clearCart(cartEntity);
        return ResponseResult.getSuccessResult(ordersEntity);
    }
}
```

与用户相关的 Service 实现类 BUserServiceImpl 的核心代码如下：

```java
@Service
@SuppressWarnings("all")
public class BUserServiceImpl extends ServiceImpl<BUserMapper, BUserEntity> implements BUserService{
    @Autowired
    private JwtTokenUtil jwtUtil;
    @Autowired
```

第 16 章　电子商务平台的设计与实现（Spring Boot+ Vue.js 3+ MyBatis Plus）

```java
    private RedisUtil redisUtil;
    @Autowired
    private ConfigurationBean config;
    @Override
    public ResponseResult<Map<String, String>> register(BUserEntity bUserEntity) {
        //对明文加密
        bUserEntity.setBpwd(MD5Util.MD5(bUserEntity.getBpwd()));
        long n = query().eq("bemail", bUserEntity.getBemail()).count();
        if(n > 0) {                                  //邮箱已注册
            return ResponseResult.getMessageResult(null, "A001");
        } else if(save(bUserEntity)) {               //注册成功
            return ResponseResult.getMessageResult(null, "A002");
        } else {                                     //注册失败
            return ResponseResult.getMessageResult(null, "A003");
        }
    }
    @Override
    public ResponseResult<Map<String, String>> login(BUserEntity bUserEntity) {
        String rand = (String)redisUtil.get("code");
        if(!rand.equalsIgnoreCase(bUserEntity.getCode())) {
            //验证码错误
            return ResponseResult.getMessageResult(null, "A000");
        }
        //链式 query
        long res = query().eq("bemail", bUserEntity.getBemail()).count();
        if(res == 0) {
            //用户名不存在
            return ResponseResult.getMessageResult(null, "A001");
        }
        Map<String, String> mapparam = new HashMap<>();
        bUserEntity.setBpwd(MD5Util.MD5(bUserEntity.getBpwd()));
        mapparam.put("bemail", bUserEntity.getBemail());
        mapparam.put("bpwd", bUserEntity.getBpwd());
        List<BUserEntity> mu = query().allEq(mapparam).list();
        if(mu.size() > 0) {                          //登录成功
            String token = jwtUtil.createToken(bUserEntity.getBemail());
            //签名时验证是否过期
             redisUtil.set("login_" + bUserEntity.getBemail(), bUserEntity.getBemail(),
                config.getRedisExpiration());
            Map<String, String> myres = new HashMap<>();
            myres.put("buserauthtoken", token);
            myres.put("bemail", bUserEntity.getBemail());
            myres.put("bid", mu.get(0).getId()+"");
            return ResponseResult.getSuccessResult(myres);
        }else{                                       //密码错误
            return ResponseResult.getMessageResult(null, "A002");
        }
    }
}
```

与购物车相关的 Service 实现类 CartServiceImpl 的核心代码如下：

```java
@Service
@SuppressWarnings("all")
public class CartServiceImpl extends ServiceImpl < CartMapper, CartEntity > implements CartService {
    @Override
    public ResponseResult<Map<String, Object>> addCart(CartEntity cartEntity) {
        Long n = query()
                .eq("busertable_id", cartEntity.getBusertableId())
                .eq("goodstable_id", cartEntity.getGoodstableId())
```

```java
                    .count();
            boolean b = false;
            if(n > 0){                              //购物车中用户已购买该商品,更新购物车
                //更新的条件构造器
                UpdateWrapper<CartEntity> updateWrapper = new UpdateWrapper<CartEntity>();
                updateWrapper.setSql("shoppingnum = shoppingnum + " + cartEntity.
getShoppingnum());
                updateWrapper.last(" where busertable_id = " + cartEntity.getBusertableId() +
                        " and goodstable_id = " + cartEntity.getGoodstableId());
                b = update(updateWrapper);
            } else {
                b = save(cartEntity);
            }
            if(b)                                   //成功
                return ResponseResult.getMessageResult(null, "A001");
            else
                return ResponseResult.getMessageResult(null, "A002");
        }
        //批量更新购物车
        @Override
        public ResponseResult<Map<String, Object>> bupDateCart(CartEntity cartEntity) {
            List<Integer> bcid = cartEntity.getBcid();
            List<Integer> bshoppingnum = cartEntity.getBshoppingnum();
            List<CartEntity> bCarts = new ArrayList<CartEntity>();
            for (int i = 0; i < bcid.size(); i++) {
                CartEntity ce = new CartEntity();
                ce.setId(bcid.get(i));
                ce.setShoppingnum(bshoppingnum.get(i));
                bCarts.add(ce);
            }
            updateBatchById(bCarts);                 //批量更新
            return ResponseResult.getMessageResult(null, "A001");
        }
        //清空购物车
        @Override
        public ResponseResult<Map<String, Object>> clearCart(CartEntity cartEntity) {
            QueryWrapper<CartEntity> queryWrapper = new QueryWrapper<CartEntity>();
            queryWrapper.eq("busertable_id", cartEntity.getBusertableId());
            remove(queryWrapper);
            return ResponseResult.getMessageResult(null, "A001");
        }
        //删除购物车
        @Override
        public ResponseResult<Map<String, Object>> removeCart(CartEntity cartEntity) {
            removeById(cartEntity);
            return ResponseResult.getMessageResult(null, "A001");
        }
}
```

与商品关注相关的 Service 实现类 FocusServiceImpl 的核心代码如下:

```java
@Service
@SuppressWarnings("all")
public class FocusServiceImpl extends ServiceImpl< FocusMapper, FocusEntity> implements
FocusService{
    @Override
    public ResponseResult<Map<String, Object>> focus(FocusEntity focusEntity) {
        long n = query()
                .eq("goodstable_id", focusEntity.getGoodstableId())
                .eq("busertable_id", focusEntity.getBusertableId())
                .count();
```

```
        if(n > 0)
            return ResponseResult.getMessageResult(null, "A001");
        focusEntity.setFocustime(new Date());
        if (save(focusEntity))
            return ResponseResult.getMessageResult(null, "A002");
        return ResponseResult.getMessageResult(null, "A003");
    }
}
```

▶16.1.7 创建控制器层

在本章前后端系统中使用 Hutool 的 JWTUtil 进行 Token 签名，并使用拦截器判断是否签名，在不需要签名的控制器方法上标注自定义注解@AuthIgnore。

在应用 ch16 的 src/main/java 目录下创建 com.ch.ch16.controller 包，并在该包中创建管理员功能和用户功能对应的控制器。

管理员登录功能对应的控制器类 AUserController 的核心代码如下：

```
@RestController
@RequestMapping("/api/admin")
@SuppressWarnings("all")
public class AUserController {
    @Autowired
    private AUserService aUserService;
    @AuthIgnore
    @PostMapping("/login")
    public ResponseResult<Map<String, String>> login(@RequestBody AUserEntity aUserEntity){
        return aUserService.login(aUserEntity);
    }
}
```

商品管理相关功能对应的控制器类 GoodsController 的核心代码如下：

```
@RestController
@RequestMapping("/api/admin/goods")
@SuppressWarnings("all")
public class GoodsController {
    @Autowired
    private GoodsService goodsService;
    private static String fileName;
    private static byte[] filecontent;
    @AuthIgnore
    @PostMapping("/fileInit")
    public void fileInit(@RequestBody MultipartFile file) {
        //MultipartFile 对象不能在另一个方法中使用,因此把文件对象变成字节数组
        try {
            filecontent = file.getBytes();
        } catch (IOException e) {
            throw new RuntimeException(e);
        }
        fileName = file.getOriginalFilename();
    }
    @PostMapping("/getGoods")
    public ResponseResult<Map<String, Object>> getGoods(@RequestBody GoodsEntity goodsEntity){
        return goodsService.getGoods(goodsEntity);
    }
    @PostMapping("/add")
```

```java
    public ResponseResult<Map<String, String>> add(@RequestBody GoodsEntity 
goodsEntity) {
        goodsEntity.setFileName(fileName);
        goodsEntity.setLogoFile(filecontent);
        return goodsService.add(goodsEntity);
    }
    @PostMapping("/delete")
    public ResponseResult<Map<String, String>> delete(@RequestBody GoodsEntity 
goodsEntity) {
        return goodsService.delete(goodsEntity);
    }
    @AuthIgnore
    @PostMapping("/getAdvGoods")
    public ResponseResult<List<GoodsEntity>>getAdvGoods(){
        return goodsService.getAdvGoods();
    }
    @AuthIgnore
    @PostMapping("/getGoodsIndex")
     public ResponseResult < List < GoodsEntity > > getGoodsIndex (@ RequestBody GoodsEntity 
goodsEntity) {
        return goodsService.getGoodsIndex(goodsEntity);
    }
    @AuthIgnore
    @PostMapping("/getGoodsById")
    public ResponseResult<GoodsEntity>getGoodsById(@RequestBody GoodsEntity 
goodsEntity) {
        return goodsService.getGoodsById(goodsEntity);
    }
}
```

商品类型管理相关功能对应的控制器类 GoodsTypeController 的核心代码如下：

```java
@RestController
@RequestMapping("/api/admin/type")
@SuppressWarnings("all")
public class GoodsTypeController {
    @Autowired
    private GoodsTypeService goodsTypeService;
    @AuthIgnore                                     //因为before的导航页需要查询商品类型
    @PostMapping("/getAllGoodsType")
    public ResponseResult<List<GoodsTypeEntity>> getAllGoodsType(){
        //{result:goodsTypeService.list()}
        return ResponseResult.getSuccessResult(goodsTypeService.list());
    }
    @PostMapping("/getGoodsType")
    public ResponseResult<Map<String, Object>> getGoodsType(@RequestBody GoodsTypeEntity 
goodsTypeEntity) {
        return goodsTypeService.getGoodsType(goodsTypeEntity);
    }
    @PostMapping("/add")
     public ResponseResult < Map < String, Object > > add ( @ RequestBody GoodsTypeEntity 
goodsTypeEntity) {
        return goodsTypeService.add(goodsTypeEntity);
    }
    @PostMapping("/update")
     public ResponseResult < Map < String, Object > > update ( @ RequestBody GoodsTypeEntity 
goodsTypeEntity) {
        return goodsTypeService.update(goodsTypeEntity);
    }
    @PostMapping("/delete")
```

第16章 电子商务平台的设计与实现（Spring Boot+ Vue.js 3+ MyBatis-Plus）

```
    public ResponseResult<Map<String, Object>> delete(@RequestBody GoodsTypeEntity
goodsTypeEntity){
        return goodsTypeService.delete(goodsTypeEntity);
    }
}
```

订单管理相关功能对应的控制器类 OrdersController 的核心代码如下：

```
@RestController
@RequestMapping("/api/admin/orders")
@SuppressWarnings("all")
public class OrdersController {
    @Autowired
    private OrdersService ordersService;
    @PostMapping("/getAllOrders")
    public ResponseResult<Map<String, Object>> getAllOrders(@RequestBody OrdersEntity
ordersEntity){
        return ordersService.getAllOrders(ordersEntity);
    }
    @PostMapping("/getOrdersDetail")
    public ResponseResult<List<Map<String, Object>>> getOrdersDetail(@RequestBody
OrdersEntity ordersEntity){
        return ordersService.getOrdersDetail(ordersEntity);
    }
    @PostMapping("/goPay")
    public ResponseResult<Map<String, String>> goPay(@RequestBody OrdersEntity
ordersEntity){
        return ordersService.goPay(ordersEntity);
    }
    @PostMapping("/selectOrderByMonth")
    public ResponseResult<List<OrdersEntity>> selectOrderByMonth(@RequestBody
OrdersEntity ordersEntity){
        return ordersService.selectOrderByMonth(ordersEntity);
    }
    @PostMapping("/selectOrderByType")
    public ResponseResult<List<Map<String, Object>>> selectOrderByType(){
        return ordersService.selectOrderByType();
    }
    @PostMapping("/submitOrder") ResponseResult<OrdersEntity> submitOrder(@RequestBody
OrdersEntity ordersEntity){
        return ordersService.submitOrder(ordersEntity);
    }
    @PostMapping("/getOrdersByUid")
    public ResponseResult<Map<String, Object>> getOrdersByUid(@RequestBody OrdersEntity
ordersEntity){
        return ordersService.getOrdersByUid(ordersEntity);
    }
}
```

用户注册登录相关功能对应的控制器类 BUserController 的核心代码如下：

```
@RestController
@RequestMapping("/api/before")
@SuppressWarnings("all")
public class BUserController {
    @Autowired
    private BUserService bUserService;
    @AuthIgnore
    @PostMapping("/register")
    public ResponseResult<Map<String, String>> register(@RequestBody BUserEntity
bUserEntity){
```

```
        return bUserService.register(bUserEntity);
    }
    @AuthIgnore
    @PostMapping("/login")
    public ResponseResult<Map<String, String>> login(@RequestBody BUserEntity 
bUserEntity){
        return bUserService.login(bUserEntity);
    }
    @AuthIgnore
    @GetMapping("/getcode")
    public void getcode(HttpServletResponse response) throws IOException {
        CircleCaptcha circleCaptcha = CaptchaUtil.createCircleCaptcha(116, 30, 4, 10);
        redisUtil.set("code",circleCaptcha.getCode(), config.getRedisExpiration());
        ServletOutputStream outputStream = response.getOutputStream();
        circleCaptcha.write(outputStream);
        outputStream.close();
    }
}
```

购物车相关功能对应的控制器类 CartController 的核心代码如下：

```
@RestController
@RequestMapping("/api/before/cart")
@SuppressWarnings("all")
public class CartController {
    @Autowired
    private CartService cartService;
    @Autowired
    private GoodsService goodsService;
    @PostMapping("/add")
    public ResponseResult<Map<String, Object>> addCart(@RequestBody CartEntity 
cartEntity){
        return cartService.addCart(cartEntity);
    }
    //查询我的购物车
    @PostMapping ("/myCart")
    public ResponseResult<List<Map<String, Object>>> myCartGoods(@RequestBody 
CartEntity cartEntity){
        return goodsService.myCartGoods(cartEntity);
    }
    //批量更新购物车
    @PostMapping("/bupDateCart")
    public ResponseResult<Map<String, Object>> bupDateCart(@RequestBody CartEntity 
cartEntity){
        return cartService.bupDateCart(cartEntity);
    }
    //清空购物车
    @PostMapping("/clearCart")
    public ResponseResult<Map<String, Object>> clearCart(@RequestBody CartEntity 
cartEntity){
        return cartService.clearCart(cartEntity);
    }
    //删除购物车
    @PostMapping("/removeCart")
     public ResponseResult < Map < String, Object > > removeCart (@ RequestBody CartEntity 
cartEntity){
        return cartService.removeCart(cartEntity);
    }
}
```

商品收藏相关功能对应的控制器类 FocusController 的核心代码如下：

第 16 章 电子商务平台的设计与实现（Spring Boot+ Vue.js 3+ MyBatis-Plus）

```
@RestController
@RequestMapping("/api/before/focus")
@SuppressWarnings("all")
public class FocusController {
    @Autowired
    private FocusService focusService;
    @Autowired
    private GoodsService goodsService;
    @PostMapping("/add")
    public ResponseResult<Map<String, Object>> add(@RequestBody FocusEntity focusEntity){
        return focusService.focus(focusEntity);
    }
    //查询我的收藏
    @PostMapping("/getMyFocusGoods")
    public ResponseResult<Map<String, Object>> getMyFocusGoods(@RequestBody GoodsEntity goodsEntity){
        return goodsService.iPageMyFocusGoods(goodsEntity);
    }
}
```

▶16.1.8 创建跨域响应头设置过滤器

跨域访问涉及请求域名、请求方式、发送的内容类型以及携带证书式访问等问题。在后端系统中将这些设置放在应用程序的主类中完成，主类 Ch16Application 的核心代码如下：

```
@SpringBootApplication
@MapperScan(basePackages={"com.ch.ch16.mapper"})
public class Ch16Application {
    public static void main(String[] args) {
        SpringApplication.run(Ch16Application.class, args);
    }
    //跨域设置
    private CorsConfiguration corsConfig() {
        CorsConfiguration corsConfiguration = new CorsConfiguration();
        //允许跨域请求的域名
        corsConfiguration.addAllowedOrigin("*");
        //允许发送的内容类型
        corsConfiguration.addAllowedHeader("*");
        //跨域请求允许的请求方式
        corsConfiguration.addAllowedMethod("*");
        corsConfiguration.setMaxAge(3600L);
        return corsConfiguration;
    }
    @Bean
    public CorsFilter corsFilter() {
        UrlBasedCorsConfigurationSource source = new UrlBasedCorsConfigurationSource();
        source.registerCorsConfiguration("/**", corsConfig());
        return new CorsFilter(source);
    }
}
```

▶16.1.9 创建工具类

在后端系统 ch16 中使用工具类 MyUtil 的 getNewFileName 方法对文件进行重命名，使用工具类 MD5Util 的 MD5 方法对用户密码进行加密。工具类 MyUtil 和 MD5Util 的代码略，请读者参见本书提供的源程序 ch16。

▶16.1.10 MyBatis-Plus 分页插件、Redis 以及 Token 签名配置

在使用 MyBatis-Plus 访问数据库时需要配置分页插件 MybatisPlusInterceptor 才能使用 MyBatis-Plus 的分页功能，因此在后端系统 ch16 中需要配置分页插件 MybatisPlusInterceptor，见 com.ch.ch16.common.config 包中的 MybatisPlusConfig 配置类，具体代码略，请读者参见本书提供的源程序 ch16。

在后端系统 ch16 中使用 Redis 及 Spring Cache 缓存技术进行签名数据的存储。Redis 配置类 RedisConfig 和 Redis 工具类 RedisUtil 分别位于 com.ch.ch16.common.config 和 com.ch.ch16.common.security.utils 包中，具体代码略，请读者参见本书提供的源程序 ch16。

在后端系统 ch16 中使用 Hutool 的 JWTUtil 进行 Token 签名，并使用拦截器 AuthInterceptor 判断是否签名，在不需要签名的控制器方法上标注自定义注解@AuthIgnore。JWTUtil、AuthInterceptor 以及 AuthIgnore 的相关类位于 com.ch.ch16.common.security 包中，具体代码略，请读者参见本书提供的源程序 ch16。

16.2 使用 Vue CLI 构建前端系统

本节只是实现电子商务平台的前端系统，注重功能的实现，旨在让读者了解前后端分离的应用程序的实现原理及开发流程。

▶16.2.1 使用 Vue CLI 构建前端项目 ebusiness-vue

Vue CLI 致力于将 Vue.js 生态工具基础标准化，确保各种构建工具平稳衔接，让开发者专注于应用的撰写上，而不必纠结配置的问题。参考本书 13.3.1 节，使用 Vue CLI 搭建基于 Router 功能的前端项目 ebusiness-vue。

▶16.2.2 安装 Element Plus 和@element-plus/icons-vue

首先使用 VS Code 打开项目 ebusiness-vue，并进入 Terminal 终端，依次执行"npm install element-plus --save"和"npm install @element-plus/icons-vue"命令，进行 Element Plus 和@element-plus/icons-vue 的安装。

然后在 ebusiness-vue 的 main.js 文件中完整引入 Element Plus，并注册图标组件 ElementPlusIconsVue，main.js 的代码具体如下：

```
import { createApp } from 'vue'
import App from './App.vue'
import router from './router'
import ElementPlus from 'element-plus'
import 'element-plus/dist/index.css'
//引入图标
import * as ElementPlusIconsVue from '@element-plus/icons-vue'
const axios = require('axios')
const app = createApp(App)
axios.defaults.baseURL = 'http://localhost:8443/eBusiness'
app.config.globalProperties.$axios = axios
//注册所有图标
for (const [key, component] of Object.entries(ElementPlusIconsVue)) {
    app.component(key, component)
}
app.use(ElementPlus).use(router).mount('#app')
//eslint-disable-next-line no-unused-vars
router.beforeEach((to, from) => {
//提示未使用,ESLint 规则 no-unused vars 关闭为 eslint-disable-next-line
```

```
        //如果 before 路由器需要验证
        if(to.meta.auth && to.meta.act){
            //对路由进行验证
            if (window.sessionStorage.getItem('bemail') === null) {
                alert("您没有登录,无权访问!")
                /* 未登录则跳转到登录界面,
                query:{ redirect: to.fullPath}表示把当前路由信息传递过去,以便登录后跳转回来 */
                return {
                    path: '/login',
                    query: {redirect: to.fullPath}
                }
            }
        } else if(to.meta.auth && !to.meta.act){                //如果 admin 路由器需要验证
            if (window.sessionStorage.getItem('aname') === null) {
                alert("您没有登录,无权访问!")
                return {
                    path: '/adminLogin',
                    query: {redirect: to.fullPath}
                }
            }
        }
    })
```

经过上述两个步骤,开发者即可在.vue 页面文件中使用 Element Plus 设计界面。

▶16.2.3 安装 ECharts

ECharts 是一款基于 JavaScript 的数据可视化图表库,提供直观、生动、可交互、可个性化定制的数据可视化图表。ECharts 最初由百度团队开源,并于 2018 年初捐赠给 Apache 基金会,成为 ASF 孵化级项目。

在 ebusiness-vue 前端应用中使用 ECharts 对销量及订单进行统计和可视化展示,因此首先使用 VS Code 打开 Vue.js 项目 ebusiness-vue,然后在 Terminal 终端使用命令 npm install echarts --save 安装 ECharts。

▶16.2.4 安装 Axios 模块并设置跨域访问

Axios 是一个基于 promise 的网络请求库,可以用于浏览器和 node.js。Axios 与原生的 XMLHttpRequest 对象相比,简单、易用;与 jQuery 相比,Axios 的包尺寸小且提供了易于扩展的接口,是专注于网络请求的库。

在 ebusiness-vue 前端应用中使用 Axios 与后端程序进行 Web 数据交互,因此需要使用 VS Code 打开项目 ebusiness-vue,并进入 Terminal 终端,执行"npm install --save axios"命令安装该模块。安装成功后,在 main.js 文件中全局注册 Axios 实现跨域访问,具体代码见 16.2.2 节的 main.js。

▶16.2.5 管理员登录组件

在开发的时候,前端用前端的服务器(如 Nginx),后端用后端的服务器(如 Tomcat),在开发前端内容时,把前端的请求通过前端服务器转发给后端,即可实时观察结果,并且不需要知道后端怎么实现,只需要知道接口提供的功能,前后端的开发人员各司其职。后端系统的开发已在 16.1 节中完成,下面开发前端组件。

右击 src/views,新建一个名为 admin 的文件夹,并在该文件夹中新建 LoginView.vue 文件,即管理员登录组件(也是管理员相关功能的首页组件)。通过"http://localhost:8000/ebusiness-vue/adminLogin"运行管理员登录组件,效果如图 16.1 所示。

图 16.1 管理员登录组件

LoginView.vue 的核心代码如下：

```vue
<template>
  <el-dialog title="管理员登录" v-model="dialogVisible" width="30%">
    <div class="box">
      <el-form ref="loginFormRef" :model="loginForm" :rules="rules" style="width:100%;" label-width="20%">
        <el-form-item label="用户名" prop="aname">
          <el-input v-model="loginForm.aname" placeholder="请输入用户名"></el-input>
        </el-form-item>
        <el-form-item label="密码" prop="apwd">
          <el-input show-password v-model="loginForm.apwd" placeholder="请输入密码"></el-input>
        </el-form-item>
        <el-form-item>
          <el-button type="primary" @click="login(loginFormRef)">登录</el-button>
          <el-button type="danger" @click="cancel">重置</el-button>
        </el-form-item>
      </el-form>
    </div>
  </el-dialog>
</template>
<script setup>
//使用 setup 语法糖
import { reactive, ref } from 'vue'
import { ElMessage } from 'element-plus'
import { useRouter } from 'vue-router'
import { getCurrentInstance } from 'vue'
//获取全局变量$axios
const axios = getCurrentInstance().appContext.config.globalProperties.$axios
const router = useRouter()
let loginForm = reactive({})
let loginFormRef = reactive({})
//验证规则
const rules = reactive({
    aname: [{ required: true, message: '请输入用户名', trigger: 'blur' }],
    apwd: [{ required: true, message: '请输入密码', trigger: 'blur' }]
})
let dialogVisible = ref(true)
//登录功能
const login = async (formEl) => {                                    //formEl 为 ref 值
  if (!formEl) return
  await formEl.validate((valid) => {
    if (valid) {
      axios.post('/api/admin/login', loginForm)
       .then(res => {
         if (res.data.msgId=== "A001"){
            ElMessage.error('用户名错误!')
         } else if (res.data.msgId=== "A002"){
            ElMessage.error('密码错误!')
         }else{
```

```
          ElMessage.success({message: '登录成功',type: 'success'})
          sessionStorage.setItem("authtoken", res.data.result.authtoken);
                  sessionStorage.setItem("aname", res.data.result.aname);
          sessionStorage.setItem("aid", res.data.result.aid);
          router.replace('/home')
        }
      })
      .catch(() => {
        ElMessage.error('访问异常')
      })
    } else {
      ElMessage.error('表单验证失败')
    }
  })
}
const cancel = ()=> {
  loginFormRef.resetFields()
}
</script>
<style scoped>
.box {
  width: 100%;
  height: 150px;
}
</style>
```

16.2.6 后台管理主界面组件

在管理员登录成功后,进入 HomeView.vue 主界面组件。右击 src/views/admin 文件夹,新建 HomeView.vue 文件,即管理员管理主界面组件。在主界面组件中使用两个子组件 HeaderCom.vue 和 SidebarCom.vue 实现导航功能。主界面组件的运行效果如图 16.2 所示。

图 16.2 管理员管理主界面

HomeView.vue 的核心代码如下:

```
<template>
    <div>
        <Header />
        <Sidebar />
        <div class="content-box">
            <div class="content">
                <router-view></router-view>
            </div>
        </div>
    </div>
</template>
<script setup>
    import Header from "@/components/HeaderCom.vue";
    import Sidebar from "@/components/SidebarCom.vue";
</script>
```

对于 HeaderCom.vue 和 SidebarCom.vue 组件的代码，请读者参见本书提供的源程序 ebusiness-vue。

16.2.7 商品类型管理组件

进入管理员管理主界面，在主界面右侧中央位置显示类型管理组件，包括新增、修改、查询、删除类型等功能。右击 src/views/admin 文件夹，新建 TypeManageView.vue 文件，即类型管理组件。本章前后端系统使用 Token 签名进行权限认证，也就是说管理员只有成功登录才可以管理商品类型，因此在类型管理组件中任何与后端的数据交互都要进行 Token 签名，TypeManageView.vue 的运行效果如图 16.2 所示。

在进入 TypeManageView.vue 组件时，首先使用 onMounted 方法加载所有类型，并分页展示。对于 TypeManageView.vue 组件的代码，请读者参见本书提供的源程序 ebusiness-vue。

16.2.8 商品管理组件

进入管理员管理主界面，在主界面左侧导航栏内单击管理模块中的商品管理即可显示商品管理组件，包括新增、修改、查询与条件查询、删除商品等功能。右击 src/views/admin 文件夹，新建 GoodsManageView.vue 文件，即商品管理组件，在商品管理组件中任何与后端的数据交互都要进行 Token 签名，GoodsManageView.vue 的运行效果如图 16.3 所示。

图 16.3 商品管理界面

在进入 GoodsManageView.vue 组件时，首先使用 onMounted 方法加载所有类型（条件查询使用）及商品信息，并分页展示商品信息。对于 GoodsManageView.vue 组件的代码，请读者参见本书提供的源程序 ebusiness-vue。

16.2.9 订单管理组件

进入管理员管理主界面，在主界面左侧导航栏内单击管理模块中的订单管理即可显示订单管理组件，包括查询与条件查询、详情、删除订单等功能。右击 src/views/admin 文件夹，新建 OrderManageView.vue 文件，即订单管理组件，在订单管理组件中任何与后端的数据交互都要进行 Token 签名，OrderManageView.vue 的运行效果如图 16.4 所示。

图 16.4 订单管理界面

第 16 章　电子商务平台的设计与实现（Spring Boot+ Vue.js 3+ MyBatis-Plus）

在进入 OrderManageView.vue 组件时，首先使用 onMounted 方法加载所有订单信息，并分页展示。OrderManageView.vue 组件的核心代码如下：

```
<template>
  <el-tabs type="border-card">
    <el-tab-pane label="订单管理">
      <el-form :inline="true" :model="searchParam" class="fl">
            <el-form-item label="订单编号">
          <el-input v-model="searchParam.id" placeholder="请输入订单编号" />
            </el-form-item>
            <el-form-item>
                <el-button type="primary" @click="loadOrders()">查询</el-button>
            </el-form-item>
        </el-form>
      <el-table :data="result":default-sort="{ prop: 'orderdate', order: 'descending' }" border :key="itemKey">
          <el-table-column prop="id" label="订单编号" width="100"></el-table-column>
          <el-table-column label="订单金额" width="150">
            <template #default="scope">
              <span>{{scope.row.amount.toFixed(1)}}</span>
            </template>
          </el-table-column>
          <el-table-column prop="orderdate" label="下单时间" sortable width="200"></el-table-column>
          <el-table-column prop="status" :formatter="stateFormat" label="订单状态" width="150"></el-table-column>
          <el-table-column label="操作">
            <template #default="scope">
              <el-row>
              <el-button size="small" type="primary" @click="handleDetail(scope.row)">详情</el-button>
                <el-popconfirm v-if="scope.row.status === 0" confirm-button-text="是" cancel-button-text="否" :icon="InfoFilled" icon-color="#626aef"
                    title="真的删除吗？" @confirm="confirmEvent()" @cancel="cancelEvent">
                  <template #reference>
                    <el-button size="small" type="danger">删除</el-button>
                  </template>
                </el-popconfirm>
              </el-row>
            </template>
          </el-table-column>
        </el-table>
        <div>
          <el-pagination background
              @current-change="handleCurrentChange"
              layout="total, prev, pager, next"
              v-model:currentPage="currentPage"
              :page-size="1" :total="total" />
        </div>
    </el-tab-pane>
  </el-tabs>
  <el-dialog title="订单详情" v-model="orderDetailVisible">
    <el-table :data="detailResult" border :key="itemKey">
        <el-table-column prop="id" label="商品编号" width="120"></el-table-column>
        <el-table-column prop="gname" label="商品名称" width="120"></el-table-column>
        <el-table-column label="单价" width="120">
          <template #default="scope">
            <span>{{scope.row.grprice.toFixed(1)}}</span>
          </template>
        </el-table-column>
```

```html
            <el-table-column prop="shoppingnum" label="数量" width="120"></el-table-column>
            <el-table-column label="小计" width="120">
                <template #default="scope">
                    <span>{{scope.row.smallTotal.toFixed(1)}}</span>
                </template>
            </el-table-column>
        </el-table>
    </el-dialog>
</template>
<script setup>
import { onMounted, getCurrentInstance, reactive, ref } from 'vue'
import { ElMessage } from 'element-plus'
let result = ref([])
let detailResult = ref([])
let orderDetailVisible = ref(false)
let searchParam = reactive({})
let total = ref(0)
let currentPage = ref(1)
let itemKey = ref(0)
//获取全局变量$axios
const axios = getCurrentInstance().appContext.config.globalProperties.$axios
//组件的初始化
onMounted(() => {
    loadOrders()
})
//加载订单
const loadOrders = ()=>{
    //headers 为第三个参数
    axios.post('/api/admin/orders/getAllOrders',
    {
        id: searchParam.id,
        currentPage: currentPage.value
    },
    {
        headers: {
            'Authorization': sessionStorage.getItem('authtoken')
        }
    })
    .then(res => {
        result.value =res.data.result.allOrders;
        itemKey.value = Math.random()                           //刷新表格数据
        total.value = res.data.result.totalPage
    })
    .catch((error) => {
        ElMessage.error(error)
    })
}
//根据订单状态显示是否支付
const stateFormat = (row) => {
    if (row.status === 0) {
        return '未支付'
    } else{
        return '已支付'
    }
}
//页码变换
const handleCurrentChange = (val) => {
    currentPage.value = val
    loadOrders()
}
//详情
```

```
const handleDetail = (row) => {
  axios.post('/api/admin/orders/getOrdersDetail',
  {
    id: row.id
  },
  {
    headers: {
      'Authorization': sessionStorage.getItem('authtoken')
    }
  })
    .then(res => {
      detailResult.value = res.data.result;
      itemKey.value = Math.random()                              //刷新表格数据
      orderDetailVisible.value = true
    })
    .catch((error) => {
      ElMessage.error(error)
    })
}
//删除
const confirmEvent = () => {
  ElMessage.error('不要轻易删除订单!')
}
const cancelEvent = () => {
}
</script>
```

▶16.2.10 销量统计（按月）组件

进入管理员管理主界面，在主界面左侧导航栏内单击统计模块中的销量统计（按月）即可显示销量统计（按月）组件，统计近一年中每月的销量。右击 src/views/admin 文件夹，新建 SalesStatisticsView.vue 文件，即销量统计（按月）组件。SalesStatisticsView.vue 组件的运行效果如图 16.5 所示。

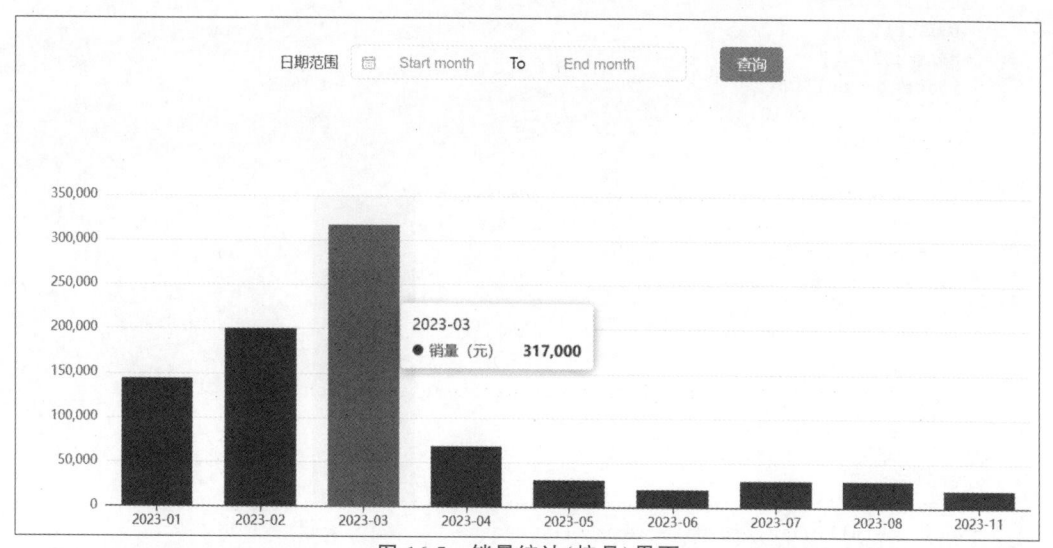

图 16.5　销量统计（按月）界面

在进入 SalesStatisticsView.vue 组件时，首先使用 onMounted 方法加载近一年的订单销量信息，并使用 ECharts 展示。SalesStatisticsView.vue 组件的核心代码如下：

```html
<template>
    <div class="demo-date-picker">
        <div class="block">
            <el-form :inline="true" :model="searchParam">
                <el-form-item label="日期范围">
                    <el-date-picker
                            v-model="searchParam.orderdate"
                            value-format="YYYY-MM"
                            type="monthrange"
                            range-separator="To"
                            start-placeholder="Start month"
                            end-placeholder="End month"
                    />
                </el-form-item>
                <el-form-item>
                    <el-button type="primary" @click="onSubmit()">查询</el-button>
                </el-form-item>
            </el-form>
        </div>
    </div>
    <div id="myChart" :style="{width: '100%', height: '380px'}"></div>
</template>
<script setup>
import { onMounted, getCurrentInstance } from 'vue'
import * as echarts from 'echarts';
import { ElMessage } from 'element-plus'
import { reactive} from 'vue'
//获取全局变量$axios
const axios = getCurrentInstance().appContext.config.globalProperties.$axios
let searchParam = reactive({})
const onSubmit = () => {                        //查询
    const datev = searchParam.orderdate
    let start1 = '';
    let end1 = '';
    let data11 = []
    let data22 = []
    if(datev != null){
        start1 = datev[0]
        end1 = datev[1]
    }
    axios.post('/api/admin/orders/selectOrderByMonth',
    {
        startDate: start1,
        endDate: end1
    },
    {
        headers: {
            'Authorization': sessionStorage.getItem('authtoken')
        }
    })
    .then(res => {
        for (let i = 0; i < res.data.result.length; i++){
            data11.push(res.data.result[i].months)
            data22.push(res.data.result[i].totalamount)
        }
        mydraw(data11, data22)
    })
    .catch((error) => {
```

```
            ElMessage.error(error)
        })
    }
    const mydraw = (datav1, datav2) => {
        const myChart = echarts.init(document.getElementById('myChart'))
        const option = {
            tooltip: {
                trigger: 'axis',
                axisPointer: {
                    type: 'shadow'
                }
            },
            grid: {
                left: '3%',
                right: '4%',
                bottom: '3%',
                containLabel: true
            },
            xAxis: {
                type: 'category',
                data: datav1,
                axisTick: {
                    alignWithLabel: true
                }
            },
            yAxis: {
                type: 'value'
            },
            series: [
                {
                    data: datav2,
                    type: 'bar',
                    name: '销量(元)',
                }
            ]
        }
        myChart.setOption(option)
    }
    onMounted (() => {
        onSubmit()
    })
</script>
```

▶16.2.11 订单统计（按类型）组件

进入管理员管理主界面，在主界面左侧导航栏内单击统计模块中的订单统计（按类型）即可显示订单统计（按类型）组件，统计近一年中每种商品类型的销量。右击 src/views/admin 文件夹，新建 OrderStatisticsView.vue 文件，即订单统计（按类型）组件。OrderStatisticsView.vue 组件的运行效果如图 16.6 所示。

在进入 OrderStatisticsView.vue 组件时，首先使用 onMounted 方法加载近一年的订单统计（按类型）信息，并使用 ECharts 展示。OrderStatisticsView.vue 组件的核心代码如下：

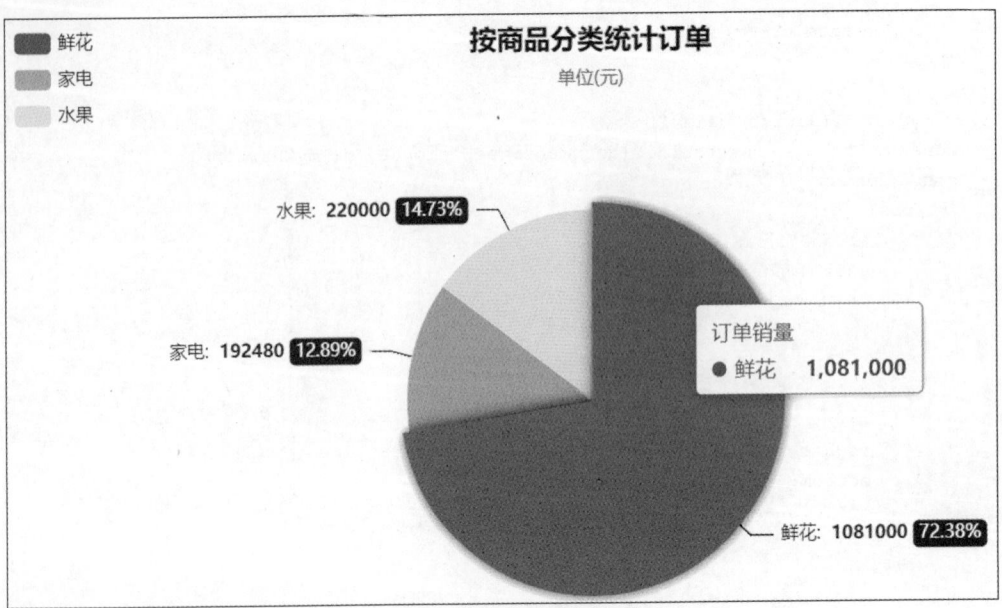

图 16.6　订单统计(按类型)界面

```
<template>
    <div id="myChart" :style="{width: '80%', height: '500px'}"></div>
</template>
<script setup>
import { onMounted, getCurrentInstance } from 'vue'
import { ElMessage } from 'element-plus'
import * as echarts from 'echarts';
//获取全局变量$axios
const axios = getCurrentInstance().appContext.config.globalProperties.$axios
const mydraw = (result) => {
    const myChart = echarts.init(document.getElementById('myChart'))
    const option = {
        title: {
            text: '按商品分类统计订单',
            subtext: '单位(元)',
            left: 'center'
        },
        tooltip: {
            trigger: 'item'
        },
        legend: {
            orient: 'vertical',
            left: 'left'
        },
        series: [
            {
                name: '订单销量',
                type: 'pie',
                radius: '50%',
                label: {
                    formatter: "{b}:{c|{c}} {per|{d}%}",
                    rich: {
                        c: {
                            color: '#4c5058',
                            fontSize: 12,
                            fontWeight: 'bold',
                            lineHeight: 33
                        },
```

```
                    per: {
                        color: '#fff',
                        backgroundColor: '#4c5058',
                        padding: [3, 4],
                        borderRadius: 4
                    }
                }
            },
            data: result,
            emphasis: {
                itemStyle: {
                    shadowBlur: 10,
                    shadowOffsetX: 0,
                    shadowColor: 'rgba(0, 0, 0, 0.5)'
                }
            }
        }
    ]
}
myChart.setOption(option)
}
onMounted (() => {
    axios.post('/api/admin/orders/selectOrderByType',{},
    {
        headers: {
        'Authorization': sessionStorage.getItem('authtoken')
        }
    })
    .then(res => {
        mydraw(res.data.result)
    })
    .catch((error) => {
        ElMessage.error(error)
    })
})
</script>
```

▶16.2.12 前端首页组件

右击 src/views，新建一个名为 before 的文件夹，并在该文件夹中新建 IndexView.vue 文件，即前端首页组件。在前端首页组件中使用子组件 HeaderView.vue 实现导航功能。通过"http://localhost:8000/ebusiness-vue"运行首页组件，效果如图 16.7 所示。

图 16.7　前端首页界面

在进入 IndexView.vue 组件时,首先使用 onMounted 方法加载最新商品信息,并展示在首页界面。IndexView.vue 的核心代码如下:

```
<template>
  <div>
    <HeaderView @goIndex="goToIndex" @searchIndex="searchToIndex"/>
  </div>
  <div>
    <el-row>
      <el-col
      v-for="(item, index) in goodslists" :key="item.id" :span="4"
      :offset="index > 0 && (index ==1 || (index !=1 && index % 5 != 0))? 1 : 0">
      <!--offset 为左侧的间隔栅数,一共 24 栅-->
        <el-card :body-style="{ padding: '0px' }">
          <el-link :underline="false" @click="goToGoodsDetail(item)">
            <img :src="require('../../assets/' + item.gpicture)" class="image"/>
          </el-link>
          <div style="padding: 5px">
            <el-link :underline="false" @click="goToGoodsDetail(item)"><span class="myfont">{{item.gname}}</span></el-link>
            <br>
            <span class="myfont">&yen;<strike>{{item.goprice.toFixed(1)}}</strike></span>  
            <span class="yourfont">&yen;{{item.grprice.toFixed(1)}}</span>
          </div>
        </el-card>
      </el-col>
    </el-row>
  </div>
</template>
<script setup>
import { onMounted, ref, getCurrentInstance } from 'vue'
import HeaderView from '@/components/HeaderView.vue'
import {useRouter} from 'vue-router'
import { ElMessage} from 'element-plus'
const router = useRouter()
const axios = getCurrentInstance().appContext.config.globalProperties.$axios
let goodslists = ref([])
onMounted (()=> {
    goToIndex(0)
})
//typeid 子组件传递过来的数据
const goToIndex = (typeid) => {
  axios.post('/api/admin/goods/getGoodsIndex',
  {
    goodstypeId: typeid
  }
  ).then(res => {
      goodslists.value =res.data.result;
  })
  .catch((error) => {
      ElMessage.error(error)
  })
}
//searchV 子组件传递过来的数据
const searchToIndex = (searchV) => {
  axios.post('/api/admin/goods/getGoodsIndex',
  {
    gname: searchV
  }
```

```
    }).then(res => {
        goodslists.value =res.data.result;
    })
    .catch((error) => {
        ElMessage.error(error)
    })
}
const goToGoodsDetail = (goods) => {
    //从 Vue Router 的 2022-8-22 更新后,弃用 params 传参
    //使用 History API 方式传递和接收,在跳转前的页面使用 state 参数
    router.push({name: 'goodsDetail', state: goods})
}
</script>
```

对于导航组件 HeaderView.vue 的代码,请读者参见本书提供的源程序 ebusiness-vue。

▶16.2.13 用户注册组件

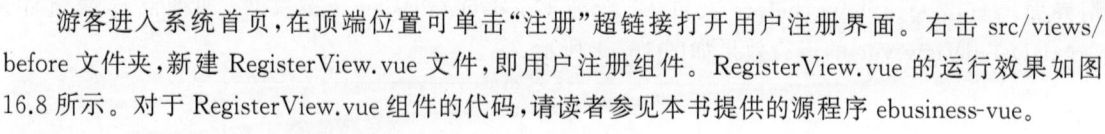

游客进入系统首页,在顶端位置可单击"注册"超链接打开用户注册界面。右击 src/views/before 文件夹,新建 RegisterView.vue 文件,即用户注册组件。RegisterView.vue 的运行效果如图 16.8 所示。对于 RegisterView.vue 组件的代码,请读者参见本书提供的源程序 ebusiness-vue。

图 16.8　用户注册界面

▶16.2.14 用户登录组件

注册用户在顶端位置可单击"登录"超链接打开用户登录界面。右击 src/views/before 文件夹,新建 LoginView.vue 文件,即用户登录组件。LoginView.vue 的运行效果如图 16.9 所示。对于 LoginView.vue 组件的代码,请读者参见本书提供的源程序 ebusiness-vue。

图 16.9　用户登录界面

▶16.2.15 个人信息组件

注册用户成功登录后,可在顶端位置单击"个人信息"超链接打开个人信息界面。右击 src/

views/before 文件夹,新建 MyselfInfoView.vue 文件,即个人信息组件。MyselfInfoView.vue 的运行效果如图 16.10 所示。对于 MyselfInfoView.vue 组件的代码,请读者参见本书提供的源程序 ebusiness-vue。

图 16.10 个人信息界面

▶16.2.16 商品详情组件

游客或用户进入系统首页后,可在广告区或最新商品区单击"商品图片"超链接打开商品详情界面。右击 src/views/before 文件夹,新建 GoodsDetailView.vue 文件,即商品详情组件。GoodsDetailView.vue 的运行效果如图 16.11 所示。

图 16.11 商品详情界面

在进入 GoodsDetailView.vue 组件时,首先使用 onMounted 方法根据商品 ID 加载商品信息,并展示在商品详情界面。GoodsDetailView.vue 的核心代码如下:

```
<template>
<el-dialog v-model="dialogVisible" @close="gogo(1) ">
    <div class="box1">
      <div class="box2">
        <img :src="getpath(goods.gpicture)" class="image"/>
      </div>
      <div class="box3">
          <p class="myfont">商品名:<span>{{goods.gname}}</span></p>
          <p class="myfont">原价:<span>&yen;<strike>{{goods.goprice}}</strike></span></p>
          <p> <span class="myfont">折扣价:</span><span style="color: rgb(249, 7, 7); font-size: 10pt;">&yen;{{goods.grprice}}</span></p>
          <p class="myfont">库存:<span>{{goods.gstore}}</span></p>
          <p> <el-input v-model="inputvalue" @input="handleEdit" class="w-50 m-2" size="small" placeholder="请输入购买量" /></p>
          <p>
            <el-button type="primary" :icon="ShoppingCart" class="button" @click="gogo(2)" size="small">加入购物车</el-button>
            <el-button type="warning" :icon="Shop" class="button" size="small" @click="gogo(3)">立刻购买</el-button>
            <el-button type="success" :icon="CirclePlusFilled" class="button" size="small" @click="gogo(4)">收藏</el-button>
         </p>
     </div>
  </div>
```

第 16 章　电子商务平台的设计与实现（Spring Boot+ Vue.js 3+ MyBatis-Plus）

```
</el-dialog>
</template>
<script setup>
import { useRouter } from 'vue-router'
import { onMounted, ref, getCurrentInstance } from 'vue'
import { ElMessage, ElMessageBox } from 'element-plus'
import { ShoppingCart, CirclePlusFilled, Shop } from '@element-plus/icons-vue'
const axios = getCurrentInstance().appContext.config.globalProperties.$axios
const router = useRouter()                    //相当于 this.$router,一般具有功能性,例如路由跳转
const dialogVisible = true
const inputvalue = ref('1')
//接收传递过来的数据
const historyParams = history.state
let goods = ref({})
onMounted (()=> {
  axios.post('/api/admin/goods/getGoodsById',
  {
      id: historyParams.id
  }).then(res => {
      goods.value =res.data.result;
  })
  .catch((error) => {
      ElMessage.error(error)
  })
})
//报错 Cannot find module './undefined'
const getpath = (path) => {
      return path ? require('../../assets/' + path) : ''
                                                      //path 不为 null 就返回对应的路径
}
//只能输入正整数
const handleEdit = (e) => {
    let value = e.replace(/^(0+)|[^\d]+/g,'');       //以 0 开头或者输入非数字,会被替换成空
    value = value.replace(/(\d{15})\d * /, '$1')     //最多保留 15 位整数
    inputvalue.value = value
}
const gogo = (myValue) => {
   //关闭
   if(myValue === 1){
     router.go(-1)
     return
   } else {
     if(sessionStorage.getItem('bemail') === null) {
       alert("您没有登录,请登录!")
       router.replace('/login')
       return false
     }
     if(myValue === 2 || myValue === 3){              //加入购物车或立刻购买
       if(inputvalue.value === ''){
         ElMessageBox.alert(
           '<span style="color: rgb(249, 7, 7); font-size: 12pt;">请输入购买量!</span>',
           '',
           {
              dangerouslyUseHTMLString: true,
           }
         )
         return false
       }
       if(myValue === 2){                              //加入购物车
         axios.post('/api/before/cart/add',
         {
```

```js
          busertableId: sessionStorage.getItem('bid'),
          goodstableId: goods.value.id,
          shoppingnum: inputvalue.value
        },
        {
          headers: {
            'Authorization': sessionStorage.getItem('buserauthtoken')
          }
        })
        .then(res => {
          if(res.data.msgId=== "A001"){
            ElMessage.success({message: '成功加入购物车',type: 'success'})
            router.push({name: 'mycart'})
          } else {
            ElMessage.error('加入失败!')
          }
        })
        .catch((error) => {
            ElMessage.error(error)
        })
      } else {                                  //立刻购买,就是直接提交订单
        let gids = [goods.value.id]
        let shoppingnums = [1]
        axios.post('/api/admin/orders/submitOrder',
        {
          bgid: gids,
          bshoppingnum: shoppingnums,
          busertableId: sessionStorage.getItem('bid'),
          amount: goods.value.grprice
        },
        {
          headers: {
            'Authorization': sessionStorage.getItem('buserauthtoken')
          }
        }).then(res => {
            if(res.data.msgId === "A001"){
              ElMessage.success({message: '订单提交成功,请付款!',type: 'success'})
              router.push({name: 'index'})
            } else
              ElMessage.error('订单提交失败!')
        })
        .catch((error) => {
            ElMessage.error(error)
        })
      }
    } else {                                    //收藏
      //headers 为第三个参数
      axios.post('/api/before/focus/add',
      {
        busertableId: sessionStorage.getItem('bid'),
        goodstableId: goods.value.id
      },
      {
        headers: {
          'Authorization': sessionStorage.getItem('buserauthtoken')
        }
      })
      .then(res => {
        if(res.data.msgId=== "A001"){
          ElMessage.error('您已关注该商品!')
        } else if (res.data.msgId=== "A002") {
```

```
                ElMessage.success({message:'成功关注该商品',type:'success'})
            } else{
                ElMessage.error(res.data.msgId + ':添加失败')
            }
        })
        .catch((error) => {
            ElMessage.error(error)
        })
      }
    }
  }
</script>
```

▶16.2.17 我的购物车组件

用户成功登录后,可在顶端位置单击"我的购物车"超链接打开购物车界面。右击src/views/before文件夹,新建MyCartView.vue文件,即我的购物车组件。MyCartView.vue的运行效果如图16.12所示。

图16.12 我的购物车界面

在MyCartView.vue组件中,首先使用onMounted方法加载登录用户的购物车商品信息,并展示在购物车界面。MyCartView.vue的核心代码如下:

```
<template>
    <el-dialog title="我的购物车" v-model="myfocusVisible" width="62%" @close="goClose(1)">
        <el-table :data="goodslists" border :key="itemKey">
        <el-table-column label="图片" width="80">
            <template #default="scope">
                <el-link @click="goToGoodsDetail(scope.row)">
                    <el-image :src="require('../../assets/' + scope.row.gpicture)" style="width: 50px; height: 50px;"/>
                </el-link>
            </template>
        </el-table-column>
        <el-table-column label="商品名称" width="155">
            <template #default="scope">
                <el-link @click="goToGoodsDetail(scope.row)" :underline="false">
                    <span>{{scope.row.gname}}</span>
                </el-link>
            </template>
        </el-table-column>
        <el-table-column label="商品实价" width="105">
            <template #default="scope">
                <span>{{scope.row.grprice.toFixed(1)}}</span>
            </template>
        </el-table-column>
```

```html
            <el-table-column label="购买量" width="150">
                <template #default="scope">
                    <el-button size="small" type="success" @click="reduce(scope.row)" :disabled="scope.row.shopnum === 1">-</el-button>
                    <span> {{scope.row.shoppingnum}} </span>
                    <el-button size="small" type="success" @click="add(scope.row)">+</el-button>
                </template>
            </el-table-column>
            <el-table-column label="小计" width="150">
                <template #default="scope">
                    <span>{{(scope.row.grprice * scope.row.shoppingnum).toFixed(1)}}</span>
                </template>
            </el-table-column>
            <el-table-column label="删除" width="100">
              <template #default="scope">
                <el-row>
<el-button size="small" type="danger" :icon="Delete" circle @click="remove(scope.row)"/>
                </el-row>
              </template>
            </el-table-column>
        </el-table>
        <br>
        <div v-if="goodslists.length > 0">总价:￥{{ totalPrice.toFixed(1) }}  
          <el-button type="success" :icon="ShoppingBag" @click="goClose(2)">去结算</el-button>
          <el-button type="danger" :icon="Delete" @click="removeAll">清空</el-button>
        </div>
  </el-dialog>
</template>
```

```vue
<script setup>
import { ref, computed, onMounted, getCurrentInstance } from 'vue'
import {useRoute, useRouter} from 'vue-router'
import { Delete, ShoppingBag } from '@element-plus/icons-vue'
import { ElMessage } from 'element-plus'
const axios = getCurrentInstance().appContext.config.globalProperties.$axios
const router = useRouter()
const route = useRoute()
const myfocusVisible = ref(true)
const goodslists = ref([])
let itemKey = ref(0)
//组件初始化
onMounted(() => {
  loadGoods()
})
//加载商品信息
const loadGoods = () => {
  axios.post('/api/before/cart/myCart',
  {
    busertableId: sessionStorage.getItem('bid')
  },
  {
    headers: {
      'Authorization': sessionStorage.getItem('buserauthtoken')
    }
  })
  .then(res => {
      goodslists.value =res.data.result;
      itemKey.value = Math.random()           //刷新表格数据
  })
  .catch((error) => {
```

第 16 章　电子商务平台的设计与实现（Spring Boot+ Vue.js 3+ MyBatis Plus）

```javascript
        ElMessage.error(error)
    })
}
const goToGoodsDetail = (goods) => {
    router.push({name: 'goodsDetail', state: goods })
}
const goClose = (n) => {
    //在修改完购物车后,关闭对话框时批量更新
    let cids = []
    let shoppingnums = []
    for (let i = 0; i < goodslists.value.length; i++) {
        let item = goodslists.value[i]
        cids[i] = item.cid
        shoppingnums[i] = item.shoppingnum
    }
    axios.post('/api/before/cart/bupDateCart',
    {
      bcid: cids,
      bshoppingnum: shoppingnums
    },
    {
      headers: {
        'Authorization': sessionStorage.getItem('buserauthtoken')
      }
    })
    //跳转到前一个页面
    let path = route.query.redirect
    if(n === 1)                              //关闭
        router.replace({ path: path === '/' || path === undefined ? '/' : path })
    else                                     //去结算
        router.push({name: 'goOrder'})
}
//减
const reduce = (goods) => {
    if(goods.shoppingnum === 1)
        return
    goods.shoppingnum--
}
//加
const add = (goods) => {
    goods.shoppingnum++
}
//删除购物车
const remove = (goods) => {
  axios.post('/api/before/cart/removeCart',
    {
      id: goods.cid
    },
    {
      headers: {
        'Authorization': sessionStorage.getItem('buserauthtoken')
      }
    }).then(res => {
        console.log(res.data.msgId)
        ElMessage.success({message: '成功删除购物车!',type: 'success'})
        loadGoods()
    })
}
//使用计算属性计算总额
const totalPrice = computed( ()=> {
    let total = 0
```

```
        for (let i = 0; i < goodslists.value.length; i++) {
            let item = goodslists.value[i]
            total = total + item.grprice * item.shoppingnum
        }
        return total
    })
    //清空购物车
    const removeAll = () => {
      axios.post('/api/before/cart/clearCart',
        {
          busertableId: sessionStorage.getItem('bid')
        },
        {
          headers: {
            'Authorization': sessionStorage.getItem('buserauthtoken')
          }
        }).then(res => {
          console.log(res.data.msgId)
          ElMessage.success({message: '已清空购物车!',type: 'success'})
          loadGoods()
        })
    }
</script>
```

▶16.2.18 我的订单组件

用户成功登录后,可在顶端位置单击"我的订单"超链接打开我的订单界面。右击 src/views/before 文件夹,新建 MyOrderView.vue 文件,即我的订单组件。MyOrderView.vue 的运行效果如图 16.13 所示。

图 16.13 我的订单界面

在 MyOrderView.vue 组件中首先使用 onMounted 方法加载登录用户的所有历史订单,并分页展示在我的订单界面。对于 MyOrderView.vue 组件的代码,请读者参见本书提供的源程序 ebusiness-vue。

▶16.2.19 我的收藏组件

用户成功登录后,可在顶端位置单击"我的收藏"超链接打开我的收藏界面。右击 src/views/before 文件夹,新建 MyFocusView.vue 文件,即我的收藏组件。MyFocusView.vue 的运行效果如图 16.14 所示。对于 MyFocusView.vue 组件的代码,请读者参见本书提供的源程序 ebusiness-vue。

▶16.2.20 订单确认组件

用户成功登录后,单击图 16.12 中的"去结算"按钮,即可打开订单确认界面。右击 src/views/before 文件夹,新建 GoOrderView.vue 文件,即订单确认组件。GoOrderView.vue 的运行效果如图 16.15 所示。对于 GoOrderView.vue 组件的代码,请读者参见本书提供的源程序 ebusiness-vue。

第 16 章　电子商务平台的设计与实现（Spring Boot+ Vue.js 3+ MyBatis-Plus）

图 16.14　我的收藏界面

图 16.15　订单确认界面

▶16.2.21　配置路由

在前端系统 ebusiness-vue 的 src/router/index.js 文件中配置路由信息，具体配置内容如下：

```
import { createRouter, createWebHistory } from 'vue-router'
//before
import IndexView from '../views/before/IndexView.vue'
import GoodsDetailView from '../views/before/GoodsDetailView.vue'
import RegisterView from '../views/before/RegisterView.vue'
import LoginView from '../views/before/LoginView.vue'
import MyselfInfoView from '../views/before/MyselfInfoView.vue'
import MyOrderView from '../views/before/MyOrderView.vue'
import MyFocusView from '../views/before/MyFocusView.vue'
import MyCartView from '../views/before/MyCartView.vue'
import GoOrderView from '../views/before/GoOrderView.vue'
//admin
import HomeView from '../views/admin/HomeView.vue'
import TypeManage from '../views/admin/TypeManageView.vue'
import GoodsManage from '../views/admin/GoodsManageView.vue'
import OrderManage from '../views/admin/OrderManageView.vue'
import AdminLoginView from '../views/admin/LoginView.vue'
import SalesStatistics from '../views/admin/SalesStatisticsView.vue'
import OrderStatistics from '../views/admin/OrderStatisticsView.vue'
import MyTestView from '../views/admin/MyTestView.vue'
const routes = [
  {
    path: '/',
    name: 'index',
    component: IndexView
  },
  {
    path: '/goodsDetail',
```

```
    name: 'goodsDetail',
    component: GoodsDetailView
},
{
    path: '/register',
    name: 'register',
    component: RegisterView
},
{
    path: '/login',
    name: 'login',
    component: LoginView
},
{
    path: '/myselfinfo',
    name: 'myselfinfo',
    component: MyselfInfoView,
    meta: {auth:true, act:true}              //需要验证登录权限
},
{
    path: '/myorder',
    name: 'myorder',
    component: MyOrderView,
    meta: {auth:true, act:true}              //需要验证登录权限
},
{
    path: '/myfocus',
    name: 'myfocus',
    component: MyFocusView,
    meta: {auth:true, act:true}              //需要验证登录权限
},
{
    path: '/mycart',
    name: 'mycart',
    component: MyCartView,
    meta: {auth:true, act:true}              //需要验证登录权限
},
{
    path: '/goOrder',
    name: 'goOrder',
    component: GoOrderView,
    meta: {auth:true, act:true}              //需要验证登录权限
},
//admin
{
    path: '/adminLogin',
    name: 'adminLogin',
    component: AdminLoginView
},
{
    path: '/home',
    name: 'home',
    component: HomeView,
    redirect: '/home/typemanage',
    meta:{auth:true, act:false},             //需要验证登录权限
    children: [
        {
            path: '/home/typemanage',
            component: TypeManage
        },
        {
```

```
                path: '/home/goodsmanage',
                component: GoodsManage
            },
            {
                path: '/home/ordermanage',
                component: OrderManage
            },
            {
                path: '/home/salesstatistics',
                component: SalesStatistics
            },
            {
                path: '/home/orderstatistics',
                component: OrderStatistics
            }
            ,
            {
                path: '/home/mytest',
                component: MyTestView
            }
        ]
    }
]
const router = createRouter({
    //推荐使用 HTML 5 模式
    history: createWebHistory(process.env.BASE_URL),
    routes
})
export default router
```

16.3　测试运行

首先运行后端系统 ch16 的主类 Ch16Application，启动后端系统 ch16，然后在 Terminal 终端运行 npm run serve 命令启动前端系统 ebusiness-vue。

在前后端系统同时启动后，即可通过"http://localhost:8000/ebusiness-vue/"测试运行。

本章小结

本章通过电子商务平台的设计与开发讲述了前后端分离开发的具体过程，旨在让读者了解 Vue.js ＋ Spring Boot 实现前后端分离开发的流程。

习题 16

1. 在 .vue 组件文件中使用 setup 语法糖时，不能使用 this 获得 main.js 中注册的全局变量（如 this.$axios），请问在使用 setup 语法糖时如何获得 main.js 中注册的全局变量？
2. 在 Vue 项目中如何安装并引入 Element Plus？

图书资源支持

感谢您一直以来对清华版图书的支持和爱护。为了配合本书的使用,本书提供配套的资源,有需求的读者请扫描下方的"书圈"微信公众号二维码,在图书专区下载,也可以拨打电话或发送电子邮件咨询。

如果您在使用本书的过程中遇到了什么问题,或者有相关图书出版计划,也请您发邮件告诉我们,以便我们更好地为您服务。

我们的联系方式:

清华大学出版社计算机与信息分社网站: https://www.shuimushuhui.com/

地　　址:北京市海淀区双清路学研大厦 A 座 714

邮　　编:100084

电　　话:010-83470236　010-83470237

客服邮箱:2301891038@qq.com

QQ:2301891038(请写明您的单位和姓名)

资源下载: 关注公众号"书圈"下载配套资源。

资源下载、样书申请	图书案例	
书圈	清华计算机学堂	观看课程直播